# Sources and Fates
# of Aquatic Pollutants

ADVANCES IN CHEMISTRY SERIES **216**

# Sources and Fates of Aquatic Pollutants

**Ronald A. Hites,** EDITOR
*Indiana University*

**S. J. Eisenreich,** EDITOR
*University of Minnesota*

Developed from a symposium sponsored by
the Division of Environmental Chemistry
at the 190th Meeting
of the American Chemical Society,
Chicago, Illinois,
September 8–13, 1985

American Chemical Society, Washington, DC 1987

**Library of Congress Cataloging-in-Publication Data**

Sources and fates of aquatic pollutants.
(Advances in chemistry series, ISSN 0065-2393; 216)

"Developed from a symposium sponsored by the
Division of Environmental Chemistry at the 190th
Meeting of the American Chemical Society, Chicago,
Illinois, September 8-13, 1985."

Includes bibliographies and indexes.

1. Water—Pollution—Congresses. 2. Organic water
pollutants—Congresses. 3. Water chemistry—
Congresses. 4. Hydrodynamics—Congresses.
5. Chemical reaction, Rate of—Congresses.

I. Hites, Ronald A. II. Eisenreich, S. J.
III. American Chemical Society. Division of
Environmental Chemistry. IV. American Chemical
Society. Meeting (190th: Chicago, Ill.) V. Series.

QD1.A355      no. 216      540 s      [628.1′68]      87-1290
[TD425]
ISBN 0-8412-0983-9

# FOREWORD

The ADVANCES IN CHEMISTRY SERIES was founded in 1949 by the American Chemical Society as an outlet for symposia and collections of data in special areas of topical interest that could not be accommodated in the Society's journals. It provides a medium for symposia that would otherwise be fragmented because their papers would be distributed among several journals or not published at all. Papers are reviewed critically according to ACS editorial standards and receive the careful attention and processing characteristic of ACS publications. Volumes in the ADVANCES IN CHEMISTRY SERIES maintain the integrity of the symposia on which they are based; however, verbatim reproductions of previously published papers are not accepted. Papers may include reports of research as well as reviews, because symposia may embrace both types of presentation.

# ABOUT THE EDITORS

RONALD A. HITES is Professor of Public and Environmental Affairs and Professor of Chemistry at Indiana University. He received a B.A. in chemistry in 1964 from Oakland University and a Ph.D. in analytical chemistry in 1968 from the Massachusetts Institute of Technology. He began his academic career as an Assistant Professor of Chemical Engineering at the Massachusetts Institute of Technology. Since 1979, he has been at Indiana University. Professor Hites is an editor of *Biomedical and Environmental Mass Spectrometry* and *Chemosphere* and has been a member of numerous review committees at various national laboratories. He has been a member of two National Academy of Sciences committees and is a member of the American Chemical Society, the American Society for Mass Spectrometry, Sigma Xi, the Geochemical Society, and the International Association for Great Lakes Research. He is currently president-elect of the American Society for Mass Spectrometry. Professor Hites created a 1-week intensive course on "Environmental Applications of Gas Chromatographic Mass Spectrometry". This course has been offered since 1979 and has been taken by more than 250 people. Professor Hites is the author of more than 100 papers on mass spectrometry and organic environmental chemistry.


S. J. EISENREICH is Professor of Environmental Engineering Sciences in the Department of Civil and Mineral Engineering at the University of Minnesota in Minneapolis. He received a B.S. degree in chemistry from the University of Wisconsin—Eau Claire; an M.S. degree in analytical chemistry from the University of Wisconsin—Milwaukee; and a Ph.D. degree in water chemistry from the University of Wisconsin—Madison. Dr. Eisenreich joined the faculty of the University of Minnesota in 1975. He has published numerous articles on the transport and fate of nonpolar organic contaminants in large lakes, the impact of atmospheric deposition of trace metals and organic contaminants on aquatic ecosystems, and the biogeochemistry of wetlands. He has edited or coedited two books on atmospheric deposition to lakes and on polychlorinated biphenyls (PCBs) in the Great Lakes. He serves on the editorial advisory board of *Environmental Science & Technology* and on the editorial board of *Water, Air and Soil Pollution*, and is Associate Editor of the *Journal of Great Lakes Research*. He is a frequent consultant to government and industry on the fate and transport of toxic organic contaminants in surface water and groundwater. Current research programs include the study of benthic processes in the Great Lakes using a deep-diving submersible, aquifer thermal energy storage, the chemical limnology of PCBs, interactions of organic chemicals with solids, biogeochemistry of bogs, and atmospheric chemistry and deposition of organic contaminants.

# CONTENTS

ix

CASE STUDIES

INDEXES

# PREFACE

OUR UNDERSTANDING OF THE PROCESSES controlling the transport and fate of inorganic and organic species in the limnic and marine environments significantly advanced in the past 10 years. This book examines these processes and their implications on environmental chemical dynamics. Some of the contributions contrast processes occurring in marine and freshwater systems.

This volume presents a holistic approach to the study of aquatic pollutant chemistry; the atmosphere, water, and sediment are treated as interdependent compartments of an ecosystem. For example, the water column of a lake receives inputs from the atmosphere and from surface and subsurface drainage and loses material to outflow, volatilization, and internal processes such as sedimentation and degradation. The rates of physical mixing within, and material transfer among, the atmospheric, hydrospheric, and sedimentary compartments and the rates of reactions occurring in each compartment determine the concentrations throughout the system. The dynamic phenomena occurring at interfaces drive the chemical fluxes, feed the degrading reaction, and control the element and compound residence times.

The theme of this book, explicitly or implicitly, is the relationship between physical mixing and chemical reaction rates. For example, rates of transport between the atmosphere and water, rates of sedimentation and volatilization, and rates of chemical transformations are important issues that are treated. The topical coverage is summarized in Figure 1.

This book is divided into four sections: air–water processes, water column processes, water–sediment processes, and case studies. The emphasis of the first three sections is on the chemical and physical processes controlling solute behavior and fate in air and water. The case studies serve to integrate information on these processes into a system-wide picture of the cycling of inorganic and organic chemicals. Although the intent is to concentrate on processes occurring in aquatic systems and not on "popular" areas of environmental research, several of the papers describe aspects of topical interest, for example, acid rain and PCBs.

We thank the authors of this volume. Our job as editors was easy compared with that of the authors. As several of them mentioned, "Without authors, editors are superfluous." We agree and acknowledge

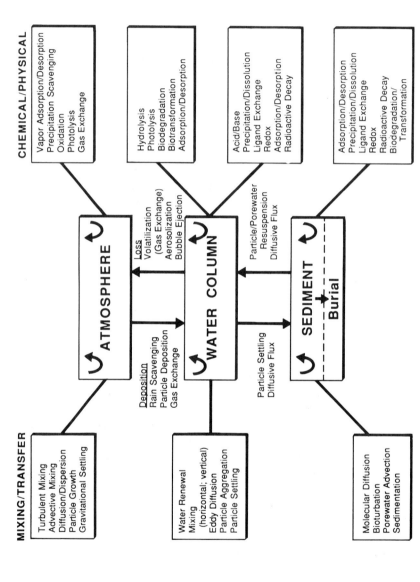

Figure 1.  The mixing and transfer processes (left) and the primarily chemical and physical processes (right) that determine compartmental concentrations and fluxes in the aquatic environment.

the first-rate contributions of all the authors involved in this book. We also thank Robin Giroux, Keith Belton, and Cara Aldridge Young of the ACS Books Department for their guidance and help.

RONALD A. HITES
School of Public and Environmental Affairs
    and Department of Chemistry
Indiana University
Bloomington, IN 47405

S. J. EISENREICH
Environmental Engineering Program
Department of Civil and Mineral Engineering
University of Minnesota
Minneapolis, MN 55455

November 1986

# AIR–WATER PROCESSES

# 1

# Methods for Estimating Solubilities of Hydrophobic Organic Compounds: Environmental Modeling Efforts

Anders W. Andren, William J. Doucette[1], and Rebecca M. Dickhut
Water Chemistry Program, University of Wisconsin—Madison, Madison, WI 53706

*A variety of environmental fate models that integrate physicochemical properties of pollutants with advective and diffusive transport equations are now available. To satisfy the demand for missing input data, scientists have begun to incorporate property estimation techniques as part of their computational procedures. The recent appearance of high quality solubility data for polyhalogenated organic compounds makes it possible to evaluate several predictive schemes for these compounds. In this work, a brief review of the thermodynamics of hydrophobic compound solubility relationships is presented. This review is followed by an examination of the UNIFAC activity coefficient prediction technique. The use of octanol–water partition coefficients, total molecular surface areas, and molecular connectivity indexes to predict aqueous solubilities is then examined, and the resulting correlations are presented.*

**M**ATHEMATICAL MODELING OF CHEMICAL FATE provides an excellent framework for sorting massive quantities of environmental data in a logical way. Parameters in a model may be varied to gain an understanding as to what processes are most important in determining the environmental behavior of a chemical. One of the most useful modeling approaches integrates data on physicochemical properties of the compound in question with hydrodynamic or aerodynamic transport models. This approach uses the results of such laboratory measurements or calculations as aque-

[1]Current address: Environmental Engineering Department, Utah State University, Logan, UT 84322–4110

0065–2393/87/0216–0003$07.00/0

ous solubility, saturation vapor pressure, liquid and vapor molecular diffusivity, Henry's law constant, UV absorption, sorption–partition coefficient, photolysis rate, chemical oxidation rate, and hydrolysis rate. These data are then incorporated into various steady-state or time-dependent fate models. Some of the models most widely used by aquatic and atmospheric scientists have been reviewed by Dickson et al. (1) and Sheehan et al. (2).

In these models, much of the aforementioned data are lacking for hydrophobic organic compounds of environmental interest. The temperature dependence of these parameters is also lacking, so most present assessments cannot be determined as a function of temperature.

Property estimation techniques have been used extensively in the chemical engineering and drug design fields (3, 4). Recently, a handbook of chemical property estimation techniques was published (5). To date, however, most techniques have been evaluated with test data sets of relatively high solubility ($>1 \times 10^{-6}$ M), and the solubility data of the most hydrophobic compounds have been of uneven quality (6, 7).

Very recent work has produced high quality aqueous solubility data for polychlorinated biphenyls (PCBs), polybrominated biphenyls (PBBs), methylated biphenyls, polychlorinated dibenzodioxins (PCDDs), and polychlorinated dibenzofurans (PCDFs) (8–10). Dickhut et al. (11) have also examined temperature effects on solubility for PCBs.

In this chapter, we briefly review a few of the most important aqueous solubility predictive methods. This property was chosen because Henry's law constants can be computed by using the approximation $P^s = HC_s$, where $P^s$ is saturation vapor pressure, $H$ is Henry's law constant, and $C_s$ is aqueous solubility. Henry's law constants may then be used to calculate liquid- and gas-phase transfer coefficients for incorporation into atmospheric flux models, for example, those models recently reviewed in Liss and Slinn (12) or presented by Mackay et al. (13).

We will first present the thermodynamic background necessary to understand some of the estimation techniques. This review is followed by an examination of the use of four activity coefficient prediction schemes with respect to their ability to predict values for chlorinated aromatic hydrocarbons ranging in solubility from $1 \times 10^{-3}$ to $1 \times 10^{-13}$ M.

## Theory

**Aqueous Solubility.**  The solubility of a substance may be considered to be an equilibrium partitioning between the pure chemical and that in solution at a specified temperature. Historically, thermodynamicists first developed the concept of "ideal solubility" that was used as an initial approximation to describe the solution behavior of solutes. However, solubility is a function of the various molecular forces that operate between solvent and solute molecules. Prausnitz et al. (*14*) classified intermolecular forces into four somewhat arbitrary types: electrostatic, induction, London dispersion, and specific chemical forces. These forces must also be incorporated into any real model that will accurately describe solubility.

Hildebrand and Scott (*15*) give a thermodynamic definition of an ideal solution: one in which the activity (*a*) equals the mole fraction (*X*) over the entire composition range and over a nonzero range of temperature and pressure. This definition establishes the ideal solution in the sense of Raoult's law. Thus, for an ideal solution:

$$X_2 = a_2 \tag{1}$$

where the subscript 2 refers to the solute.

The activity may be viewed as a measure of the difference between the substance's free energy at the state of interest and that at its standard state. Lewis and Randall (*16*) defined the activity as the ratio of the pure solute fugacity at any temperature and pressure to its fugacity at some standard state:

$$a_2 = \frac{f_2\,(T,P,X)}{f_2\,(T,P_2^o,X_2^o)} = \frac{f_2}{f_2^{ss}} \tag{2}$$

where the superscript ss refers to the standard state, and $f$, $T$, and $P$ denote fugacity, temperature, and pressure, respectively. Fugacity may be considered as the escaping tendency of a substance. The standard-state composition ($X_2^o$) and pressure ($P_2^o$) are arbitrary, but the temperature must be the same as the state of interest. For organic nonelectrolytes, the standard-state fugacity is usually defined as the fugacity (vapor pressure) of the pure liquid solute at the solution temperature.

Because equilibrium partitioning of a solute between phases is achieved when fugacities or chemical potentials are equal, the equilibrium solution condition for the ideal liquid solute is

$$X_2^L = a_2 = \frac{f_2^L}{f_2^{ss}} = \frac{f_2^L}{f_2^L} \tag{3}$$

where the superscript $L$ denotes liquid. This equilibrium condition follows because $f_2^L$, the fugacity of the pure liquid, is equal to the standard state fugacity ($f_2^{ss}$), which we chose as the pure liquid. The ideal solubility equation for liquids reduces to $X_2^L = 1$, which indicates solubility for all mole fractions (infinite miscibility) of a liquid in a liquid.

For a solid, the equation of equilibrium for an ideal solution is

$$X_2^s = a_2 = \frac{f_2^s}{f_2^{ss}} = \frac{f_2^s}{f_2^{SL}} \tag{4}$$

where $f_2^s$ is the fugacity of the pure solid solute, and $f_2^{SL}$ is the fugacity of the pure supercooled liquid. The supercooled liquid is a hypothetical state in which the solid is considered to be a liquid at a temperature below its melting point. The supercooled liquid is usually chosen as the standard state for solids. Thus, the fugacity ratio for solids is dependent on the free energy necessary to melt the solid and may be calculated from thermodynamic considerations (14) by

$$\ln \frac{f_2^{ss}}{f_2^s} = \frac{\Delta H_f}{RT_t}\left(\frac{T_t}{T} - 1\right) - \frac{\Delta C_p}{R}\left(\frac{T_t}{T} - 1\right) + \frac{\Delta C_p}{R}\ln\frac{T_t}{T} \tag{5}$$

where $\Delta H_f$ is the enthalpy of fusion (kcal $mol^{-1}$), $T_t$ is the triple point temperature (K), $\Delta C_p$ is the change in solute heat capacity (cal $deg^{-1}$ $mol^{-1}$) when changing from a solid to a liquid at constant pressure, and $R$ is the gas constant. Because the triple point and melting point temperatures are very close, the melting point temperature $(T_m)$ is usually used in equation 5. At the melting point, the phases are in equilibrium and

$$\Delta G_f = \Delta H_f - T\Delta S_f = 0 \tag{6}$$

where $\Delta G_f$ and $\Delta S_f$ are Gibbs free energy of fusion and entropy of fusion, respectively. In this case,

$$\Delta H_f = T\Delta S_f \tag{7}$$

and equation 5, together with equation 4, takes the form:

$$\ln X_2^s = -\frac{\Delta S_f}{R}\left(\frac{T_m}{T} - 1\right) + \frac{\Delta C_p}{R}\left(\frac{T_m}{T} - 1\right) - \frac{\Delta C_p}{R}\ln\frac{T_m}{T} \tag{8}$$

Very few determinations of $\Delta C_p$ for hydrophobic organic compounds have been reported. One of the following three assumptions is generally made. Either (a) the values of $\Delta C_p$ are small, and because the terms containing this parameter are of opposite sign, the last two terms on the right side cancel (14); (b) $\Delta C_p = 0$ (17); or (c) $\Delta C_p = \Delta S_f$, which is a constant (18). At the present time, not enough evidence exists to substantiate any one of these assumptions. However, equation 8 reduces to equation 9 by assumption a or b or to equation 10 by assumption c.

$$\ln X_2^s = -\frac{\Delta S_f}{R}\left(\frac{T_m}{T} - 1\right) \tag{9}$$

$$\ln X_2^s = -\frac{\Delta S_f}{R}\ln\frac{T_m}{T} \tag{10}$$

The ideal solubility equations have proved to be reliable for estimating solubilities of liquids and solids in solvents of a similar chemical nature. This reliability results from the fact that ideal solubility depends only on the properties of the solute and is independent of the nature of the solvent (*15*).

Real solutions, especially hydrophobic compounds dissolved in water, do not behave ideally. In contrast to ideal solutions, the mole fraction in a real solution is not equal to, but proportional to, the activity. On the basis of Raoult's law, the solute deviation from ideality is described by an activity coefficient, $\gamma_2$ (*16*), so that equations 1 and 2 are modified to

$$a_2 = \gamma_2 X_2 = \frac{f_2}{f_2^{ss}} \tag{11}$$

When $\gamma_2 > 1$, the fugacity of the solute is greater than in an ideal solution of the same concentration. Similarly, when $\gamma_2 < 1$, the fugacity is lower than in an ideal solution of the same concentration (*19*). In nonpolar solutions, where only dispersion forces are important, $\gamma_2$ is generally larger than unity (i.e., lower solubility than that predicted from ideal solubility). In cases where polar or specific chemical forces are important, $\gamma_2$ may be less than unity.

In terms of thermodynamic functions, $\gamma_2$ represents the excess Gibbs free energy ($G_{excess}$) associated with nonideal solutions. From phase-equilibrium thermodynamics (*14*),

$$G_{excess} = G_{real} - G_{ideal} \tag{12}$$

$$G_{real} = RT \ln f_{real} = RT \ln f_2^{ss} X_2 \gamma_2 \tag{13}$$

$$G_{ideal} = RT \ln f_{ideal} = RT \ln f_2^{ss} X_2 \tag{14}$$

Substituting equations 13 and 14 into equation 12 yields

$$G_{excess} = RT \ln \gamma_2 \tag{15}$$

The activity coefficient thus gives a quantitative measure of departure from ideal behavior.

If no solubility of the solvent in the solute occurs, the solubility model is modified by the addition of this correction term. The expression for mole fraction solubility becomes

$$X_2 = (f^\circ / f_2^{ss})\,(1/\gamma_2) \tag{16}$$

where $f^\circ$ represents the pure solute (liquid or solid) fugacity. For a liquid, the expression may conveniently be expressed in a logarithmic form:

$$\ln X_2^l = -\ln \gamma_2 \tag{17}$$

This expression is valid because $f_2^l = f_2^{ss}$ at equilibrium. The model may be similarly expressed for solids by combining equation 16 with equation 9 or 10 to yield equation 18 or 19, respectively.

$$\ln X_2^s = \frac{-\Delta S_f}{R}\,\frac{T_m}{T} - 1 - \ln \gamma_2 \tag{18}$$

$$\ln X_2^s = \frac{-\Delta S_f}{R}\,\ln \frac{T_m}{T} - \ln \gamma_2 \tag{19}$$

Equations 18 and 19 would not be particularly useful because the existence of entropy of fusion values is rather limited in the literature. However, Yalkowsky (20) found that rigid aromatic molecules exhibit a fairly constant value for this term of 13.5 eu (Walden's rule). Miller et al. (8) subsequently showed that 16 PCBs exhibited $\Delta S_f$ values ranging from 12.6 to 16.4 eu and had an average value of 13.2 eu. A reasonable approximation for the entropy term will likely be achieved with $\Delta S_f = 13.2$ for PCBs, polycyclic aromatic hydrocarbons (PAHs), furans, and dioxins. Also, if $R = 1.987$ cal/deg mol, and $T = 298$ K, equations 18 and 19 can be further simplified to equations 20 and 21, respectively.

$$\ln X_2^s = 6.64 - 0.0223\,T_m - \ln \gamma_2 \tag{20}$$

$$\ln X_2^s = 37.8 - 6.64 \ln T_m - \ln \gamma_2 \tag{21}$$

Equations 17, 20, and 21 can be very useful. The value $T_m$ is an experimentally accessible parameter, and the solubility of solids and liquids can be compared. The effect of the energy necessary to melt the solid in the overall solution process can be visualized by rearranging equation 20:

$$X_2^s \gamma_2 = e^{(6.64-0.0223\,T_m)} \qquad (22)$$

The amount of energy necessary for this melting process is zero for a substance that is a liquid at the temperature of interest, and this energy increases exponentially for solids of increasing $T_m$. Figure 1 shows this effect for solubilities at 25 °C on solids of varying $T_m$. Thus, the importance of the correction factor can easily be appreciated for a halogenated aromatic hydrocarbon whose $T_m = 400$ °C because the solid solubility is reduced approximately 4500 times when compared with the liquid solubility.

Our ability to provide a thermodynamic framework to account for the energy required to overcome the crystalline lattice forces inherent in a solid solute is important for attempts to predict aqueous solubilities. This thermodynamic framework is also necessary and must be applied to vapor pressure and Henry's law constant predictive schemes.

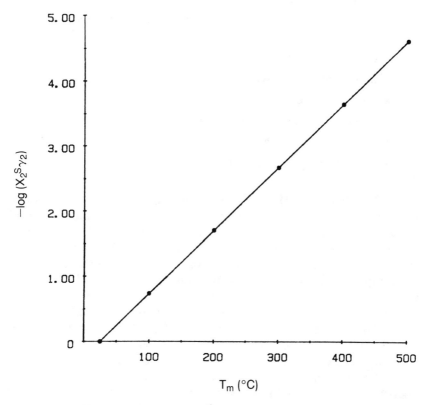

*Figure 1. Melting point correction factor as a function of* $T_m$ *on solubility at 25 °C.*

**Activity Coefficient Estimation Techniques.** Various molecular models and predictive techniques have been devised to calculate solute activity coefficients from the properties of the pure components. Reviews of efforts in this area were written by Hildebrand et al. (21), Ben-Naim (22), Pierotti (23), Prausnitz et al. (14), and Acree (24). Precise theoretical treatment of the various solution interactions has proven to be too complicated to be feasible for incorporation into our present models. A more practical and attractive approach that has found widespread application is to examine the relationships between aqueous solubility and intrinsic molecular properties that are readily determined either on a computational or experimental basis. Some of the more widely used techniques employed by workers in the field of drug design and environmental science are the following:

- universal quasi-chemical functional groups activity coefficient (UNIFAC) (25-27)
- experimental log $K_{ow}$ (octanol–water partition coefficient) (9, 28)
- calculated log $K_{ow}$ (29, 30)
- molecular connectivity (9, 31-33)
- molar volume (34-36)
- molecular weight (9, 37)
- molecular surface area (7, 9, 38-42)
- solvametric parameters (43-45)

Although the UNIFAC method and various modifications of the Scatchard–Hildebrand techniques employ sound thermodynamically derived components, several approaches employ only an appropriate amount of chemical intuition in combination with multiregressional analysis. In this review, we summarize some of the efforts that have been applied toward predicting aqueous activity coefficients of halogenated aromatic hydrocarbons of environmental interest.

UNIFAC. One of the more thermodynamically sound models for determining the activity coefficient is the UNIFAC group contribution method. This method is based on the concept that, whereas literally millions of organic chemicals are known, the number of functional fragments (groups) that constitute these compounds is limited. The activity coefficient is divided into two parts: the combinatorial contribution, due mostly to differences in molecular size and shape, and the residual contribution, accounting for differences in intermolecular forces of attraction (26, 46). For a molecule $i$ in any solution,

$$\ln \gamma_i = \ln \gamma_i^c + \ln \gamma_i^R \tag{23}$$

where $\gamma_i$ is the total activity coefficient, $\gamma_i^c$ is the combinatorial part, and $\gamma_i^R$ is the residual part. For a solute in water, the combinatorial part in equation 23 is given by equation 24 (25).

$$\ln \gamma_2^c = \ln \frac{\phi_2}{X_1} + \frac{z}{2} q_2 \ln \frac{\theta_2}{\phi_2} + l_2 - \frac{\phi_2}{X_2} (X_1 l_1 + X_2 l_2) \tag{24}$$

where $z$ is the coordination number; $\phi_2$ is the segment fraction; $\theta_2$ is the area fraction; and the subscripts 1 and 2 denote the solvent and solute, respectively. The parameters are defined as follows:

$$\phi_2 = \frac{r_2 x_2}{r_1 x_1 + r_2 x_2} \tag{25}$$

$$\theta_2 = \frac{q_2 x_2}{q_1 x_1 + q_2 x_2} \tag{26}$$

$$l_1 = \frac{z}{2} (r_1 - q_1) - (r_1 - 1) \tag{27}$$

$$z = 10 \tag{28}$$

$$l_2 = \frac{z}{2} (r_2 - q_2) - (r_2 - 1) \tag{29}$$

$$r_1 = \sum_k \nu_k^{(1)} R_k \tag{30}$$

$$r_2 = \sum_k \nu_k^{(2)} R_k \tag{31}$$

$$q_1 = \sum_k \nu_k^{(1)} Q_k \tag{32}$$

$$q_2 = \sum_k \nu_k^{(1)} Q_k \tag{33}$$

where $X_1$ and $X_2$ are the mole fractions of water and solute, respectively; $\nu_k = 1, 2 \ldots N$, in which $N$ is the number of groups in water and solute molecules; $R_k$ is the van der Waals volume for group $k$; $Q_k$ is the van der Waals surface area for group $k$; and $k$ is the group number.

For a solute in water, the residual part of equation 23 is given by

$$\ln \gamma_2^R = \sum_k \nu_k^{(2)} (\ln \Gamma_k - \ln \Gamma_k^{(2)}) \tag{34}$$

In this equation,

$$\ln \Gamma_k = Q_k \left[ 1 - \ln \left( \sum_m \theta_m \psi_{mk} \right) - \sum_m \left( \theta_m \psi_{km} / \sum_n \theta_n \psi_{nm} \right) \right] \qquad (35)$$

where $m = 1, 2 \ldots M$, and $n = 1, 2 \ldots N$ (all groups). Also,

$$\theta = \frac{Q_m X_m}{\sum\limits_n Q_n X_n} \qquad (36)$$

$$\psi_m = \frac{\sum\limits_j \nu_m^{(j)} X_j}{\sum\limits_j \sum\limits_n \nu_n^{(j)} X_j} \qquad (37)$$

where $j = 1, 2 \ldots M$, and $n = 1, 2 \ldots N$. Individual parameter definitions include the following: $j$ is the component number (1 or 2 for a binary mixture), $M$ is the total number of components, $n$ is the group number, $N$ is the total number of groups, $\theta_m$ is the group surface area fraction, $X_m$ is the group fraction, $X_j$ is the mole fraction of molecule $j$ in solution, $\nu_m^{(j)}$ is the number of groups of type $m$ in molecule $j$, and $\psi_{mn} = e^{(-a_{nm}/T)}$. The parameter $a_{nm}$ is the group interaction parameter and indicates the difference in interaction energy between various solvent and solute groups.

The various UNIFAC parameters have been determined for about 40 molecular groups (46). Grain (47) also presented a step-by-step illustration on how to calculate the UNIFAC activity coefficient. Aqueous solubilities for solids may then be calculated by using relationships already discussed in the theory section.

The appropriate parameters necessary to calculate activity coefficients for PCBs are presented in Table I. Burkhard et al. (48) tested the predictive success of the UNIFAC method by using a recently compiled

**Table I. UNIFAC Parameters for Polychlorinated Biphenyl Groups**

| Group | K | $R_k$ | $Q_k$ | $a_{nm}$ | $a_{mn}$ |
|-------|-----|--------|--------|----------|----------|
| Aromatic Cl | 54 | 1.1562 | 0.844 | | |
| Aromatic CH | 10 | 0.5313 | 0.400 | | |
| $H_2O$ | 17 | 0.92 | 1.40 | | |
| Aromatic Cl–$H_2O$ | | | | 920.4 | 678.2 |
| Aromatic CH–$H_2O$ | | | | 362.3 | 903.8 |

SOURCE: Data are from reference 46.
NOTE: Abbreviations are as follows: $K$ is a group number assigned by the original developers of UNIFAC; $R_k(\text{Å}^3)$ is the van der Waals volume for group $k$; $Q_k(\text{Å}^2)$ is the van der Waals surface area for group $k$; and $a_{nm}(K)$ is a group interaction parameter that indicates the difference between groups $n$ and $m$.

data set ($N = 24$) on PCB solubility (*9*). All measurements were made by using the generator column method (*49, 50*). Figure 2 is a plot of experimental $\gamma$ versus those values calculated via the UNIFAC method where vapor–liquid equilibria interaction parameters were used. The method is obviously unsatisfactory in that errors of as much as 4 orders of magnitude result. Upon the advice of Arbuckle (*51*), we changed the aromatic chlorine–water interaction parameter to 92.04 K. From our experimental data, we also obtained a new value for the aromatic chlorine–water interaction parameter of 526 K. The subsequent agreement between experimental and predicted activity coefficients for PCBs is shown in Figure 3. A correlation coefficient ($r^2$) of 0.97 results. Much of this error is due to the fact that the UNIFAC method cannot distinguish between isomers (*ortho, meta,* and *para* substitutions). However, PCB

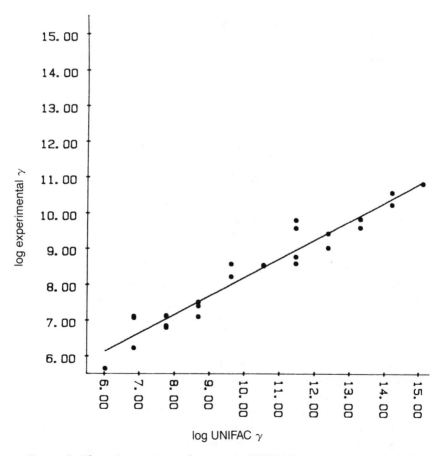

*Figure 2. Plot of experimental $\gamma$ versus UNIFAC $\gamma$ using vapor–liquid equilibria interaction parameters.*

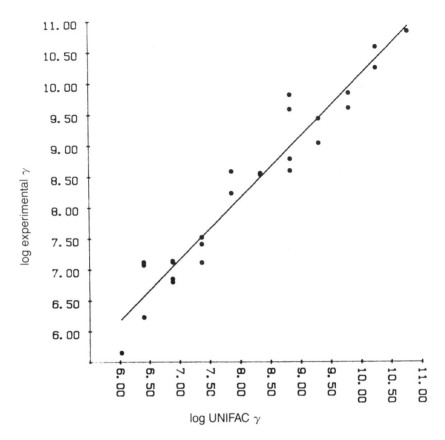

*Figure 3. Plot of experimental* γ *versus UNIFAC* γ *using an aromatic chlorine–water interaction parameter of 92.04 K and a water–aromatic chlorine interaction parameter of 526 K.*

activity coefficients may be determined within a factor of 2 or 3, and this accuracy is acceptable for many environmental applications. Application of the UNIFAC technique to other chlorinated aromatic hydrocarbons, although satisfactory as an initial activity coefficient estimate, must await development of an accurate data base.

**OCTANOL–WATER PARTITION COEFFICIENT RELATIONSHIP TO SOLUBILITY.** The octanol–water partition coefficient $(K_{ow})$ is defined as the equilibrium concentration ratio of an organic chemical partitioned between octanol and water. Hansch et al. (29) first observed the linear relationship between the logarithms of aqueous solubilities and octanol–water partition coefficients of organic liquids. Since then, many researchers have reported similar inverse relationships between aqueous solubility and $K_{ow}$ for several classes of compounds. In a recent compilation, Lyman et al. (5) cite 18 such correlations. Most notable in terms of environmental interest are correlations by Yalkowsky and Valvani (30), Mackay et al. (7), Amidon et al. (52), Chiou et al. (53), Tewari et al. (54), Banerjee et al. (28), Bowman and Sans (55), and Miller et al. (56).

Initially, the correlations were strictly empirical (29, 57) and took the form

$$\log S = A \log K_{\mathrm{ow}} + B \tag{38}$$

where $S$ is the molar solubility, and $A$ and $B$ are linear regression constants. The $S$–$K_{\mathrm{ow}}$ relationship has a thermodynamic basis, however, and Mackay et al. (7), Chiou et al. (53), and Miller et al. (56) have demonstrated the necessity to understand this relationship for the correct interpretation of available data sets. A brief derivation of the appropriate thermodynamic relationships that describe partitioning of both liquid and solid solutes between water and octanol will be presented here.

The equilibration of a liquid solute partitioned between water and octanol leads to the following expression:

$$X_o \gamma_o^* = X_w \gamma_w^* \tag{39}$$

where $X_o$ and $X_w$ are solute mole fractions in octanol and water, respectively, and $\gamma_o^*$ and $\gamma_w^*$ are activity coefficients of liquid solute in octanol and water, respectively. Because some miscibility of octanol and water occurs, the activity coefficients are noted by an asterisk (*) to distinguish them from their pure phases.

Expressing $K_{\mathrm{ow}}$ in terms of molar concentration ratios is customary. At dilute solute concentrations, each mole fraction is approximated by the product of the solute molar concentration, $C$, and molar volume, $V^*$, of each phase. Substitution into equation 39 yields

$$C_o V_o^* \gamma_o^* = C_w V_w^* \gamma_w^* \tag{40}$$

where $C_o$ and $C_w$ are the solute molar concentrations in octanol and water, respectively, and $V_o^*$ and $V_w^*$ are the solvent molar volumes in octanol saturated with water and water saturated with octanol, respectively. Because $K_{\mathrm{ow}} = C_o/C_w$, equation 40 can be used to express the partition coefficient.

$$K_{\mathrm{ow}} = \frac{C_o}{C_w} = \frac{\gamma_w^* V_w^*}{\gamma_o^* V_o^*} \tag{41}$$

For all practical purposes, the molar volumes of a sparingly soluble solute in pure water and in octanol-saturated water are approximately equal [solubility of octanol in water is $4.5 \times 10^{-3}$ M (58), that is, $V_w^* = V_w$].

To relate the aqueous molar solute solubility, S, to $C_w$ requires an expression for the water-saturated liquid solute in equilibrium with water.

$$X_s^o \gamma_s = X_w^o \gamma_w \tag{42}$$

where $X_s^o$ and $X_w^o$ are the mole fractions of solute in the organic-rich (solute) and water-rich phases, respectively, and $\gamma_s$ and $\gamma_w$ are the activity coefficients in their respective phases. Chiou et al. (53) made the reasonable assumption that very little water dissolves into the organic solute. In that case, $X_s^o \gamma_s = X_w^o \gamma_w = 1$. Using earlier arguments, $X_w^o = C_w^o V_w$, and at saturation equilibrium, $C_w^o = S$. The resulting expression for aqueous solute solubility is then

$$S = 1/\gamma_w V_w \tag{43}$$

Incorporation of equation 43 into equation 41 with $V_w^* = V_w$ and a log transformation results in

$$\log S = -\log K_{ow} - \log V_o^* - \log \gamma_o^* + \log (\gamma_w^*/\gamma_w) \tag{44}$$

This expression only holds for a liquid or supercooled liquid solute. The equivalent expression for a solid solute requires incorporation of the melting point correction. Equation 43 must thus be modified by the fugacity ratio so that equation 45 is valid for a solid.

$$S = \frac{1}{\gamma_w V_w} \frac{f_2^s}{f_2^{ss}} \tag{45}$$

Incorporation of equation 5 with the same assumptions as those used to derive equation 9 leads to

$$\log S = -\log K_{ow} - \frac{\Delta S_f}{R} \left( \frac{T_m}{T} - 1 \right)$$
$$- \log V_o^* - \log \gamma_o^* + \log \left( \frac{\gamma_w^*}{\gamma_w} \right) \tag{46}$$

An examination of equations 44 and 45 reveals several important points that should be considered when evaluating correlations between S and $K_{ow}$. These points are as follows:

1. All solid solubility data should be adjusted via the fugacity correction if comparisons are made with solute data sets containing both liquids and solids.

2. If the solute forms an ideal solution in the water-saturated octanol phase, $\log \gamma_o^\bullet = 0$.

3. If the solute solubility in the pure water and octanol-saturated water phase are equal, the last term drops out. A plot of $\log K_{ow}$ versus $\log S$ would then produce a line with a slope of $-1$ and an intercept at $-\log V_o^\bullet$. This line has the same form as equation 38, where $A = -1$.

Miller et al. (*56*) examined the effects of the various components in equation 44 on the slope in the $\log S - \log K_{ow}$ correlation. They suggest an alternative form of equation 38, for example,

$$\log K_{ow} + \log S_L = \log Q = A - BV \tag{47}$$

where $A$ and $B$ represent the intercept and slope, respectively, and $V$ is the molar volume of each chemical. Such an approach should also be fruitful for molecular surface areas or any other molecular descriptor.

To examine the $\log S - \log K_{ow}$ correlations for high molecular weight aromatic hydrocarbons, Doucette (*9*) used the expression formulated in equation 38. He concluded that interpretation is very difficult for many of the previous correlations that have appeared in the literature for these classes of compounds. Reasons for this difficulty were as follows:

1. Many of the data sets consisted of experimental values obtained from different literature sources where a variety of conditions and analytical techniques were used. This lack of consistency makes accuracy and precision assessments difficult.

2. In several studies, $\log K_{ow}$ values estimated from group contribution methods (*59, 60*) were used in place of experimental values. Group contribution methods often overestimate $\log K_{ow}$ values, especially for the highly hydrophobic compounds (*61-63*). Errors associated with these methods tend to increase with increasing $K_{ow}$ values. In addition, these methods do not account for the variation in $\log K_{ow}$ between isomers.

3. Previous data sets generally lacked experimental values for compounds having $\log K_{ow} > 6$. Thus, detailed examina-

tions of the correlations were prevented for very hydrophobic substances.

In the study by Doucette (9), only experimental generator-column-
derived $K_{ow}$ and S values were used. Also, all experimental values were
measured at a single temperature of 25 °C. Finally, the data sets contained numerous experimental values for compounds having log $K_{ow} >$
6. In all, the data set contained 55 aromatic liquids and solids representing compounds such as alkylbenzenes, halogenated benzenes, PCBs,
PBBs, PCDFs, and PCDDs.

A plot of log $K_{ow}$ versus log S is shown in Figure 4. For solids, the
experimental values were corrected to their corresponding supercooled
liquid solubility by using the fugacity term. The regression obtained is

$$\log S = -1.24 \log K_{ow} + 1.19 \qquad r^2 = 0.96 \qquad (48)$$

Note that log $K_{ow}$ values range from 2 to 8 and aqueous solubility values
range from $1 \times 10^{-2}$ to $1 \times 10^{-13}$ M. Doucette (9) also examined the
slope for five individual classes of aromatic hydrocarbons and found
that B ranged from −0.965 for chlorobenzenes to −1.68 for the PCDDs.
He also found that calculated values are unreliable for most compounds
whose log $K_{ow}$ value exceeds 6. The average error in using equation 48
was 5.1% and had a range of −14.2%–18.9% for the log $K_{ow}$ value.

TOTAL MOLECULAR SURFACE AREA.    According to the total molecular surface area (TSA) concept, the major factor in the solubilization
process is the energy required to create a cavity in the solvent into which
the solute is placed. The energy needed for the hole formation is considered to be proportional to the surface area of the solute.

TSA is an advantageous parameter for solubility correlations because
structural effects (i.e., branching) are accounted for. TSA has been
found to be linearly related to the logarithm of the hydrocarbon solubility. Work in the pharmacology field (52, 64, 65) has established that (1)
branching, cyclization, and positional isomerism can be accounted for
by using TSA without introducing additional terms; (2) the surface area
method shows considerable promise for solubility estimation of complex
organic substances in water; and (3) the generality of TSA makes it the
single best molecular parameter for estimating solubilities. However, this
evaluation included very few compounds with solubilities less than
$10^{-6}$ M.

Hermann (66) originally developed the TSA concept and method
for calculating this parameter. He described TSA as the area of the cavity surface containing the centers of the water molecules in the first layer
around the solute. Valvani et al. (67) suggested a simplified computa-

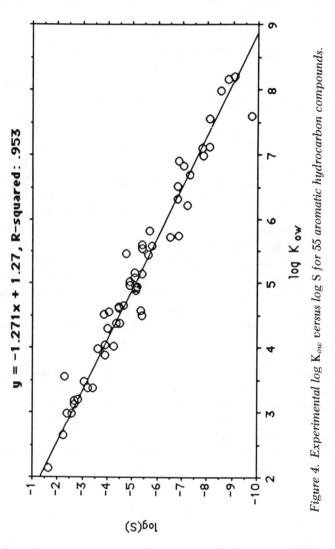

*Figure 4. Experimental log $K_{ow}$ versus log S for 55 aromatic hydrocarbon compounds.*

tional method that eliminated the need for a solvent radius. The most detailed account of various computational approaches to obtain TSA has been given by Pearlman (68).

Aqueous solubility correlations with TSA have been used by a variety of investigators for PCBs and PAHs (7, 38, 40–42). In general, these correlations take the form

$$\log S = C - B \left(\frac{T_m - T}{T}\right) - A(\text{TSA}) \tag{49}$$

where $A$, $B$, and $C$ are constants. If solid solubilities are corrected to their supercooled liquid state, the correlation takes the form

$$\log S = D - E \log (\text{TSA}) \tag{50}$$

where $D$ and $E$ are constants.

In some recent work, Doucette (9) examined the usefulness of the TSA–solubility correlations for the identical data base to the one presented in the log $K_{ow}$–solubility section. Values of TSA were calculated by using the method of Pearlman (68) with a program obtained from the Quantum Chemistry Program Exchange at Indiana University, Bloomington, IN. In this approach, each atom of a molecule is represented by a sphere centered at the equilibrium position of the nucleus. The radius of the sphere is equal to that of the van der Waals radius. The program calculates the surface area represented by all intersecting atomic spheres in the molecule. Standard geometry, standard interatomic bond lengths and bond angles, all of which were supplemented with X-ray diffraction data, were used to construct molecular topology. The substituted benzenes, PCDDs, and PCDFs were assumed to have planar conformations. A nonplanar conformation was used for halogenated biphenyls, and the phenyl rings were assumed to be perpendicular to each other. Methyl groups were approximated as single spheres of radius 2.0 Å, as suggested by Valvani et al. (67).

The relationship for 66 aromatic hydrocarbons, including chlorinated benzenes, PCBs, PBBs, PCDDs, and PCDFs, was best expressed by

$$\log S = -0.0313 \ (\text{TSA}) + 1.588 \qquad r^2 = 0.94 \tag{51}$$

The results obtained by this method compare favorably with those of previous studies that examined the relationship between S and TSA for hydrocarbons (66), alcohols (67), alkyl aromatic hydrocarbons (64), and PCBs (7). This correlation, however, was carried out with a uniform data set; for example, all solubilities were determined at the same temperature and by the generator column technique. The average error for equation 51 was found to be 7.6% and had a range of −18.4%–21.1%. All errors were incorporated in the log value of S. Of all the techniques examined, the TSA–solubility correlation is one of the most reliable and can most likely be improved by using molecular mechanics techniques (69).

MOLECULAR CONNECTIVITY INDEX.   The molecular connectivity index (MCI) is a calculated parameter that gives a quantitative assessment of molecular complexity. Methods for calculating MCIs have been given by Kier (70) and Kier and Hall (71). Briefly, the structure of the molecule is first drawn in the hydrogen atom suppressed-skeletal form. A simple connectivity value ($\delta^v$) is then assigned to each atom of the molecular skeleton. This value is expressed as $\delta^v = Z - h$, where Z is the maximum valence, and h is the number of hydrogens on the molecule.

The first-order, valence-corrected MCI, $^1X^v$, is computed as

$$^1X^v = \sum_1^n (\delta_i \delta_j) \tag{52}$$

where $\delta_i$ and $\delta_j$ are the valence-corrected connectivity values for the adjacent atoms i and j, respectively, and n is the number of bonds within the molecule. Higher order MCIs may also be calculated by including the number of paths, chains, clusters, and path clusters. Burkhard et al. (33) used 136 different MCIs for each molecule in an attempt to examine the utility of these to predict solubilities of n-alkanes and PCBs. A principal component analysis was used to obtain new sets of variables that had the same meaning for all compounds.

Significant correlations of MCIs with water solubility have also been observed for branched, cyclic, and straight-chain alcohols and hydrocarbons (31, 32). Other physical properties such as solute boiling point, density, gas–liquid chromatography (GLC) retention time, octanol-water partition coefficient, and molecular volume have also been correlated with MCIs (71).

Using the same experimental data base that was discussed in previous sections, Doucette (9) correlated solubilities with first-order MCIs. He found that the relationship is best given by the nonlinear equation

$$\log S = 0.0315 \ (MCI)^2 - 1.371 \ (MCI) + 0.844 \tag{53}$$

Of all the predictive methods examined, only experimental log $K_{ow}$ values showed a better correlation with solubility than MCI. An average error of 5.6%, ranging from −16.4% to 14.6%, was obtained from the 66 aromatic hydrocarbons.

A summary of errors for the predictive techniques discussed in this chapter is presented in Table II. Each technique may be used depending upon what data are available. Classes of compounds represented in these correlations include PCBs, PBBs, PCDDs, PCDFs, chlorinated benzenes, and methylated biphenyls. All solubilities included in the correlations were determined by the generator column method and only apply for an ambient temperature of 25 °C. The average error expected ranges from 5% to 7% in the log solubility value. This error range means that prediction of solubility for these classes of compounds within a factor of about 2 is possible when the solubility is around $1 \times 10^{-6}$ M. For compounds whose solubility is around $1 \times 10^{-13}$ M, the error increases to a factor of about 5. As bigger and better data bases appear, additional refinements will be possible. Errors may then be reduced even further.

**Table II. Predictive Errors for Selected Solubility Estimation Methods**

| Method | Number of Compounds | Average Error (%) | Range of Error (%) |
|--------|--------------------|--------------------|--------------------|
| UNIFAC | 35 | 6.1 | −8.1–12.2 |
| log $K_{ow}$ | 55 | 5.1 | −14.2–18.9 |
| TSA | 66 | 7.6 | −18.4–21.1 |
| MCI | 66 | 5.6 | −16.4–14.6 |
| log $K_{ow}$[a] | 66 | 14.2 | −29.4–43.4 |

SOURCE: Data are from reference 9.
[a]These values of log $K_{ow}$ were calculated.

## Conclusions

One of the most fruitful approaches to gain understanding of how chemicals behave in the environment involves the integration of physicochemical properties of a chemical with the appropriate transport equations in fluid media. This type of an approach permits "exposure profiles" to be performed for a variety of chemicals and environments. Such models now exist for a variety of media, for example, groundwater, surface water, and the atmosphere.

The physicochemical properties that are necessary to model and interpret the environmental behavior of hydrophobic organic chemicals include aqueous solubility, vapor pressure, Henry's law constant, and liquid- and vapor-phase molecular diffusivities. Information on photolytic and chemical reaction kinetics is also necessary. Much of this data is presently not available for chlorinated aromatic hydrocarbons, especially as a function of temperature.

The recent appearance of high quality solubility and $K_{ow}$ data for selected PCBs, PBBs, PCDDs, and PCDFs (9, 10, 56) makes testing and developing physicochemical-property predictive schemes for these compounds possible. These data may then be used in combination with vapor pressure data to derive Henry's law constants for each substance. In this chapter, we evaluated several available methods to predict aqueous solubilities (activity coefficients), including the UNIFAC method, and we made correlations with the molecular surface area, the octanol–water partition coefficient, and the molecular connectivity index. The correlations for the latter three techniques covering the aforementioned molecular classes are

$$\log = -1.24 \log K_{ow} + 1.19 \qquad\qquad r^2 = 0.96 \quad (54)$$

$$\log = -0.0313 \ (TSA) + 1.588 \qquad\qquad r^2 = 0.94 \quad (55)$$

$$\log = -0.0315 \ (MCI)^2 - 1.371 \ (MCI) + 0.844 \qquad r^2 = 0.96 \quad (56)$$

Our analysis indicates that compounds whose solubility is in the $1 \times 10^{-6}$-M range may be predicted within a factor of 2 or 3, and compounds whose solubility is around $1 \times 10^{-13}$ M can be predicted within a factor of 5 or 6. These predictive errors are often better than available errors in present experimental data bases.

## Abbreviations and Symbols

| | |
|---|---|
| $\gamma$ | activity coefficient |
| $\gamma_s$ | activity coefficient of solute-rich phase |
| $\gamma_w$ | activity coefficient of water-rich phase |
| $\delta_i$ | valence-corrected connectivity value for atom $i$ |
| $\delta_j$ | valence-corrected connectivity value for atom $j$ |
| $\nu_k$ | number of groups in water and solute molecules |
| $\nu_m^{(j)}$ | number of groups of type $m$ in molecule $j$ |
| $a_2$ | activity of solute 2 |
| $a_{nm}$ | group interaction parameter |
| $A$ | linear regression constant |
| $B$ | linear regression constant |
| $\Delta C_p$ | change in solute heat capacity |
| $f^{L}$ | fugacity of pure liquid |
| $f^{ss}$ | fugacity at steady state |
| $\Delta G_f$ | Gibbs free energy of fusion |
| $\Delta H_f$ | enthalpy of fusion |
| $i$ | component number |
| $k$ | group number |
| $K$ | difference in interaction energy between various solvent and solute groups |
| $K_{ow}$ | octanol–water partition coefficient |
| $M$ | total number of components |

MCI     molecular connectivity index
$P$     pressure
$Q_k$     van der Waals surface area for group $k$
$r^2$     regression coefficient
$R$     gas constant
$R_k$     van der Waals volume for group $k$
$\Delta S_f$     entropy of fusion
$T$     temperature
$T_m$     melting point temperature
$T_t$     triple point temperature
TSA     total molecular surface area
$V_W$     total volume of water-rich phase
$X_j$     mole fraction of molecule $j$ in solution
$X_O$     solute mole fraction of octanol
$X_W$     solute mole fraction of water
$X_S^O$     mole fraction of solute in organic-rich phase
$X_W^O$     mole fraction of solute in water-rich phase
$^1X^v$     first-order, valence-corrected molecular connectivity index

## Acknowledgments

We thank Jean Schneider and Helen Grogan for typing the drafts and the revised manuscript. Nicholas Loux wrote the computer program for our UNIFAC calculations. Much of this work was funded by the University of Wisconsin Sea Grant College Program under grants from the National Sea Grant College Program, the National Oceanic and Atmospheric Administration, the U.S. Department of Commerce, and the state of Wisconsin.

## References

1. Dickson, K. L.; Maki, A. W.; Cairns, J. *Modeling the Fate of Chemicals in the Aquatic Environment;* Ann Arbor Science: Ann Arbor, MI, 1982.
2. Sheehan, P.; Korte, F.; Klein, W.; Bourdeau, P. *Appraisal of Tests to Predict the Environmental Behaviour of Chemicals;* Wiley: New York, 1985.
3. Reid, R. C.; Prausnitz, J. M.; Sherwood, T. K. *The Properties of Gases and Liquids;* McGraw–Hill: New York, 1977.
4. Yalkowsky, S. H.; Valvani, S. C.; Mackay, D. *Residue Rev.* 1983, 85, 42–55.
5. Lyman, W. J.; Reehl, W. F.; Rosenblatt, D. H. *Handbook of Chemical Property Estimation Methods. Environmental Behavior of Organic Compounds;* McGraw–Hill: New York, 1982.
6. Mackay, D.; Shiu, W. Y. *J. Phys. Chem. Ref. Data* 1981, 10, 1175–1199.
7. Mackay, D.; Mascarenhas, R.; Shiu, W. Y.; Valvani, S. C.; Yalkowsky, S. H. *Chemosphere* 1980, 9, 257–264.
8. Miller, M. M.; Ghodbane, S.; Wasik, S. P.; Tewari, Y. B.; Martire, D. E. *J. Chem. Eng. Data* 1984, 29, 184–190.

9. Doucette, W. J. Ph.D. Dissertation, University of Wisconsin, Madison, 1985.
10. Dickhut, R. M. M.S. Thesis, University of Wisconsin, Madison, 1985.
11. Dickhut, R. M.; Andren, A. W.; Armstrong, D. E. *Environ. Sci. Technol.* **1986**, *20*, 807-810.
12. Liss, P. S.; Slinn, W. G. N. *Air-Sea Exchange of Gases and Particles;* Reidel: Boston, 1983.
13. Mackay, D.; Paterson, S.; Schroeder, W. H. *Environ. Sci. Technol.* **1986**, *20*, 810-816.
14. Prausnitz, J. M.; Lichtenthaler, R. N.; de Azevedo, E. G. *Molecular Thermodynamics of Fluid-Phase Equilibria;* Prentice-Hall: Englewood Cliffs, NJ, 1986.
15. Hildebrand, J. H.; Scott, R. L. *Regular Solutions;* Prentice-Hall: Englewood Cliffs, NJ, 1962.
16. Lewis, G. N.; Randall, M. *Thermodynamics*, 2nd ed.; McGraw-Hill: New York, 1961.
17. Yalkowsky, S. H. *J. Pharm. Sci.* **1981**, *70*, 971-973.
18. Hollenbeck, G. R. *J. Pharm. Sci.* **1980**, *69*, 1241-1242.
19. Castellan, G. W. *Physical Chemistry;* Addison-Wesley: Reading, MA, 1971.
20. Yalkowsky, S. H. *Ind. Eng. Chem. Fundam.* **1979**, *18*, 108-111.
21. Hildebrand, J. H.; Prausnitz, J. M.; Scott, R. L. *Regular and Related Solutions;* Van Nostrand Reinhold: New York, 1970.
22. Ben-Naim, A. *Water and Aqueous Solutions: Introduction to a Molecular Theory;* Plenum: New York, 1974.
23. Pierotti, R. A. *Chem. Rev.* **1976**, *76*, 717-726.
24. Acree, W. E. *Thermodynamic Properties of Nonelectrolyte Solutions;* Academic: New York, 1984.
25. Fredenslund, A.; Gmehling, J.; Rasmussen, P. *Vapor-Liquid Equilibria Using UNIFAC, a Group-Contribution Method;* Elsevier Scientific: New York, 1977.
26. Gmehling, J. G.; Anderson, T. F.; Prausnitz, J. M. *Ind. Eng. Chem. Fundam.* **1978**, *17*, 269-273.
27. Campbell, J. R.; Luthy, R. G. *Environ. Sci. Technol.* **1985**, *19*, 980-985.
28. Banerjee, S.; Yalkowsky, S. H.; Valvani, S. C. *Environ. Sci. Technol.* **1980**, *14*, 1227-1229.
29. Hansch, C.; Quinlan, J. E.; Lawrence, G. L. *J. Org. Chem.* **1968**, *33*, 347-350.
30. Yalkowsky, S. H.; Valvani, S. C. *J. Pharm. Sci.* **1980**, *69*, 912-922.
31. Hall, L. H.; Kier, L. B.; Murry, W. J. *J. Pharm. Sci.* **1975**, *64*, 1974-1977.
32. Edward, J. T. *Can. J. Chem.* **1982**, *60*, 2573-2578.
33. Burkhard, L. P.; Andren, A. W.; Armstrong, D. E. *Chemosphere* **1983**, *12*, 935-943.
34. McAuliffe, C. *J. Phys. Chem.* **1966**, *70*, 1267-1275.
35. Leinonen, P. J.; Mackay, D.; Phillips, C. R. *Can. J. Chem. Eng.* **1971**, *49*, 288-290.
36. Lande, S. S.; Banerjee, S. *Chemosphere* **1981**, *10*, 751-759.
37. Horvath, C. *Halogenated Hydrocarbons;* Marcel Dekker: New York, 1982.
38. Burkhard, L. P.; Armstrong, D. E.; Andren, A. W. *Environ. Sci. Technol.* **1985**, *19*, 590-596.
39. Yalkowsky, S. H.; Valvani, S. C. *J. Chem. Eng. Data* **1979**, *24*, 127-129.
40. Yalkowsky, S. H.; Orr, R. J.; Valvani, S. C. *Ind. Eng. Chem. Fundam.* **1979**, *18*, 351-353.
41. Amidon, G. L.; Anik, S. T. *J. Chem. Eng. Data* **1981**, *26*, 28-33.
42. Pearlman, R. S.; Yalkowsky, S. H.; Banerjee, S. J. *J. Phys. Chem. Ref. Data* **1984**, *13*, 555-562.

43. Kamlet, M. J.; Abbound, J. L. M.; Abraham, M. H.; Taft, R. W. *J. Org. Chem.* **1983**, *48*, 2877-2887.
44. Kamlet, M. J.; Abraham, M. H.; Doherty, R. M.; Taft, R. W. *J. Am. Chem. Soc.* **1984**, *106*, 464-466.
45. Taft, R. W.; Abraham, M. H.; Doherty, R. M.; Kamlet, M. J. *Nature* **1985**, *313*, 384-386.
46. Gmehling, J.; Rasmussen, P.; Fredenslund, A. *Ind. Eng. Chem. Process Des. Dev.* **1982**, *21*, 118-127.
47. Grain, C. F. In *Handbook of Chemical Property Estimation Methods. Environmental Behavior of Organic Compounds;* Lyman, W. J.; Reehl, W. F.; Rosenblatt, D. H., Eds.; McGraw-Hill: New York, 1982.
48. Burkhard, L. P.; Andren, A. W.; Loux, N. T.; Armstrong, D. E. *Environ. Sci. Technol.* **1986**, *20*, 527-528.
49. May, W. E.; Wasik, S. P.; Freeman, D. H. *Anal. Chem.* **1978**, *50*, 997-1000.
50. Stolzenburg, T. R.; Andren, A. W. *Anal. Chim. Acta* **1983**, *151*, 271-274.
51. Arbuckle, W. B. *Environ. Sci. Technol.* **1986**, *20*, 527.
52. Amidon, G. L.; Yalkowsky, S. H.; Leung, S. *J. Pharm. Sci.* **1974**, *63*, 1858-1866.
53. Chiou, C. T.; Schmedding, D. W.; Maines, M. *Environ. Sci. Technol.* **1982**, *16*, 4-10.
54. Tewari, Y. B.; Martire, D. E.; Wasik, S. P.; Miller, M. M. *J. Solution Chem.* **1982**, *11*, 435-445.
55. Bowman, B. T.; Sans, W. W. *J. Environ. Sci. Health* **1983**, *B18* (2), 221-227.
56. Miller, M. M.; Wasik, S. P.; Huang, G.; Shiu, W.; Mackay, D. *Environ. Sci. Technol.* **1985**, *19*, 529-537.
57. Chiou, C. T.; Freed, V. H.; Schmedding, D. W.; Kohnert, R. L. *Environ. Sci. Technol.* **1977**, *11*, 475-478.
58. Collander, R. *Acta Chem. Scand.* **1951**, *5*, 774.
59. Nys, G. G.; Rekker, R. F. *Eur. J. Med. Chem.* **1974**, *9*, 361-375.
60. Hansch, C.; Leo, A. J. *Substituent Constants for Correlation Analysis in Chemistry and Biology;* Wiley: New York, 1979.
61. Tulp, M. T. M.; Hutzinger, O. *Chemosphere* **1978**, *10*, 849-860.
62. Valvani, S. C.; Yalkowsky, S. H.; Roseman, T. J. *J. Pharm. Sci.* **1981**, *70*, 502-507.
63. Woodburn, K. B.; Doucette, W. J.; Andren, A. W. *Environ. Sci. Technol.* **1984**, *18*, 457-459.
64. Amidon, G. L.; Yalkowsky, S. H.; Valvani, S. C. *J. Phys. Chem.* **1975**, *79*, 2239-2246.
65. Amidon, G. L.; Anik, S. T. *J. Pharm. Sci.* **1976**, *65*, 801-806.
66. Hermann, R. B. *J. Phys. Chem.* **1972**, *76*, 2754-2759.
67. Valvani, S. C.; Yalkowsky, S. H.; Amidon, G. L. *J. Phys. Chem.* **1976**, *80*, 829-835.
68. Pearlman, R. S. In *Physical Chemical Properties of Drugs;* Yalkowsky, S. H.; Sinkula, A. A.; Valvani, S. C., Eds.; Marcel Dekker: New York, 1980.
69. Burkert, U.; Allinger, N. L. *Molecular Mechanics;* ACS Monograph 177; American Chemical Society: Washington, DC, 1982.
70. Kier, L. B. In *Physical Chemical Properties of Drugs;* Yalkowsky, S. H.; Sinkula, A. A.; Valvani, S. C., Eds.; Marcel Dekker: New York, 1980.
71. Kier, L. B.; Hall, L. H. In *Medicinal Chemistry. A Series of Monographs;* DeStevens, G., Ed.; Academic: New York, 1976.

RECEIVED for review June 16, 1986. ACCEPTED October 10, 1986.

$A = vapour$
$F = particle$

# 2

# Vapor–Particle Partitioning of Semivolatile Organic Compounds

Terry F. Bidleman[1] and William T. Foreman[2,3]

[1]Department of Chemistry, Marine Science Program, and Belle W. Baruch Institute for Marine Biology and Coastal Research, University of South Carolina, Columbia, SC 29208
[2]Department of Chemistry, University of South Carolina, Columbia, SC 29208

*Pesticides, polychlorinated biphenyls (PCBs), and other semivolatile organic compounds (SOCs) exist in air as vapors and are adsorbed to particulate matter. Estimates of the vapor–particle distribution and factors influencing it were obtained from high-volume sampling experiments using a glass fiber filter to collect particles and an adsorbent trap to collect vapors. Measurements of airborne organochlorine pesticides and PCBs in four cities over a wide temperature range were used to determine the partition coefficient A(TSP)/F, where A and F are the adsorbent- and filter-retained SOC concentrations (ng/m³), respectively, and TSP is the total suspended particle concentration (μg/m³). Laboratory determinations of A(TSP)/F were carried out by equilibrating particle-loaded filters at 20 °C in an airstream containing controlled SOC vapor concentrations. Field and laboratory A(TSP)/F values agreed well for most organochlorines, with the exception of hexachlorobenzene (HCB). The A(TSP)/F value was closely correlated with the subcooled liquid vapor pressure (p$_L^o$) of the SOCs, but not with the solid-phase vapor pressure (p$_S^o$). Implications of vapor–particle partitioning to the atmospheric deposition of SOCs are discussed.*

$\mathbf{S}$EMIVOLATILE ORGANIC COMPOUNDS (SOCs) such as pesticides, polychlorinated biphenyls (PCBs), and polycyclic aromatic hydrocarbons (PAHs) are present in the atmosphere as vapors and are attached to suspended particles. The vapor–particle distribution $(V/P)$ influences

[3]Current address: Cooperative Institute for Research in Environmental Sciences, University of Colorado, Campus Box 449, Boulder, CO 80309

0065–2393/87/0216–0027$08.50/0

atmospheric removal of SOCs. For example, heavier PCBs having greater fractions in the particle phase are preferentially deposited by washout and dry deposition (1, 2). Understanding $V/P$ is also important in designing control systems for stacks because removing particles is easier than removing vapors. In this article, results of field and laboratory high-volume sampling experiments are presented to investigate the distribution of SOCs between the vapor phase and urban air particulate matter.

## Characteristics of Urban Air Particulate Matter

Urban aerosols are derived from many sources including auto exhaust, incineration, industrial emissions, photochemical gas–particle conversion, and eolian weathering of soils. Major constituents of urban particulate matter include organic and elemental carbon, sulfate, nitrate, ammonium, silicates, alkali and alkaline earth metals, aluminum, iron, and lead (3–5). Quantifying the contributions of various sources to urban aerosol composition was the subject of a U.S. Environmental Protection Agency (EPA) workshop in 1982 (6).

The particle size spectrum ranges roughly from 0.005 to 20 $\mu$m in diameter. The particle range is limited on the low end by rapid coagulation of smaller particles and on the high end by gravitational settling (7). The distribution of particle sizes is approximately log normal, and a convenient measure of size is the mass median effective diameter (MMED). For a polydispersed aerosol having MMED = 1 $\mu$m, 50% of the total suspended particle (TSP) mass is contained on particles larger and smaller than 1 $\mu$m. The dispersion is expressed by the geometric standard deviation ($\sigma_g$). Discussions of particle statistics and general characteristics of particles in air were given by Cadle (8), Corn (9), and Whitby (10). The MMED and $\sigma_g$ values are remarkably similar for particles in different U.S. cities. A 1970 survey in six cities showed the average MMED between 0.4 and 0.8 $\mu$m and $\sigma_g$ between 5 and 10 $\mu$m (11). However, these yearly averages blur large short-term differences. For example, in Cincinnati, MMEDs were 0.44–0.48 $\mu$m during noninversion periods and 1.3–2.0 $\mu$m during inversion periods (11).

Surface areas of particles collected in Pittsburgh have been measured by Corn et al. (12) by using a nitrogen adsorption method. Surface areas varied from 1.9 $m^2/g$ in summer to 3.0 $m^2/g$ in winter. To put these into perspective, surface areas of solid adsorbents used to sample organic vapors are in the 20–800-$m^2/g$ range (13). The data of Corn et al. (12) were for TSPs collected on glass fiber filters and thus may differ from surface areas of particles suspended in air.

The TSP concentrations are determined by their weight on a glass fiber filter following high-volume sampling. The current primary and secondary 24-h standards for TSP concentrations in ambient air are 75 and 50 $\mu g/m^3$, respectively (14). In addition to TSP, two other measures of particle concentration in ambient air are frequently used. A size-selective inlet can be placed over the high-volume sampler to screen out the very large particles and provide a measure of inhalable particulate matter less than 15 $\mu m$. Dichotomous samplers split the particle mass into coarse (>2.5 $\mu m$) and fine (<2.5 $\mu m$) fractions.

The carbonaceous fraction of urban particulate matter has been the subject of many investigations. Only an overview will be given here; several reviews (15-21) present detailed information. In 1975, annual average particulate carbon (organic plus elemental) concentrations in 46 U.S. cities ranged from 4 to 20 $\mu g/m^3$ with an overall mean of 10.3 $\mu g/m^3$ (21). The range and mean at 20 rural sites were 1.5-6.0 and 3.7 $\mu g/m^3$, respectively (21). Carbon typically constitutes 11%-21% of TSP (Table I).

Several methods are used to determine unspeciated carbon. Extractable organic matter (EOM) or extractable organic carbon is obtained by extracting the particle-loaded filters with an organic solvent and weighing the residues on a microbalance after solvent evaporation (22-24) by combustion of the extract to $CO_2$ in a carbon analyzer (25-28), or less commonly by gas chromatographic analysis (29, 30). Extraction yields depend on solvent polarity (22-30), and binary mixtures of a polar and nonpolar solvent are recommended for highest efficiency (25). Sometimes EOM is fractionated by successive extraction with solvents of different polarities. This scheme can be used to distinguish "primary" organic compounds, which are released directly into the atmosphere in the particle phase or which condense shortly after introduction, from "secondary" organic compounds, which result from gas-particle conversion following chemical reactions of gaseous precursors. Higher proportions of polar fraction EOM resulting from the oxidation of hydrocarbons have been found during pollution episodes and have been correlated with ozone levels (26-28, 31-32).

Total organic carbon (TOC) is determined by dry (21, 25, 31, 32, 34) or wet (35) oxidation of a small portion of a particle-loaded filter in a carbon analyzer. Free or elemental carbon is determined by combustion in a carbon analyzer after prior wet or dry oxidative removal of organic carbon (21, 31, 34, 36, 37) or solvent extraction of EOM (26-28), and by absorption or reflectance measurements (38-40).

Table I gives some illustrative values of EOM, TOC, and elemental carbon as percentages of airborne particulate matter. Several of these were taken from Shah et al. (21), who analyzed more than 1300 filters from the National Air Surveillance Network filter bank in 1975. The study covered 46 cities and 20 rural sites. Although average EOM contents among cities appear similar, large short-term differences can occur. For example, during a pollution episode in New Jersey in February

Table I. Mean Carbon Percentages in Urban Particulate Matter

| Place and Time Period | Particle Type[a] | EOM (%) | TOC (%) | Elemental Carbon (%) | Ref. |
|---|---|---|---|---|---|
| Pasadena, CA | | | | | |
| Annual, 1973 | TSP | 23 | | | 28 |
| Annual, 1975 | TSP | | 9.9 | 4.9 | 21 |
| Denver, CO | | | | | |
| Winter, 1978 | TSP | | 11.3 | 8.2 | 3 |
| Winter, 1978 | FPM | | 19.6 | 16.3 | 3 |
| Annual, 1975 | TSP | | 7.0 | 3.9 | 21 |
| New York, NY | | | | | |
| Unspecified | | | 8.1 | 10.2 | 3 |
| Winter, 1976, 1978 | TSP | 16 | | | 23 |
| Summer, 1976–1978 | TSP | 13 | | | 23 |
| Annual, 1975 | TSP | | 11.8 | 8.8 | 21 |
| Newark, NJ | | | | | |
| Summer, 1981 | IPM | 23 | 10.8 | 4.5 | 33 |
| Annual, 1975 | TSP | | 9.3 | 7.3 | 21 |
| Elizabeth, NJ | | | | | |
| Summer, 1981 | IPM | 22 | 8.0 | 3.3 | 33 |
| Annual, 1975 | TSP | | 9.7 | 5.7 | 21 |
| Camden, NJ | | | | | |
| Summer, 1981 | IPM | 22 | 10.0 | 5.9 | 33 |
| Annual, 1975 | | | 8.0 | 5.2 | 21 |
| New Jersey, northeast | | | | | |
| Winter, 1983 | IPM | 33 | | | 41 |
| Detroit, MI | | | | | |
| Summer, 1981 | FPM | | 22 | 9.4 | 48 |
| Annual, 1975 | TSP | | 6.8 | 3.7 | 21 |
| Chicago, IL | | | | | |
| Annual, 1975 | TSP | | 6.3 | 4.5 | 21 |
| Miami, FL | | | | | |
| Annual, 1975 | TSP | | 6.9 | 3.6 | 21 |
| Phoenix, AZ | | | | | |
| Annual, 1975 | TSP | | 10.4 | 3.6 | 21 |
| Seattle, WA | | | | | |
| Annual, 1975 | TSP | | 11.5 | 8.0 | 21 |
| Columbia, SC | | | | | |
| Annual, 1975 | TSP | | 10.7 | 7.5 | 21 |
| Boston, MA | | | | | |
| Annual, 1975 | TSP | | 7.3 | 6.2 | 21 |

[a] Abbreviations are as follows: TSP is total suspended particles, IPM is inhalable particulate matter (<15 $\mu$m), and FPM is fine particulate matter (<2.5 $\mu$m).

1983, EOM rose to 30%–55% of inhalable particulate matter (*41*). Grosjean and Friedlander (*28*) found that EOM in Pasadena aerosols varied in the range 5%–54%. Also, carbon and individual organic compounds are enriched on the fine fraction from dichotomous samplers (*3*) and on the smaller aerosols separated by cascade impaction (*42–47*).

In addition to the above nonspecific measures of particulate carbon, a large number of individual organic compounds have been identified and quantified in aerosols by gas and liquid chromatography and gas chromatography–mass spectrometry (GC–MS). The scope of this work can be appreciated from several reviews (*15–20*). A detailed knowledge of the organic composition of airborne particulate matter is important in understanding pollutant transport, health risks, and natural geochemical cycles.

## Vapor–Particle Distribution of Semivolatile Organic Compounds in the Atmosphere

**Exchangeable and Nonexchangeable Compounds.** Although much remains to be learned about the nature of $V/P$ interactions, SOCs bound to particles appear to be of two types: a nonexchangeable fraction that is strongly adsorbed to active sites on the particles or imbedded within the particle matrix and is not in equilibrium with its vapor phase in the atmosphere, and an exchangeable fraction that is more loosely bound and appears to be controlled by the concentration of the SOC vapor in air.

Indications that some of the SOCs may be nonexchangeable are provided by studies showing that the yield of PAHs from the extraction of atmospheric particulate matter depends on the choice of solvent. Some investigators found that aromatic and polar solvents recover more PAHs than nonpolar solvents like hexane or cyclohexane (*49–51*). Others found no significant differences (*52*). Griest et al. (*53*) reported difficulty recovering PAHs adsorbed on fly ash. Eiceman and Vandiver (*54*) used gas chromatographic experiments to investigate the equilibrium between PAH vapors and fly ash. Irreversible adsorption occurred at low PAH concentrations, but once a layer of PAHs had become attached to the particles, additional PAHs were adsorbed reversibly.

Nearly a decade ago, Junge (*55*) considered the situation of the exchangeable fraction and presented a theoretical model of physical adsorption to aerosols. The fraction of particle-bound SOCs ($\phi$), solute vapor pressure ($p^\circ$), and the particle surface area per unit volume of air ($\theta$) available for adsorption were related by

$$\phi = c\theta/(p^\circ + c\theta) \tag{1}$$

where $c$ is a constant. From this model, Junge prepared plots of $\phi$ versus $\theta$ for compounds having $p°$ values in the range $10^{-4}$–$10^{-8}$ torr. Over the range of $\theta$ in urban air, a compound having $p° = 10^{-6}$ torr might be expected to be 20%–80% in the particle fraction. As we will show, this range is similar to experimental estimates in urban air.

**Problems with Determining the Exchangeable Fraction.** Direct determination of $V/P$ in ambient air is an experimentally challenging problem that has not yet been solved. Aerial concentrations of SOCs are usually low, on the order of nanograms or picograms per cubic meter, so large volumes of air must be sampled to obtain enough material for analysis. The common technique for preconcentrating SOCs is to draw hundreds to thousands of cubic meters of air through a glass or quartz fiber filter followed by a solid adsorbent trap (13), and the apparent $V/P$ is operationally defined by the adsorbent- to filter-retained ratio $(A/F)$.

How closely $A/F$ represents $V/P$ is uncertain. Many investigators feel that $A/F$ overestimates $V/P$ because of the "blow-off effect": stripping of SOCs from particles on the filter by the flowing airstream. The particle mass on the filter may also act as an adsorbent and lead to an underestimation of $V/P$. Field collections using different sampling times and different types of filters have provided evidence for each process, although blow-off losses are more commonly reported. Several studies (24, 45, 56–59) are reviewed briefly in reference 13.

A limitation of field investigations is that aerial concentrations and temperatures change during the normal 24-h or longer sampling period. The temperature effect is especially important because vapor pressures of SOCs nearly double with a 5 °C increase. Whether blow-off losses or adsorption gains are observed depends on how and when these variables change during a sampling period. For example, what is deposited on the filter at night may be desorbed again in the heat of the next day. The situation with PAHs is complicated by the fact that degradative losses may also occur (60–66).

## Experimental Measurement of the Adsorbent-to-Filter-Retained Ratio

**Field Investigations.** Despite the problems mentioned, field $A/F$ measurements provide insights into the factors influencing vapor–particle interactions in the atmosphere. Investigations of many classes of SOCs (PAHs, n-alkanes, fatty acids, phthalates, and organochlorines) have shown that, within a homologous series, $A/F$ decreases with increasing molecular weight (1, 45, 57, 67–75). Higher proportions of filter-retained SOCs have also been noted for samples collected in winter compared with summer (*57, 68, 70–73, 76*). Volatility is therefore a major factor controlling adsorption of SOCs to atmospheric particulate matter.

Yamasaki et al. (*73*) were the first to quantitatively relate $A/F$ to temperature and particle concentration. Samples were taken over 24-h periods for 1 year in Tokyo with a glass fiber filter to collect particles followed by a polyurethane foam column to adsorb vapors. The $A/F$ of individual PAHs were related to average sampling temperature, $T$, and the TSP concentration by

$$\log A(\text{TSP})/F = m/T + b \tag{2}$$

where $T$ is in Kelvin units. This equation, which is essentially the same as Junge's relationship (equation 1, *see* Appendix in this chapter), was successful in describing $A/F$ of PAHs in urban air. Regression fits to equation 2 yielded $r^2 \geq 0.8$ for most PAHs. The Tokyo samples are the best available data on the adsorption of PAHs to ambient air particulate matter. Temperatures were fairly constant, varying by only 2.2–2.5 °C over each collection period; nonequilibrium effects caused by changes in vapor pressure with temperature were minimized.

Keller and Bidleman (*71*) collected EOM and PAHs in Columbia, SC, under conditions where the diurnal temperature changes were much larger, on the order of 10–20 °C. Nevertheless, $A/F$ partitioning of EOM, fluoranthene, and pyrene followed equation 2, and the parameters $m$ and $b$ were reasonably close to those of the Tokyo samples.

Recently, we carried out field and laboratory investigations of PCBs and organochlorine pesticides partitioning between the vapor phase and urban particulate matter on glass fiber filters. Results of the urban air collection experiments are discussed here, and laboratory equilibration studies are presented in a later section. Experimental details of both are given elsewhere (*76, 77*).

High-volume samples were taken in four cities using a glass fiber filter backed up by a polyurethane foam, diphenyl-*p*-phenylene oxide (Tenax), or styrene-divinylbenzene (XAD-2) trap. The cities and dates of sampling were Columbia, SC (1977–1980); Denver, CO (January 1980); New Bedford, MA (June 1980); and Stockholm, Sweden (1983–1985). Sampling periods were usually 24–48 h. Experiments in the U.S. cities were done to also evaluate adsorbent collection efficiencies for PCBs and organochlorine pesticide vapors, and these results, along with analytical methods, have been presented (*68, 76*). A summary of organochlorine and TSP concentrations in the four cities is given in Table II.

Partitioning of organochlorines between the vapor phase and the particle mass on the glass fiber filter was strongly influenced by volatility. A chromatogram of the PCB-containing fraction of a Stockholm air sample is shown in Figure 1. Higher proportions of the less volatile SOCs

**Table II. Average Organochlorine and Total Suspended Particle Concentrations**

| Location | PCB[a] | DDE | DDT | Chlordane[b] | α-HCH | HCB | TSP |
|---|---|---|---|---|---|---|---|
| Columbia, SC | 1.5 | 0.093 | 0.048 | 1.3 | 0.9 | 0.29 | 53 |
| Denver, CO | 0.45 | 0.021 | 0.043 | 0.063 | 0.22 | 0.24 | 175 |
| New Bedford, MA (landfill) | 9.3 | | | | | | 193 |
| Stockholm, Sweden | 0.067 | 0.0034 | 0.0061 | 0.010 | 0.40 | 0.065 | 34 |

NOTE: All results are given in nanograms per cubic meter, except for TSP values, which are given in micrograms per cubic meter.
[a] The PCB is Aroclor 1254.
[b] Chlordane is a mixture of *cis*-chlordane, *trans*-chlordane, and *trans*-nonachlor.

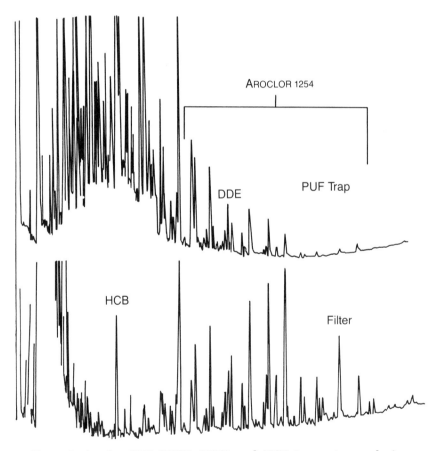

*Figure 1. Aroclor 1254 (PCB), DDE, and HCB in an air sample from Stockholm taken in January 1985. Most organochlorine pesticides have been removed by silicic acid fractionation. Column: 25-m fused silica BP-1 bonded phase (SGE Corp.).*

can be seen on the glass fiber filter compared to the polyurethane foam trap, and even within the Aroclor 1254 (PCB) pattern, the heavier components are enriched on the glass fiber filter.

Plots of equation 2 are shown in Figure 2 for Aroclor 1254 and $\alpha$-hexachlorocyclohexane ($\alpha$-HCH). Aroclor 1254 $A/F$ values were obtained in all four cities. The $\alpha$-HCH was measurable on the glass fiber filter only at low temperatures, and so all $\alpha$-HCH results are from Denver and Stockholm. Plots of equation 2 were also made for 1,1-dichloro-2,2-bis (p-chlorophenyl)ethylene (DDE), 1,1,1-trichloro-2,2-bis(p-chlorophenol)ethane (DDT), chlordane, and hexachlorobenzene (HCB) (76). The DDE, DDT, and chlordane data were from Columbia, Denver, and

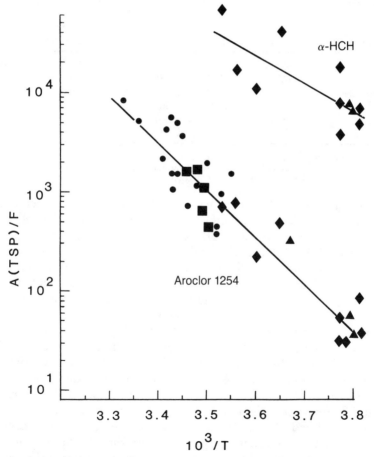

*Figure 2. Equation 2 plots for Aroclor 1254 and $\alpha$-HCH. Key: ●, Columbia, SC; ■, New Bedford, MA; ▲, Denver, CO; and ◆, Stockholm, Sweden.*

Stockholm, but $A/F$ values of the more volatile HCB were determined only in the latter two cities.

Regression parameters for the organochlorines are given in Table III along with best-fit $A(TSP)/F$ values at 20 and 0 °C. The individual points in Figure 2 and similar plots show some variability, but the $r^2$ values in the 0.88–0.90 range for Aroclor 1254, DDE, DDT, and chlordane compare well with the PAH $r^2$ values of Yamasaki et al. (73). Lower $r^2$ values were obtained for HCB and $\alpha$-HCH because of the smaller number of data points and greater scatter. As mentioned, HCB and $\alpha$-HCH residues on the glass fiber filters approached the detection limits except at the coldest temperatures.

### Table III. Regression Parameters for Organochlorines

| | | | | | log A(TSP)/F[a] | |
|---|---|---|---|---|---|---|
| Organochlorine | m | b | r² | n | 0 °C | 20 °C |
| α-HCH | −2755 | 14.286 | 0.574 | 11 | 4.194 | 4.883 |
| HCB | −3328 | 16.117 | 0.687 | 8 | 3.925 | 4.758 |
| Aroclor 1254 | −4686 | 19.428 | 0.885 | 34 | 2.265 | 3.436 |
| Chlordane | −4995 | 21.010 | 0.901 | 18 | 2.711 | 3.960 |
| p,p'-DDE | −5114 | 21.048 | 0.881 | 15 | 2.320 | 3.598 |
| p,p'-DDT | −5870 | 22.824 | 0.885 | 18 | 1.320 | 2.788 |

NOTE: All regression parameters are based on equation 2.
[a] Values of $A(TSP)/F$ were calculated by dividing nanograms of SOCs per cubic meter of air by nanograms SOCs per microgram of particles.

Several other factors may have contributed to the scatter in the equation 2 plots: day–night temperature fluctuations over the collection periods may have caused blow-off losses or adsorption gains to particles on the glass fiber filter during individual experiments. Also, organochlorine concentrations in the air probably varied over an experiment. Data in the equation 2 plots were obtained from measurements in several cities over 8 years and were influenced by local differences in particle size distribution, surface area, and content of carbonaceous material. The influence of relative humidity on SOC adsorption to atmospheric particles is unknown, and although not measured, relative humidity probably varied substantially over the course of these experiments. Adsorption of nonpolar SOCs might be expected to decrease at high humidity because of the displacing effect of water molecules. Chiou and Shoup (78) found that the uptake of chlorobenzene vapors by soil was greatly diminished at high humidities. Spencer et al. (79, 80) incorporated lindane and dieldrin into dry soil and measured an increase in their partial pressures when the soil was hydrated. However, urban aerosols have a much higher organic content than most soils, and humidity effects in soil adsorption studies may not apply to atmospheric particulate matter. Considering the large list of variables, the correlations are surprisingly good. Volatility thus stands out as the primary factor governing SOCs partitioning to urban aerosols.

If the interaction between SOC vapors and atmospheric particulate matter is relatively weak physical adsorption, the slope ($m$) of equation 2 plots should be the same as the slope ($m'$) of log $p°$ versus $1/T$ (*see* Appendix in this chapter). Equation 2 slopes for PCBs, DDE, DDT, and chlordane were in the $-4700$ to $-5900$ range (Table III). These slopes are similar to $m'$ for most high molecular weight SOCs. Heats of adsorption ($\Delta H_A$), calculated from $m$ values, were at most 2–4 kcal/mol greater than heats of vaporization of the subcooled liquid phase ($\Delta H_{V,L}$) for PCBs, DDE, DDT, and chlordane (*76*). Yamasaki et al. (*81*) also found that $\Delta H_A$ values were only 2–4 kcal/mol greater than $\Delta H_{V,L}$ values for PAHs.

The strong dependence of $A/F$ on temperature suggested a general relationship based on SOC volatility. Vapor pressures of the organochlorines in this work and the PAHs investigated by Yamasaki et al. (*73, 81*) span 4 orders of magnitude (Table IV). Figure 3 shows a log–log plot of the best-fit $A(TSP)/F$ values at 20 °C (Table III) versus the subcooled liquid vapor pressure ($p_L°$) of the SOCs. Differences between $p_L°$ and the solid-phase vapor pressure $p_S°$ will be discussed in the next section. The $A(TSP)/F$ values have been plotted for the organochlorines, and for PAHs by using data from Yamasaki et al. (*73*) and Keller and Bidleman (*71*).

Because the uncertainties in the equation 2 parameters for $\alpha$-HCH and HCB were larger than for the other organochlorines, we also extrapolated $A(TSP)/F$ values at 20 °C from data obtained at subzero temperatures (where filter-retained quantities were easily measured), assuming the slope ($m'$) of the log $p_L°$ versus $1/T$ relationship (*76*) (*see* Appendix in this chapter). The bars in Figure 3 thus represent the range of $A(TSP)/F$ at 20 °C obtained by using two different slopes in equation 2.

The $A/F$ of both classes of pollutants were closely related to $p_L°$, but different lines were obtained for organochlorines and PAHs. At a given $p_L°$, the proportion of PAHs on particulate matter was higher than for organochlorines. This behavior is currently under investigation, but as of now we suggest two possible explanations: PAHs are planar molecules, and may be more strongly adsorbed than most of the organochlorines, only one of which is flat (HCB). Also, urban particulate matter may contain some nonexchangeable PAHs bound to highly active sites or trapped within the particles. These nonexchangeable PAHs are extracted with solvent during analysis and counted with the exchangeable PAHs.

Comparing field $A/F$ values with those predicted by Junge (*55*) is interesting. From Figure 3, $A(TSP)/F$ values in urban air are approxi-

### Table IV. Vapor Pressures of Organochlorines and Polycyclic Aromatic Hydrocarbons

| Compound | mp (°C) | $p_S^{\circ a}$ | $p_L^{\circ b}$ |
|---|---|---|---|
| Organochlorines | | | |
| α-HCH | 159 | $2.5 \times 10^{-5}$ | $6.3 \times 10^{-4}$ |
| γ-HCH | 112 | $2.8 \times 10^{-5}$ | $2.4 \times 10^{-4}$ |
| HCB | 230 | $1.1 \times 10^{-5}$ | $1.4 \times 10^{-3}$ |
| Aroclor 1254 | | | $1.4 \times 10^{-5}$ |
| trans-Chlordane | 106 | $3.9 \times 10^{-6}$ | $2.9 \times 10^{-5}$ |
| cis-Chlordane | 106 | $3.0 \times 10^{-6}$ | $2.2 \times 10^{-5}$ |
| trans-Nonachlor | | | $1.7 \times 10^{-5}$ |
| p,p'-DDE | 89 | $2.6 \times 10^{-6}$ | $1.3 \times 10^{-5}$ |
| p,p'-DDT | 109 | $1.6 \times 10^{-7}$ | $1.3 \times 10^{-6}$ |
| Polycyclic aromatic hydrocarbons | | | |
| Phenanthrene | 101 | $9.5 \times 10^{-5}$ | $6.2 \times 10^{-4}$ |
| Anthracene | 216 | $4.3 \times 10^{-6}$ | $4.1 \times 10^{-4}$ |
| Fluoranthene | 111 | $5.1 \times 10^{-6}$ | $4.2 \times 10^{-5}$ |
| Pyrene | 156 | $3.1 \times 10^{-6}$ | $7.3 \times 10^{-5}$ |
| Benzo[k]fluoranthene | 217 | $3.9 \times 10^{-10}$ | $3.7 \times 10^{-8}$ |
| Benzo[e]pyrene | 179 | $2.4 \times 10^{-9}$ | $9.6 \times 10^{-8}$ |
| Benzo[a]pyrene | 177 | $2.4 \times 10^{-9}$ | $9.2 \times 10^{-8}$ |

NOTE: All vapor pressure values were determined at 20 °C.
[a]The $p^{\circ}$ values are averages of literature values compiled as an appendix to reference 82, except for cis- and trans-chlordanes. Values of $p_S^{\circ}$ for these two compounds were estimated from $p_L^{\circ}$ by using equation 3.
[b]The $p_L^{\circ}$ values were calculated from literature $p_S^{\circ}$ values by using equation 3 except for Aroclor 1254 (which is a liquid at 20 °C), cis- and trans-chlordanes, and trans-nonachlor. The $p_L^{\circ}$ values of the last three compounds were determined by using the GC method described in reference 82. The $p_L^{\circ}$ value of Aroclor 1254 was estimated by summing the partial pressures of PCB congeners (83) that were included in the air sample analysis.

mately 500 and 50 for an organochlorine and a PAH, respectively, having $p_L^{\circ} = 10^{-6}$ torr. If TSP = 60 $\mu$g/m$^3$, then $\phi = F/(A + F) = 0.11$–$0.55$ (11%–55%). These values agree reasonably well with $\phi = 20\%$–$80\%$, which is estimated from Junge's diagram (55) over the particle concentration range in urban air.

**Laboratory Equilibration Experiments.** Some of the problems with field experiments can be circumvented by investigating vapor adsorption to particles in the laboratory where temperature, relative humidity, and vapor concentrations can be controlled. We designed a system for equilibrating particle-loaded filters with SOC vapors that is similar to a high-volume sampling train (77) (Figure 4). Clean air was supplied by drawing laboratory air through a glass fiber filter and adsorbent cartridge. The SOC vapors were bled into a stainless steel chamber from coated-sand generator columns fitted with 0.2-$\mu$m filters (Acrodisc CR) on the outlet ends. The vapors were mixed with the high-

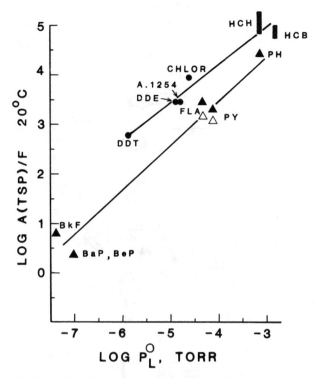

*Figure 3. Relationships between A(TSP)/F at 20 °C and $p_L^o$. Key: ● and ■, organochlorines; ▲, PAH in Tokyo (73); and △, PAH in Columbia (71). PH is phenanthrene, FLA is fluoranthene, PY is pyrene, BkF is benzo[k]-fluoranthene, and BaP and BeP are benzo[a]pyrene and benzo[e]pyrene, respectively. ■ for HCH and HCB span the range of A(TSP)/F at 20 °C obtained by extrapolating equation 2 with regression slope m (Table III) and slope m' for log $p_L^o$ versus 1/T.*

volume air stream (0.3–0.5 m³/min), passed through a particle-loaded filter backed up by a clean filter, and finally through a polyurethane foam trap that stripped the vapors from the airstream and preconcentrated them for analysis.

Equilibration experiments were carried out by exposing two 20- × 25-cm glass fiber filters to SOC vapors at 20 °C and a relative humidity of 58%–73% for varying times and air volumes. The front filter was loaded with approximately 100–250-mg urban air particulate matter by high-volume sampling in Columbia for 48 h. The back filter was a blank to correct for vapor adsorption to the filter matrix itself. Because only a small fraction of the total SOC vapor input was retained by the particles, the vapor concentration in the back filter was considered the same as that in the front filter. The polyurethane foam trap behind the filters was

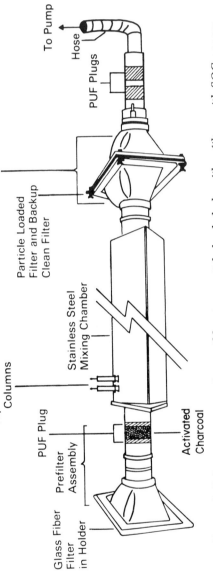

*Figure 4. Laboratory system for equilibrating particle-loaded glass fiber filters with SOC vapors.*

changed every 2–3 h and analyzed to determine changes in vapor concentration with time. At the end of the experiment, the filters were extracted with dichloromethane and analyzed for organochlorines by gas chromatography–electron-capture detection (GC–ECD). Details of the analytical methods have been given (77). The experiments yielded vapor–particle partition coefficients at 20 °C and A(TSP)/F values, which were calculated by dividing nanograms of SOCs per cubic meter of air by nanograms of SOCs per microgram of particles.

Eight runs were carried out with organochlorine pesticides over a range of concentrations and equilibration times (air volumes). Conditions are summarized in Table V. Three lots of particulate matter were used. The lots were collected from Columbia air on October 31–November 1, 1984; April 2–5, 1985; and August 26–28, 1985.

Table V. Range of Conditions and Mean A(TSP)/F Values for Laboratory Vapor–Particle Equilibration Experiments

| Compound | Experiments | Vapor Concentration ($ng/m^3$) | Air Volume ($m^3$) | log A(TSP)/F |
|----------|-------------|-------------------------------|--------------------|--------------|
| α-HCH | 7 | 53–160 | 230–790 | 5.362 |
| γ-HCH | 7 | 1–252 | 150–790 | 4.908 |
| HCB | 7 | 30–87 | 150–830 | 5.833 |
| *trans*-Chlordane | 7 | 17–56 | 230–790 | 4.041 |
| *cis*-Chlordane | 8 | 9–40 | 150–790 | 4.000 |
| *p,p′*-DDE | 8 | 0.1–27 | 150–1360 | 3.653 |
| *p,p′*-DDT | 6 | 0.013–1.5 | 230–1360 | 2.748 |

NOTE: All experiments were done at 20 °C. Particle weights on the glass fiber filter ranged from 100 to 250 mg.

Changes in vapor concentration behind the exposed filters for one run of 770 $m^3$ (about 30 h) are shown in Figure 5. Concentrations of the more volatile organochlorines remained constant over the entire run, whereas a lag time of several hours was noted before the heavier organochlorines reached stable levels. In these cases, A(TSP)/F values were calculated using plateau vapor concentrations (e.g., the mean of the last four *p,p*-DDT points in Figure 5).

Experiments carried out in the 9–54-h (150–1360 $m^3$ of air) range showed that a steady state was reached quickly between the particles on the glass fiber filter and vapors in the airstream (Figure 6). Residues on backup glass fiber filters averaged ≤10% of those on the particle-loaded filters. Thus, adsorption on the glass fiber filter itself was slight compared to adsorption by the particles. Backup glass fiber filter residues were subtracted from those on the front filter when calculating A(TSP)/F. Vapor concentrations of some SOCs (Table V) were varied by 1–2 orders of magnitude without affecting A(TSP)/F. The constancy

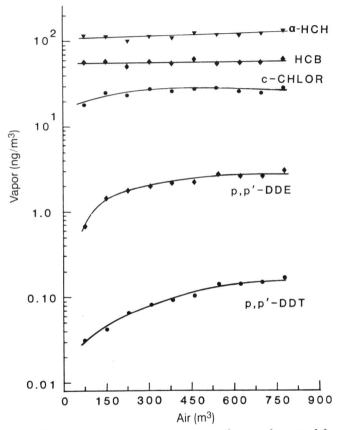

*Figure 5. Changes in vapor concentration with air volume in laboratory equilibration system. These changes were monitored by polyurethane foam plugs behind glass fiber filters.*

of $A(TSP)/F$ is rather remarkable, considering that three different lots of urban particulate matter were used for the experiments.

An important question is whether vapor–particle partitioning is controlled by the vapor pressure of the subcooled liquid ($p_L^o$) or the crystalline solid ($p_S^o$). The two vapor pressures can be interconverted by using the expression

$$\ln (p_L^o/p_S^o) = \Delta S_f (T_m - T)/RT \tag{3}$$

where $T_m$ and $T$ are the melting point and ambient temperatures (K), respectively (84). Estimates of the entropy of fusion ($\Delta S_f$) for rigid molecules range from 10 to 17 cal/deg·mol, with a most likely value of 13.5 cal/deg·mol (85). The range of experimentally determined $\Delta S_f$ for

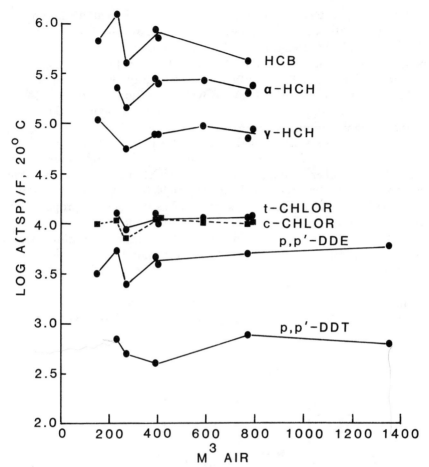

*Figure 6. Apparent vapor–particle partition coefficients, A(TSP)/F, at different air volumes. Equilibration times ranged from about 9 h at 150 m³ to 54 h at 1360 m³.*

16 PCB congeners was 9.8–16.6 cal/deg·mol (*86*). The mean for this range was 13.1 cal/deg·mol. We used 13.5 cal/deg·mol ($\Delta S_f/R = 6.79$) in equation 3.

Differences between $p_S^o$ and $p_L^o$ increase rapidly with melting point. At 20 °C, a low melting point substance like p,p'-DDE (mp 89 °C) has a $p_L^o$ value 5 times higher than $p_S^o$, whereas the difference is a factor of 130 for HCB (mp 230 °C) (Table IV). Vapor pressure is reduced over a solid relative to its subcooled liquid because of the crystal lattice energy. When individual solute molecules condense on an indifferent surface (i.e., physical adsorption), no solute crystal lattice is present and the adsorption might be expected to be controlled by $p_L^o$ rather than $p_S^o$. Several lines of evidence indicate that some environmentally important phase distributions are controlled by liquid-phase physical properties. Yamasaki et al. (81) found that $\Delta H_A$ for PAHs onto urban air particulate matter were close to $\Delta H_{V,L}$ for PAHs. Breakthrough of PAHs and organo-chlorine vapors on solid adsorbent collection traps is governed by $p_L^o$ rather than by $p_S^o$ (87, 88). Octanol–water partition coefficients ($K_{ow}$) and bioconcentration factors are inversely related to the water solubility of the subcooled liquid phase (89).

Figures 7 and 8 show log–log plots of A(TSP)/F versus $p_S^o$ or $p_L^o$, respectively. The points in Figure 7 show a good deal of scatter, espe-cially for the higher melting compounds $\alpha$-HCH and HCB, whereas the correlation in Figure 8 is excellent ($r^2 = 0.998$). This laboratory data, along with field data (Figure 3), support $p_L^o$ as the relevant property for describing vapor–particle equilibria in the atmosphere.

The A(TSP)/F values from field and laboratory experiments are compared in Figure 9. The agreement is very good for the chlordanes, DDE, and DDT, but less so for $\alpha$-HCH and HCB. As mentioned earlier, filter-retained quantities of these two volatile organochlorines were dif-ficult to determine except at very cold temperatures, and the uncertain-ties in extrapolating equation 2 plots to 20 °C were large. These uncer-

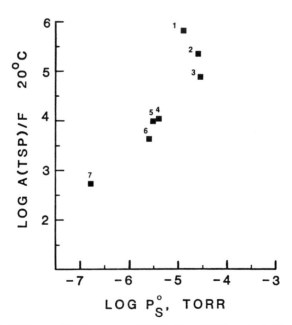

*Figure 7. Relationship between* A(TSP)/F *at 20 °C and* $p_S^o$. *Key: 1, HCB; 2, $\alpha$-HCH; 3, $\gamma$-HCH; 4, trans-chlordane; 5, cis-chlordane; 6, DDE; and 7, DDT.*

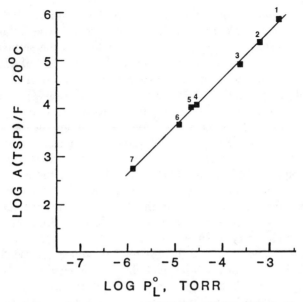

*Figure 8. Relationship between A(TSP)/F at 20 °C and $p_L^o$. Key: same as Figure 7.*

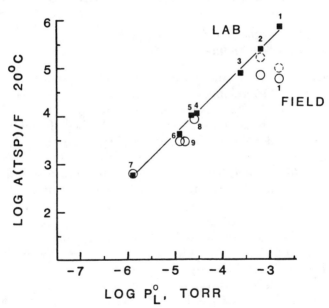

*Figure 9. Comparison of field and laboratory A(TSP)/F values at 20 °C. Ranges of field A(TSP)/F values are shown for HCB and HCH by assuming equation 2 slopes m (solid circles) and m' (broken circles). Key: 1, HCB; 2, α-HCH; 3, γ-HCH; 4, trans-chlordane; 5, cis-chlordane; 6, sum of chlordanes (field samples); 7, DDE; 8, Aroclor 1254; and 9, DDT.*

tainties may in part account for the field and laboratory differences seen in Figure 9. The agreement for $\alpha$-HCH can be improved by assuming a slope for its equation 2 plot equal to $m'$, which is the slope of the $p_L^o$ temperature dependence (76). However, the situation of HCB is not greatly improved. An alternative explanation may lie in the different ways that SOCs become bound to particles in urban air versus laboratory equilibration experiments. Vapors introduced in the latter experiments probably condensed on the surface of the particles on the glass fiber filters. The SOCs on filters from field collections also included material trapped within the particle matrix or strongly adsorbed to active sites (i.e., nonexchangeable SOCs). Differences between the laboratory and field $A(TSP)/F$ might be expected to be greatest for SOCs that become incorporated into the particles at the time of formation and less for SOCs that condense onto the particles later in their lifetime. The SOCs in the former category include combustion-derived compounds such as PAHs and HCB, and we found the greatest difference between laboratory and field $A(TSP)/F$ for HCB. Organochlorine pesticides probably enter urban air as vapors via transport from agricultural areas or by evaporation from treated areas within the city (e.g., chlordane used in structural termite control) and then condense onto urban aerosols. For these compounds, laboratory and field $A(TSP)/F$ values agree quite well.

## Relationship of Vapor–Particle Partitioning to Precipitation Scavenging

Atmospheric deposition contributes substantially to pollutant loadings in the Great Lakes. Several investigators have documented aerial transport and flux of SOCs (PCBs, pesticides, and PAHs) to the Great Lakes (2, 90–105). Hydrocarbons and fatty acids have also been found in precipitation from the midwestern United States (106). Analysis of PCBs in ice cores from Lake Huron provided an integrated measure of wintertime deposition (107). Atmospheric deposition is the most likely explanation for the presence of PCBs, pesticides, and dioxins in fish and sediments from Lake Siskiwit, a landlocked lake located on Isle Royale in Lake Superior (100, 108), and for DDT (109) and toxaphene (110) in peat cores from Minnesota, Maine, and southeastern Canada.

Atmospheric fluxes of SOCs occur by wet and dry deposition of gases and particles and also by exchange of gases across the air–water interface. An excellent overview of these processes and their role in contributing SOCs to the Great Lakes has been given by Eisenreich et al. (101). The $V/P$ value has an important influence on the atmospheric removal of SOCs. Vapor washout is governed by Henry's law constant ($H$), which is an air–water partition coefficient that may be calculated from the ratio of vapor pressure to water solubility. Other SOCs that have substantial fractions in the particle phase or Henry's law constants unfavorable for vapor scavenging are removed by particle washout.

A model for SOC washout incorporating vapor and particle scavenging was presented by Pankow et al. (*111*) and Ligocki et al. (*112, 113*). The overall washout ratio, $W$, which is the mass of SOCs per volume rain divided by the mass of SOCs per volume air, is related to the washout ratios of vapors, $W_v$ or $RT/H$, the washout ratio of particles, $W_p$, and the fraction of particulate SOC, $\phi$, by

$$W = (RT/H)(1 - \phi) + W_p\phi \qquad (4)$$

Equation 4 is a simple equilibrium model that ignores the complications of meteorology. Several articles (*115–119*) report details of the physical process of precipitation scavenging. The dominant rain scavenging mechanism depends on the relative magnitudes of $W_v$, $W_p$, and $\phi$. Ligocki et al. (*112, 113*) collected rain and concurrent air samples in Portland, OR, and found that two- to four-ring PAHs were washed out as vapors. Field $W_v$ values were 3–6 times higher than those predicted from $H$ at 25 °C, but if $H$ at the appropriate rain temperatures was used, the agreement between field and equilibrium values was a factor of 2 or better. Other compounds removed mainly by vapor scavenging are HCH (*112, 113, 120  121*) and phenols (*122*). Some SOCs for which particle scavenging dominates are PCBs and DDT (*1, 2*), PAHs of five or more rings (*113*), and *n*-alkanes (*113, 123*). Within the suite of PCB congeners, those that are less volatile and have greater fractions on particles are preferentially deposited by rain and dry deposition (*1, 2*).

We used equation 4 to estimate the likely mode of wet deposition for toxaphene, an organochlorine pesticide of concern in the Great Lakes. In the 3 years before its ban in November 1982, more than 30,000 metric tons of toxaphene was applied (*124*) to soybeans, cotton, grain, and as a cattle dip (*125*). Toxaphene is translocated from heavy application areas in the southern United States to the Great Lakes (*103*) and eastern Canada (*126*) and has been found in peat cores from the north central and eastern United States and southeastern Canada (*110*). Toxaphene is a widespread contaminant in fish from the Great Lakes (*127*) and eastern Canadian marine waters (*128*). Because toxaphene is carcinogenic in laboratory animals (*129*), the U.S. Environmental Protection Agency (EPA) cited potential risks to fish-consuming populations in the Great Lakes and Mississippi Delta regions in their decision to ban toxaphene (*125, 130*). Toxaphene is one of the 11 critical pollutants recommended for study of sources and transport by the Great Lakes Water Quality Board (*131*).

Toxaphene is a complex mixture of chlorinated bornanes and camphenes and has a wide range of poorly defined physical properties.

Nearly all physical data refer to the technical toxaphene mixture, a waxy semisolid with a melting range of 70–90 °C. A brief review of these properties has been given (132). Seiber et al. (133) and Murphy et al. (134) reported average vapor pressures for the mixture in the range $3 \times 10^{-7}$–$6 \times 10^{-6}$ torr at 20 °C. Murphy et al. (134) also determined the water solubility of toxaphene (0.55 mg/L, or $1.3 \times 10^{-3}$ mol/m$^3$ assuming that the average molecular composition is $C_{10}H_{10}Cl_8$). From the solubility and vapor pressure, $H = 6.1 \times 10^{-6}$ atm m$^3$/mol.

If toxaphene behaves like other organochlorines (Figure 3), about 2%–17% should be filter-retained at 20 °C, and TSP = 40 $\mu$g/m$^3$. This TSP concentration is the geometric mean found by Andren and Strand (92) over Lake Michigan. For reasons discussed earlier in this chapter, the filter-retained percentage may or may not represent the true particulate fraction in ambient air, but it is the only estimate we have.

Gatz (135) reported $W_p$ for trace metals, and these values generally increase with MMED. The $W_p$ values for Pb and Zn (MMED $\sim$ 0.7–1.2 $\mu$m) were $6.3 \times 10^4$ and $1.4 \times 10^5$, respectively. The SOCs are preferentially concentrated on the smaller particles, and MMEDs for particles containing PAHs, alkanes, and organic acids are typically 1–2 $\mu$m or less (42–44, 136). Therefore, $W_p$ for SOCs might be expected to be on the order of $10^5$. Actual field measurements of $W_p$ for SOCs are few and variable. Ligocki et al. (113) found $W_p$ for $n$-alkanes in the 1.3–2.2 $\times 10^4$ range, and lower values for particulate PAHs. The $W_p$ values for particulate alkanes found by Farmer and Wade (123) were much higher, ranging from 6.0–8.6 $\times 10^5$. According to Scott (114), hydrophilic aerosols precipitated from warm clouds have $W_p \sim 10^6$. Values of $W_p \leq 10^5$ can occur during precipitation from cold clouds, or if aerosols are not efficiently scavenged. Young aerosols of submicrometer size that are not sufficiently hydrophilic to grow by water vapor condensation would fall into the latter category.

Assuming $W_p = 1 \times 10^5$ for toxaphene, $\phi = 0.02$–0.17, and $H = 6.1 \times 10^{-6}$ atm m$^3$/mol, then $(RT/H)(1-\phi) = 3.2$–$3.9 \times 10^3$, $W_p\phi = 0.2$–1.7 $\times 10^4$, and $W = 0.6$–2.0 $\times 10^4$. Vapor scavenging thus accounts for a major share of toxaphene deposition. Particle scavenging could be 85% or only 33% of the total, depending on the choice of toxaphene vapor pressure (controlling $\phi$). The overall $W$ value is also sensitive to the choice of $W_p$, and as discussed earlier, a great deal of uncertainty exists concerning the magnitude of $W_p$ for particulate SOCs.

The predicted $W$ from toxaphene physical properties and V/P partitioning may be compared with field values. Between 1976 and 1978, we measured the rainfall flux of toxaphene and other organochlorines to the North Inlet Estuary, a high salinity marsh on the South Carolina coast that receives little freshwater input other than through rainfall (1, 137). Toxaphene was the most abundant pesticide in precipitation; loading to the estuary was estimated at 4.6 kg/ha from June to September 1977 (*137*). Concurrent air samples were taken with 16 rain events at North Inlet, and two air–rain sample pairs were collected from shipboard about 300 km off the coast of Delaware. Toxaphene $W$ ranged from $<0.13–7.2 \times 10^5$. In our original articles (*1, 137*), the arithmetic mean for $W$ was given as $2.3 \times 10^5$. This mean is too high for two reasons. First, only positive samples (13 of 18) were averaged. Second, the distribution of $W$ is log normal, and a better central measure is the geometric mean (*116*). The geometric mean gives less weight to a few large values than does the arithmetic mean. The geometric mean of the 18 North Inlet and shipboard $W$ values, including upper limits for samples below the detection limit, was $6.6 \times 10^4$. This value is higher than the predicted $W$ of $0.6–2.0 \times 10^4$. A better knowledge of toxaphene physical properties as well as rain scavenging mechanisms for particulate SOCs would be helpful in explaining this difference.

The preceding exercise shows how a model incorporating vapor–particle partitioning can be used to predict precipitation removal mechanisms. Mackay et al. (*138*) recently presented a thorough discussion of wet and dry SOC deposition that includes vapor–particle equilibria.

Some mechanisms for concentrating SOCs in rainfall may not be included in the equation 4 model. Rain contains surface-active material (*20*), and the importance of surface films in modifying SOC vapor exchange across the air–drop interface has not been examined. Glotfelty et al. (*139*) collected fog droplets with a rotary impactor in the California San Joaquin Valley, a heavy pesticide use area. The filtered fog water and concurrent air samples were analyzed for organophosphate insecticides and a dinitroaniline herbicide. Pesticide concentrations in the fog water were enriched 88–2400 times above equilibrium values predicted from Henry's law. The enrichment mechanism has not yet been elucidated, but SOC solubility enhancement due to dissolved organic material or partitioning of SOCs into surface films on the droplets might be a factor.

## Conclusions

Many uncertainties still remain in our knowledge of vapor–particle interactions in ambient air. A major question, unanswered by either our field or laboratory experiments, is how closely does $A/F$ represent the true $V/P$ in the atmosphere. The process of collecting particles on a filter will likely alter the particle size and surface area distribution from the situation in which the particles were floating freely in the atmosphere. Therefore, the $A(TSP)/F$ partition coefficients in this work are between vapors and urban air particulate matter on filters, not airborne particles.

Despite this shortcoming, field and laboratory experiments have provided some valuable information about the nature of vapor–particle interactions. Particles on glass fiber filters reach a steady state with SOC vapors within a few hours. The $A/F$ partitioning of nonpolar compounds appears to be due to simple physical adsorption controlled by SOC volatility, and the apparent partition coefficient, $A(TSP)/F$, is better correlated with $p_L^o$ than $p_S^o$.

Implications of the $p_L^o$ hypothesis are important for high melting compounds. For example, $p_S^o$ of 2,3,7,8-tetrachlorodibenzo-$p$-dioxin (TCDD) has recently been measured by the gas saturation method ($140$). At 20 °C, $p_S^o = 6.1 \times 10^{-10}$ torr, and if $p_S^o$ controlled $V/P$ in the atmosphere, all the TCDD would be expected to be particle bound. However, the melting point of TCDD is 305 °C, and $p_L^o$ at 20 °C is 745 times $p_S^o$, or $4.5 \times 10^{-7}$ torr. Based on this $p_L^o$ value, and assuming that TCDD behaves like other organochlorines (Figures 3 and 8), a substantial proportion of the TCDD may exist in the atmosphere in the vapor phase. A better knowledge of vapor–particle equilibria and the controlling thermodynamic factors is thus important to understand SOC atmospheric chemistry.

### Abbreviations and Symbols

| | |
|---|---|
| $\theta$ | particle surface area per unit volume of air |
| $\sigma_g$ | geometric standard deviation |
| $\phi$ | fraction of particle-bound SOCs |
| $A(TSP)/F$ | partition coefficient where $A$ is adsorbent-retained SOC concentration (mg/m³), $F$ is filter-retained SOC concentration (mg/m³), and TSP is total suspended particle concentration (μg/m³) |
| EOM | extractable organic matter |
| FPM | fine particulate matter |
| $H$ | Henry's law constant |
| $\Delta H_A$ | heat of adsorption |
| $\Delta H_{V,L}$ | heat of vaporization of a subcooled liquid |
| IPM | inhalable particulate matter |
| $K_{ow}$ | octanol–water partition coefficient |
| MMED | mass median effective diameter |
| $p_L^o$ | subcooled liquid vapor pressure |
| $p_S^o$ | solid-phase vapor pressure |
| $R$ | gas constant |
| $\Delta S_f$ | entropy of fusion |
| $T$ | temperature |
| $T_m$ | melting point temperature (K) |
| $W$ | overall washout ratio |

$W_p$           washout ratio of particles
$W_v$           washout ratio of vapor
$V/P$           vapor–particle distribution

## Acknowledgments

Support for field collection studies was provided by the U.S. EPA, Grant 807048; by the National Science Foundation, Grant INT 8317424; and by the National Environmental Protection Board of Sweden (SNV). We thank the Colorado State Department of Health and the Special Analytical Laboratory of SNV for their assistance in providing sampling sites and analytical facilities. The laboratory portion of this work was supported by the U.S. Department of Agriculture, Grant 58-32U4-4-750. Gene Slice and John Cooper of the South Carolina Department of Health and Environmental Control generously made their high-volume stations available for collection of particulate matter. We also thank Mark Zaranski for help with sampling, analysis, and design of the laboratory equilibration apparatus. This chapter is contribution 638 of the Belle W. Baruch Institute.

## Appendix: Relationship between (Junge's) Equation 1 and (Yamasaki's) Equation 2

Assume that $\phi$, the fraction of aerosol-bound SOCs, is equal to the fraction of filter-retained SOCs:

$$\phi = F/(A + F) = c\theta/(p° + c\theta)$$

$$1 + A/F = 1 + p°/c\theta, \text{ or } A/F = p°/c\theta \tag{1}$$

where $\theta$ is the surface area of suspended particulate matter per cubic centimeter of air, and therefore $\theta = k(\text{TSP})$.

$$A/F = p°/ck(\text{TSP})$$

$$\log A(\text{TSP})/F = \log p° - \log ck$$

The temperature dependence of $p°$ over the range where the heat of vaporization can be considered approximately constant is

$$\log p° = m'/T + b'$$

Combining these last two equations yields

$$\log A(\text{TSP})/F = m'/T + b' - \log ck$$

Setting $m = m'$ and $b = b' - \log ck$ yields equation 2.

$$\log A(TSP)/F = m/T + b \tag{2}$$

In the case of physical adsorption, the slope $(m)$ of equation 2 is the same as the slope $(m')$ of $\log p°$ versus $1/T$.

## References

1. Bidleman, T. F.; Christensen, E. J. *J. Geophys. Res.* **1979**, *84*, 7857–7862.
2. Murphy, T. J.; Rzeszutko, C. P. *J. Great Lakes Res.* **1977**, *3*, 305–312.
3. Countess, R. J.; Wolff, G. T.; Cadle, S. H. *J. Air Pollut. Control Assoc.* **1980**, *30*, 1194–1200.
4. Kowalczyk, G.; Gordon, G. E.; Rheingrover, S. W. *Environ. Sci. Technol.* **1982**, *16*, 79–90.
5. Johnson, D. L.; Davis, B. L.; Dzubay, T. G.; Hasan, H.; Crutcher, E. R.; Courtney, W. J.; Jaklevic, J. M.; Thompson, A. C.; Hopke, P. K. *Atmos. Environ.* **1984**, *18*, 1539–1553.
6. Stevens, R. K.; Pace, T. G. *Atmos. Environ.* **1984**, *18*, 1499–1506.
7. Junge, C. E. *Air Chemistry and Radioactivity;* Academic: New York, 1963; pp 111–112.
8. Cadle, R. D. *The Measurement of Airborne Particles;* Wiley: New York, 1975.
9. Corn, M. In *Air Pollution*, 3rd ed.; Stern, A. C., Ed.; Academic: New York, 1976; Vol. 1, Chapter 3, pp 77–168.
10. Whitby, K. T. *Atmos. Environ.* **1978**, *12*, 135–159.
11. Lee, R. E. *Science (Washington, D.C.)* **1972**, *178*, 567–575.
12. Corn, M.; Montgomery, T. L.; Esman, N. A. *Environ. Sci. Technol.* **1971**, *5*, 155–158.
13. Bidleman, T. F. In *Trace Analysis;* Lawrence, J. F., Ed.; Academic: New York, 1985; Vol. 4, pp 51–100.
14. *Code of Federal Regulations 1984,* Title 40, Protection of the Environment, Subchapter C, Air Programs, Parts 50.6 and 50.7, p 514.
15. Daisey, J. M. *Ann. NY Acad. Sci.* **1980**, *338*, 50–69.
16. Lamb, S. I.; Petrowski, C.; Kaplan, I. R.; Simoneit, B. R. T. *J. Air Pollut. Control Assoc.* **1980**, *30*, 1098–1115.
17. Simoneit, B. R. T.; Mazurek, M. A. *CRC Crit. Rev. Environ. Control* **1981**, *11*, 219–276.
18. Duce, R. A.; Mohnen, V. A.; Zimmerman, P. R.; Grosjean, D.; Cautreels, W.; Chatfield, R.; Jaenicke, R.; Ogren, J. A.; Pellizzari, E. D.; Wallace, G. T. *Rev. Geophys. Space Phys.* **1983**, *21*, 921–952.
19. Wolff, G. T.; Klimisch, R. L. *Particulate Carbon: Atmospheric Life Cycle;* Plenum: New York, 1982.
20. Gill, P. S.; Graedel, T. E. *Rev. Geophys. Space Phys.* **1983**, *21*, 903–920.
21. Shah, J. J.; Johnson, R. L.; Heyerdahl, E. K.; Huntzicker, J. J. *J. Air Pollut. Control Assoc.* **1986**, *36*, 254–257.
22. Daisey, J. M.; Leyko, M. A.; Kleinman, M. T.; Hoffman, E. *Ann. NY Acad. Sci.* **1979**, *322*, 125–142.
23. Daisey, J. M.; Hershman, R. J.; Kneip, T. J. *Atmos. Environ.* **1982**, *16*, 2161–2168.

24. Schwartz, G. P.; Daisey, J. M.; Lioy, P. J. *J. Am. Ind. Hyg. Assoc.* **1981,** *42,* 258–263.
25. Grosjean, D. *Anal. Chem.* **1975,** *47,* 797–805.
26. Appel, B. R.; Hoffer, E. M.; Kothny, E. L.; Wall, S. M.; Haik, M.; Knights, R. L. *Environ. Sci. Technol.* **1979,** *13,* 98–104.
27. Appel, B. R.; Colodny, P.; Weslowski, J. J. *Environ. Sci. Technol.* **1976,** *10,* 359–363.
28. Grosjean, D.; Friedlander, S. K. *J. Air Pollut. Control Assoc.* **1975,** *25,* 1038–1048.
29. Ketseridis, G.; Hahn, J. *Fresenius' Z. Anal. Chem.* **1975,** *273,* 257–261.
30. Ketseridis, G.; Hahn, J.; Jaenicke, R.; Junge, C. E. *Atmos. Environ.* **1976,** *10,* 603–610.
31. Daisey, J. M.; Morandi, M.; Lioy, P. J.; Wolff, G. T. *Atmos. Environ.* **1984,** *18,* 1411–1419.
32. Pratsinis, S.; Novakov, T.; Ellis, E. C.; Friedlander, S. K. *J. Air Pollut. Control Assoc.* **1984,** *34,* 643–650.
33. Lioy, P. J.; Daisey, J. M.; Reiss, N. M.; Harkov, R. *Atmos. Environ.* **1983,** *17,* 2321–2330.
34. Cadle, S. H.; Groblicki, P. J.; Stroup, D. P. *Anal. Chem.* **1980,** *52,* 2201–2206.
35. Hoffman, E. J.; Duce, R. A. *J. Geophys. Res.* **1974,** *79,* 4474–4477.
36. Pimenta, J.; Wood, G. R. *Environ. Sci. Technol.* **1980,** *14,* 556–561.
37. Ogren, J. A.; Charlson, R. J.; Groblicki, P. J. *Anal. Chem.* **1983,** *55,* 1569–1572.
38. Heintzenberg, J. *Atmos. Environ.* **1982,** *16,* 2461–2469.
39. Delumyea, R. G.; Chu, L. C.; Macias, E. S. *Atmos. Environ.* **1980,** *14,* 647–652.
40. Edwards, J. D.; Ogren, J. A.; Weiss, R. E.; Charlson, R. J. *Atmos. Environ.* **1983,** *17,* 2337–2341.
41. Lioy, P. J.; Daisey, J. M.; Greenberg, A.; Harkov, R. *Atmos. Environ.* **1985,** *19,* 429–436.
42. Pierce, R. C.; Katz, M. *Environ. Sci. Technol.* **1975,** *9,* 347–353.
43. Katz, M.; Chan, C. *Environ. Sci. Technol.* **1980,** *14,* 838–843.
44. Van Vaeck, L.; Van Cauwenberghe, K. *Atmos. Environ.* **1978,** *12,* 2229–2239.
45. Van Vaeck, L.; Van Cauwenberghe, K.; Janssens, J. *Atmos. Environ.* **1984,** *18,* 417–430.
46. Van Vaeck, L.; Broddin, G.; Van Cauwenberghe, K. *Atmos. Environ.* **1979,** *13,* 1494–1502.
47. Miguel, A. H.; Friedlander, S. K. *Atmos. Environ.* **1978,** *12,* 2407–2413.
48. Wolff, G. T.; Korsog, P. E.; Kelly, N. A.; Ferman, M. A. *Atmos. Environ.* **1985,** *19,* 1341–1349.
49. Cautreels, W.; Van Cauwenberghe, K. *Water, Air, Soil Pollut.* **1976,** *6,* 103–110.
50. Sternberg, U. R.; Alsberg, T. E. *Anal. Chem.* **1981,** *53,* 2067–2072.
51. Grimmer, G.; Naujack, K. W.; Schneider, D. *Fresenius' Z. Anal. Chem.* **1982,** *311,* 475–484.
52. Greenberg, A.; Darack, F.; Harkov, R.; Lioy, P. J.; Daisey, J. M. *Atmos. Environ.* **1985,** *19,* 1325–1339.
53. Griest, W. H.; Caton, J. E.; Guerin, M. R.; Yeatts, L. B.; Higgins, C. E. *Anal. Chem.* **1980,** *52,* 199–201.
54. Eiceman, G. A.; Vandiver, V. J. *Atmos. Environ.* **1983,** *17,* 461–465.
55. Junge, C. E. In *Fate of Pollutants in the Air and Water Environments,* Part I; Suffet, I. H., Ed.; Wiley: New York, 1977, pp 7–26.

56. König, J.; Funcke, W.; Balfanz, E.; Grosch, G.; Potts, F. *Atmos. Environ.* **1980**, *14*, 609–613.
57. Broddin, G.; Cautreels, W.; Van Cauwenberghe, K. *Atmos. Environ.* **1980**, *14*, 895–910.
58. Spitzer, T., Dannecker, W. *Anal. Chem.* **1983**, *55*, 2226–2228.
59. Grosjean, D. *Atmos. Environ.* **1983**, *17*, 2565–2573.
60. Korfmacher, W. A.; Natusch, D. F. S.; Taylor, D. R.; Mamantov, G.; Wehry, E. L. *Science (Washington, D.C.)* **1980**, *207*, 763–765.
61. Butler, J. D.; Crossley, P. *Atmos. Environ.* **1981**, *15*, 91–94.
62. Van Vaeck, L.; Van Cauwenberghe, K. *Atmos. Environ.* **1984**, *18*, 323–328.
63. Brorström, E.; Grennfelt, P.; Lindskog, A. *Atmos. Environ.* **1983**, *17*, 601–605.
64. Brorström-Lunden, E.; Lindskog, A. *Environ. Sci. Technol.* **1985**, *19*, 313–316.
65. Pitts, J. N., Jr.; Zielinska, B.; Sweetman, J. A.; Atkinson, R.; Winer, A. M. *Atmos. Environ.* **1985**, *19*, 911–915.
66. Pitts, J. N., Jr.; Sweetman, J. A.; Zielinska, B.; Winer, A. M.; Atkinson, R. *Atmos. Environ.* **1985**, *19*, 1601–1608.
67. Cautreels, W., Van Cauwenberghe, K. *Atmos. Environ.* **1978**, *12*, 1133–1141.
68. Billings, W. N.; Bidleman, T. F. *Atmos. Environ.* **1983**, *17*, 383–391.
69. Eichmann, R.; Ketserides, G.; Schebeske, G.; Jaenicke, R.; Hahn, J.; Junge, C. E. *Atmos. Environ.* **1980**, *13*, 587–599.
70. Galasyn, J. F.; Hornig, J. F.; Soderberg, R. H. *J. Air Pollut. Control Assoc.* **1984**, *34*, 57–59.
71. Keller, C. D.; Bidleman, T. F. *Atmos. Environ.* **1984**, *18*, 837–845.
72. Thrane, K.; Mikalsen, A. *Atmos. Environ.* **1981**, *15*, 909–918.
73. Yamasaki, H.; Kuwata, K.; Miyamoto, H. *Environ. Sci. Technol.* **1982**, *16*, 189–194.
74. Duce, R. A.; Gagosian, R. B. *J. Geophys. Res.* **1982**, *87*, 7192–7200.
75. Van Vaeck, L.; Broddin, G.; Van Cauwenberghe, K. *Biomed. Mass Spectrom.* **1980**, *7*, 473–483.
76. Bidleman, T. F.; Billings, W. N.; Foreman, W. T. *Environ. Sci. Technol.* **1986**, *20*, 1038–1043.
77. Foreman, W. T.; Bidleman, T. F., submitted for publication in *Environ. Sci. Technol.*
78. Chiou, C. T.; Shoup, T. D. *Environ. Sci. Technol.* **1985**, *19*, 1196–1200.
79. Spencer, W. F.; Cliath, M. M.; Farmer, W. J. *Soil Sci. Soc. Am. Proc.* **1969**, *33*, 509.
80. Spencer, W. F.; Cliath, M. M. *Soil Sci. Soc. Am. Proc.* **1970**, *34*, 574.
81. Yamasaki, H.; Kuwata, K.; Yoshio, K. *Nippon Kagaku Kaishi* **1984**, 1324–1329; *Chem. Abstr.* **1984**, *101*, 156747.
82. Bidleman, T. F. *Anal. Chem.* **1984**, *56*, 2490–2496.
83. Foreman, W. T.; Bidleman, T. F. *J. Chromatogr.* **1985**, *330*, 203–216.
84. Mackay, D.; Bobra, A.; Chan, D. W.; Shiu, W. Y. *Environ. Sci. Technol.* **1982**, *16*, 645–649.
85. Yalkowsky, S. H. *Ind. Eng. Chem. Fundam.* **1979**, *18*, 108–111.
86. Miller, M. M.; Ghodbane, S.; Wasik, S. P.; Tewari, Y. B.; Martire, D. E. *J. Chem. Eng. Data* **1984**, *27*, 184–190.
87. Feng, Y.; Bidleman, T. F. *Environ. Sci. Technol.* **1984**, *18*, 330–333.
88. Bidleman, T. F.; Simon, C. G.; Burdick, N. F.; Feng, Y. *J. Chromatogr.* **1984**, *301*, 448–453.

89. Mackay, D. *Environ. Sci. Technol.* **1982**, *16*, 274–278.
90. Eisenreich, S. J.; Hollod, G. J.; Johnson, T. C. In *Atmospheric Pollutants in Natural Waters;* Eisenreich, S. J., Ed.; Ann Arbor Science: Ann Arbor, MI, 1977; Chapter 21, pp 425–444.
91. Murphy, T. J.; Schinsky, A.; Paolucci, G.; Rzeszutko, C. P. Ibid., Chapter 22, pp 445–458.
92. Andren, A. W.; Strand, J. W. Ibid., Chapter 23, pp 459–479.
93. Eisenreich, S. J.; Looney, B. B.; Hollod, G. J. In *Physical Behavior of PCBs in the Great Lakes;* Mackay, D.; Paterson, S.; Eisenreich, S. J.; Simmons, M. S., Eds.; Ann Arbor Science: Ann Arbor, MI, 1983; Chapter 7, pp 115–125.
94. Eisenreich, S. J.; Looney, B. B. Ibid., Chapter 9, pp 141–156.
95. Murphy, T. J.; Pokojowczyk, J. C.; Mullin, M. D. Ibid., Chapter 3, pp 49–58.
96. Strachan, W. M. J.; Huneault, H. *J. Great Lakes Res.* **1979**, *5*, 61–68.
97. Strachan, W. M. J. *Environ. Toxicol. Chem.* **1985**, *4*, 677–683.
98. Strachan, W. M. J.; Huneault, H. *Environ. Sci. Technol.* **1984**, *18*, 127–130.
99. Strachan, W. M. J.; Huneault, H.; Schertzer, W. M.; Elder, F. C. In *Hydrocarbons and Halogenated Hydrocarbons in the Aquatic Environment;* Afghan, B. K.; Mackay, D., Eds.; Plenum: Toronto, Ontario, Canada, 1980; pp 387–396.
100. Swain, W. R. *J. Great Lakes Res.* **1978**, *4*, 398–407.
101. Eisenreich, S. J.; Looney, B. B.; Thornton, J. D. *Environ. Sci. Technol.* **1981**, *15*, 30–38.
102. Doskey, P. V.; Andren, A. W. *J. Great Lakes Res.* **1981**, *7*, 15–20.
103. Rice, C. P.; Samson, P. J.; Noguchi, G. *Environ. Sci. Technol.* **1986**, *20*, 1109–1116.
104. Rice, C. P.; Eadie, B. J.; Erstfeld, K. M. *J. Great Lakes Res.* **1982**, *8*, 265–270.
105. Murphy, T. J. In *Toxic Contaminants in the Great Lakes;* Nriagu, J. O.; Simmons, M. S., Eds.; Wiley: New York, 1984.
106. Meyers, P. A.; Hites, R. A. *Atmos. Environ.* **1982**, *16*, 2169–2175.
107. Murphy, T. J.; Schinsky, A. W. *J. Great Lakes Res.* **1983**, *9*, 92–96.
108. Czuczwa, J. M.; Niessen, F.; Hites, R. A. *Chemosphere* **1985**, *14*, 623.
109. Rapaport, R. A.; Urban, N. R.; Capel, P. D.; Baker, J. E.; Looney, B. B.; Eisenreich, S. J. *Chemosphere* **1985**, *14*, 1167–1173.
110. Rapaport, R. A.; Eisenreich, S. J. *Atmos. Environ.* **1986**, *20*, 2367–2379.
111. Pankow, J. F.; Isabelle, L. M.; Asher, W. E. *Environ. Sci. Technol.* **1984**, *18*, 310–318.
112. Ligocki, M. P.; Leuenberger, C.; Pankow, J. F. *Atmos. Environ.* **1985**, *19*, 1609–1617.
113. Ligocki, M. P.; Leuenberger, C.; Pankow, J. F. *Atmos. Environ.* **1985**, *19*, 1619–1623.
114. Scott, B. C. In *Atmospheric Pollutants in Natural Waters;* Eisenreich, S. J., Ed.; Ann Arbor Science: Ann Arbor, MI, 1981; Chapter 1, pp 3–21.
115. Slinn, W. G. N. *Water, Air, Soil Pollut.* **1977**, *7*, 513–543.
116. Slinn, W. G. N. In *Air–Sea Exchange of Gases and Particles;* Liss, P. S.; Slinn, W. G. N., Eds.; NATO Advanced Science Institute Series; Reidel: Boston, 1983; pp 299–405.
117. Slinn, W. G. N.; Hasse, L.; Hicks, B. B.; Hogan, A. W.; Lal, D.; Liss, P. S.; Munnich, K. O.; Sehmel, G. A.; Vittori, O. *Atmos. Environ.* **1978**, *12*, 2055–2087.

118. Peters, L. K. In *Air–Sea Exchange of Gases and Particles;* Liss, P. S.; Slinn, W. G. N., Eds.; NATO Advanced Science Institute Series; Reidel: Boston, 1983; pp 173–240.
119. Topol, L. E.; Vijayakumar, R.; McKinley, C. M.; Waldron, T. L. *J. Air Pollut. Control Assoc.* **1986,** *36,* 393–398.
120. Bidleman, T. F.; Leonard, R. *Atmos. Environ.* **1982,** *16,* 1099–1107.
121. Atlas, E. L.; Giam, C. S. *Science (Washington, D.C.)* **1981,** *211,* 163–165.
122. Leuenberger, C.; Ligocki, M. P.; Pankow, J. F. *Environ. Sci. Technol.* **1985,** *19,* 1053–1058.
123. Farmer, C. T.; Wade, T. L. *Water, Air, Soil Pollut.* **1986,** *29,* 439–452.
124. *Production Yearbooks;* Food and Agricultural Organization of the United Nations: Rome, 1984–1985; Vol. 37–38.
125. *Toxaphene: Decision Document;* U.S. Environmental Protection Agency. Office of Pesticide Programs: Washington, DC, 1982.
126. Bidleman, T. F.; Christensen, E. J.; Billings, W. N.; Leonard, R. *J. Mar. Res.* **1981,** *39,* 443–464.
127. Rice, C. P.; Evans, M. S. In *Toxic Contaminants in the Great Lakes;* Nriagu, J. O.; Simmons, M. S., Eds.; Wiley: New York, 1984; Chapter 8, pp 163–194.
128. Musial, C. J.; Uthe, J. F. *Int. J. Environ. Anal. Chem.* **1983,** *14,* 117–126.
129. Reuber, M. D. *J. Toxicol. Environ. Health* **1979,** *5,* 729–748.
130. *Fed. Regist.* **1982,** *47* (229), 53784.
131. Kingman, J. O.; Adamkus, V. V. Report to the International Joint Commission on Great Lakes Water Quality; Great Lakes Water Quality Board: Windsor, Ontario, Canada, 1985.
132. Bidleman, T. F.; Zaranski, M. T.; Walla, M. D. *Proc. 2nd World Conf. Large Lakes,* in press.
133. Seiber, J. N.; Woodrow, J. E.; Sanders, P. F. Presented at the 182nd National Meeting of the American Chemical Society, New York, 1981.
134. Murphy, T. J.; Mullin, M. D.; Meyer, J. A., submitted for publication in *Environ. Sci. Technol.*
135. Gatz, D. F. *Water, Air, Soil Pollut.* **1975,** *5,* 239–251.
136. Van Vaeck, L.; Broddin, G.; Van Cauwenberghe, K. *Environ. Sci. Technol.* **1979,** *12,* 1494–1502.
137. Harder, H. W.; Christensen, E. J.; Matthews, J. R.; Bidleman, T. F. *Estuaries,* **1980,** *3,* 142–147.
138. Mackay, D.; Paterson, S.; Schroeder, W. H. *Environ. Sci. Technol.* **1986,** *20,* 810–816.
139. Glotfelty, D. E.; Seiber, J. N.; Liljedahl, L. A. *Nature (London),* in press.
140. Schroy, J. M.; Hileman, F.; Cheng, S. C. *Chemosphere* **1985,** *14,* 877–880.

RECEIVED for review May 6, 1986. ACCEPTED October 13, 1986.

# Lake Sediments as Historic Records of Atmospheric Contamination by Organic Chemicals

Jeffrey W. Astle, Frank A. P. C. Gobas, Wan-Ying Shiu, and Donald Mackay

Department of Chemical Engineering and Applied Chemistry, University of Toronto, Toronto, Ontario M5S 1A4, Canada

*A mathematical model is presented that describes the variation in lake sediment concentration of a persistent organic chemical with time and depth as a function of atmospheric concentrations. This model may be used to estimate historic atmospheric concentrations from analyses of sectioned sediments. The model includes wet and dry air particulate deposition, rain dissolution, volatilization and absorption, water column particulate deposition and resuspension, sediment-water diffusion, degradation reactions in water and sediment, and sediment burial. If a chemical's properties are known, the dominant transfer processes can be identified, and it can be determined if the sediment can provide a reasonable record of atmospheric concentration. Defining surface sediment-air and water-air concentration ratios for organic chemicals may be feasible. The model is applied to dioxin concentration data from Siskiwit Lake on Isle Royale in Lake Superior.*

THE ATMOSPHERE IS A SIGNIFICANT SOURCE of organic contaminants to lakes, and realization of this fact has increased in the last decade. Swain (*1*) showed that the pristine Siskiwit Lake on Isle Royale in Lake Superior had concentrations of polychlorinated biphenyls (PCBs) that could have been derived only from the atmosphere. Eisenreich and co-workers (*2–4*) later reviewed and quantified the atmospheric sources of organic compounds to the Great Lakes. More recently, Czuczwa et al. (*5*) showed that polychlorinated dibenzodioxins (PCDDs) and dibenzofurans were present in the sediments of Siskiwit Lake and that the sediment record suggests their source as being combustion of wastes

0065–2393/87/0216–0057$06.25/0

containing chlorinated compounds since about 1940. A satisfying correspondence existed between air particulate and sediment distributions of isomers. A similar study was reported for Swiss lakes by Czuczwa et al. (6).

Sediments can thus provide an invaluable record of atmospheric contamination. Sediments not only provide samples for analysis derived from sources that are now unobtainable, but they also presumably integrate concentrations over a long period and thus eliminate much of the temporal variation that can confound atmospheric sampling. Sediments can give a qualitative indication of previous contamination, but can they give a quantitative indication of contamination? If a link could be established between a specific amount of a chemical in air and a specific amount in the corresponding sediments, information that has been unavailable could be accessed. Presumably, each chemical has a unique ratio of sediment–atmosphere concentrations that reflects a variety of properties of the lake and the chemical. In some cases, this ratio may be impossible to discern because of degradation reactions or other processes such as volatilization.

This possibility, quantitative indication of contamination, will be explored by using a simple mathematical model based on the fugacity concept. The questions that will be answered by using the model are as follows:

1. Is a linear concentration relationship expected between sediment and atmosphere (i.e., does a constant concentration ratio apply)?

2. What are the factors influencing the magnitude of this ratio?

3. Can the ratio be quantified for a specific lake and contaminant system?

4. Are any existing data consistent with these answers?

Another purpose of this chapter is to discuss the factors that control the processes that influence sediment–air concentration relationships in the hope that more effort may be devoted to quantifying these various processes.

## Model Description

Consider a simple one-dimensional system consisting of an atmosphere, a water column, and sediment (Figure 1). In this system, the time period of interest is assumed to be 1 year, and the air and water column are assumed to be well mixed. This last assumption may preclude considera-

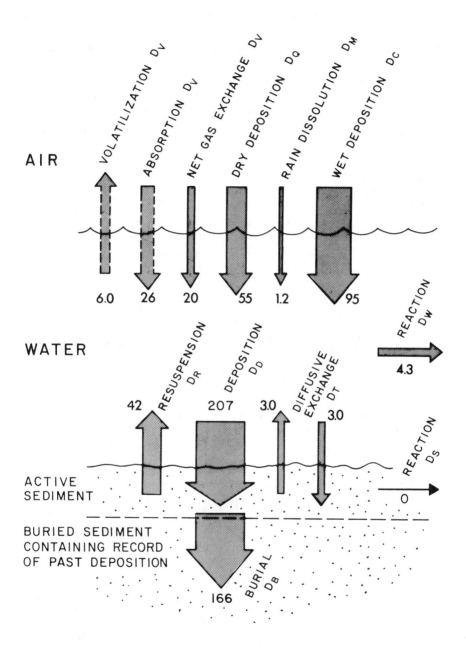

Figure 1. Simple one-dimensional system showing the relative transport and transformation values ($10^9$ mol/h) calculated for a tetrachlorinated dibenzodioxin (tetra-CDD) in Siskiwit Lake.

tion of very deep, stratified lakes; however, if sedimentation is rapid, as appears to occur with PCBs in Lake Superior (7), this model may be applicable. Steady state, also assumed to occur, implies that the capacity of the phases is small compared with the fluxes. Equivalently, the response time of the water and sediment to concentration changes is small compared with 1 year, or with the period of historic record. The sediment is viewed as a well-mixed layer of perhaps 1-mm depth underlain with sediments that are essentially inaccessible, unchanged, and unmixed with time. Thus, the "memory" of sedimentation history is contained in this active surface layer. As deposition occurs, the previously active sediment layer becomes buried to progressively greater depths.

The fugacity approach was used to formulate the model because this approach simplifies the equations and facilitates interpretation. Details of the assumptions are given elsewhere (8, 9).

In the model, fugacity, $f$ (Pa), is used instead of concentration, $C$ (mol/m$^3$). These quantities are linearly related by the fugacity capacity, $Z$ (mol/m$^3$ Pa), which is specific to the chemical and the medium. The fluxes of chemical between air, water, and sediment are expressed as products of fugacity in the source phase and a transport parameter, $D$ (mol/Pa h). A horizontal area, $A$ (m$^2$), is assumed to apply to this model. A reaction $D$ value is also included for the water and sediment. The depths of water and sediment are $h_W$ and $h_S$ (m), respectively. Thus, $V_W = h_W A$, and $V_S = h_S A$, where $V_W$ and $V_S$ have units of cubic meters.

**Transport Parameter Estimates.**   In a previous paper (8), $D$ values were defined for all relevant air–water–sediment transfer and transformation processes. The $D$ values can be estimated from one of four approaches.

REACTION.   The reaction rate is conventionally the volume, concentration, and rate constant product, $VCk$, or in fugacity terms, $VZkf$. The first-order rate constant is $k$ (h$^{-1}$). Because the rate is also expressed as $Df$, $D = VZk$.

The parameters for the water column and for the sediment are $D_W$ and $D_S$, respectively.

ADVECTIVE OR BULK FLOW.   For advective or bulk flow, the rate is $GC$, where $G$ is the volumetric flow rate of some phase, and $C$ is the concentration in that phase. When this rate is expressed as $Df$, $D = GZ$. This approach is used for dissolution in rain, deposition of particulates, burial of sediment, and suspended-matter deposition and resuspension.

MASS TRANSFER BY DIFFUSION.   For mass transfer by diffusion, the rate is $K'A\Delta C$, where $K'$ (m/h) is a mass-transfer coefficient, and $\Delta C$ is the concentration departure from equilibrium causing the diffusion. In this case, $D = K'AZ$. Two resistances in series may occur in this approach. For example, resistances may occur at the air–water interface during volatilization.

DIFFUSION IN PORE WATER.  For diffusion in pore water, the rate is $BA\Delta C/Y$, where $B$ (m$^2$/h) is an effective diffusivity, and $Y$ (m) is the diffusion path length. The value $B$ contains a correction for void fraction. The ratio $B/Y$ is thus equivalent to $K'$ as discussed previously. In this case, $D = BAZ/Y$.

The principal advantage of this approach is that these diverse processes are expressed in common units and thus can be grouped and added.

Steady-State Relationships.  If the system in Figure 1 is considered, steady-state mass balance equations for the water and sediment, respectively, can be written as follows:

$$D_R f_s + D_{Tf} s + D_V f_A + D_Q f_A + D_C f_A + D_M f_A = D_{Tf} w + D_D f w + D_W f w + D_V f w \quad (1)$$

$$D_{Tf} w + D_D f w = D_B f s + D_S f s + D_R f s + D_{Tf} s \quad (2)$$

where the subscripts A, B, C, M, Q, R, S, T, V, and W represent air, burial, wet deposition, rain dissolution, dry deposition, resuspension, sediment reaction, sediment–water transfer, volatilization, and water reaction, respectively.

Rearrangement of these equations yields equations 3 and 4.

$$f_w = f_s (D_B + D_S + D_R + D_T)/(D_T + D_D) \quad (3)$$

$$f_A = f_w[(D_T + D_D + D_W + D_V) - f_s(D_R + D_T)]/(D_V + D_Q + D_M + D_C)$$

$$= f_w \frac{\left[(D_T + D_D + D_W + D_V) - \dfrac{(D_R + D_T)(D_D + D_T)}{(D_B + D_S + D_R + D_T)}\right]}{(D_V + D_Q + D_M + D_C)} \quad (4)$$

These equations demonstrate a simple linear relationship between $f_S$, $f_W$, and $f_A$. For example, if the air concentration (and fugacity) doubles, the water and sediment concentrations should also double, albeit after some delay until steady state is reached. Each fugacity value can be replaced by concentration in equations 3 and 4 to give equations 5 and 6, respectively.

$$C_W = C_S(\overline{Z_W}/Z_S) \ (D_B+D_S+D_R+D_T)/(D_T+D_D) \tag{5}$$

$$C_A = \cfrac{C_W(\overline{Z_A}/\overline{Z_W}) \ \left[(D_T+D_D+D_W+D_V) - \cfrac{(D_R+D_T) \ (D_D+D_T)}{(D_B+D_S+D_R+D_T)}\right]}{(D_V+D_Q+D_M+D_C)} \tag{6}$$

The bars over the fugacity capacities in equations 5 and 6 are used to discriminate between the Z values for pure air and water, ($Z_A$ and $Z_W$, respectively), and those of the particulate-laden phases that actually exist in the system ($\overline{Z_A}$ and $\overline{Z_W}$). Each fugacity ratio can also be replaced by introducing simple steady-state, but not equilibrium, intermedia concentration ratios:

$$K_{SW} = (Z_S/\overline{Z_W}) \ (D_T+D_D)/(D_B+D_S+D_R+D_T) \tag{7}$$

$$K_{WA} = \cfrac{(\overline{Z_W}/\overline{Z_A}) \ (D_V+D_Q+D_M+D_C)}{\left[(D_T+D_D+D_W+D_V) - \cfrac{(D_R+D_T) \ (D_D+D_T)}{(D_B+D_S+D_R+D_T)}\right]} \tag{8}$$

$$K_{SA} = \cfrac{(Z_S/Z_A) \ (D_T+D_D) \ (D_V+D_Q+D_M+D_C)}{[(D_B+D_S+D_R+D_T) \ (D_T+D_D+D_W+D_V) - (D_R+D_T) \ (D_D+D_T)]} \tag{9}$$

After replacement of these concentration ratios, equations 10–12 result.

$$C_S = C_W K_{SW} \tag{10}$$

$$C_W = C_A K_{WA} \tag{11}$$

$$C_A = C_S/(K_{SW}K_{WA}) = C_S/K_{SA} \tag{12}$$

where $K_{SA} = K_{SW}K_{WA}$.

The presence of particles reduces the fugacity for a given total concentration. The expressions relating particle-laden phases to pure phases are

$$\overline{Z_A} = Z_A + x_A Z_Q \tag{13}$$

$$\overline{Z_W} = Z_W + x_W Z_P \tag{14}$$

where $x_A$ and $x_W$ are the volume fractions of particulates in the air and water, respectively, and $Z_Q$ and $Z_P$ are the Z values of the corresponding particulate phases. Later, values of $10^{-11}$ for $x_A$ ($\approx 20 \ \mu g/m^3$) and $10^{-5}$ for $x_W$ ($\approx 15 \ mg/L$) are assumed. The amounts of chemical dissolved in water and sorbed to particulates are proportional to $Z_W$ and $x_W Z_P$, respectively. This relationship is similar for air.

These expressions look complex, but they reduce to relatively simple and intuitively obvious expressions under certain conditions.

If the reaction terms in water $(D_W)$ and sediment $(D_S)$ are large compared with the other $D$ values, all the $K$ terms approach zero, and $C_S$ and $C_W$ will be nearly zero or a very small fraction of $C_A$. Physically, the chemical is reacting rapidly, and little is left in the water or sediment. No historic record results.

If deposition $(D_D)$ and burial $(D_B)$ are large compared to the atmospheric deposition processes $(D_C + D_M + D_Q)$, any chemical that enters the water will rapidly scavenge to the sediment, and the water concentration will be low. The rate of burial will then approximate the rate of atmospheric deposition; a situation results that is believed to apply to PCBs in Lake Superior as discussed by Eisenreich and Hollod (4). In practice, the deposition rate term, $D_D$, will likely be similar in magnitude to the burial rate term, $D_B$, because burial is the net difference between deposition and resuspension. Indeed, if no resuspension occurs, if deposition and burial are equal, and if no sediment reaction occurs $(D_S = 0)$, $K_{SW} = Z_S/\overline{Z_W}$, and the sediment layer achieves equilibrium with the water.

Using fish as monitors of water concentration, or fugacity, may be convenient. Thus, in small fish, the chemical's fugacity, which equilibrates rapidly with the water and is not subject to appreciable biomagnification, will be expected to equal that of the water. An additional term, $K_{FW}$, or $Z_F/\overline{Z_W}$, which is a bioconcentration factor, could then be introduced, where the subscript $F$ denotes fish.

Notable omissions from this one-dimensional (vertical) model are processes of advective inflow and outflow of water. These processes are treated as being either zero or equal. The chemical buried in an area of sediment is assumed to be derived only from the atmosphere vertically above it. In many cases, appreciable input may come from horizontally remote areas, for example, by runoff of chemical deposited on soils in the lake watershed. In such cases, the area of atmospheric input is enlarged. Quantifying this effect requires knowledge of the extent of runoff, or the "capturing efficiency" of the soil and the relevant areas of soil and water. Sediment "focusing", which is the phenomenon of sediment deposition in only some areas of a lake, is also assumed not to occur.

**Time Response.** The equations presented are applicable for steady-state conditions and imply a fast response of the system to changes in air concentrations. This implication may not always apply. For example, the net atmospheric deposition rate for a particular area may be 10 mol/year, and the prevailing capacity of the water column may be 1 mol. If the air concentration doubles to give a deposition rate of 20 mol/year, the water concentration and amount will double to 2 mol. This change will occur in a time period of 0.1 year because the water capacity is small compared to the annual flux. If the water capacity was 100 mol, the response time would be 10 years, and the detail of year-to-year concentration variations would be lost. The water–sediment sensing system thus behaves like an electrical amplifier with defined frequency transmission characteristics. Knowing these characteristic times is useful because certain sediments may respond too slowly to be useful.

For water, the capacity is $V_W \overline{Z_W} f_W$, and the rate of loss (and gain) is $f_W(D_D+D_V+D_W+D_T)$. Thus, the characteristic time, $t$, is defined as

$$t_W = V_W \overline{Z_W}/(D_D+D_V+D_W+D_T) \tag{15}$$

Similarly, for sediment, the characteristic time is defined as

$$t_S = V_S Z_S/(D_R+D_B+D_S+D_T) \tag{16}$$

Order-of-magnitude estimates of these times can be readily obtained. Interestingly, these times are dependent not only on the dynamics of the lake, but also on the partitioning properties of the chemical. Thus, a lake sediment may provide a record of one chemical but not another. The time $t_W$ is also of interest because it is the response time of the lake water to changes in loading. The key quantity is often the deposition term, $D_D$, which equals $G_D Z_P$. By ignoring the other $D$ terms and replacing $Z_P/\overline{Z_W}$ by a partition coefficient, $K_P$, $t_W$ can be found to approximate $V_W/(G_D K_P)$. The implication is that the ratio of water volume to the product of deposited sediment volume and the partition coefficient controls the time. For hydrophobic organic compounds, $K_P$ can be of the order of $10^6$. Thus, if only $10^{-6}$ volumes of sediment are deposited per volume of lake water, the response time will be 1 year. This value could represent deposition at a rate of 1 mm/year at a lake depth of 1000 m. The lake response time to a loading change may be totally unrelated to its hydraulic retention time. As Eisenreich (7) has pointed out, Lake Superior, which has a retention time of centuries, responds to PCB loading changes in 1 year. The importance from a regulatory and remedial viewpoint is that loading reductions can result in rapid improvements in water and fish quality due to sedimentation.

## Parameterization

If the model is valid, a linear relationship exists between the sediment, water, and air concentrations of a particular substance, and the corresponding intermedia concentration ratios should be constant. These concentration ratios are specific for each chemical, depending on the $Z$ values and reactivities in the respective phases. As a result, the relative proportions of chemicals (such as PCBs and PCDDs) simultaneously introduced as a mixture in a particular lake's ecosystem will be different in the various compartments (water, air, and sediment).

For every chemical, a concentration ratio can be quantified if physicochemical data are available to permit calculation of $Z$ values; if the chemical's reactivities in air, water, and sediment are known; and if the appropriate environmental data for the particular lake's ecosystem are available. We can thus explore in an order-of-magnitude fashion the likely distribution of PCDDs and quantify the concentrations in the various compartments of Siskiwit Lake by using the data of Czuczwa et al. (5) as a guide.

**Fugacity Capacity Calculation.** For this purpose, PCDDs are treated as five pseudocompounds, or classes of equally chlorinated isomers: tetrachlorinated dibenzodioxins (tetra-CDD), pentachlorinated dibenzodioxins (penta-CDD), hexachlorinated dibenzodioxins (hexa-CDD), heptachlorinated dibenzodioxins (hepta-CDD), and octachlorinated dibenzodioxins (octa-CDD). For each class of chlorinated dioxin congeners, a set of $Z$ values was calculated from the physicochemical properties of one "average" congener. Table I gives approximate, estimated physicochemical properties of dioxins at 15 °C. These properties include the subcooled liquid solubility, $C_L$ (mol/m³); the subcooled liquid vapor pressure, $P_L$ (Pa); and the octanol–water partition coefficient, $K_{OW}$. The estimates were based on the experimental data reported by Shiu et al. (10), Burkhard and Kuehl (11), Friesen et al. (12), Rordorf (13, 14), Marple et al. (15), and Schroy et al. (16, 17). The $Z$ values were

Table I. Estimated Physicochemical Properties of
Polychlorinated Dibenzodioxins

| Congener Group | Molecular Mass | $C_L{}^a$ ($\cdot 10^6$ mol/m³) | $P_L{}^b$ ($\cdot 10^6$ Pa) | log $K_{OW}{}^c$ |
|---|---|---|---|---|
| Tetra-CDD | 322 | 63 | 126 | 7.0 |
| Penta-CDD | 356.4 | 12.6 | 15.8 | 7.6 |
| Hexa-CDD | 391 | 2.5 | 2.0 | 8.0 |
| Hepta-CDD | 425.2 | 0.50 | 0.25 | 8.2 |
| Octa-CDD | 460 | 0.10 | 0.03 | 8.3 |

NOTE: All values were estimated assuming a temperature of 15 °C.
$^a$$C_L$ is solubility of liquid phase.
$^b$$P_L$ is vapor pressure of liquid phase.
$^c$$K_{OW}$ is the partition coefficient between octanol and water.

then calculated from the relationships for air, water, bottom and suspended sediment, and air particles by using equations 17–20.

$$Z_A = 1/RT \tag{17}$$

$$Z_W = 1/H = C_L/P_L \tag{18}$$

$$Z_S = Z_P = 0.41\theta K_{OW} Z_W \cdot 1.5 \tag{19}$$

$$Z_Q = 6 \cdot 10^6 Z_A/P_L \tag{20}$$

where $R$ is the gas constant (8.314 Pa m$^3$/mol K), $T$ is temperature (288 K), $H$ is Henry's law constant (m$^3$ Pa/mol), $\theta$ is the organic carbon content of the sediment (assumed to be 0.1 g/g), and the subscript $L$ represents the liquid phase. The constant in equation 20 is an assumed value taken from Mackay et al. (18) and may be in error. Thus, care must be taken in interpreting air–particulate partitioning. The Z values are listed in Table II.

**Table II. Calculated Fugacity Capacity Values for the Polychlorinated Dibenzodioxins, and Percentages in Sorbed Form**

| Congener Group | $Z_A$ | $Z_W$ | $Z_S$ | $Z_Q$ | $\overline{Z_A}$ | $\overline{Z_W}$ | %Sorbed Air | %Sorbed Water |
|---|---|---|---|---|---|---|---|---|
| Tetra-CDD | 0.00042 | 0.5 | $3.1 \cdot 10^5$ | $2.0 \cdot 10^7$ | .0006 | 3.6 | 32 | 86 |
| Penta-CDD | 0.00042 | 0.8 | $2.0 \cdot 10^6$ | $1.6 \cdot 10^8$ | .002 | 21 | 79 | 96 |
| Hexa-CDD | 0.00042 | 1.3 | $8.1 \cdot 10^6$ | $1.3 \cdot 10^9$ | .013 | 82 | 97 | 98 |
| Hepta-CDD | 0.00042 | 2.0 | $2.2 \cdot 10^7$ | $1.0 \cdot 10^{10}$ | .10 | 220 | 100 | 99 |
| Octa-CDD | 0.00042 | 3.2 | $3.9 \cdot 10^7$ | $7.9 \cdot 10^{10}$ | .79 | 390 | 100 | 99 |

NOTE: All values of Z are in moles per cubic meter pascal.

## Sediment–Deposition Flux Calculation.

The lake is assumed to have an area of $1.5 \cdot 10^7$ m$^2$, a mean water depth of 20 m, and an active sediment depth of 1 mm. The water volume is thus $3.0 \cdot 10^8$ m$^3$, and the sediment volume is $1.5 \cdot 10^4$ m$^3$.

The deposition rate in Siskiwit Lake is approximately 8 cm/50 years, or 1.6 mm/year. If the sediment is 50% solids, this rate corresponds to a net volumetric deposition, or burial rate, of $8.0 \cdot 10^{-4}$ m$^3$/m$^2$ year, or 1.4 m$^3$/h for the entire lake. If one-fifth of the deposited sediment is resuspended ($G_R = 0.34$ m$^3$/h), and four-fifths is permanently buried ($G_B = 1.4$ m$^3$/h), then the sediment deposition flux, $G_D$, is 1.7 m$^3$/h.

TRANSPORT PARAMETER CALCULATION. The following procedures, which are described in more detail by Mackay et al. (8, 18), were used to estimate the $D$ values at the assumed temperature of 15 °C.

**DRY DEPOSITION.** A dry deposition velocity, $V_D$, of 0.3 cm/s, or 10.8 m/h, is assumed to apply to the particles present at a concentration, $x_A$, of $10^{-11}$. Thus, the rate of dry deposition, $G_Q$, over the total area, $A$, is $V_D x_A A$, or $1.62 \cdot 10^{-3}$ m³/h. The $D_Q$ value is obtained by multiplying $G_Q$ by $Z_Q$.

**WET DEPOSITION.** Each volume of rain is assumed to scavenge particles from 200,000 volumes of air ($Q$). A rainfall rate, $G_M$, of 1400 m³/h or 0.82 m/year, is assumed. Thus, the particulate wet deposition rate ($G_c$) is $Q G_M x_A$, or 0.0028 m³/h. Multiplying this rate by $Z_Q$ yields $D_C$.

**RAIN DISSOLUTION.** The transport parameter for rain dissolution, $D_M$, is the product of $G_M$ and $Z_W$ and is negligible compared with $D_Q$ and $D_C$.

**VOLATILIZATION.** To calculate $D_V$, the two-resistance theory is applied with a water-phase mass-transfer coefficient, $K'_W$, of 0.01 m/h, and an air-phase value, $K'_A$, of 3 m/h (*19*). The individual $D$ values are $K'_W A Z_W$ and $K'_A A Z_A$. The overall $D$ value is the reciprocal of the sum of the reciprocals because the resistances are in series.

**WATER TRANSFORMATION.** From a decrease in the total amount of 2,3,7,8-tetrachlorodibenzo-*p*-dioxin in an outdoor pond with time (*20*) and a corresponding increase in metabolites, an environmental half-life of 395 days, or a degradation rate constant of $7.3 \cdot 10^{-5}$ h$^{-1}$, was calculated. Degradation was mainly attributed to photochemical conversion in the water phase, where algae are believed to produce photosensitizers.

Assuming that relative degradation rate constants of the five chlorinated dioxin congeners are the same in the environment and in laboratory photolysis experiments as performed by Nestrick et al. (*21*), rate constants of $9 \cdot 10^{-6}$ h$^{-1}$ for penta-CDD, $1.1 \cdot 10^{-5}$ h$^{-1}$ for hexa-CDD, $2.3 \cdot 10^{-6}$ h$^{-1}$ for hepta-CDD, and $2.8 \cdot 10^{-6}$ h$^{-1}$ for octa-CDD are estimated. The value of $D_W$ can then be calculated by multiplying the degradation rate constant by the water volume and $Z_W$. This approach implicitly assumes that all photodegradation occurs to the dissolved PCDDs and none occurs to the particle-bound material.

**SEDIMENT DEPOSITION.** As discussed earlier, $G_D = 1.7$ m³/h for the entire lake. The $D$ value is the product of $G_D$ and $Z_P$.

**SEDIMENT RESUSPENSION.** The sediment resuspension parameter, $D_R$, is assumed to be one-fifth of the deposition rate, $D_D$.

**SEDIMENT BURIAL.** The sediment burial transport parameter, $D_B$, is assumed to be four-fifths of the deposition rate, $D_D$.

SEDIMENT TRANSFORMATION.   Because pond data for 2,3,7,8-tetra-chlorodibenzo-p-dioxin (20) suggest that degradation predominantly occurs by photochemical conversion in the water phase, sediment trans-formation rate constants are considered to be extremely low. Because the sediment record contains isomers that date from 40 years ago, the rate is likely to be very small, and the half-life likely exceeds 10 years. A zero rate constant is assumed. If a value had been adopted, $D_S$ would be the product of the sediment fugacity capacity, the sediment volume, and the rate constant.

SEDIMENT–WATER DIFFUSION.   A diffusivity of $1.0 \cdot 10^{-6}$ m²/h is assumed to apply over a 1-mm path length, and an effective mass-transfer coefficient, $K'_T$, of $1.0 \cdot 10^{-3}$ m/h results. Thus, $D_T$ is the prod-uct of $K'_T$, $Z_W$, and $A$. This parameter value proves to be small com-pared with the sediment deposition and resuspension parameter values.

In many cases, the results are insensitive to the assumed parameter values; thus, not all values need to be known accurately. A problem that recurs when modeling complex environmental systems is which data should be known most accurately in advance. The best approach to this problem is to assume a reasonable value first, and then perform a sensi-tivity analysis later.

Of particular relevance are rates of sediment deposition $(D_D)$ and resuspension $(D_R)$. Eisenreich (7) has suggested that deposition and resuspension rates may be 2–10 times the burial rate (ie., there is a region of very active sediment–water exchange near the bottom). From a modeling viewpoint, quantifying these rates and the vertical distance over which they apply is essential. In this case, the results are fairly insensitive to the assumed deposition rate $(D_D)$, provided that the rate exceeds a certain value.

## Discussion

A short computer program was written (a copy can be obtained on request from the authors) to calculate the Z values, D values, and the air and water concentrations from an input sediment concentration. In this case, the concentrations for Siskiwit Lake reported by Czuczwa et al. (6) were used. The output for a tetra-CDD is presented here. The physi-cochemical properties of tetra-CDD and lake properties of Siskiwit Lake are listed on page 69, the estimated D values and molar fluxes for tetra-CDD are shown (Table III), and the concentrations of tetra-CDD in Siskiwit Lake are presented (Table IV). The data are shown pictorially in Figure 1.

The Z values in Table II show that appreciable fractions of the PCDDs are sorbed to atmospheric particulates and are thus susceptible

---

### Physicochemical Properties of Tetra-CDD

Molecular weight: 322 g/mol
Henry's law constant: 2.0 $m^3$ Pa/mol
Subcooled liquid vapor pressure: $1.3 \cdot 10^{-4}$ Pa
Subcooled liquid solubility: 0.02 $g/m^3$
Water mass-transfer coefficient: 0.01 m/h
Air mass-transfer coefficient: 3.0 m/h
Log of the octanol–water partition coefficient: 7.0
Sediment pore water diffusivity: $1.0 \cdot 10^{+6}$ $m^2$/h

### Physical Properties of Siskiwit Lake

Ambient temperature: 15 °C
Water volume: $3.0 \cdot 10^8$ $m^3$
Surface area: $1.5 \cdot 10^7$ $m^2$
Air particulate concentration: $1.0 \cdot 10^{-11}$ $m^3$ particles/$m^3$ air
Water particulate concentration: $1.0 \cdot 10^{-5}$ $m^3$ particles/$m^3$ water

---

to wet and dry deposition. Similarly, appreciable fractions are sorbed to suspended matter in the water and are thus susceptible to sediment deposition.

The estimated $D$ values listed in Table V show clearly which processes are most important. Wet and dry deposition from the atmosphere ($D_C$ and $D_Q$, respectively) exceed gaseous exchange ($D_V$), and rain dissolution ($D_M$) is negligible. In the water column, deposition ($D_D$) is much faster than diffusion ($D_T$) or reaction ($D_W$). An advantage of the fugacity approach is that these diverse process rates can be compared directly and the dominant processes identified.

The water column response times vary from 70 to 80 days, and the sediment response times are approximately 1 year. Thus, this system is capable of recording changes in air concentration that occur during a period of a few years.

The estimated concentration ratios and the calculated air concentrations are given in Figure 2 and Table VI. The model is linear, and thus these ratios apply at any concentration below saturation. The only possible nonlinearity would arise from a nonlinear biodegradation rate, but this does not apply in this case. These ratios have two interesting features. First, as shown in Figure 2, the ratios are fairly constant for this series because the dominant transport mechanisms involve deposition in association with particles, and almost all the dioxin deposited on the water eventually reaches the sediment. Second, the air and sediment fugacities only differ by a small factor. This system is fairly close to

**Table III. Estimated Transport–Transformation Parameters and Molar Fluxes for Tetra-CDD in Siskiwit Lake**

| Physicochemical Property | Kinetic Term | $Z$ $(mol/m^3/Pa)$ | $D$ $(mol/h/Pa)$ | $f$ $(Pa)$ | $Flux^a$ $(mol/h)$ |
|---|---|---|---|---|---|
| Air | | $Z_A=4.2\cdot10^{-4}$ | | | |
| Sediment Burial | $G_B=1.4$ | $Z_S=3.0\cdot10^5$ | $D_B=4.2\cdot10^5$ | $f_S$ | $1.66\cdot10^{-7}$ |
| Sediment Transformation | $k_S=0.0$ | $Z_S=3.0\cdot10^5$ | $D_S=0.0$ | | $0.00$ |
| Sediment Resuspension | $G_R=3.4\cdot10^{-1}$ | $Z_S=3.0\cdot10^5$ | $D_R=1.0\cdot10^5$ | | $4.15\cdot10^{-8}$ |
| Sediment–Water Diffusion | $K_T=1.0\cdot10^{-3}$ | $Z_W=4.9\cdot10^{-1}$ | $D_T=7.4\cdot10^3$ | $4.00\cdot10^{-13}$ | $2.95^2\cdot10^{-9}$ |
| Water–Sediment Diffusion | $K_T=1.0\cdot10^{-3}$ | $Z_W=4.9\cdot10^{-1}$ | $D_T=7.4\cdot10^3$ | $f_W$ | $2.95\cdot10^{-9}$ |
| Sediment Deposition | $G_D=1.7$ | $Z_P=3.0\cdot10^5$ | $D_D=5.2\cdot10^5$ | | $2.07\cdot10^{-7}$ |
| Water Transformation | $k_W=7.3\cdot10^{-5}$ | $Z_W=4.9\cdot10^{-1}$ | $D_W=1.1\cdot10^{-4}$ | | $4.31\cdot10^{-9}$ |
| Water–Air Volatilization | $K_V=2.0\cdot10^{-3}$ | $Z_W=3.9\cdot10^{-1}$ | $D_V=1.5\cdot10^4$ | $4.00\cdot10^{-13}$ | $5.98\cdot10^{-9}$ |
| Air–Water Absorption | $K_V=2.0\cdot10^{-3}$ | $Z_W=4.9\cdot10^{-1}$ | $D_V=1.5\cdot10^4$ | $f_A$ | $2.55\cdot10^{-8}$ |
| Dry Atmospheric Deposition | $G_Q=1.6\cdot10^{-3}$ | $Z_Q=2.0\cdot10^7$ | $D_Q=3.2\cdot10^4$ | | $5.48\cdot10^{-8}$ |
| Wet Atmospheric Deposition | $G_C=2.8\cdot10^{-3}$ | $Z_Q=2.0\cdot10^7$ | $D_C=5.6\cdot10^4$ | | $9.47\cdot10^{-8}$ |
| Rain Dissolution | $G_M=1.4\cdot10^3$ | $Z_W=4.9\cdot10^{-1}$ | $D_M=6.9\cdot10^2$ | $1.70\cdot10^{-12}$ | $1.17\cdot10^{-9}$ |
| Bulk Air | | $\overline{Z_A}=6.2\cdot10^{-4}$ | | | |
| Bulk Water | | $\overline{Z_W}=3.5$ | | | |

$^a$The flux was calculated from the product of $D$ and $f$.

Table IV. Concentrations of Tetra-CDD in Siskiwit Lake

| Concentration Type | $mol/m^3$ | $g/m^3$ |
|---|---|---|
| Air | | |
| Bulk concentration | $1.05 \cdot 10^{-15}$ | $3.38 \cdot 10^{-13}$ |
| Concentration in gas | $7.10 \cdot 10^{-16}$ | $2.29 \cdot 10^{-13}$ |
| Concentration on particulates | $3.38 \cdot 10^{-16a}$ | $1.09 \cdot 10^{-13a}$ |
| | $3.38 \cdot 10^{-5b}$ | $1.09 \cdot 10^{-2b}$ |
| | | $5.45 \cdot 10^{-9c}$ |
| Water | | |
| Bulk concentration | $1.41 \cdot 10^{-12}$ | $4.53 \cdot 10^{-10}$ |
| Dissolved concentration | $1.97 \cdot 10^{-13}$ | $6.34 \cdot 10^{-11}$ |
| Concentration on particles | $1.21 \cdot 10^{-12d}$ | $3.90 \cdot 10^{-10d}$ |
| | $1.21 \cdot 10^{-7b}$ | $3.90 \cdot 10^{-5b}$ |
| | | $2.60 \cdot 10^{-11c}$ |
| Sediment | | |
| Bulk concentration | $1.21 \cdot 10^{-7e}$ | $3.90 \cdot 10^{-5e}$ |
| | | $2.60 \cdot 10^{-11f}$ |

NOTE: The air and water concentrations were estimated, and the sediment concentration was reported. [a]Concentration is in units of mass per volume air. [b]Concentration is in units of mass per volume particles. [c]Concentration is in units of grams tetra-CDD per gram of particles. [d]Concentration is in units of mass per volume water. [e]Concentration is in units of mass per volume sediment. [f]Concentration is in units of grams tetra-CDD per gram of sediment.

equilibrium because the reaction rates are small compared with the transport rates.

These features indicate that obtaining a first estimate of sediment–air concentration ratios for conservative compounds may be possible by computing the ratio of Z values in sediment and bulk air. Inspection of the defining equations shows that this ratio is independent of vapor pressure provided that a substantial fraction of the chemical is in sorbed form in the air. Furthermore, the ratio is proportional to the product $K_{OW}C_L$, which is fairly constant because $K_{OW}$ and $C_L$ are inversely related. Relating sediment to air concentrations may be easier than has been expected, at least for conservative chemicals of low volatility. Fish may also give a useful indication of recent air concentrations if the chemical is known to be introduced only by the atmosphere.

The air and air particle concentrations estimated from the reported sediment concentrations are in fair order-of-magnitude agreement with available air concentration data. For example, PCDD concentrations on air particles reported by Czuczwa et al. (5) were 170–200 ng/g for octa-CDD, 21–25 ng/g for hepta-CDD, 1.2–1.6 ng/g for hexa-CDD, 0.2–6.4 ng/g for penta-CDD, and 0.5–1.1 ng/g for tetra-CDD. The model-predicted values are 130 (octa-CDD), 16 (hepta-CDD), 2.3 (hexa-CDD), 2.7 (penta-CDD), and 5.4 (tetra-CDD) ng/g. These values are all in fair

Table V. Estimated Transport–Transformation Parameters for the Polychlorinated Dibenzodioxins

| Congener Group | $D_T$ | $D_D$ | $D_B$ | $D_S$ | $D_R$ | $D_V$ | $D_Q$ | $D_M$ | $D_C$ | $D_W$ |
|---|---|---|---|---|---|---|---|---|---|---|
| Tetra-CDD | $7.5 \cdot 10^3$ | $5.3 \cdot 10^5$ | $4.2 \cdot 10^5$ | 0 | $1.1 \cdot 10^5$ | $1.5 \cdot 10^4$ | $3.2 \cdot 10^4$ | $7.0 \cdot 10^2$ | $5.6 \cdot 10^4$ | $1.1 \cdot 10^4$ |
| Penta-CDD | $1.2 \cdot 10^4$ | $3.4 \cdot 10^6$ | $2.7 \cdot 10^6$ | 0 | $6.8 \cdot 10^5$ | $1.6 \cdot 10^4$ | $2.6 \cdot 10^5$ | $1.1 \cdot 10^3$ | $4.4 \cdot 10^5$ | $2.1 \cdot 10^3$ |
| Hexa-CDD | $1.9 \cdot 10^4$ | $1.4 \cdot 10^7$ | $1.1 \cdot 10^7$ | 0 | $2.8 \cdot 10^6$ | $1.7 \cdot 10^4$ | $2.0 \cdot 10^6$ | $1.8 \cdot 10^3$ | $3.5 \cdot 10^6$ | $4.2 \cdot 10^3$ |
| Hepta-CDD | $3.0 \cdot 10^4$ | $3.7 \cdot 10^7$ | $3.0 \cdot 10^7$ | 0 | $7.5 \cdot 10^6$ | $1.8 \cdot 10^4$ | $1.6 \cdot 10^7$ | $2.8 \cdot 10^3$ | $2.8 \cdot 10^7$ | $1.4 \cdot 10^3$ |
| Octa-CDD | $4.7 \cdot 10^4$ | $6.6 \cdot 10^7$ | $5.3 \cdot 10^7$ | 0 | $1.3 \cdot 10^7$ | $1.8 \cdot 10^4$ | $1.3 \cdot 10^8$ | $4.4 \cdot 10^3$ | $2.2 \cdot 10^8$ | $2.7 \cdot 10^3$ |

NOTE: All values are in moles per hour pascal.

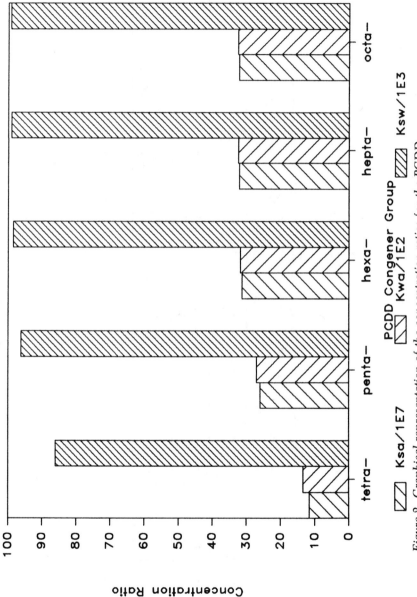

*Figure 2. Graphical representation of the concentration ratios for the PCDD congener groups.*

**Table VI. Estimated Water and Air Concentration Ratios Derived from Reported Surface Sediment Concentrations**

| Congener Group | Reported Sediment Concentration $(g/g)$ | Reported Sediment Concentration $(g/m^3)$ | Estimated Concentrations Water $(g/m^3)$ | Estimated Concentrations Air $(g/m^3)$ | $K_{SW}$ | $K_{SA}$ | $K_{WA}$ |
|---|---|---|---|---|---|---|---|
| Tetra-CDD | $26 \cdot 10^{-12}$ | $3.9 \cdot 10^{-5}$ | $4.5 \cdot 10^{-10}$ | $34 \cdot 10^{-14}$ | $8.6 \cdot 10^4$ | $1.2 \cdot 10^8$ | $1.3 \cdot 10^3$ |
| Penta-CDD | $12 \cdot 10^{-12}$ | $1.8 \cdot 10^{-5}$ | $1.9 \cdot 10^{-10}$ | $6.9 \cdot 10^{-14}$ | $9.6 \cdot 10^4$ | $2.6 \cdot 10^8$ | $2.7 \cdot 10^3$ |
| Hexa-CDD | $10 \cdot 10^{-12}$ | $1.5 \cdot 10^{-5}$ | $1.5 \cdot 10^{-10}$ | $4.8 \cdot 10^{-14}$ | $9.8 \cdot 10^4$ | $3.1 \cdot 10^8$ | $3.2 \cdot 10^3$ |
| Hepta-CDD | $70 \cdot 10^{-12}$ | $10 \cdot 10^{-5}$ | $11 \cdot 10^{-10}$ | $33 \cdot 10^{-14}$ | $9.9 \cdot 10^4$ | $3.2 \cdot 10^8$ | $3.2 \cdot 10^3$ |
| Octa-CDD | $560 \cdot 10^{-12}$ | $84 \cdot 10^{-5}$ | $85 \cdot 10^{-10}$ | $260 \cdot 10^{-14}$ | $9.9 \cdot 10^4$ | $3.2 \cdot 10^8$ | $3.2 \cdot 10^3$ |

agreement. Bulk air concentration reported by the Ontario Ministry of the Environment (*22*) for tetra-CDD (2.6 pg/m$^3$) compares well with the predicted value (0.34 pg/m$^3$), considering that the Ministry value represents a maximum ambient tetra-CDD concentration value near an incinerator source of PCDDs. The Ministry's concentrations for octa-CDD (0.75 pg/m$^3$), hepta-CDD (1.0 pg/m$^3$), hexa-CDD (2.3 pg/m$^3$), and penta-CDD (2.4 pg/m$^3$) compare well with the predicted values (2.6, 0.3, 0.05, and 0.07 pg/m$^3$) given by the model. Clearly, the model is giving a fair representation of the air–sediment concentration relationships, especially considering the uncertainties in physicochemical properties, analyses, and lake process parameters.

Thus, the sediment record from pristine lakes can be used to estimate historic air concentrations, at least for some chemicals. Clearly, the model needs to be validated or calibrated by comparing current and documented air, water, sediment, and possibly fish concentrations. This validation will require acquisition of reliable physicochemical property data, reliable expressions for reactivity, partitioning in air and water phases, as well as prevailing environmental data such as mass-transfer coefficients and the various deposition rate quantities.

Comparing the sediment record from lakes, in which the atmosphere is known to be the only source of contamination, with nearby lakes experiencing other sources such as industrial or municipal discharges, may enable discrimination between source contributions by amount and by nature (using, for example, congener distributions as "fingerprints"). Assigning sources to in situ contaminants and suggesting appropriate source reduction strategies would then be possible.

### Conclusions

An advantage of the model described is that it clearly demonstrates the relative magnitudes of the various transport and transformation processes and identifies the important processes. Sediments can thus provide an invaluable record of past concentrations, especially for chemicals with reaction half-lives that exceed the time of interest in the record, have low vapor pressures, and are highly hydrophobic (ie., low water solubility and large octanol–water partition coefficients). Testing the model approach on other chemicals such as PCBs and toxaphenes with similar properties and ascertaining the model's applicability to the range of chemicals present in the environment would be an interesting future study.

### Abbreviations and Symbols
### Quantity

$\theta$    sediment organic carbon content (g/g)

$A$    area (m$^2$)

| | |
|---|---|
| $B$ | effective diffusivity ($m^2$/h) |
| $C$ | concentration or solubility (mol/$m^3$) |
| $D$ | transport–transformation parameter (mol/h·Pa) |
| $f$ | fugacity (Pa) |
| $G$ | flow rate ($m^3$/h) |
| $h$ | depth (m) |
| $H$ | Henry's law constant (Pa·$m^3$/mol) |
| $k$ | rate constant ($h^{-1}$) |
| $K'$ | mass-transfer coefficient (m/h) |
| $K$ | partition coefficient (dimensionless) |
| $P$ | vapor pressure (Pa) |
| $R$ | gas constant (8.314 Pa·$m^3$/mol K) |
| $t$ | characteristic time (h) |
| $T$ | temperature (K) |
| $V$ | volume ($m^3$) |
| $x$ | volume fraction of particulate matter (dimensionless) |
| $Y$ | diffusion path length (m) |
| $Z$ | fugacity capacity (mol/$m^3$·Pa) |

## Subscripts

| | |
|---|---|
| $A$ | air |
| $B$ | burial |
| $C$ | wet deposition |
| $D$ | sediment deposition |
| $F$ | fish |
| $L$ | liquid phase |
| $M$ | rain dissolution |
| $P$ | suspended sediment |
| $Q$ | air particles |
| $R$ | resuspension |
| $S$ | sediment |
| $T$ | sediment–water |
| $V$ | volatilization |
| $W$ | water |

## *References*

1. Swain, W. R. *J. Great Lakes Res.* **1978**, *4(3-4)*, 398–407.
2. Eisenreich, S. J.; Looney, B. B.; Thornton, J. D. *Environ. Sci. Technol.* **1981**, *15(1)*, 30–38.
3. Eisenreich, S. J. *Atmospheric Pollutants in Natural Waters;* Ann Arbor Science: Ann Arbor, MI, 1981.
4. Eisenreich, S. J.; Hollod, G. J. *Environ. Sci. Technol.* **1979**, *13(5)*, 569–573.
5. Czuczwa, J. M.; McVeety, B. D.; Hites, R. A. *Chemosphere* **1985**, *14(6-7)*, 623–626.

6. Czuczwa, J. M.; Niessen, F.; Hites, R. A. *Chemosphere* **1985**, *14(9)*, 1175-1179.
7. Eisenreich, S. J., University of Minnesota, personal communication, 1986.
8. Mackay, D.; Joy, M.; Paterson, S. *Chemosphere* **1983**, *12(7-8)*, 981-997.
9. Mackay, D.; Paterson, S. *Environ. Sci. Technol.* **1982**, *16(12)*, 654A-660A.
10. Shiu, W. Y.; Gobas, F. A. P. C.; Mackay, D. In *QSAR in Environmental Toxicology II*; Kaiser, K. L. E., Ed.; Reidel: Dardrecht, in press.
11. Burkhard, L. P.; Kuehl, D. W. *Chemosphere* **1986**, *15*, 163-167.
12. Friesen, K. J.; Sarna, L. P.; Webster, G. R. B. *Chemosphere* **1985**, *14*, 1267-1274.
13. Rordorf, B. F. *Chemosphere* **1985**, *14(6-7)*, 885-893.
14. Rordorf, B. F. *Thermochim. Acta* **1985**, *85*, 435-438.
15. Marple, L.; Brunck, R.; Throop, L. *Environ. Sci. Technol.* **1986**, *20(2)*, 180-182.
16. Schroy, J. M.; Hileman, F. D.; Cheng, S. C. *Chemosphere* **1985**, *14(6-7)*, 877-880.
17. Schroy, J. M.; Hileman, F. D.; Cheng, S. C. In *Aquatic Toxicology and Hazard Assessment, Eighth Symposium;* Bahner, R. C.; Hansen, D. J., Eds.; ASTM Special Technical Publication 891; American Society for Testing and Materials: Philadelphia, PA, 1985.
18. Mackay, D.; Paterson, S.; Schroeder, W. H. *Environ. Sci. Technol.* **1986**, *20(8)*, 810-816.
19. Mackay, D.; Yeun, A. *Environ. Sci. Technol.* **1983**, *17(4)*, 211-217.
20. Tsushimoto, G.; Matsumura, F.; Sago, R. *Environ. Toxicol. Chem.* **1982**, *1*, 61-63.
21. Nestrick, T. J.; Lamparshi, L. L.; Townsend, D. J. *Anal. Chem.* **1980**, *52*, 1865-1880.
22. "Polychlorinated Dibenzodioxins and Polychlorinated Dibenzofurans— PCDDs and PCDFs"; Scientific Criteria Document for Standard Development No. 4-84; Ontario Ministry of the Environment: Toronto, Ontario, Canada, 1985.

RECEIVED for review June 10, 1986. ACCEPTED August 11, 1986.

# 4

# Depositional Aspects of Pollutant Behavior in Fog and Intercepted Clouds

Jed M. Waldman[1] and Michael R. Hoffmann[2]

Environmental Engineering Science, W. M. Keck Laboratories, California Institute of Technology, Pasadena, CA 91125

*Droplet deposition during fog is shown to play an important role in the removal of anthropogenic pollutants from the atmosphere. Relevant theoretical principles are reviewed. The in-cloud scavenging of aerosols and soluble gases coupled with the small size of fog droplets results in higher chemical concentrations in fog water than in rainwater. In the urban regions of southern California and the southern San Joaquin Valley, fog water chemistry is dominated by sulfate, nitrate, and ammonium ions, which are measured at millimolar levels. The formation of fog is shown to accelerate deposition rates for water-scavenged atmospheric constituents. During stagnation episodes, pollutant removal by ventilation of valley air requires at least 5 days, while the enhancement of deposition by fog formation leads to pollutant lifetimes on the order of 6–12 h. Thus, in an environment characterized by flat, open landscape and low wind speed, droplet sedimentation can be the dominant removal mechanism of pollutants during prolonged stagnation episodes with fog.*

T HE REMOVAL OF ANTHROPOGENIC EMISSIONS and windblown material to ground surfaces occurs by processes known as wet and dry deposition. The deposition of airborne pollutants is essential for cleansing the

[1]Current address: Environmental and Community Medicine, University of Medicine & Dentistry of New Jersey—Robert Wood Johnson Medical School, Piscataway, NJ 08854
[2]To whom correspondence should be addressed

0065-2393/87/0216-0079$21.00/0
© 1987 American Chemical Society

atmosphere, but some deposited pollutants may have the potential for significant environmental impact when they reach the surface. Key deposition pathways are shown schematically in Figure 1.

Scavenging and deposition of ambient gases and aerosols by raindrops is more rapid and efficient than removal under dry conditions. However, precipitation is an intermittent event. The dry flux of pollutants due to the cumulative effect of long dry periods may contribute a greater mass of material than rainfall. Liljestrand (1) found this situation to be the case in southern California. A small fraction of total $NO_x$ and $SO_2$ emissions in the region was accounted for by wet deposition monitoring; Liljestrand estimated a 10-fold greater removal by dry versus wet deposition, and the majority of emissions was advected away from the region. In contrast, the net deposition in the eastern United States has been associated primarily with precipitation. Bischoff et al. (2) have estimated that wet fluxes constitute 75% of $SO_2$ and more than 90% of $NO_x$ emissions.

In addition, the potential exists for appreciable deposition during fog episodes. Even though the occurrence of fog is often infrequent, fog-induced deposition may be important in several specific environments. In coastal regions, fogs often serve as a dominant source of moisture and nutrient input to local ecosystems (3). Similarly, many mountainous regions are frequently intercepted by clouds and the accompanying chemical input (4). Urban areas, subject to higher ambient pollutant concentrations, may have substantial fog-induced deposition as well.

In many respects, fog is simply a ground-level cloud in which processes such as nucleation, gas scavenging, and droplet growth are important. At the same time, fogs consist of discrete water droplets, and sedimentation and inertial impaction are dominant transport mechanisms because of the relatively large sizes of these particles. However, because fog droplets may change size continually, their transport behavior may be altered.

Fog may make important contributions to pollutant deposition or impacts in selected environments for the following reasons:

(1) Fog and cloud droplets are important chemical reactors (5) that can modify the nature of pollutant material in the atmosphere (6). They act as sinks for many gaseous pollutants such as nitric acid, ammonia, and sulfur dioxide that are appreciably soluble in aqueous solutions. Fog and cloud droplets are sufficiently small such that gas scavenging is not limited by mass transport in most cases (7). Their effect on pollutant speciation, in turn, will have impacts on the chemical balance of depositing components.

(2) Under dry conditions, much of the aerosol mass of pollutant species is found in the size range 0.1–1.0-$\mu$m diameter. Deposition velocities ($V_d$) of these small particles are extremely low (8). Scavenging of

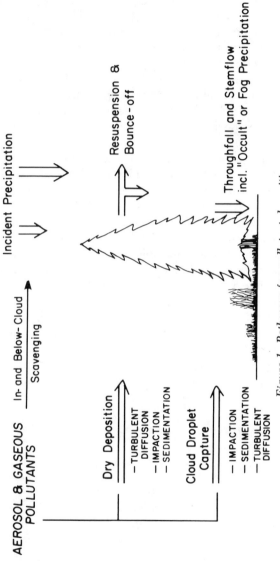

*Figure 1. Pathways for pollutant deposition.*

ambient aerosols by fog droplets leads to the association of solute mass with larger sized particles. Hence, when dissolved in fog droplets, solute species are more efficiently deposited by particle impaction and sedimentation.

(3) Surface moisture deposited by fog and dew can significantly increase the $V_d$ of aerosols by reducing rebound and resuspension (9). In addition, moisture on surfaces also serves as a sink for soluble gases (10).

(4) The capture of fog droplets by foliar and ground surfaces can be a significant component of the water and nutrient fluxes to an ecosystem (3, 4). Lovett (11) reported water flux at rates of 0.1–0.3 mm h$^{-1}$ from advected clouds to a subalpine forest canopy. This range agreed with his numerical model, which indicated that sedimentation and inertial impaction, especially to the upper 3 m of the canopy, were the dominant deposition mechanisms.

(5) The analyses of fog and cloud water show at least an order-of-magnitude greater concentration for ambient solutes compared with rainwater. Therefore, droplets can make a greater contribution to pollutant flux per volume of water deposition.

(6) Intercepted fog and drizzle wet surfaces but do not necessarily flush them clean as does rain. Dissolution of previously accumulated material can lead to concentrated solutions of acids (12) and metals (13) at collection surfaces. Waldman et al. (14) found that droplets, after depositing on pine needles, were in general more concentrated than ambient cloud water except when the needles had been previously rinsed by measurable rainfall. The acidity of drops removed from foliar surface was as high as the ambient cloud water despite enrichment with calcareous material.

(7) Damage to sensitive receptor surfaces, as measured in field and laboratory studies (15–17), has been caused by the exposure to acidic fog droplets.

## Theoretical Considerations of the Chemistry and Microphysics of Fog

**Fog Microphysics.**   Fog and cloud droplets are formed by condensation of water vapor in saturated air. Droplets form with diameters between 2 and 100 $\mu$m, although in nonprecipitating clouds and fog, the majority of droplet mass occurs in the range of 5–30 $\mu$m. In contrast, raindrops are far larger, mostly in the range of 200–2000 $\mu$m. The mass of liquid water is typically 0.01–0.5 g m$^{-3}$; the number concentration of droplets is generally ten to hundreds per cubic centimeter. Droplet interactions, such as coagulation or differential settling, have negligible impact on their dynamics except when intensive convection leads to rain or drizzle (18).

Vertical motions lead to the formation of most clouds with the exception of fogs. Generally, extensive ground contact suppresses net vertical motion. Fogs are primarily classified as either advection- or radiation-type, depending on their mechanism of formation (19). Advection fogs are formed by large-scale, horizontal air movement. For example, in the case of marine fogs, a warm, moist air mass comes in contact with a cooler surface. Radiation fog forms by radiative cooling near the ground to a clear, nighttime sky. This cooling causes an intense thermal gradient within the ground layer that may lead to fog or dew formation. Turbulent eddy transfer of momentum and heat play important roles under these conditions. A high level of turbulence inhibits fog (20), but some eddy mixing is essential to fog formation (21). Finally, fogs also occur on mountain slopes by the interception with a mesoscale cloud deck or fog formed locally by the upward sloping wind component.

The presence of condensation nuclei, which are composed of both soluble and nonsoluble material, is essential to the formation of atmospheric water droplets. The effects of surface tension and the chemical potential of the aquated solutes are important processes that raise and lower the saturation vapor pressure near the droplet surface, respectively. Accretion or evaporation of water to the condensation nucleus or droplet is forced by the difference between the ambient and local (i.e., surface) humidities. Droplet growth equations have been derived by coupling microscale heat and mass transfer at the aerosol surface (18). The principal terms depend on atmospheric relative humidity, droplet surface tension, and solute activity. Condensation caused by the radiative cooling of water droplets may also be an important factor in the dynamics of radiation fog (22).

During growth, the droplet temperature differs from the ambient because of latent heat release, which in turn depends upon the instantaneous growth rate. Hence, the complete growth-rate equation takes an implicit form; however, simplifying assumptions can be applied for all but the very smallest nuclei (18). The equation for rate of change in droplet diameter $(D_o)$ is

$$D_o \frac{dD_o}{dt} = \frac{S_v - a_w e^{(B)}}{C + E (A - 1) a_w e^{(B)}} \tag{1}$$

where $t$ is time; $S_v$ is the ambient water vapor saturation ratio; $a_w$ is the activity of water in a droplet; and $A$, $B$, $C$, and $E$ are constants defined in the appendix at the end of this chapter.

Families of curves can be deduced by solving the growth equation for the equilibrium condition $dD_o/dt = 0$ at various saturation ratios for different dry aerosol masses. An example of these families, known as the

Köhler curves, is shown in Figure 2. The maxima occur at points known as the critical supersaturation, $SS_{cr}$, and activation size, $D_{act}$. When the ambient water vapor supersaturation, $SS_v$, is less than $SS_{cr}$, nuclei achieve stable equilibrium sizes ($D_{eq} < D_{act}$, where $D_{eq}$ is the equilibrium diameter). When $SS_v$ is sustained at values greater than $SS_{cr}$, droplets will form and continue to grow.

**Fog Chemistry.** The study of fog has traditionally remained in the domain of atmospheric physicists principally concerned with its effect on visibility or the mechanisms of formation analogous to clouds. Yet, even the data of early investigations of fog indicated that fog water can be highly concentrated with respect to a variety of chemical components (Table I). Cloud water, sampled aloft, has been found to exhibit similar composition, although not with the same extreme values (23–25). On the other hand, rainwater compositions when compared with fog water are found to be far more dilute. Fog droplets are approximately 100 times smaller than raindrops, which form partially by the further condensation of water vapor. Hence, fog droplets should be more concentrated in solute derived from the condensation nuclei. Furthermore, fog forms in the ground layer where gases and aerosols are most concentrated.

The higher aqueous concentrations and extremes in acidity found in fog water are reason for concern. Fog-derived inputs have the potential to add substantially to the burden of acid deposition caused by precipitation. This input contribution can be disproportional to the sum of additional moisture because of its higher concentrations. Perhaps more importantly, deleterious effects of fog deposition may be associated with the intensity of solution acidity. For instance, appreciable nutrient leaching (26) and damage to leaf tissue (16) have been noted with application of acid fogs or mists. Finally, a historical correlation of fog with the most severe air pollution episodes provides additional cause for concern (41). Identification of a link between urban fog events and human health injury was made even before detailed measurements of fog composition were performed (42).

**Fog Scavenging Processes.** Fog droplets are highly effective at scavenging ambient materials present in the air. The overall fraction incorporated into fog droplets depends upon two processes: nucleation scavenging (i.e., activation of aerosols) and gas dissolution. The speciation of pollutant components precursory to fog formation is therefore important. Furthermore, in situ chemical transformations may alter this speciation and the effectiveness of fog scavenging while the droplet phase is present.

The following notation has been adopted in this chapter: (C) is the

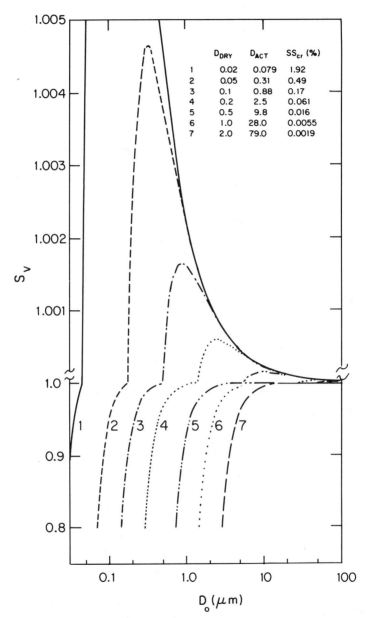

Figure 2. *Equilibrium vapor pressure ($S_v$) over aquated ammonium sulfate nuclei as a function of droplet diameter ($D_o$). The dry diameter ($D_{dry}$), activation diameter, ($D_{act}$), and critical supersaturation ($SS_{cr}$) were calculated from the equations given in the appendix.*

**Table I. Summary of Fog Water Composition Measurements**

| Location | Date | Type | Number of Samples | pH | $Na^+$ | $NH_4^+$ | $Ca^{2+}$ | $Mg^{2+}$ | $Cl^-$ | $NO_3^-$ | $SO_4^{2-}$ | Ref. |
|---|---|---|---|---|---|---|---|---|---|---|---|---|
| Mt. Washington, NH (1900 m) | 1930–1940 | I | 35° | 4.5 | | | | | 4 | | 150 | 27 |
| Coastal MA and ME | 1930–1940 | M | 37° | 3.0–5.9 | | | | | 0–34 | | 4–1100 | |
| | | | | 4.7 | | | | | 940 | | 380 | |
| | | | | 3.5–6.3 | | | | | 0–5800 | | 60–2600 | |
| Germany | | | | | | | | | | | | |
| Baltic Sea | 1955–1965 | M | 42° | 3.8 | 1500 | 2300 | 750 | | 700 | 900 | 1900 | 28 |
| Dresden | 1955–1965 | R | 12° | 4.2 | | 2100 | 3200 | | 590 | 450 | 780 | |
| Harz Mtn. (1150 m) | 1955–1965 | I | 18° | 5.1 | 300 | 710 | 220 | | 200 | | | |
| Japan | | | | | | | | | | | | |
| Mt. Noribura (3026 m) | July 1963 | I | 10+ | 3.9 | 87 | 175 | | | 110 | 36 | 3300 | 29 |
| | | | | 3.4–4.3 | 45–165 | 115–260 | | | 75–230 | 25–175 | 230–1250 | |
| Mt. Tsukaba (876 m) | November 1963 | I | 5+ | 5.9 | 290 | 880 | | | 800 | 17 | 1600 | |
| | | | | 5.6–6.5 | 180–435 | 110–965 | | | 295–1290 | 5–37 | 360–2100 | |
| Whiteface Mtn., NY (1500 m) | August 1976 | I# | 28° | 3.7 | 11 | 89 | 17 | 6 | 31 | 90 | 140 | 30 |
| | August 1980 | I | 50 | 3.2–4.0 | 1–7 | 1–200 | | | 1–14 | 7–190 | 32–800 | 31 |
| Los Angeles, CA foothills (780 m) | Spring 1982 and 1983 | I | 120+ | 2.9 | 240 | 580 | 140 | 80 | 190 | 1510 | 840 | 14 |
| | | | | 2.1–3.9 | 135–8700 | 62–7400 | 5–3000 | 1–1800 | 15–9650 | 160–16300 | 130–9300 | |

| Location | Date | Type | n | pH | | | | | | | | Ref. |
|---|---|---|---|---|---|---|---|---|---|---|---|---|
| Nova Scotia, Canada | August 1975 | M | 14° | — | 1040 (600–1530) | 33 (3–94) | 45 (20–69) | 100 (13–130) | 87 (3–450) | 115 (24–235) | 250 (50–500) | 32 |
| California Central Coast | Fall 1976 | M | 8° | — | 320 (80–950) | 190 (0–580) | 55 (9–100) | 68 (23–175) | 400 (95–1240) | 110 | 200 (77–490) | 33 |
| Los Angeles area, CA | Fall–Winter 1980–1982 | M, I | 11+ | 3.3 (2.7–7.1) | 139 (30–620) | 1580 (420–4260) | 168 (0–460) | 54 (22–310) | 223 (68–423) | 580–2980 | 584 (354–1875) | 34 |
| Los Angeles area, CA | Fall–Winter 1981–1982 | M | 24 | 2.3–5.8 | 12–2180 | 370–7960 | 190–4350 | 7–1380 | 56–1110 | 130–12000 | 62–5000 | 35 |
| Pt. Reyes, CA | August 1982 | M | 17+ | 4.5 (3.5–5.0) | 190 (21–4700) | 64 (28–330) | 10 (0–240) | 36 (5–1200) | 215 (34–7000) | 23 (2–526) | 186 (36–1281) | 36 |
| San Nicholas Island, CA | August 1982 | M | 7ˣ | 3.9 | 6100 | 450 | 450 | 1500 | 5300 | 1580 | 1080 | 37 |
| San Diego Area, CA | January 1983 | M | 5ˣ | 2.9 | 510 | 780 | 49 | 130 | | 1850 | 470 | 38 |
| Albany, NY | October 1982 | R | 24+ | 5.8 (4.3–6.4) | 36 (10–100) | 215 (70–350) | 120 (65–350) | 13 (6–47) | 47 (18–175) | 85 (11–220) | 155 (21–1360) | 39 |
| Bakersfield, CA | Winter 1983 | R | 108+ | 4.2 (2.6–7.0) | 20 (1–325) | 1440 (490–1330) | 47 (7–3500) | 6 (1–430) | 47 (1–980) | 850 (200–6800) | 1160 (10–9400) | 40 |
| Po Valley, Italy | February and November 1984 | R | 5° | 3.5–4.3 | 10–110 | 580–1620 | 60–130 | 10–50 | 20–120 | 290–1100 | 400–990 | |

NOTE: All concentrations are in $\mu$equiv L$^{-1}$. Abbreviations are as follows: I is intercepted stratiform cloud, M is marine or coastal fog, and R is radiation fog. + denotes average number of samples or events. ° denotes median number of samples or events. # denotes nonprecipitating stratiform cloud data only. ˣ denotes average number of n samples. ° denotes volume-weighted mean of n samples. ° denotes average, range, or both average and range of n events.

total concentration of species $C$ in the atmosphere, and $(C)_f$, $(C)_g$, and $(C)_a$ are the concentrations of $C$ corresponding to fog water, gas, and nonactivated aerosol phases expressed as moles per cubic meter of air. Aqueous concentrations in fog water are expressed in moles per liter with the notation $[C]$. Multiplication of these concentrations by the liquid water content (LWC) yields $(C)_f = \text{LWC}\ [C]$, where LWC is expressed as liters per cubic meter. For components in the gas phase, partial pressure $(P_C)$ values may be converted to give the same units. The mass balance may then be written as follows:

$$(C) = \text{LWC}\ [C] + P_C\ (RT)^{-1} + (C)_a \qquad (2)$$

where $R$ is the gas constant (atm $m^3$ $mol^{-1}$ $K^{-1}$) and $T$ is absolute temperature (K). The components for which we are most concerned are sulfur dioxide, sulfuric acid, nitric acid, ammonia, and the neutralized salts of these compounds because of their abundance and effect on acidification in urban-impacted environments. The terms S(IV), S(VI), N(V), and N(−III) are used to refer to all species of that oxidation state in any phase.

NUCLEATION SCAVENGING.    The pollutant species of concern are often present as hygroscopic aerosols (e.g., ammonium sulfate and ammonium nitrate). These aerosols will deliquesce to form aquated condensation nuclei at high relative humidity (RH), growing to larger equilibrium sizes as RH approaches 100%. The Köhler curves show the range of aerosols that can become activated into droplets. For soluble nuclei, the following proportionalities are valid: $D_{eq} : D_{dry}^{1.5} : SS_{cr}^{-1}$, where $D_{dry}$ is the dry diameter for a hygroscopic aerosol. Achievement of greater $SS_v$ can lead progressively to activation of smaller nuclei. The presence of a nonsoluble fraction in particles of a given size can significantly raise their $SS_{cr}$ (43). The chemical composition of nuclei also plays a role, although the differences between aerosols composed of pure solute species are minor, as shown here, where $K$ is the proportionality coefficient (see equation A12 in the appendix at the end of this chapter).

| Solute | K for $D_{dry}$ ($\mu m$) at 10 °C |
|--------|-----------------------------------|
| $(NH_4)_2SO_4$ | $4.93 \times 10^{-5}$ |
| $NH_4HSO_4$ | $4.59 \times 10^{-5}$ |
| $NH_4NO_3$ | $4.82 \times 10^{-5}$ |
| $H_2SO_4$ | $4.21 \times 10^{-5}$ |
| NaCl | $3.68 \times 10^{-5}$ |

Figure 3a shows three hypothetical particle size distributions formed by superimposing the two modes generally observed in the aerosol size spectra of sulfate (*44*). The smaller (0.05 $\mu$m) mode corresponds to particles formed by condensation of gases; the larger (0.5 $\mu$m) mode is known as the accumulation mode, which is formed primarily by coagulation and combustion. A third, coarser fraction is also routinely observed, although the contribution from anthropogenic sources may vary, especially with regard to nitrate aerosol (*45, 46*). The activation of these distributions, corresponding to $(NH_4)_2SO_4$ particles, is shown in Figure 3b as a function of $SS_v$. The nonactivated fraction is expressed as $f_C^a = (C)_a/(C)$, and $1 - f^a$ represents the aerosol scavenging efficiency in the fog.

Measurements of $SS_v$ in the atmosphere are not widely available. Gerber (*47*) found that rapid oscillations of ambient humidities occurred

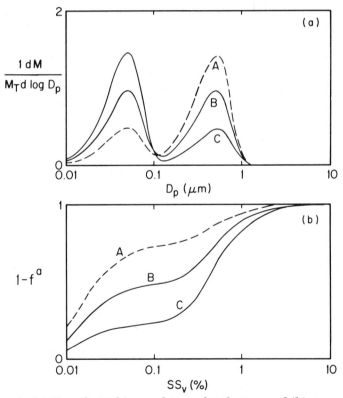

*Figure 3. (a) Hypothetical fog nuclei size distributions and (b) scavenging efficiency ($1-f^a$) as a function of supersaturation ($SS_v$). Curves A, B, and C represent 75%, 50%, and 25% of total mass in the larger size mode. Abbreviations are as follows: M is fog water mass, $M_T$ is total mass, and $f^a$ is the nonactivated fraction.*

in radiation fog with excursions about a mean near to RH = 100%; $SS_v$ rarely exceeded 0.1%–0.2%. Hudson (48) measured interstitial aerosol in fog and cloud environments; he found a systematic removal (i.e., activation) of larger nuclei. These cutoff values were used to calculate an effective supersaturation for the ambient atmosphere. In fogs and ground-based sampling of stratus clouds, Hudson's measurements indicated $SS_v$ values were in the range of 0.03%–0.2%. Comparing these measured or calculated values to Figure 3b shows that little of the smaller size mode is likely to be scavenged in these environments. Conversely, solutes that are associated with a coarser (>1 $\mu$m) fraction can be readily activated.

At the same time, the fraction of these nuclei (i.e., $D_{eq} > D_{act}$) that will be activated or the droplet size that they will achieve is not known. The relative increase in droplet size is faster for smaller diameters, and modeling would indicate that droplets of any size can grow from nuclei of any mass, depending upon the $SS_v$ history they have experienced, mixing, and the aerosol size spectrum (49). Limited field data address this controversy. Measurements of solute mass for individual fog droplets have been attempted. On the Japanese coast, Naruse and Maruyama (50) found correlations between droplet size and nucleus mass, in which the biggest droplets generally had large sea salt nuclei, and the smaller droplets formed on smaller, ammonium sulfate aerosols. Hudson and Rogers (51) indirectly measured condensation nuclei within stratus cloud droplet spectra, selectively removing droplets above a certain size. They reported an increasing fraction of low $SS_{cr}$ (i.e., large) nuclei in the larger droplets and a vanishing fraction of these nuclei in the interstitial aerosol spectrum. These relationships are at best qualitative. This area of fog microphysics—the size–composition relationship within droplet spectra—may strongly affect the chemistry and deposition of fog-borne material and remains in need of further research.

EQUILIBRIUM DISSOLUTION OF SOLUBLE GASES. Dissolution of gaseous species in fog water is dependent on gas–aqueous equilibria and the quantity of liquid water present. Temperature and pH are the most important parameters that affect phase equilibria (e.g., aqueous, gas-aqueous, etc.). Ionic strength effects are not as important for the aqueous concentrations generally found (see Table I). For highly soluble species, LWC may control the overall partitioning because the droplets can provide a sufficient solution volume to deplete the gas phase. Thermodynamic data for important gas–aqueous equilibria for $SO_2$, $HNO_3$, and $NH_3$ are given in Table II.

Sulfur dioxide is fairly insoluble at low pH. At higher pH (7 and above), dissociation of $SO_2 \cdot H_2O$ to $HSO_3^-$ and $SO_3^{2-}$ can lead to much greater S(IV) solubility. With increasing LWC, $(S(IV))_f$ rises and can

**Table II. Thermodynamic Constants**

| Constant[a] | Reaction | $pK$ | $\Delta H$ (kcal/ mol) | Ref. |
|---|---|---|---|---|
| $H_s$ | $SO_2(g) + H_2O \rightleftharpoons SO_2 \cdot H_2O(aq)$ | $-0.095$ | $-6.25$ | 52 |
| $K_{s1}$ | $SO_2 \cdot H_2O(aq) \rightleftharpoons H^+ + HSO_3^-$ | 1.89 | $-4.16$ | 52 |
| $K_{s2}$ | $HSO_3^- \rightleftharpoons H^+ + SO_3^{2-}$ | 7.22 | $-2.23$ | 52 |
| $K_s$ | $HSO_4^- \rightleftharpoons H^+ + SO_4^{2-}$ | 2.20 | $-4.91$ | 52 |
| $H_N^\bullet$ | $HNO_3(g) \rightleftharpoons H^+ + NO_3^-$ | $-6.51$ | $-17.3$ | 53 |
| $H_A$ | $NH_3(g) + H_2O \rightleftharpoons NH_3 \cdot H_2O(aq)$ | $-1.77$ | $-8.17$ | 52 |
| $K_B$ | $NH_3 \cdot H_2O(aq) \rightleftharpoons NH_4^+ + OH^-$ | 4.77 | 0.9 | 52 |
| $H_F$ | $CH_2O(g) + H_2O \rightleftharpoons CH_2O \cdot H_2O(aq)$ | $-3.8$ | $-12.85$ | 54 |
| $K_F$ | $HMSA \rightleftharpoons CH_2O + HSO_3^-$ | $-5.0$ | | 55 |
| $K_w$ | $H_2O \rightleftharpoons H^+ + OH^-$ | 14.00 | 13.35 | 52 |

[a] Abbreviations are as follows: $H_s$, $H_A$, and $H_F$ are the corresponding Henry's law constants for sulfur(IV), ammonia, and formaldehyde; $K_{s1}$ and $K_{s2}$(M) are the first and second dissociation constants for $SO_2$(aq), respectively; $H_N^\bullet$(M atm$^{-1}$) is the modified Henry's law constant for nitrogen; $K_B$ and $K_w$(M) are the dissociation constants for ammonia and water, respectively; and $K_F$(M) is the formation constant for HMSA.

lead to depletion of the gas phase (Figure 4a). The fraction of solute partitioning into the droplet phase is expressed as $f_C = (C)_f/(C)_f + (C)_g$. Appreciable [S(IV)$_{(aq)}$] may be supported in the aqueous phase even at low $P_{SO_2}$, where $P_{SO_2}$ is the partial pressure of $SO_2$. Figure 4b shows the aqueous concentration as a function of prefog $P_{SO_2}$ at high and low LWC. Presence of gaseous aldehydes has been shown to substantially increase S(IV) solubility by the formation of stable bisulfite adducts (56).

Nitric acid is completely scavenged to droplets for fogs even at minimal LWC because it has such a high Henry's law coefficient [for $HNO_3(g) + H_2O \rightleftharpoons HNO_3 \cdot H_2O_{(aq)}$, $H_N = 2.1 \times 10^5$ M atm$^{-1}$ (53)]. Ammonia scavenging is variable in the ranges of pH and LWC found. Similar to the case of $SO_2(g)$, the atmosphere may exist as a reservoir for $NH_3(g)$ that can be dissolved or released from fog water as a function of the relative amount of atmospheric acidity and liquid water. Reservoirs of N(V) and N(−III), which also exist in the aerosol phase (e.g., ammonium nitrate), are discussed in the next section.

Gas transfer to fog-sized droplets by molecular diffusion is sufficiently rapid for phase equilibria to be achieved on the order of seconds or less under most conditions (57, 58). Greater time may be required in the cases where high solubility strongly favors gas-phase partitioning. The characteristic times required to supply the total amount of solute necessary to attain this equilibrium were evaluated by Schwartz and Freiberg (7). They found that times of the order 10 s were sufficient for

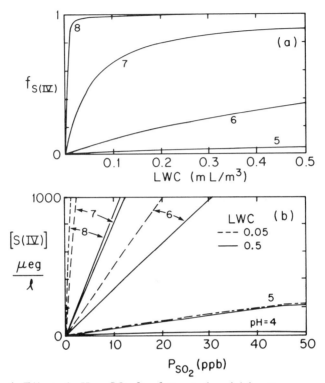

*Figure 4. Effect of pH on $SO_2$ dissolution in fog: (a) fraction scavenged as function of LWC; (b) fog water concentration versus partial pressure of $SO_2$ ($P_{SO_2}$).*

$SO_2(g)$ to dissolve into droplets even at high pH. Similarly, the absorption of highly soluble gases such as $HNO_3$ and $NH_3$ in fog droplets is not instantaneous. However, phase equilibria are established relatively rapidly—times of the order 10–100 s (*104*)—compared to the lifetimes of droplets.

**OVERALL FOG SCAVENGING.** Nonactivated aerosol fractions of nitrate, sulfate, and ammonium coexist with fog water and gaseous constituents in the atmosphere. Solutes incorporated within the droplet spectrum are subject to gas–aqueous equilibria; nonactivated aerosols appear to remain inert in this regard. In the case of nonvolatile species such as sulfate, partitioning is exclusively between nonactivated aerosol and droplet phases. Applying mass balances for $HNO_3(g)$ and $NH_3(g)$, the overall fractions, $F_C = (C)_f/(C)$, may be calculated as functions of gas–aqueous equilibria and nonactivated aerosol fraction. The mass balances for $S(VI)$, $N(V)$, $N(-III)$, and $S(IV)$ are given by equations 3–6.

$$(S(VI)) = LWC \, [S(VI)] + (S(VI))_a \tag{3}$$

$$(N(V)) = \frac{P_{HNO_3}}{RT} + (LWC) \, [NO_3^-] + (N(V))_a \tag{4}$$

$$(N(-III)) = \frac{P_{NH_3}}{RT} + (LWC) \, ([NH_3 \cdot H_2O] + [NH_4^+]) + (N(-III))_a \tag{5}$$

$$(S(IV)) = \frac{P_{SO_2}}{RT} + (LWC) \, ([SO_2 \cdot H_2O] +$$

$$[HSO_3^-] + [SO_3^-] + [HMSA]) \tag{6}$$

where HMSA is hydroxymethylsulfonic acid, and $(S(IV))_a$ is assumed to be negligible. The resulting scavenging efficiencies ($F$) for equations 3–6 are given by equations 7–10.

$$F_{S(VI)} = (S(VI))_f / (S(IV)) = 1 - f_{S(VI)}^a \tag{7}$$

$$F_{N(V)} = (N(V))_f / (N(V))$$

$$= \frac{H_N^\circ (LWC) RT}{H_N^\circ (LWC) RT + [H^+]} (1 - f_{N(V)}^a) \tag{8}$$

$$F_{N(-III)} = (N(-III))_f / (N(-III))$$

$$= \frac{H_A (LWC) RT \, (1 + K_B \, [H^+]/K_w)}{1 + H_A \, (LWC) RT \, (1 + K_B [H^+]/K_w)} (1 - f_{N(-III)}^a) \tag{9}$$

$$F_{S(IV)} = (S(IV))_f / (S(IV))$$

$$= 1 - \left[ 1 + (LWC) H_s RT \left( 1 + \frac{K_{s1}}{[H^+]} \right. \right.$$

$$\left. \left. + \frac{K_{s1} K_{s2}}{[H^+]^2} + \frac{K_{s1} K_F H_F P_{CH_2O}}{[H^+]} \right) \right]^{-1} \tag{10}$$

where $K_B$ is the base dissociation constant; $K_W$ is the dissociation constant for water; $H_A$, $H_S$, $H_N$, and $H_F$ are the corresponding Henry's law constants for ammonia, sulfur(IV), nitrogen, and formaldehyde; $K_{s1}$ and $K_{s2}$ are the first and second dissociation constants for $SO_2(aq)$, respectively; $K_F$ is the formation constant for HMSA; and $H_N^\circ$ is the modified Henry's law constant for nitrogen.

Without specific knowledge about solute size–composition relationships, a more general model of fog scavenging can be proposed based on solute speciation and the relative abundance of ambient acids and bases. Aerosols are generally found as neutral salts in the system dominated by $H_2SO_4$, $HNO_3$, and $NH_3$. Nitrate and sulfate are counterbalanced by ammonium ion, while excess acids or bases reside in the gas phase as $HNO_3(g)$ or $NH_3(g)$ (*25, 59*). Only for the cases where sulfate acidity exceeds the ambient ammonia [i.e., $(S(VI)) > (N(-III))$] would an acidic aerosol be thermodynamically stable in the dry atmosphere (*60, 61*). Consider two cases for these constituents:

$$\text{case i: } (N(V)) + 2(S(VI)) = (acids) > (N(-III))$$

$$\text{case ii: } (N(-III)) > (acids).$$

where acids is the total amount of sulfuric and nitric acids.

Nucleation scavenging is assumed to depend on LWC. For example, a linear dependence to a maximum of $F_{S(VI)} = 90\%$ at LWC $= 0.3$ mL m$^{-3}$ is assumed for sulfate in Figure 5a. This maximum approximates the progressive activation of a spectrum dominated by the large aerosol fraction. We also assume that scavenging of aerosol $N(V)$ and $N(-III)$ follows this same relationship. Solute-specific size spectra are not addressed at this point; actual differences in the proportion of fine versus coarse particles would be reflected in the nucleation scavenging efficiencies for each solute.

For case i, available $N(-III)$ is immediately exhausted in neutralizing acidic sulfate and nitrate. In the atmosphere, the dissociation of ammonium nitrate aerosol [$NH_4NO_3 \rightleftharpoons NH_3(g) + HNO_3(g)$] is highly dependent upon ambient temperature and RH (*60*). Ammonium nitrate aerosol volatilizes at low RH and high temperature. However, at high RH, the formation of aerosols leads to equimolar depletion of the gases until one is present at a negligible level; essentially, the gases do not coexist. Hence, $N(V)$ in excess of $N(-III)$ remains in the gas phase. When a droplet phase forms, $HNO_3(g)$ is 100% scavenged, and only nonactivated $N(V)$ aerosol is left outside the droplet spectrum. Figure 5b shows the effect of $N(-III)$/acid and $N(V)$/acid equivalent ratios based on the assumption that aerosol scavenging for nitrate proceeds similar to sulfate. Overall, ambient $N(V)$ will be more efficiently scavenged when the total acids are in greater excess of $N(-III)$ and other bases, and when $S(VI)$ makes a greater contribution to the net acidity in the precursor atmosphere.

In case ii, we assume $(NH_3(g)) = (N(-III)) - (acids)$ in the humid atmosphere, while the remaining material forms neutral ammonium aerosol. Furthermore, assuming ammonium aerosol nucleation proceeds as in Figure 5a, the total $N(-III)$ scavenging depends on the gas-aqueous equilibrium, fog water pH, and LWC (Figure 5c). In the alkaline regime, the presence of soluble, weak-acid gases can play an important role in determining the resultant pH of fog water that forms, and this pH affects the scavenging of $N(-III)$. Dissolution of formic and acetic acids $pK_a = 3.8$ and 4.8, respectively (*62*), for example, provides base-neutralizing capacity to the droplet. Also, the acidity of the droplet

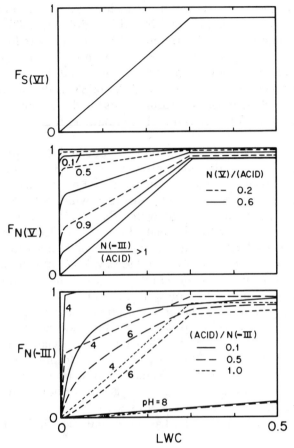

*Figure 5. Fog scavenging (F) for S(VI), N(V), and N(−III) as a function of liquid water content (LWC). Aerosol scavenging for N(V) and N(−III) is similar to the form shown for S(VI). For $F_{N(V)}$, the 0.1, 0.5, and 0.9 values represent N(−III)/(ACID).*

can be increased by the formation of S(IV)–formaldehyde adducts (e.g., HMSA), which increase S(IV) solubility, or in situ S(VI) production. Each process will lead to N(−III) scavenging that can eventually deplete $NH_3(g)$.

Nitrate and sulfate aerosols may be found within different size fractions. In southern California, Appel et al. (63) reported that nitrate is found predominantly in a coarser fraction than sulfate. In the San Joaquin Valley, Heisler and Baskett (64) found a similar relationship except during the wintertime; a substantial increase in the coarse sulfate fraction (>2.5 μm) was observed during wintertime stagnation. Ammonium was found to occur in the same fraction as nitrate. However, the measurements of Heisler and Baskett did not include gaseous ammonia, which can be a sizeable portion of total N(−III) in the region. Therefore, the true partitioning was not determined. Differences in predominant sizes for these three major constituents can significantly alter results of the simplified model given above. Most significant would be the presence of a coarse particle fraction enriched with N(V) or S(VI) that would be readily scavenged. Simultaneous measurements of these components in the three phases have not been satisfactorily made in fog.

**Fog Deposition.**  Early studies documented the hydrological and chemical importance of intercepted cloud and fog water inputs to mountain slope and coastal ecosystems (65). For instance, open collectors placed below trees exposed to fog-laden wind on the San Francisco pennisula measured an average water flux of 50 cm in one rainless month (66). Natural and artificial "fog-drip" collectors exposed for extended periods have shown that these "occult" inputs can have comparable magnitudes to annual precipitation values for water and nutrient capture (3, 4). However, the true relationship of these measurements to the capture by actual vegetation was uncertain.

Fog deposition from the ambient atmosphere to ground surfaces can be viewed as a two-step process (67). First, a droplet in the ambient atmosphere must be transported through an aerodynamic layer toward the ground. Next, once within a canopy layer, droplets must be deposited to canopy surfaces. The resistance to mass transport may reside in either layer, primarily depending on the geometry of the canopy and the degree of atmospheric turbulence. Research in the subject area of droplet or large particle deposition has addressed the specific mechanisms that can control their flux.

From measurements of droplet capture by monitored trees, Hori (68) and co-workers reported deposition rates on the order of 0.5 mm h$^{-1}$ for a forest intercepted by dense coastal fog. Yosida and Kuroiwa (69) found that momentum and droplet transport coefficients were of similar order of magnitude, based on measurements of wind force and droplet-capture rate by a small conifer tree. However, although the drag coefficient for the canopy elements decreased with increasing wind speed, the authors found that capture efficiency rose because of impaction. In the same field area, Oura (70) determined that interception near the leading edge of the forest was about 3 times more effective than for interior locations.

Lovett (11) adapted a resistance model to cloud droplet capture by a balsam fir forest canopy. He included turbulent transport and impaction to the canopy elements, conceived as a set of 1-m-thick height strata in a 10-m-high forest, and edge effects were not included. The model predicted a nearly linear correspondence between water flux and canopy-top wind speeds greater than 2 m s$^{-1}$. Droplet impaction, primarily to the upper 3 m of strata, was the dominant deposition mechanism. Wind speeds between 2 and 10 m s$^{-1}$ yielded water fluxes of 0.2–1.1 mm h$^{-1}$ (equivalent to $V_d = 10$–70 cm s$^{-1}$ for the simulated conditions). For lesser wind speeds, Lovett found that sedimentation controlled the water capture.

Legg and Price (71) calculated that the sedimentation flux of large particles to vegetation with a large leaf-area index (i.e., total leaf area per surface area of ground) would also increase with wind speed. This increase would be caused by wind-driven turbulence bringing particles to the lower leaves where sedimentation would lead to additional removal. Their model did not account for a vertical profile caused by depletion of particles, especially by impaction to the top of the canopy. Lovett (11) showed that the depletion reduced the net sedimentation flux nearer to the ground as wind speeds increased; nonetheless, greater leaf-area index caused sedimentation fluxes slightly higher than by terminal fall velocities alone.

Davidson and Friedlander (67) identified particle size regimes where different transport mechanisms dominated for a short-grass canopy. Under moderate wind conditions, the filtration efficiencies in calculations for $D_o > 10$ μm were high enough that deposition was effectively limited by turbulent transport or sedimentation to the canopy from above.

Droplet precipitation measured to flat plates in radiation fog averaged 0.03 mm h$^{-1}$, and this value agreed with calculated terminal fall velocities; measured rates for grass-model collectors (<1 m high) indicated approximately a twofold greater deposition (72). In another field study of radiation fog, Roach et al. (21) calculated that sedimentation removed as much as 90% of water that condensed to droplets during fog. Brown and Roach (73) parameterized the fog deposition rate as a linear function of LWC in their companion modeling paper. However, Brown (74) later determined that the gravitational flux was overestimated by this relationship for higher LWC. Jiusto and Lala (75) also found a linear fit between sedimentation rate and LWC as measured from droplet size spectra in radiation fog. Corrandini and Tonna (76) evaluated this parameterization for droplet size spectra given in the literature for different types of fog and found it did not fit well for advection and valley fogs.

Dollard and Unsworth (77) made direct measurements of turbulent fluxes for wind-driven fog drops above a grass surface. Their technique relied on precise determinations of LWC and wind speed made simultaneously at several heights. From the gradient of these parameters, the authors calculated turbulent transport for droplets and momentum. Their experimental data gave values of turbulent droplet flux 1.8 ± 0.9

times the sedimentation rate when wind speeds were 3–4 m s$^{-1}$. At wind speeds less than 2 m s$^{-1}$, their measurements showed that total fluxes were no more than 50% greater than by sedimentation alone.

**Transport Parameters and Processes.** The difference between deposition in dry and fog-laden air is primarily due to the increase in particle size for the latter case. Fog droplets are of the size where inertial impaction and sedimentation dominate their deposition to collection surfaces (78). In the general case, impaction to surface elements occurs when particles diverge, because of their inertia, from the airflow streamlines where they curve around the obstacle. The efficiency of impaction depends on the radius of curvature of the impaction surface ($R_f$) and the particle inertia and is characterized by the Stokes number (St):

$$St = \frac{\rho_w \, D_o^2 \, U_s}{18 \, \mu \, R_f} \tag{11}$$

where $\rho_w$ is the density of the particle (i.e., water for droplets); $D_o$ is the droplet diameter; $\mu$ is the dynamic viscosity of air; and $U_s$ is the relative velocity between the particle and the obstacle. Fog and cloud droplet sedimentation or terminal fall velocity ($V_s$) follows the form of Stokes law:

$$V_s = \frac{\rho_w \, g \, D_o^2}{18 \, \mu} \tag{12}$$

where $g$ is the gravitational acceleration. At larger diameters, such as raindrops, Stokes law no longer holds because of viscosity and drop deformation effects (18).

IMPACTION. The droplet impaction to cylinders can be viewed as an idealized analog of particle capture by grasses or conifer leaves. Impaction efficiencies have been theoretically derived for potential flow as a function of Stokes number (79) and extended for higher Reynolds number (Re = $\rho_a U_s D_o / \mu$) (80).

Experimental data have also been applied to calculations of particle or droplet flux to receptor surfaces. For example, Davidson and Friedlander (67) used a least-squares fit to data for impaction on cylinders. The efficiency, $\eta$, represented the fraction of particles that impact compared to the total number of particles passing through the projected area of the obstacle. For a single cylinder, the empirical expression was given as

$$\eta = \frac{St^3}{St^3 + 0.75 \, St^2 + 2.80 \, St - 0.20} \tag{13}$$

The flux by impaction to a canopy of cylinders of diameter $d_f$ was calculated by integrating over the length of the cylinder and multiplying by the number of cylinders per unit area of ground ($N$):

$$J = - d_f N \int_{z_s}^{H} \eta(z) U(z) C(z) dz \qquad (14)$$

where $J$ is the flux of particles per unit area of ground, $z$ is the vertical scale, $U$ is the horizontal wind speed, $C$ is the concentration of particles, $H$ is the canopy height, and $z_s$ is the particle sink [i.e., the level at which either $C$ or $U$ is assumed to vanish to zero, and impaction no longer occurs (67)]. The convention is that $J$ is positive for upward flux.

Alternatively, Lovett (11) used the experimental results of Thorne et al. (81). These results were measured specifically for components of the balsam fir canopy that they studied. The efficiency was given as

$$\eta = e^{[-1.84 + 0.90(\ln St) - 0.11(\ln St)^2 - 0.04(\ln St)^3]} \qquad (15)$$

The needles were oriented randomly, and the effect of interferences in airflow caused by neighboring canopy elements was accounted for empirically. For 50% efficiency, $St = 4$ according to Thorne et al. (81) and $St = 2$ in equation 13. Lovett (11) calculated the matrix of boundary-layer resistances for droplet capture as a function of horizontal wind speed and canopy structure (e.g., leaf-area index) for the different levels within the canopy.

Because the wind speed and droplet concentration profiles are also strongly dependent on the canopy structure, solutions for fog deposition due to impaction are not readily generalized. Davidson et al. (78) showed that the range of large particle deposition rates was 1 order of magnitude among five wild grass canopies that were studied. As stated in the previous section, the overall efficiency was found to be limited by turbulent transport to the top of the canopy for large particles ($\geq 10$ $\mu m$). In the case of the balsam fir canopy studied by Lovett (11), no such limitations were found to occur, and impaction led to very high deposition rates, also mentioned previously.

In practice, the collection efficiency of particles may be different from impaction efficiency. This difference largely depends on the surface properties of the collector element. In wind tunnel experiments (9, 82) and field measurements (83), particle deposition to dry surfaces was far below that to wet surfaces. As wind speeds increased, so did particle rebound. For wet surfaces, collection efficiencies of dry particles were more in accord with impaction theory. Experimental results for droplet collection efficiencies were generally in good agreement with theory

(*84*). The surface tension of water droplets was found to provide adequate adhesion to ensure near-perfect retention for $D_o < 50$ $\mu$m (*85*). Wind-induced shear may cause some droplet removal from foliar elements, and this removal was observed in wind tunnel tests with glycerol droplets when St $> 10$ (*81*). Nonetheless, sheared drops will generally fall because of their large size rather than be resuspended. In the wind tunnel experiments of Merriam (*86*), LWC was found to be a more important factor in determining total droplet capture than variations in canopy element geometry.

DEPOSITION VELOCITY. Although an overall deposition may be modeled for a specific canopy geometry and elements, wind speed and turbulence profiles, and particle size distributions (*11, 78, 87*), the flux to the canopy can be parameterized by the quantity known as deposition velocity, $V_d$. In this approach, the depositional flux, $J_d$, is scaled to the ambient concentration of some component at a reference height, $H$, above the canopy:

$$J_d = - V_d C (H) \tag{15a}$$

The deposition velocity may be calculated in terms of droplet fluxes or the specific chemical elements contained within the droplets.

TURBULENCE VERSUS SEDIMENTATION. The flux of droplets through the aerodynamic layer ($J_a$) can be expressed by

$$J_a = - K_p(z) \frac{dC}{dz} - V_s C \tag{16}$$

where $K_p$ is the eddy diffusivity for particles, and $V_s$ is the sedimentation velocity. In the steady state, $J_a$ is constant with height, so $C$ and $dC/dz$ are functions of $z$.

The relative importance of turbulence versus sedimentation transport may be expressed by $E = V_d/V_s$, the ratio of deposition to sedimentation velocities. With substitution ($J_a = J_d$), equation 16 becomes

$$K_p(z) \frac{dC}{dz} + V_s C = E V_s C (H) \tag{17}$$

This equation can be readily solved for $C(z)$ to give the concentration profile:

$$\frac{C(z)/C(H) - E}{1 - E} = e^{\left[ - \int_{H}^{z_s} \frac{V_s}{K_p(z)} dz \right]} \tag{18}$$

For the momentum exchange between the air and the ground, a logarithmic wind profile can be assumed for thermally neutral conditions (88):

$$U(z) = \frac{U^\bullet}{k} \ln \left(\frac{z}{z_o}\right) \tag{19}$$

where $U^\bullet$ is the friction velocity, $k$ is von Karman's constant (0.4), and $z_o$ is a roughness scale. In cases where an appreciable canopy structure exists, the profile will be displaced by some height, $d$, above the ground surface, and $z$ is replaced by $(z - d)$. Above the canopy, an analogy between turbulent exchange of particles and momentum is often assumed for neutral atmospheric stability (9, 78, 88):

$$K_p (z) = k \, U^\bullet \, z \tag{20}$$

Again, when displacement of the wind profile is observed, $z - d$ replaces $z$. The presumption of neutral stability is not always warranted; with daytime heating, vertical motions are enhanced by buoyancy. However, fog occurs during periods where minimal isolation and thermal neutrality or slight stability predominate (21, 89).

The analogy between momentum and particle transport may sometimes be inappropriate. For example, in the viscous boundary layer, momentum transfer to the canopy elements is augmented by the bluff-body (or normal pressure) forces. There is no analogy in heat or mass transport. Hence, the resistance to momentum exchange is generally less than for the other entities (90). The failure of large particles to follow fluid streamlines will also reduce their effective turbulent diffusivity, although this reduction is only important for $D_o > 30$ $\mu$m, based on the calculations of Csanady (91). However, for the conditions of fog, the analogy between momentum transfer with turbulent transport of particles has given satisfactory results (77).

In applying measurements made in radiation fog, we were interested in identifying the relative importance of sedimentation and turbulent transport for various degrees of turbulence (i.e., $U^\bullet$) and sizes of particles ($V_s$). We solved equation 18 by using equation 20 for eddy diffusivity and the boundary condition of $C \to 0$ at $z = z_s$, the particle sink (67). Figure 6 shows the results of calculations for $z_s/H$ between 0.05 and 0.005 (e.g., for $H = 3$ m, particle sink at 15 or 1.5 cm). This range corresponds to a shallow canopy where impaction is effective enough that transport is limited by turbulence or sedimentation (78). The enhancement of deposition by turbulent-driven droplet transport is seen to be rather limited for low and moderate wind speeds. The rea-

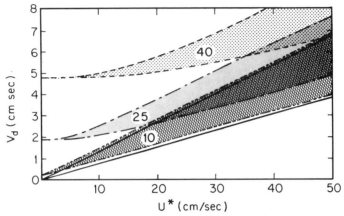

*Figure 6. Fog deposition curves. Deposition velocity ($V_d$) as a function of friction velocity of wind ($U°$) for the droplet diameters ($\mu m$) indicated. The indicated range for $z_s/H$ is between 0.05 (upper) and 0.005 (lower), where $z_s$ is the particle sink and H is the reference height for $V_d$.*

sons stated for less-effective turbulent exchange for particles vis-a-vis momentum would require that the indicated rates could be even lower. These calculations identified the same range of turbulent transport values given by Dollard and Unsworth (77).

**DROPLET LIFETIMES.** Changes in particle size will strongly affect depositional processes. The rate at which hygroscopic aerosols achieve equilibrium in the humid region over wet surfaces was recently studied in wind tunnel experiments by Jenkin (92). If equilibria were attained as particles approached the wet surface, an order-of-magnitude enhancement in deposition would be expected. Jenkin's experiments indicated that the growth rate was not sufficiently rapid; a twofold increase was the maximum observed. Hence, the residence of depositing particles within the humid region was not long enough for growth to equilibrium sizes.

An alternative concern is for the converse case in which fog droplets are exposed to lower humidity in the region near warmer-than-air surfaces. A rapid shift of the droplet distribution to smaller sizes would significantly alter both the chemistry and deposition rate. Even under nighttime, radiative conditions, the ground may remain warmer than the overlying air during fog because of the ground's high heat capacity. However, the vertical extent of conductive warming is limited to several centimeters in fog until insolation becomes important (75).

The lifetimes for fog droplets instantaneously exposed to drier air were calculated from the growth equation (equation 1) as a function of relative humidity and nucleus mass. These calculation results are shown in Figure 7 for several initial diameters. Mature fog droplets ($D_o \geq 20$ $\mu$m) are very resistant to rapid evaporation until RH drops well below 100%. Solute concentration has little effect on the shrinkage rate for these larger droplets. On the other hand, the rates at which smaller droplets evaporate could be quite rapid; those with greater solute mass change size less rapidly. Because sedimentation alone will transport larger droplets downward at 1–3 cm s$^{-1}$, droplet evaporation would not be expected to alter size-dependent depositional processes until a drier region extends several meters above the canopy surface. This situation is what happens when a fog starts to "lift", generally within several hours after sunrise. Even before the fog dissipates, evaporation from wetted surfaces can be important to the net water flux (*11*), but until the atmosphere dries sufficiently, the flux of fog water solutes will continue.

POLLUTANT SCAVENGING.   An essential facet of fog deposition is the scavenging of ambient aerosols and gaseous constituents into droplets. Partitioning of species between phases determines the relative importance of respective removal pathways. As particulates are incorporated into droplets, their deposition rate will increase with enhanced sedimentation and impaction efficiency. Simple models of fog deposition

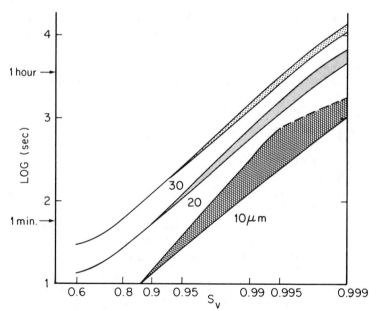

*Figure 7. Droplet lifetimes are the times required for droplets of a given initial diameter to shrink to equilibrium size for the indicated humidity ($S_V$ = RH/100). The indicated range is from 0.1 (lower) to 1.0 (upper) $\mu$m for the nucleus diameter.*

presume that fog leads to an increase for all particle sizes. However, when only a portion of aerosol mass achieves droplet sizes, the actual increase in deposition is reflected in this proportionality. Furthermore, fog may effectively scavenge important pollutant gases such as $SO_2(g)$, $HNO_3(g)$, and $NH_3(g)$. Turbulent diffusional processes are predominantly responsible for deposition of gases, and deposition velocities of about 1 cm s$^{-1}$ have been reported for $SO_2(g)$ (93) and 2 cm s$^{-1}$ for $HNO_3(g)$ (94). However, gas scavenging by droplets would cause removal of these species to be dominated by the sedimentation or impaction flux.

## Field Experiments

**Experimental Results.** Atmospheric pollutant behavior was studied in the southern San Joaquin Valley of California during periods of dense fog and stagnation (90). Fluxes to the ground of water-soluble species were determined by surrogate-surface collectors while simultaneous fog water and aerosol composition measurements were made. Repetitive, widespread fogs were observed only when the base of the temperature inversion was 150–400 m above the valley floor. Dense fogs (visual range < 200 m) lasted 10–17 h at sampling locations. Atmospheric loadings of water-soluble species in the droplet phase were composed almost entirely of $NH_4^+$, $NO_3^-$, and $SO_4^{2-}$ (Figure 8). Substantial concentrations of free acidity ($H^+$) and S(IV) species were occasionally measured in fog water samples (Table III). Concurrent with fogs, appreciable gaseous ammonia was often present, but only when atmospheric acidity was absent. Gaseous nitric acid concentration was generally lower than detection limits during all fog periods. Deposition samples (Figure 9) were similarly dominated by the same major ions as fog water and aerosol phases, although slightly greater contributions were made by soil dust species. These contributions were still far lower than the major species.

Substantially higher deposition rates for aerosol species occurred during fogs compared to rates during nonfog periods. Rates for major ions were enhanced by factors of 5–20 as shown in Figure 9. Deposition velocities, $V_d$ and $V_{d,fog}$, were calculated by normalizing measured deposition rates by the total and droplet-phase atmospheric loadings, respectively (Figure 10). The proportions of deposited solute for major ions were closely matched to the fog water composition. That is, despite large differences between $V_d$ values, there was close agreement for $V_{d,fog}$ values among major ions (Figure 11). The median value of $V_{d,fog}$ was approximately 2 cm s$^{-1}$, and measurements were in the range of 1–5 cm s$^{-1}$. Calculations of $V_{d,fog}$ were sensitive to LWC data, which have large uncertainties (±50%) for absolute values. An operationally defined

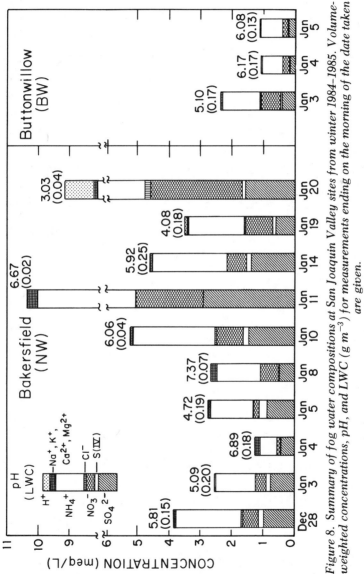

*Figure 8. Summary of fog water compositions at San Joaquin Valley sites from winter 1984–1985. Volume-weighted concentrations, pH, and LWC (g m⁻³) for measurements ending on the morning of the date taken are given.*

**Table III. Fog Water Composition Summary**

| Location, Date, and Time | N | LWC | pH | $Na^+$ ($\mu equiv\ L^{-1}$) | $K^+$ | $NH_4^+$ | $Ca^{2+}$ | $Mg^{2+}$ | $Cl^-$ | $NO_3^-$ | $SO_4^{2-}$ | $S(IV)$ ($\mu mol\ L^{-1}$) | $CH_2O$ | $-/+^{a,b}$ | $\Delta$ | % |
|---|---|---|---|---|---|---|---|---|---|---|---|---|---|---|---|---|
| **Bakersfield Airport** | | | | | | | | | | | | | | | | |
| December 28, 1984 | | | | | | | | | | | | | | | | |
| 00:35 to 08:35 | 6 | 0.149 | 5.81 | 14 | 4 | 2100 | 46 | 9 | 41 | 505 | 1064 | 153 | 98 | 0.81 | +410 | 10 |
| January 2–3, 1985 | | | | | | | | | | | | | | | | |
| 20:15 to 10:15 | 10 | 0.202 | 5.09 | 10 | 3 | 1260 | 25 | 3 | 13 | 335 | 764 | 128 | 78 | 0.95 | +69 | 3 |
| January 3–4, 1985 | | | | | | | | | | | | | | | | |
| 20:00 to 01:55 | 5 | 0.176 | 6.89 | 5 | 3 | 530 | 171 | 11 | 6 | 96 | 442 | 18 | 32 | 0.78 | +156 | 12 |
| January 4–5, 1985 | | | | | | | | | | | | | | | | |
| 20:10 to 09:15 | 13 | 0.191 | 4.72 | 21 | 4 | 1350 | 33 | 4 | 19 | 190 | 889 | 231[c] | 105 | 0.93 | +102 | 4 |
| January 8, 1985 | | | | | | | | | | | | | | | | |
| 09:00 to 10:00 | 1 | 0.070 | 7.37 | 57 | 8 | 1380 | 109 | 24 | 6 | 570 | 501 | 16 | 27 | 0.69 | +483 | 18 |
| January 10, 1985 | | | | | | | | | | | | | | | | |
| 07:15 to 08:50 | 2 | 0.035 | 6.06 | 25 | 9 | 2600 | 54 | 5 | 50 | 830 | 1619 | 180 | | 0.99 | +18 | 0 |
| January 11, 1985 | | | | | | | | | | | | | | | | |
| 06:35 to 07:30 | 1 | 0.016 | 6.67 | 58 | 24 | 4920 | 248 | 9 | 20 | 2115 | 2920 | | | 0.96 | +206 | 2 |
| January 14, 1985 | | | | | | | | | | | | | | | | |
| 01:30 to 04:30 | 1 | 0.249 | 5.92 | 26 | 3 | 2350 | 37 | 5 | | 630 | 1400 | 145 | 123 | 0.90 | +246 | 5 |
| January 18–19, 1985 | | | | | | | | | | | | | | | | |
| 20:05 to 09:00 | 10 | 0.182 | 4.08 | 7 | 7 | 1780 | 21 | 4 | 32 | 870 | 650 | 89 | 147 | 0.86 | +263 | 7 |
| January 20, 1985 | | | | | | | | | | | | | | | | |
| 07:15 to 09:30 | 3 | 0.040 | 3.03 | 26 | 17 | 3280 | 64 | 10 | 183 | 2900 | 1610 | 84 | 165 | 1.10 | −461 | 5 |
| **Buttonwillow** | | | | | | | | | | | | | | | | |
| January 2–3, 1985 | | | | | | | | | | | | | | | | |
| 20:20 to 06:50 | 6 | 0.174 | 5.10 | 7 | 3 | 120 | 38 | 4 | 17 | 640 | 440 | 36 | 55 | 0.89 | +135 | 6 |
| January 3–4, 1985 | | | | | | | | | | | | | | | | |
| 17:25 to 10:30 | 7 | 0.173 | 6.17 | 5 | 2 | 600 | 27 | 3 | 40 | 225 | 195 | 9 | 28 | 0.73 | +172 | 15 |
| January 4–5, 1985 | | | | | | | | | | | | | | | | |
| 19:30 to 09:00 | 11 | 0.133 | 6.08 | 3 | 2 | 670 | 19 | 3 | 14 | 190 | 245 | 22 | 33 | 0.68 | +224 | 19 |

NOTE: Concentration values are volume-weighted mean values collected for indicated time periods. Abbreviations are as follows: $N$ is the number of individual fog water samples taken, LWC is liquid water content, $-/+ = \Sigma$ anions/$\Sigma$ cations, $\Delta = \Sigma$ anions $- \Sigma$ cations, and % is sum of all ions.
[a]S(IV) is assumed to be in monovalent form.
[b]Correction for bicarbonate is not included.
[c]The S(IV) concentration for six samples was >300 $\mu mol\ L^{-1}$, which is the upper limit for the colorimetric method.

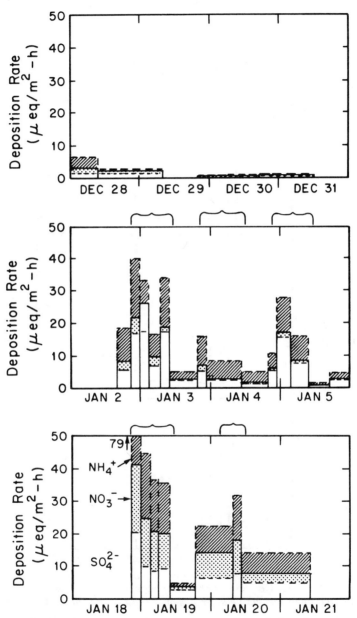

*Figure 9. Deposition rates of major ions to Petri dish collectors at the Bakersfield Airport (NW) site. Braces above figures indicate periods of fog.*

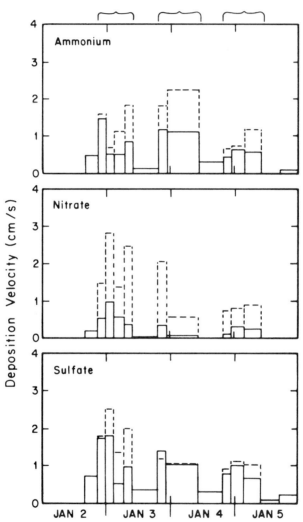

*Figure 10. Deposition velocities for major ions measured to Petri dish collectors at NW site. Concurrent values of* $V_d$ *and* $V_{d,fog}$ *are shown as solid and dashed lines, respectively. Braces over figures indicate periods of fog.*

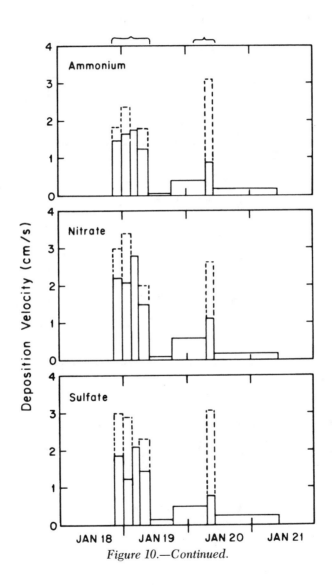

Figure 10.—Continued.

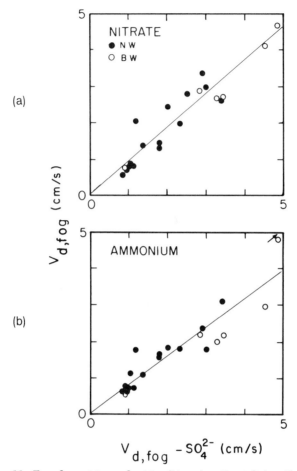

*Figure 11. Fog deposition velocities ($V_{d,fog}$) to Petri dish collectors at NW site. The graphs are (a) nitrate and (b) ammonium versus sulfate. Lines indicate best fit of data. Key: ● is Bakersfield Airport (NW) data, and ○ is Buttonwillow (BW) data.*

LWC was used for which the relative error was less (±20%). For the data, no correlation with LWC was found for droplet solute deposition. The rates were comparable to the terminal settling velocities of typical fog droplets.

Volatile loss of ammonium ion from the fog-wetted deposition surfaces was indicated. This loss may have been enhanced by the fallout of calcareous dust onto the hydrophobic collector surface. However, under normal conditions, $NH_3$ loss via volatilization of fog-deposited ammonium aerosol can also be expected to some degree.

Substantially greater deposition rates ($V_d$) for sulfate were found

compared with nitrate in nonacidic fogs. This difference has been attributed to more efficient scavenging of soluble sulfur species by fog
droplets. Sulfate deposition during fog was in the range of 0.5–2.0
cm s$^{-1}$, and had a median value of about 1 cm s$^{-1}$; the nitrate rate was
generally 50% less than that for sulfate. In nonacidic fogs, as on January
2–3 and 4–5 (Figures 12 and 13, respectively), nitrate scavenging was
uniformly low, and the fraction of sulfate incorporated into the droplet
phase rose with increasing LWC. For the acidic fog on January 18–19
(pH < 4), the fraction of ambient nitrate scavenged by the fog droplet
was much higher and increased with LWC (Figure 14). On this later
date, higher atmospheric acidity is believed to have altered N(V) partitioning prior to fog formation. Higher acidity would support higher
HNO$_3$(g) concentrations in the prefog air, followed by subsequent nitric
acid scavenging once the fog formed. However, measured HNO$_3$(g)
was not as great as the observed enhancement of N(V) scavenging.
Depletion of gaseous ammonia accompanied the period of higher
atmospheric acidity and low pH fog, and this depletion is believed to
have caused a reduction in the formation of the smaller NH$_4$NO$_3$
aerosol. In the absence of detectable NH$_3$(g), newly formed N(V) was
apparently incorporated into a coarser aerosol fraction and readily
scavenged in fog.

**Mass Balance Analysis.**    When widespread stagnation suppresses
convective transport out of the basin, the accumulation of pollutants
may proceed. The processes that control the fates of primary emissions
in the atmosphere are varied and complex (*96*). Nonetheless, profiles of
concentrations versus time have been reasonably interpreted based on
continuous-flow, stirred-tank considerations of pollutant inputs and removal pathways (*97*). This accumulation of atmospheric constituents is
governed by primary emissions, in situ transformations (production or
loss terms), intrabasin circulation, ventilation, and removal by deposition
to ground surfaces. The mixing height, $H$, controls the volume in which
these processes occur. For example, mass balances for S(IV) and S(VI)
may be described by

$$\frac{d(S(IV))}{dt} = \frac{E_{SO_2}}{H} - k_s (SO_2) - \nabla \cdot ((SO_2)\vec{V}) - \frac{(SO_2)}{\tau_v} - V'_d \frac{(SO_2)}{H} \qquad (21)$$

$$\qquad\quad \text{(a)} \qquad\quad \text{(b)} \qquad\quad \text{(c)} \qquad\quad \text{(d)} \qquad\quad \text{(e)}$$

$$\frac{d(S(VI))}{dt} = \qquad k_s (SO_2) - \nabla \cdot ((SO_4^{2-})\vec{V}) - \frac{(SO_4^{2-})}{\tau_v} - V_d \frac{(SO_4^{2-})}{H} \qquad (22)$$

where $E$ is an areal emission rate (mol m$^{-2}$ h$^{-1}$); $k_s$ is a pseudo-first-order
rate constant for S(IV) oxidation (h$^{-1}$); $V$ is the horizontal transport vector
(m h$^{-1}$); $\tau_v$ (h) is the characteristic time for vertical ventilation; and $\nabla$ is

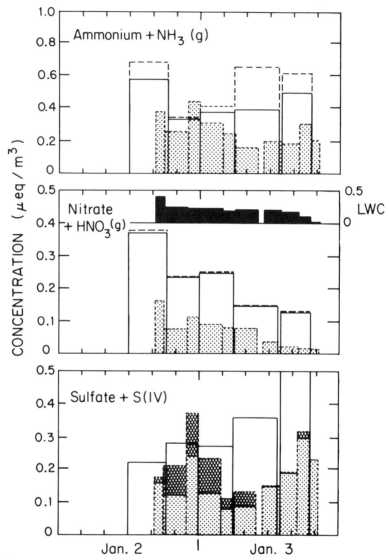

*Figure 12. Total (line) and fog water (shaded) concentrations of N(−III), N(V), and sulfur species at Bakersfield Airport site on January 2-3. Gaseous species (dashed line) and liquid water content (black) are also shown. Fog water S(IV) is indicated by heavy shading.*

the divergence vector operator. Deposition velocities (m h$^{-1}$) will depend on the species and the phase: gas, aerosol, or droplet. Similar expressions may be formulated for nitrogen species, although the chemical transformations involving NO$_x$ species are far more complex.

Under wintertime stagnation conditions in the southern San Joaquin

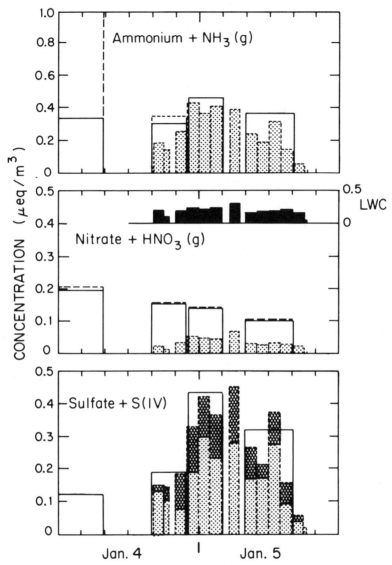

*Figure 13. Same as Figure 12, but for January 4-5.*

Valley, $\tau_v = 3$–$5$ days or more (*98*). Time scales characteristic of emission and deposition rates are strongly dependent on mixing height. Mixing heights during stagnation episodes are generally 200–800 m above the valley floor. In the years we conducted studies, widespread fog occurred when the mixing height was ≤400 m (*39, 97*). For low wind speeds, deposition velocities can range from ~0.05 cm s$^{-1}$ for submicrometer aerosols to ~2 cm s$^{-1}$ for fog droplets or reactive gases. Hence, $\tau_d$, given as $H/V_d$, can range from >3 days to <3 h. For sulfate aerosol, this range is largely dependent on the presence or absence of fog.

A characteristic time for S(IV) oxidation may be given as $k_s^{-1}$. In reality, in situ transformations are rarely simple first-order processes dependent on reactant concentration alone. These rates will depend on the nature and concentration of oxidants, metals, and other catalytic components, in addition to pH and LWC (99). The pseudo-first-order

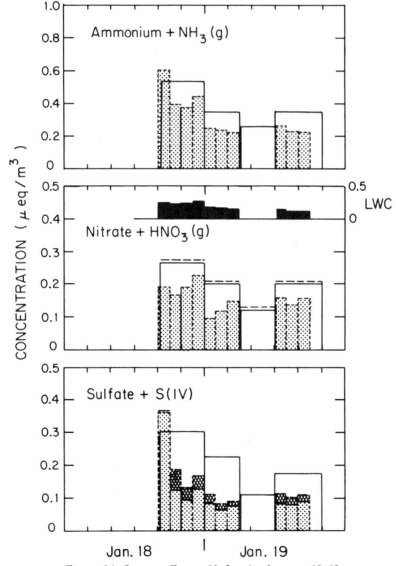

Figure 14. Same as Figure 12, but for January 18-19.

rate constant is a convenient means to parameterize the observed atmospheric kinetics. The oxidation reactions are thought to proceed more rapidly in fog, although interpretations of our prior San Joaquin Valley field measurements have not statistically verified this assertion (39, 97). Using values in a range given by Jacob et al. ($1\% < k_s < 10\%$ h$^{-1}$), we estimated the time for S(IV) oxidation between 2 and 24 h (to lower the concentration by a factor of $e$).

Emission inventories for $SO_2$, $NO_x$, and $NH_3$ have been determined by Jacob et al. (97) for the southern San Joaquin Valley as 192, 190, and 79 tons day$^{-1}$. These emissions translate to 29, 20, and 22 $\mu$equiv m$^{-2}$ h$^{-1}$ when expressed as area-wide averages. These units correspond to the secondary products; therefore, $SO_2$ yields 2 equiv mol$^{-1}$. A characteristic time, taken here as the time required for emissions to replace a given atmospheric loading of pollutant $C$, would be expressed as $\tau_E = (C)H/E$. Given ($SO_2$) $\sim$ 10 ppb and $H \sim$ 400 m, the calculated $\tau_E$ for $SO_2$ is on the order of 12 h; for $H \sim$ 200 m, $\tau_E$ is only half of this value. This term is useful to assess whether a balance of sources and sinks has been achieved. For example, when the time scales for loss terms are longer than $\tau_E$, atmospheric concentrations will increase. Higher concentrations of $NO_x$ are a clear indication that $NO_x$ depletion rates are slower than either $SO_2$ or $NH_3$, because sources are comparable.

During dense fog, deposition becomes the predominant loss term for secondary aerosol species. Flux measurements to surrogate surfaces demonstrated that removal from the atmosphere can be very rapid. In Table IV, characteristic times have been calculated for deposition during dense fog. These values were determined from the total solute fluxes, mixing heights, and average aerosol concentrations measured during the individual events. The characteristic removal times were calculated to be 6–12 h for these periods with the exception of N(V). As discussed previously, during the sampling periods of acidic atmospheric condition (January 19 and 20), there was a distinct increase in the relative scavenging efficiency for nitrate species into fog water. The deposition rate for N(V) also increased for these events, and this can be seen in much lower $\tau_d$ values for nitrate compared to the earlier fog events when pH $\geq$ 5. Between the occurrences of fog, aerosol deposition was substantially reduced. Deposition velocities were generally an order of magnitude below in-fog values. Fogs persisted more than 50% of the time during January 2–5. In the absence of production terms, aerosol components would be more than 90% depleted during protracted fog episodes. However, such a net depletion was not observed. By inference, in situ production rates must have at least equaled the deposition rates.

Advection in this environment is difficult to assess. Wind directions are found to be erratic at valley stations during stagnation episodes (100). Resultant winds for the early January period were <1 m s$^{-1}$ at all stations; therefore, net cross-valley transport would require $\sim$1 day. Before Jan-

Table IV. Characteristic Removal Times and Production Rates in San Joaquin Valley Fogs

| Location and Date | Duration (h) | H (m) | $\tau_d$ (h) | | | Production Rate (ppb h$^{-1}$) | | |
|---|---|---|---|---|---|---|---|---|
| | | | $NH_4^+$ | $NO_3^-$ | $SO_4^{2-}$ | $E_A$ | $E_N$ | $E_S$ |
| Bakersfield Airport (NW) | | | | | | | | |
| Dec. 28, 1984 | 4 | 200 | 6 | 15 | 6 | 1.5 | 0.3 | 0.5 |
| Jan. 2–3, 1985 | 14 | 240 | 6 | 10 | 6 | 1.5 | 0.4 | 0.7 |
| Jan. 3–4, 1985 | 12 | 210 | 6 | 42 | 6 | 0.6 | 0.1 | 0.2 |
| Jan. 4–5, 1985 | 12 | 230 | 11 | 27 | 7 | 0.8 | 0.1 | 0.5 |
| Jan. 14, 1985 | 3 | 500 | 12 | 22 | 7 | 0.7 | 0.2 | 0.4 |
| Jan. 18–19, 1985 | 14 | 300 | 6 | 4 | 5 | 1.5 | 1.1 | 0.5 |
| Jan. 20, 1985 | 7 | 350 | 11 | 9 | 12 | 0.9 | 0.7 | 0.3 |
| Buttonwillow (BW) | | | | | | | | |
| Jan. 2–3, 1985 | 17 | 290 | 7 | 6 | 7 | 2.3 | 1.4 | 0.3 |
| Jan. 3–4, 1985 | 17 | 260 | 10 | 17 | 9 | 0.5 | 0.2 | 0.1 |
| Jan. 4–5, 1985 | 15 | 230 | 7 | 18 | 6 | 0.9 | 0.2 | 0.2 |

NOTE: The characteristic time for pollutant removal ($\tau_d$) was calculated from the equation $\tau_d = H/V_d$, where $V_d$ is the flux divided by the ambient concentration. The production or emission rate ($E_A$, $E_N$, or $E_S$) was determined as the deposition divided by the product of duration and height above ground level ($H$).

uary 1, the concentrations of sulfate were uniformly low on both sides of the valley; after January 1, higher sulfate values were monitored in Bakersfield than in areas to the west or north. The spatial gradient for sulfate across the valley indicated that advection of sulfate aerosol was a less prominent term during January 2–5. Clearly, deposition of sulfate was more rapid than its transport away from the source region.

However, nitrate aerosol concentrations were uniform throughout the sampling network at that same time. Similar temporal variations were also observed at all valley sites except McKittrick, which was above the insertion base throughout that period. Because the deposition rates measured for nitrate were much lower than for sulfate during that period, more complete mixing could have occurred. This uniformity was probably aided by more widely dispersed emissions of $NO_x$.

For the two sites in the Bakersfield area, Bakersfield Airport (NW) and downtown Bakersfield (BA), concurrently measured aerosol concentrations agreed within 20% in most cases (12 for $NO_3^-$ and 10 for $SO_4^{2-}$ for 13 sampling periods), even though temporal variations spanned a factor of 5. Simultaneous peaks in $NH_3$ concentrations were further indication of the spatial homogeneity at the two sites. However, it is impossible to state unequivocally that short-term changes at a particular site were due to in situ transportation rather than localized transport. Without a network of greater spatial resolution, the interpretation of sequential S(VI) or N(V) concentrations in terms of a generalized continuity-equation analysis would be moot. Nonetheless, advection of pollutant species away from this source region must represent a sink over longer time scales.

As a lower limit, we calculated production rates necessary to balance removal rates of aerosol species measured during fog (i.e., the production rate is proportionate to the deposition flux divided by $H$). Essentially, this calculation equates terms (b) and (e) in equation 22 and neglects the rest. Production rates have been calculated in units of the primary emissions, $NH_3$, $NO_x$, and $SO_2$ (Table IV). Sulfur dioxide values measured at the fog study sites were mostly ≤10 ppb, although spatial variability of $SO_2$, especially near the oil fields, makes the calculation of an areal average concentration questionable. Assuming 10 ppb for the gaseous concentration, the pseudo-first-order S(IV) oxidation rates were calculated to be 2%–7% $h^{-1}$. Considering that advection represented a loss term for the Bakersfield area, the total sinks were likely to have been even greater than measured by deposition alone. The calculated S(IV) oxidation rates at Buttonwillow (BW) were not as great; assuming that $SO_2$ concentrations were a factor of 2 less at the BW site compared with Bakersfield, the rates of S(IV) oxidation at the two locations in dense fog were similar. The S(VI) production and deposition rates should have been comparable at these sites because the meteorological conditions were similar (i.e., widespread nighttime fog and afternoon haze).

**Caveats.** Fog deposition as measured to flat surfaces or buckets represented a lower limit to rates occurring in the southern San Joaquin Valley. The NW site was chosen for its open and featureless terrain. In many ways, this site was characteristic of wintertime land use in this region. The majority of the southern San Joaquin Valley is cropland (43%) or rangeland (44%), and the remaining portions of the valley are orchard (8%), residential–commercial (4%), or forest (1%). In areas of tall (dense or sparse) canopy or buildings, wind profiles would be more turbulent, and canopy-top wind speeds would be closer to the National Weather Service sensor (10 m) values rather than to the low values measured close to the open field. Simply the presence of bare trees would substantially increase the surface area for pollutant deposition. Droplet impaction becomes important under these circumstances (*11*). In such cases, $V_{d,fog}$, and hence, $V_d$, would be adjusted upwards. Fog deposition may therefore lead to even greater pollutant influx to these areas. Although this influx may be of significance for certain receptor areas, we believe the effect is relatively unimportant to the conclusion regarding the dominance of sedimentation and the range of deposition values we have reported.

Finally, these measurements were specific to particles and droplets. Although we are convinced that dry deposition of $SO_2$, for example, did not significantly contribute to our measurements, this sink for S(IV) may itself be substantial. Moreover, scavenging of $SO_2$ by droplets aided in its deposition during fog. In addition to the chemical transformations of S(IV), the removal of $SO_2$ to surfaces, although not studied here, is an important term in the overall sulfur budget. Also, $HNO_3$ deposition may have an important role in N(V) removal under certain circumstances.

**Precipitation Scavenging.** In the case of the rainfall event, the total amount of material brought to the ground was 2–5 times greater than the overburden of pollutants in the air (i.e., the product of ambient concentration and mixing height) prior to the rain (Figure 15). After the initial loading of accumulated pollutants was washed (or ventilated) out of the air mass, additional nitrate and sulfate must come from the transformation of primary emissions. Low $SO_2$ concentrations were measured during the rainfall period; simultaneously, sulfate deposition was large. This measurement clearly indicated a period in which rapid S(IV) oxidation must have occurred in the presence of an aqueous phase. The smaller fluxes of nitrate suggested either a slower rate of N(V) formation for the same period or less efficient precipitation scavenging of secondary nitrate. However, without a detailed knowledge of atmospheric mixing and below-cloud processes, in situ oxidation rates cannot be determined from these rainfall measurements.

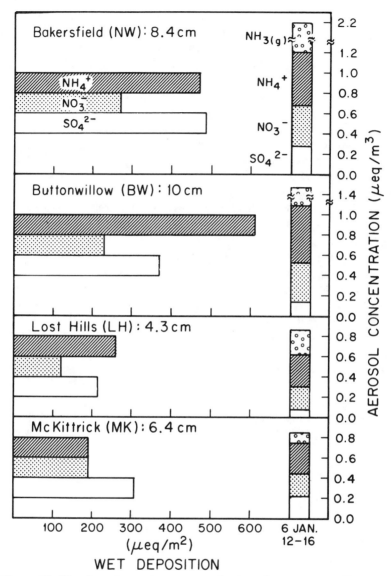

*Figure 15. Wet deposition of major ions for four San Joaquin Valley sites measured on January 6-8, 1985. Filter data for the sampling interval immediately preceding rainfall are also shown.*

## Discussion

The efficacy of fog in scavenging gaseous and aerosol species and enhancing their removal can cause the fog to be the dominant factor controlling ambient levels reached by pollutants during wintertime stagnation episodes. The measurements of depositional flux during the stagnation periods in the winter of 1985 illustrated the differences between episodes with dense fog and those when fog was absent. Detailed fog water and air quality data were also collected for the winters of 1983 and 1984 as part of our earlier field programs. These former studies included periods with and without dense, widespread fog. Concentrations of the major species in the fog water, aerosols, and gas phases are compared in Table V. The comparison is limited to parameters measured in the Bakersfield area. The values presented were for periods of stagnation when mixing heights were low and concentrations in the air had achieved apparent steady-state levels. Given the characteristic time for fog deposition and the emission rates of sulfur species, a steady state for sulfate could be attained in less than 1 day. Nonfoggy intervals require much longer times, and true steady-state levels may not have been fully reached before a change in meteorological conditions occurred. This type of comparison is readily acknowledged to be far from conclusive because other factors (e.g., daytime insolation, oxidant concentrations, and wind trajectories) have not been evaluated. We used this comparison primarily to point out the relationships that these unique data sets have provided.

Mixing heights were comparable during the six episodes. A disparity between two regimes is readily apparent with respect to aerosol nitrate and sulfate concentrations. The highest values were associated with January 2-6, 1983, and January 2-6, 1984. These were periods of low clouds when fog was absent. Average concentrations of particulate sulfate exceeding California air quality standards (25 $\mu$g m$^{-3}$ or 520 nequiv m$^{-3}$) were monitored during both episodes. Periods of dense fog led to lower concentrations of particulate loading, which leads to air of better quality despite lesser visibility.

The January 2-5, 1985 samples complemented a data set of pre-

**Table V. Fog Water and Air Quality**

| | | | Fog Water | | | |
|---|---|---|---|---|---|---|
| Date | H (m) | pH | $NH_4^+$ (mequiv L$^{-1}$) | $NO_3^-$ (mequiv L$^{-1}$) | $SO_4^{2-}$ (mequiv L$^{-1}$) | %S(IV) |
| Dec. 30–31, 1982[a] | 200 | 4.1 | 1600 | 850 | 800 | 62 |
| Jan. 2–6, 1983[a] | 400 | | | | | |
| Jan. 6–8, 1983[a] | 350 | 3.9 | 2600 | 900 | 1600 | 38 |
| Jan. 11–15, 1983[a] | 300 | 3.6 | 1400 | 700 | 900 | 57 |
| Jan. 2–6, 1984[a] | 350 | | | | | |
| Jan. 2–5, 1985[b] | 200 | 4.9 | 1200 | 250 | 800 | 32 |
| Jan. 18–20, 1985[b] | 300 | 3.9 | 1900 | 1000 | 600 | 22 |

[a]Measurements made at downtown Bakersfield (BA) site.
[b]Measurements made at Bakersfield Airport (NW) site. Parentheses denote BA values for same time period.

dominantly acidic fogs measured in January 1983. Simultaneous fog water and aerosol measurements were not made during the 1982–1983 study, but the ionic ratios indicate that N(V) scavenging was as great or greater than for sulfate during these low pH fogs. Both N(V) and S(VI) production were important sources of atmospheric acidity. Emission inventories for all producers during the selected periods are not available, but data on the east side support an assumption that year-to-year differences in primary emitter operations were not a factor (D. Anderson, Texaco, Inc., private communication). No obvious effect of acidity on $SO_2$ levels was observed; gaseous concentrations were uniformly low for both years. Interestingly, dissolution of S(IV) in fog water was not reduced in the acidic regime of the earlier year (Table V), as might be expected from gas–aqueous equilibria.

The primary determinant for acidity is most likely the relative N(−III) abundance as discussed in Jacob et al. (97) with respect to spatial patterns in the region. In an ammonia source region, factors can suppress or accelerate $NH_3$ release. Temperature, moisture, and land cover may be primary factors. Dawson (101) showed that there are strong dependencies for soil release of ammonia; release increases with temperature and with soil moisture until 20% saturation and then decreases. Uptake by vegetation can drastically reduce the net release of ammonia (102). Hutchinson et al. (103) measured ammonia release from a large cattle feedlot and found lower fluxes when the surface was wet from rain. However, this lowering was offset by greater-than-average rates while the surface dried.

The January 2–5, 1985 episode distinguished itself in one important regard. During other episodes, no daytime clearing occurred. Heavy overcase generally continued after the fog lifted. On the afternoons of the early January 1985 episode, there were periods of hazy sunshine and appreciable warming. This warming would promote ammonia release,

**during Winter Stagnation Episodes**

| Aerosol | | | Gas | | | | |
|---|---|---|---|---|---|---|---|
| $NH_4^+$ (nequiv m³) | $NO_3^-$ (nequiv m³) | $SO_4^{2-}$ (nequiv m³) | $NH_3$ (ppb) | $NO_x$ (ppb) | $SO_2$ (ppb) | Dense Fog | Ref. |
| 600 | 280 | 280 | | 100 | 20 | Yes | 39 |
| 900 | 300 | 900 | | 100 | 25 | No | 39 |
| 600 | 150 | 450 | | 150 | 20 | Yes | 39 |
| 300 | 40 | 200 | | 200 | 40 | Yes | 39 |
| 1000 | 400 | 700 | 1 | 100 | 20 | No | 97 |
| 350 | 190 | 210 | 9 | 25 | 10 | Yes | this study |
| (380) | (210) | (270) | (6) | (100) | (10) | | |
| 510 | 280 | 320 | <0.5 | (50) | (10) | Yes | this study |

especially from feedlots and agricultural soils. Despite this cooling, the ground would maintain higher temperature at nighttime as well. On the other hand, the acidification of the San Joaquin Valley atmosphere would be promoted by cooler and steady overcast conditions. Ammonia release also may be reduced in postrainfall periods. At the same time, this moisture is often important in sustaining widespread fog.

## Summary

The shallow and poorly ventilated mixed layer of the San Joaquin Valley under the intense temperature inversion represented a reactor of limited volume. A comparison was made between the mass deposited and the overburden of pollutants (i.e., the product of mixing height and ambient loading). For S(VI) and N(−III) species, 2–3 times the apparent overburden was deposited during prolonged fogs. This deposition should have caused substantial depletion of atmospheric concentrations; however, this depletion was not observed. Steady aerosol sulfate concentrations required S(IV) oxidation to proceed rapidly. A pseudo-first-order constant for $SO_2$ oxidation was calculated to be in the range of 2%–7% $h^{-1}$. The true rate may be several times higher because (a) the reactant concentrations were frequently below the value (10 ppb) used to make the calculation, (b) the depositional flux measurements tacitly neglected the sink due to droplet impaction, and (c) advective loss terms in the mass balance equation were neglected in the source region. In similar fashion, ammonia emissions were calculated to be approximately 1 ppb $h^{-1}$ in order to balance solute removal in fog.

The two processes important for the determination of ambient sulfate concentration were removal by deposition and production by S(IV) oxidation. Measurements made during three wintertime fog–aerosol studies were summarized to consider the overall mass balance during stagnation episodes. This data set supported the hypothesis that fog deposition lowered the ambient concentrations of aerosol sulfate and nitrate during stagnation periods compared with periods of no fog. However, the importance of dry deposition of $SO_2$, sulfate production in haze aerosol, and the contribution of impaction to droplet fluxes needs to be more fully investigated. Finally, the conditions under which fog water acidity was high were related to factors favoring ammonia release from sources (e.g., higher soil moisture and temperature).

### Appendix—The Droplet Growth Equation

The droplet growth equation is

$$D_o \frac{dD_o}{dt} = \frac{S_v - a_w\, e^B}{C + E\, (A-1)\, a_w\, e^B} \tag{A1}$$

where

$$A = \frac{M_w L}{RT} \qquad (A2)$$

$$B = \frac{4 M_w \sigma}{\rho_w D_o RT} \qquad (A3)$$

$$C = \frac{\rho_s RT}{4 D_m M_w e_s (T)} \qquad (A4)$$

$$E = \frac{\rho_s L}{4kT} \qquad (A5)$$

$$a_w = e^{\left(-\phi \frac{n_s}{n_w}\right)} \qquad (A6)$$

For the activation point,

$$\frac{dD_o}{dt} = 0 \qquad (A7)$$

$$D_o = D_{act} \qquad (A8)$$

$$S_v = 1 + SS_{cr} \qquad (A9)$$

These equations yield

$$D_{act} = \left(\frac{3 \, LNA}{B}\right)^{1/2} \qquad (A10)$$

$$SS_{cr} = \frac{2}{3\sqrt{3}} \left(\frac{B^3}{LNA}\right)^{1/2} \qquad (A11)$$

$$= K \, D_{dry}^{-3/2} \qquad (A12)$$

In equation A10, LNA is defined as

$$LNA = \frac{M_w \, \rho_s}{M_s \, \rho_w} \, D_{dry}^3 \qquad (A13)$$

Parameters for these equations are defined as follows:

$D_o, D_{act}$     droplet diameter and diameter at activation point, respectively

$D_{dry}$     dry diameter for hygroscopic aerosol

$e, e_s (T)$     ambient and saturation water vapor pressure at $T$, respectively

| $S_v$ | ambient water vapor saturation ratio, $e/e_s$ $(T)$ |
| $SS_{cr}$ | critical supersaturation |
| $\rho_w$, $\rho_s$ | density of pure water and solute salt, respectively |
| $R$ | gas constant |
| $T$ | temperature (K) |
| $L$ | latent heat of evaporation |
| $D_m$ | molecular diffusivity of water vapor in air |
| $k$ | heat conductivity of air |
| $\sigma$ | surface tension of water |
| $M_w$, $M_s$ | molecular weight of water and dissociated solute ions, respectively |
| $n_w$, $n_s$ | number of moles of water and dissociated solute in droplet, respectively |
| $a_w$ | activity of water in droplet |
| $\phi$ | osmotic coefficient of water in solution (assumed to equal 1) |

## Abbreviations and Symbols

| $\nabla$ | gradient |
| $\mu$ | dynamic viscosity of air |
| $\eta$ | fraction of particles that impact receptor surface |
| $\rho_s$ | density of solute salt |
| $\rho_w$ | density of pure water |
| $\sigma$ | surface tension of water |
| $\tau_E$ | time for emissions to replace a given atmospheric loading of pollutant $C$ |
| $\tau_v$ | characteristic time for vertical ventilation |
| $\phi$ | osmotic coefficient of water in solution |
| $a_w$ | activity of water in a droplet |
| $A$ | constant |
| $B$ | constant |
| $C$ | concentration of particles; constant |
| $(C)$ | total concentration of species $C$ in the atmosphere |
| $[C]$ | aqueous concentration of species $C$ in fog water |
| $(C)_a$ | concentration of species $C$ in the nonactivated aerosol phase |
| $(C)_f$ | concentration of species $C$ in the fog water phase |
| $(C)_g$ | concentration of species $C$ in the gas phase |
| $d$ | height |
| $d_f$ | diameter of cylinder |
| $D_{act}$ | activation size (diameter) |
| $D_{dry}$ | dry diameter |
| $D_{eq}$ | equilibrium diameter |
| $D_m$ | molecular diffusivity of water vapor in air |

| | |
|---|---|
| $D_o$ | droplet diameter |
| $e$ | the exponential; ambient water vapor pressure |
| $e_s$ | saturation water vapor pressure |
| $E$ | ratio of deposition to sedimentation velocities; areal emission rate |
| $f^a$ | nonactivated fraction |
| $f_c$ | fraction of solute partitioning into the droplet phase |
| $F_c$ | overall fraction of species $C$ |
| $g$ | gravitational acceleration |
| $H$ | canopy height |
| $H_A$ | Henry's law constant for ammonia |
| $H_F$ | Henry's law constant for formaldehyde |
| $H_N^*$ | modified Henry's law constant for nitrogen |
| $H_S$ | Henry's law constant for sulfur (IV) |
| $J$ | flux of particles per unit area of aerosol |
| $J_a$ | aerodynamic layer flux |
| $J_d$ | depositional flux |
| $k$ | heat conductivity of air; von Karman's constant |
| $K$ | proportionality coefficient |
| $K_B$ | dissociation constant for ammonia |
| $K_F$ | formation constant for HMSA |
| $K_p$ | eddy diffusivity |
| $K_s$ | dissociation constant for sulfur |
| $K_{s1}$ | first dissociation constant for $SO_2$ (aq) |
| $K_{s2}$ | second dissociation constant for $SO_2$ (aq) |
| $K_w$ | dissociation constant for water |
| $L$ | latent heat of evaporation |
| LWC | liquid water content |
| $M$ | fog water mass |
| $M_s$ | molecular weight of dissociated solute ions |
| $M_T$ | total mass |
| $M_w$ | molecular weight of water |
| $n_s$ | number of moles of solute in droplet |
| $n_w$ | number of moles of water in droplet |
| $N$ | number of cylinders per unit area of ground |
| p$K$ | $-$ log dissociation constant |
| $P_c$ | partial pressure of species $C$ |
| $R$ | gas constant |
| $R_f$ | radius of cylinder |
| Re | Reynolds number |
| RH | relative humidity |
| St | Stokes number |
| $S_v$ | ambient water vapor saturation ratio |
| $SS_{cr}$ | critical supersaturation |

$t$   time
$T$   temperature
$U$   horizontal wind speed
$U^*$   frictional velocity
$U_s$   relative velocity between particle and obstacle
$V$   horizontal transport vector
$V_d$   depositional velocity
$V_{d,\text{fog}}$   depositional velocity due to fog
$V_s$   sedimentation velocity
$Z_o$   roughness scale
$z_s$   particle sink

## Acknowledgments

We are grateful to the California Air Resources Board (CARB) for their financial support (CARB A4-075-32) and their assistance in the field aspects of this project. We appreciate the guidance and assistance provided by the program manager, Eric Fujita. We are also indebted to our colleagues, Daniel Jacob and J. William Munger, who spent many long hours in the field working on aspects of this research.

## References

1. Liljestrand, H. M. Ph.D. Thesis, California Institute of Technology, Pasadena, 1980.
2. Bischoff, W. D.; Paterson, V. L.; Mackenzie, F. T. In *Geological Aspects of Acid Deposition;* Bricker, O. P., Ed.; Acid Precipitation Series; Butterworth: Boston, 1984; Vol. 7, pp 1-21.
3. Azevedo, J.; Morgan, D. L. *Ecology* 1974, 55, 1135-1141.
4. Schlesinger, W. H.; Reiners, W. A. *Ecology* 1974, 55, 378-386.
5. Peterson, T. W.; Seinfeld, J. H. In *Advances in Environmental Science and Technology;* Pitts, J. N.; Metcalf, R. L., Eds.; Wiley: New York, 1980; Vol. 10, pp 125-180.
6. Barrett, E.; Parungo, F. P.; Pueschl, R. F. *Meteorol. Res.* 1979, 32, 136-149.
7. Schwartz, S. E.; Freiberg, J. E. *Atmos. Environ.* 1981, 15, 1129-1144.
8. Slinn, W. G. N. *Atmos. Environ.* 1981, 16, 1785-1794.
9. Chamberlain, A. C. *Proc. R. Soc.* 1967, 296, 45-70.
10. Brimblecombe, P. *Tellus* 1978, 30, 151-157.
11. Lovett, G. M. *Atmos. Environ.* 1984, 18, 361-371.
12. Wisniewski, J. *Water, Air, Soil Pollut.* 1982, 17, 361-377.
13. Lindberg, S. E.; Harriss, R. C.; Turner, R. R. *Science (Washington, D.C.)* 1982, 215, 1609-1611.
14. Waldman, J. M.; Munger, J. W.; Jacob, D. J.; Hoffmann, M. R. *Tellus* 1985, 37B, 91-108.
15. Thomas, M. D.; Hendricks, R. H.; Hill, G. R. In *Air Pollution: Proceedings of U.S. Technical Conference;* McCabe, L., Ed.; McGraw-Hill: New York, 1952; pp 41-47.
16. Haines, B.; Stefani, M.; Hendrix, F. *Water, Air, Soil Pollut.* 1980, 14, 403-407.

17. Granett, A. L.; Musselman, R. C. *Atmos. Environ.* **1984**, *18*, 887-891.
18. Pruppacher, H. R.; Klett, J. D. In *Microphysics of Clouds and Precipitation;* Reidel: Amsterdam, 1978; pp 9-27, 136-148, 412-421.
19. Myers, J. N. *Sci. Am.* **1968**, *219*, 75-82.
20. Taylor, G. I. *Q. J. R. Meteorol. Soc.* **1917**, *43*, 241-268.
21. Roach, W. T.; Brown, R.; Caughey, S. J.; Garland, J. A.; Readiness, C. J. *Q. J. R. Meteorol. Soc.* **1976**, *102*, 313-333.
22. Roach, W. T. *Q. J. R. Meteorol. Soc.* **1976**, *102*, 361-372.
23. Oddie, B. C. V. *Q. J. R. Meteorol. Soc.* **1962**, *88*, 535-538.
24. Hegg, D. A.; Hobbs, P. V. *Atmos. Environ.* **1982**, *16*, 2663-2668.
25. Daum, P. H.; Schwartz, S. E.; Newman, L. *J. Geophys. Res.* **1984**, *89*, 1447-1458.
26. Scherbatskoy, T.; Klein, R. M. *J. Environ. Qual.* **1983**, *12*, 189-195.
27. Houghton, H. G. *J. Meteorol.* **1955**, *12*, 355-357.
28. Mrose, H. *Tellus* **1966**, *18*, 266-270.
29. Okita, T. *J. Meteorol. Soc. Jpn.* **1968**, *46*, 120-126.
30. Castillo, R. A.; Jiuso, J. E.; McLaren, E. *Atmos. Environ.* **1983**, *17*, 1497-1505.
31. *Cloud Chemistry and Meteorological Research at Whiteface Mountain: Summer 1980;* Falconer, P. D., Ed.; Publication No. 806; Atmospheric Sciences Research Center, State University of New York: Albany, 1981.
32. Mack, E. J.; Katz, U. "The Characteristics of Marine Fog Occurring off the Coast of Nova Scotia"; Report CJ-5756-M-1; Calspan: Buffalo, NY, 1976.
33. Mack, E. J. et al. "An Investigation of the Meteorology, Physics and Chemistry of Marine Boundary Layer Processes"; Report CJ-6017-M-1; Calspan: Buffalo, NY, 1977.
34. Brewer, R. L.; Ellis, E. C.; Gordon, R. J.; Shepard, L. S. *Atmos. Environ.* **1983**, *17*, 2267-2271.
35. Munger, J. W.; Jacob, D. J.; Waldman, J. W.; Hoffmann, M. R. *J. Geophys. Res.* **1983**, *88C*, 5109-5121.
36. Jacob, D. J. Ph.D. Thesis, California Institute of Technology, Pasadena, 1985.
37. Jacob, D. J.; Waldman, J. M.; Munger, J. W.; Hoffmann, M. R. *Environ. Sci. Technol.* **1985c**, *19*, 730-735.
38. Fuzzi, S.; Castillo, R. A.; Jiusto, J. E.; Lala, G. G. *J. Geophys. Res.* **1984**, *89D*, 7159-7164.
39. Jacob, D. J.; Waldman, J. M.; Munger, J. W.; Hoffmann, M. R. *Tellus* **1984b**, *36B*, 272-285.
40. Fuzzi, S., unpublished manuscript.
41. Hoffmann, M. R. *Environ. Sci. Technol.* **1984**, *18*, 61-64.
42. Firket, J. *Trans. Faraday Soc.* **1936**, *32*, 1192-1197.
43. Saxena, V. K.; Fisher, G. F. *Aerosol Sci. Technol.* **1984**, *3*, 335-344.
44. Whitby, K. T. *J. Aerosol Sci.* **1978**, *12*, 135-159.
45. Wolfe, G. T. *Atmos. Environ.* **1984**, *18*, 977-981.
46. Manane, Y.; Noll, K. E. *Atmos. Environ.* **1985**, *19*, 611-622.
47. Gerber, H. E. *J. Atmos. Sci.* **1981**, *38*, 454-458.
48. Hudson, J. G. *J. Clim. Appl. Meterol.* **1984**, *23*, 42-51.
49. Lee, I-Y.; Pruppacher, H. R. *Pure Appl. Geophys.* **1977**, *115*, 523-545.
50. Naruse, H.; Maruyama, H. *Pap. Meteorol. Geophys.* **1971**, *22*, 1-21.
51. Hudson, J. G.; Rogers, C. F. *Proc. 9th Int. Cloud Phys. Conf.* 1984.
52. Smith, R. M.; Martell, A. E. *Critical Stability Constants;* Plenum: New York, 1976; Vol. 4.
53. Schwartz, S. E.; White, W. H. In *Advances in Environmental Science and*

*Engineering;* Pfafflin, J. R.; Ziegler, E. N., Eds.; Wiley–Interscience: New York, 1981; pp 1–45.
54. Ledbury, W.; Blair, E. W. *J. Am. Chem. Soc.* **1925**, *127*, 2832–2839.
55. Dasgupta, P. K.; DeCesare, K.; Ullrey, J. C. *Anal. Chem.* **1980**, *52*, 1912–1922.
56. Munger, J. W.; Jacob, D. J.; Hoffmann, M. R. *J. Atmos. Chem.* **1984**, *1*, 335–350.
57. Baboolal, B.; Pruppacher, H. R.; Topalian, J. H. *J. Atmos. Sci.* **1981**, *38*, 856–870.
58. Schwartz, S. E. In *SO$_2$, NO and NO$_2$ Oxidation Mechanisms: Atmospheric Considerations;* Calvert, J. G., Ed.; Acid Precipitation Series; Butterworth: Boston, 1984; Vol. 3, pp 173–208.
59. Jacob, D. J.; Waldman, J. M.; Munger, J. W.; Hoffmann, M. R. *J. Geophys. Res.* **1986**, *91*, 1089–1096.
60. Stelson, A. W.; Seinfeld, J. H. *Atmos. Environ.* **1982**, *16*, 983–992.
61. Bassett, M. E.; Seinfeld, J. H. *Atmos. Environ.* **1983**, *17*, 2237–2252.
62. Martell, A. E.; Smith, R. M. *Critical Stability Constants;* Plenum: New York, 1977; Vol. 3.
63. Appel, B. R.; Kothny, E. L.; Hoffer, E. M.; Hidy, G. M.; Wesolowski, J. J. *Environ. Sci. Technol.* **1978**, *12*, 418–425.
64. Heisler, S.; Baskett, R. "Particle Sampling and Analysis in the California San Joaquin Valley"; Report CARB-RR-81-14; California Air Resources Board: Sacramento, 1981.
65. Kerfoot, O. *For. Abstr.* **1968**, *29*, 8–20.
66. Oberlander, G. T. *Ecology* **1956**, *37*, 851–852.
67. Davidson, C. I.; Friedlander, S. K. *J. Geophys. Res.* **1978**, *83*, 2342–2352.
68. *Studies on Fog in Relation to Fog-Preventing Forest;* Hori, T., Ed.; Tanne Trading: Sapporo, Japan, 1953; p 399.
69. Yosida, Z.; Kuroiwa, D. Ibid., pp 261–278.
70. Oura, H. Ibid., pp 239–252.
71. Legg, B. J.; Price, R. I. *Atmos. Environ.* **1980**, *14*, 305–309.
72. Wattle, B. J.; Mack, E. J.; Pilie, R. J.; Harley, J. T. "The Role of Vegetation in the Low-Level Water Budget in Fog"; Report 7096-M-1; Arvin/Calspan Advanced Technology Center: Buffalo, NY, 1984.
73. Brown, R.; Roach, W. T. *Q. J. R. Meteorol. Soc.* **1976**, *102*, 335–354.
74. Brown, R. *Q. J. R. Meteorol. Soc.* **1980**, *106*, 781–802.
75. Jiusto, J. E.; Lala, G. G. *Radiation Fog Field Programs—Recent Studies;* Publication No. 869; Atmospheric Sciences Research Center, State University of New York: Albany, 1983.
76. Corrandini, C.; Tonna, G. *J. Atmos. Sci.* **1980**, *37*, 2535–2539.
77. Dollard, G. J.; Unsworth, M. H. *Atmos. Environ.* **1983**, *17*, 775–780.
78. Davidson, C. I.; Miller, J. M.; Pleskow, M. A. *Water, Air, Soil Pollut.* **1982**, *18*, 25–43.
79. Brun, R. J.; Lewis, W.; Perkins, P. J.; Serafini, J. S. "Impingement of Cloud Droplets on a Cylinder and Procedure for Measuring Liquid Water Content and Droplet Sizes in Supercooled Clouds by Rotating Multicylinder Method"; Report 1215; National Agricultural Chemicals Association: Washington, DC, 1955.
80. Israel, R.; Rosner, D. E. *Aerosol Sci. Technol.* **1983**, *2*, 45–51.
81. Thorne, P. G.; Lovett, G. M.; Reiners, W. A. *J. Appl. Meteorol.* **1982**, *21*, 1413–1416.
82. Chamberlain, A. C.; Chadwick, R. C. *Ann. Appl. Biol.* **1972**, *71*, 141–158.

83. Davidson, C. I. Ph.D. Thesis, California Institute of Technology, Pasadena, 1977.
84. May, K. R.; Clifford, R. *Ann. Occup. Hyg.* **1967**, *10*, 83–95.
85. Hartley, G. S.; Brunskill, R. T. In *Surface Phenomena in Chemistry and Biology;* Danielli, S. F. et al., Eds.; Pergamon: Oxford, 1958.
86. Merrian, R. A. *Water Resour. Res.* **1973**, *9*, 1591–1598.
87. Bache, D. H. *Atmos. Environ.* **1979**, *13*, 1681–1687.
88. Thom, A. S. In *Vegetation and the Atmosphere;* Monteith, J. L., Ed.; Academic: London, 1975; pp 57–110.
89. Pilie, R. J.; Mack, E. J.; Kolmund, W. C.; Eadie, W. J.; Rogers, C. W. *J. Appl. Meteorol.* **1975**, *14*, 364–374.
90. Chamberlain, A. C. In *Vegetation and the Atmosphere;* Monteith, J. L., Ed.; Academic: London, 1975; pp 155–203.
91. Csanady, G. T. *J. Atmos. Sci.* **1963**, *20*, 201–208.
92. Jenkin, M. E. *Atmos. Environ.* **1984**, *18*, 1017–1024.
93. Sehmel, G. A. *Atmos. Environ.* **1980**, *14*, 883–1011.
94. Huebert, B. J.; Robert, C. H. *J. Geophys. Res.* **1985**, *90D*, 2085–2090.
95. Waldman, J. M. Ph.D. Thesis, California Institute of Technology, Pasadena, 1986.
96. McRae, G. J. Ph.D. Thesis, California Institute of Technology, Pasadena, 1981.
97. Jacob, D. J.; Munger, J. W.; Waldman, J. M.; Hoffmann, M. R. *J. Geophys. Res.* **1986**, *91*, 1073–1088.
98. Reible, D. D. Ph.D. Thesis, California Institute of Technology, Pasadena, 1982.
99. Hoffmann, M. R.; Jacob, D. J. In *SO₂, NO and NO₂ Oxidation Mechanisms: Atmospheric Considerations;* Calvert, J. G., Ed.; Acid Precipitation Series; Butterworth: Boston, 1984; Vol. 3, pp 101–172.
100. Aerovironment Report AV–FR–80/603R; Aerovironment: Pasadena, CA, 1982.
101. Dawson, G. A. *J. Geophys. Res.* **1977**, *82*, 3125–3133.
102. Denmead, O. T.; Freney, J. R.; Simpson, J. R. *Soc. Biol. Chem.* **1976**, *8*, 161–164.
103. Hutchinson, G. L.; Mosier, A. R.; Andre, C. E. *J. Environ. Qual.* **1982**, *11*, 288–293.
104. Jacob, D. J. *J. Geophys. Res.* **1985**, *90D*, 5864.

RECEIVED for review May 6, 1986. ACCEPTED October 10, 1986.

# 5

# Air–Sea Transfer of Trace Elements

Richard Arimoto and Robert A. Duce

Center for Atmospheric Chemistry Studies, Graduate School of Oceanography, University of Rhode Island, Kingston, RI 02882-1197

*At Enewetak Atoll in the tropical North Pacific, the atmospheric concentrations of Al, Sc, Mn, Fe, Co, Cs, Ba, Ce, Eu, Hf, Ta, and Th are at times dominated by mineral aerosols from Asia. Present rates of air–sea transfer for several elements representative of mineral aerosols are similar to their rates of accumulation in sediments over the past 5000-10,000 years. This similarity implies that the air–sea transfer of particles is tied to the sedimentary cycle. The atmospheric concentrations of Pb, Zn, Cu, Se, and Cd are higher than those expected from mineral or sea salt aerosols. For these enriched elements, a comparison of air–sea transfer rates with their estimated inputs to surface water from vertical mixing shows that air–sea transfer processes also affect the chemistry of reactive trace elements in the ocean.*

INTEREST IN THE AIR-SEA EXCHANGE of trace elements grew out of the general awareness of pollution in the 1970s and out of the concern over environmental issues, especially the fear that the air–sea transfer of airborne pollutants could affect the health of the oceans. In 1976, the sea–air exchange (SEAREX) program was established to investigate the atmospheric transport of trace elements and other trace substances to the Pacific Ocean and to study the air–sea transfer of the materials brought to the ocean by prevailing winds. The general strategy of SEAREX has been to sample sites in the Pacific Ocean that are representative of the four major surface level wind regimes—the easterly (trade) winds and the westerly winds over the North and South Pacific.

The specific objectives of the SEAREX studies presented here were (1) to identify the sources for selected trace elements in the atmosphere, (2) to investigate the mechanisms responsible for the air–sea transfer of the elements, and (3) to estimate the fluxes of various elements to the Pacific Ocean. We used the results obtained during the experiments at

0065-2393/87/0216-0131$06.00/0
© 1987 American Chemical Society

Enewetak to illustrate the sources for and the air–sea transfer of particulate trace elements. The role of air–sea exchange processes on the geochemistry of the oceans is also discussed.

## Experimental Details

Three experiments have been conducted on remote islands for SEAREX: Enewetak Atoll in the easterlies of the tropical North Pacific, American Samoa in the easterlies of the tropical South Pacific, and New Zealand in the westerlies of the South Pacific (Figure 1). The fourth surface level wind field over the Pacific, the North Pacific westerlies, was sampled during a cruise in 1986. Towers were erected on each of the islands and on the ship to ensure that the samples were representative of the clean atmosphere over the ocean and not contaminated by materials from the islands or from the ship.

Samples of airborne particles were collected with four different types of samplers at Enewetak (1). Two of the samplers (the 7-in.-round and the 8- × 10-in.-rectangular) were high-volume systems that collected bulk aerosol samples on cellulose filters (Whatman 41 filters, Whatman Inc.). A third type of sampler, the cascade impactor (model 235, Sierra Instruments, Inc.), separated the aerosols by size. These three high-volume systems operated at ~68 m$^3$ h$^{-1}$. The fourth sampler, a 47-mm-diameter, 0.4-$\mu$m-porosity filter (Nuclepore Corp.) inserted in a polycarbonate open-faced filter holder, was a low-volume system that operated at ~1 m$^3$ h$^{-1}$. A total of 47 aerosol samples was collected during two sampling periods at Enewetak: from April to June and from July to August. Detailed

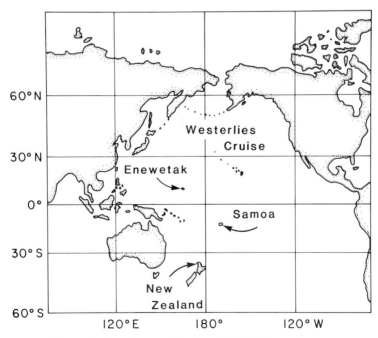

*Figure 1. Locations of the four SEAREX experiments.*

information on the collection efficiencies of these filters for particles of various sizes at different flow rates has recently been published (2). For the cellulose filters, the collection efficiency for particles from 0.035 to 1.0 $\mu$m in diameter ranges from 43% to 99.5%. For 0.4-$\mu$m-porosity filters, the filter efficiency range is from 78% to 99.99%.

Because of the extremely low concentrations anticipated for some of the elements, precautions were taken during the SEAREX studies to keep the samples free from contamination. These precautions included monitoring of the concentrations of condensation nuclei and ozone, both of which exist in clean air but are useful indicators of air containing contaminants generated at the sampling site by pumps, generators, and vehicles (1). The sampling pumps automatically shut off at low wind speeds ($\leq$1.3 m s$^{-1}$) and whenever the number of condensation nuclei exceeded 750 cm$^{-3}$. Wind direction also was continuously monitored to ensure that the air being sampled had not recently passed over land or any other sources of local contaminants. At Enewetak, the sampling sector was from 35° to 170° because air from that sector was directly from the ocean. Aerosol sampling also was stopped during precipitation.

Rainwater samples were collected on an event basis from the tops of the towers with a conventional polyethylene funnel, which was ~1 m in diameter and precleaned in a series of acid baths. The rainwater samples were collected in precleaned polyethylene bottles and then frozen shortly after collection. Most of the rainwater samples from Enewetak were aliquots of those collected by Settle and Patterson (3), and further details of the methods used for the rainwater sampling may be found in their paper.

Dry deposition samples were collected with a circular, conventional polyethylene plate that had a slightly concave collection surface and an area of 2780 cm$^2$. The sampler was covered when the wind was not from the predetermined sampling sector, when ships or airplanes passed upwind of the sampler, and when precipitation occurred. The deposits were removed from the plate with ultrapure 4 N HNO$_3$, and a quartz rod was used with the acid to strip the deposits from the surface of the plate. Further details of the dry deposition sampling may be found in reference 3.

The aerosol samples and samples of wet and dry deposition were analyzed by atomic absorption spectroscopy (AAS) and instrumental neutron activation. The methods used for chemical analyses have been described elsewhere (1, 5). These references also include descriptions of the numerous blanks that were taken along with the samples to assess the potential for contamination during the collection and processing of the samples.

## Results and Discussion

**Sources for Particulate Trace Elements.** The results obtained with the four independent aerosol sampling systems used at Enewetak agreed well; in Figure 2, the data for Al are used to illustrate the consistency of results. Similar agreement among samplers was observed for other elements (1), and interlaboratory calibrations were conducted to verify the accuracy of the data. The concentration of Al in the remote marine atmosphere is directly proportional to the concentration of mineral aerosols, which are produced by the erosion of soils and the weathering of continents. Thus, Al frequently is used as an indicator of

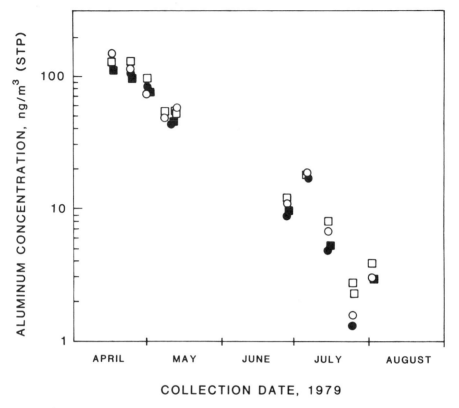

*Figure 2. Concentrations of Al at Enewetak as determined by four independent sampling devices. Key:* ●, *7-in.-round sampler;* ■, *cascade impactor;* ○, *8- ×  10-ir.-rectangular sampler; and* □, *filter.*

this weathered crustal material. Similarly, Na is used as an indicator of atmospheric sea salt, which is produced as bubbles burst at the surface of the ocean. One of the most significant early results of the experiments at Enewetak was that the concentrations of mineral aerosol or dust decreased exponentially from April to August 1979. In contrast, the atmospheric sea salt concentration decreased less dramatically (Figure 3) (4).

The source for the mineral aerosols collected at Enewetak was identified in Asia through analyses of air mass trajectories (4, 6) and through analysis of lead isotope source signatures (3). More specifically, meteorological analyses (6) indicated that arid and semiarid regions in China were the source for the material that was transported thousands of kilometers through the atmosphere to the open North Pacific. In addition, most of the transport of dust was determined to occur during the spring because strong surface winds at that time, combined with

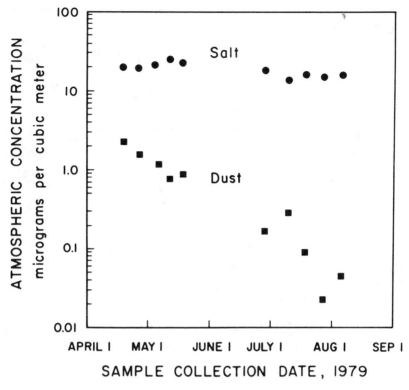

Figure 3. *Changes in the concentrations of sea salt and atmospheric dust at Enewetak.*

low rainfall and the plowing of fields in China, cause frequent and intense dust storms. Mechanisms for injecting dust into the atmosphere exist at that time of the year, and the synoptic flow also is favorable for the long-range transport of materials from Asia to the open Pacific (*6*).

Analyses of trace element data from Enewetak (*1*) indicated that the trace elements could be classified into three groups: (1) those from the ocean due to the production of sea salt particles, (2) those from the continents as a result of the weathering of rock and the erosion of soils, and (3) those enriched relative to both bulk sea water and mineral aerosols. The enrichments of the third group of elements may be due to emissions from high temperature natural and anthropogenic sources or from biologically driven reactions.

One technique that has proven useful as a first step for identifying the source of a trace element is to normalize the trace element concentration with respect to a reference element. As previously mentioned, Al and Na frequently are used as indicators of mineral aerosols and atmospheric sea salt, respectively, and enrichment factors calculated relative to average crustal rock and bulk sea water are often used to identify likely sources for trace elements. The enrichment factor for a hypothetical element Z relative to crustal material ($EF_{crust,Z}$) is defined as

$$EF_{crust,Z} = (Z/Al)_{sample}/(Z/Al)_{crust} \qquad (1)$$

where $(Z/Al)_{sample}$ and $(Z/Al)_{crust}$ are the respective concentration ratios in the sample and in average crustal rock according to the compilation of Taylor (7). Similarly, the enrichment factor relative to bulk sea water ($EF_{sea,Z}$) is

$$EF_{sea,Z} = (Z/Na)_{sample}/(Z/Na)_{sea\ water} \qquad (2)$$

where $(Z/Na)_{sample}$ and $(Z/Na)_{sea\ water}$ are the respective concentration ratios in the sample and in bulk sea water. If the $EF_{crust,Z}$ or the $EF_{sea,Z}$ value for an element approaches unity, then one may infer that unfractionated crustal rock or bulk sea water, respectively, is the predominant source for that element. Upper limits of 3–5 often are used for the interpretation of the enrichment factors owing to uncertainties in the actual composition of the mineral aerosol precursor and in the concentrations of some trace elements in bulk sea water.

The enrichment factors calculated relative to sea water showed that in the aerosol samples, the concentrations of Br, Cl, Mg, K, and Ca relative to Na were similar to those in bulk sea water (Figure 4). All of these elements except Br exhibited similar changes in concentration from the local dry season (February–June) to the local wet season (July–January), with a mean decrease of 28% ± 15% (Table I). The mean concentration of Br did not change between seasons, however, and the cause for this lack of temporal variability has not been determined.

The enrichment factors also showed that the concentrations of 11 elements (Sc, Mn, Fe, Co, Cs, Ba, Ce, Eu, Hf, Ta, and Th) relative to Al were similar to those in average crustal rock. These crustal-derived elements exhibited more pronounced temporal variability, and the mean percent change from the dry season to the wet season was 91% ± 4.2% (Table I). The concentrations of four elements, V, Cr, Rb, and Cu, were dominated by mineral aerosols during the high dust season. When the concentrations of dust subsided, however, enrichments of these elements were observed, particularly on the largest particles as a result of the processes involved in the formation of sea salt aerosols and on the smallest particles, which are produced by high temperature reactions or gas–particle conversion.

A third group of elements, I, Zn, Pb, Ag, Se, Sb, and Cd, exhibited concentrations in the aerosols that were clearly higher than those expected from sea water or from unfractionated mineral aerosols (Figure 4). This third group of elements is thus considered enriched, for even though their concentrations are low, they are higher than expected from the production of sea salt aerosols or the dispersal of weathered crustal material.

Much of the research on the atmospheric chemistry of Pb has been done by C. C. Patterson and his colleagues at the California Institute of Technology. Their work has shown that Pb is enriched in the atmosphere by emissions from automobile exhausts and from smelters (8). The concentrations of other elements, such as Zn and Cd, also may be affected by pollution sources because the industrial emissions of these elements apparently exceed their inputs from natural sources (9). Still other elements may be enriched in marine aerosols by gas–particle conversions of vapor-phase species, and these reactions may be driven by processes in the oceans themselves. For example, Mosher and Duce (10) proposed that the oceans may be a significant source for vapor-phase Se, and they suggested that Se concentrations may be elevated in aerosols by gas–particle conversions or the sorption of vapor-phase species by aerosol particles. Similarly, enrichments of iodine in marine aerosols may be due to either sorption or heterogeneous reactions because the oceans evidently are a significant source for gas-phase iodine species (11, 12).

**Atmospheric Residence Times for Aerosols.**   The atmospheric residence times of particulate trace elements are controlled by three processes: coagulation, precipitation scavenging, and dry deposition. The relative importance of these three processes is dependent upon the size of the aerosols of interest. According to Jaenicke (13), the residence times for particles $\sim$0.01 $\mu$m in radius are on the order of 1 day due to coagulation with other particles. That is, these particles cease to exist as individuals in about 1 day, and the behavior of the aggregate of particles then controls the residence times of the trace elements. Atmospheric residence times for very large particles (radii $>$ 10 $\mu$m) also are limited to about 1 day because the high settling velocities of these particles make dry deposition an efficient removal mechanism. Particles with radii of $\sim$0.3 $\mu$m have the longest residence times, and the removal of these particles is mainly through precipitation scavenging. In the following sections, the removal of particulate trace elements by dry and wet deposition is discussed.

DRY DEPOSITION MEASUREMENTS.   Direct measurements of the dry deposition of particles to the surface of the ocean are not technically feasible at present, and thus surrogate surfaces have been used to study

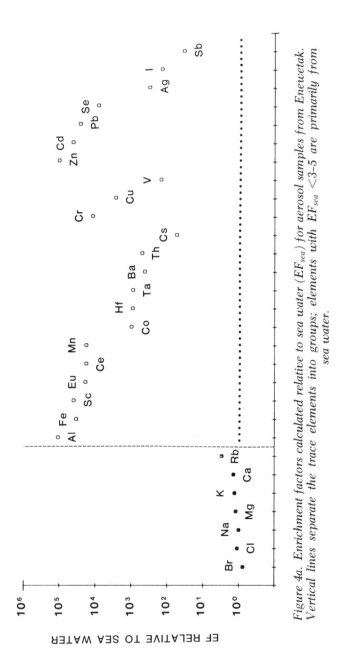

Figure 4a. Enrichment factors calculated relative to sea water ($EF_{sea}$) for aerosol samples from Enewetak. Vertical lines separate the trace elements into groups; elements with $EF_{sea}$ <3–5 are primarily from sea water.

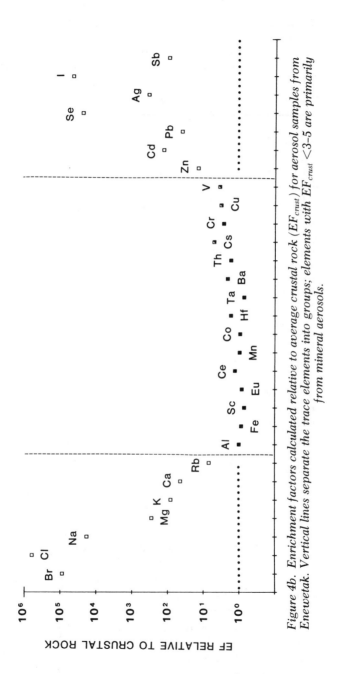

*Figure 4b. Enrichment factors calculated relative to average crustal rock ($EF_{crust}$) for aerosol samples from Enewetak. Vertical lines separate the trace elements into groups; elements with $EF_{crust}$ <3–5 are primarily from mineral aerosols.*

Table I. Geometric Mean Concentrations for Trace Elements in Aerosols
from Enewetak

| Element | Units | Geometric Mean | | | Change[a] (%) |
|---|---|---|---|---|---|
| | | All Samples | Feb.–June | July–Jan. | |
| Na | $\mu$g m$^{-3}$ | 4.8 | 5.6 | 4.1 | −27 |
| Br | ng m$^{-3}$ | 20 | 20 | 20 | 0 |
| Ca | ng m$^{-3}$ | 230 | 300 | 170 | −43 |
| Cl | $\mu$g m$^{-3}$ | 8.3 | 9.5 | 7.2 | −24 |
| K | ng m$^{-3}$ | 220 | 270 | 160 | −41 |
| Mg | $\mu$g m$^{-3}$ | 0.66 | 0.80 | 0.56 | −30 |
| Rb | pg m$^{-3}$ | 120 | 220 | 60 | −73 |
| Al | ng m$^{-3}$ | 21 | 79 | 6.1 | −92 |
| Ba | ng m$^{-3}$ | 0.53 | 0.11 | 0.30 | −79 |
| Ce | pg m$^{-3}$ | 21 | 89 | 4.8 | −95 |
| Co | pg m$^{-3}$ | 7.6 | 21 | 1.7 | −92 |
| Cs | pg m$^{-3}$ | 5.1 | 13 | 1.2 | −91 |
| Eu | pg m$^{-3}$ | 0.54 | 1.0 | 0.09 | −91 |
| Fe | ng m$^{-3}$ | 17 | 50 | 3.3 | −93 |
| Hf | pg m$^{-3}$ | 1.9 | 3.9 | 0.45 | −89 |
| Mn | ng m$^{-3}$ | 0.29 | 0.79 | 0.063 | −92 |
| Sc | pg m$^{-3}$ | 4.7 | 16 | 0.94 | −94 |
| Ta | pg m$^{-3}$ | 0.52 | 1.2 | 0.12 | −90 |
| Th | pg m$^{-3}$ | 5.2 | 17 | 1.1 | −94 |
| Cr | pg m$^{-3}$ | 91 | 140 | 37 | −74 |
| Cu | pg m$^{-3}$ | 44 | 72 | 14 | −81 |
| V | pg m$^{-3}$ | 82 | 170 | 42 | −75 |
| Ag | pg m$^{-3}$ | 4.1 | 4.0 | 4.8 | +20 |
| Cd | pg m$^{-3}$ | 3.5 | 4.6 | 2.5 | −46 |
| I | ng m$^{-3}$ | 2.7 | 4.4 | 1.7 | −61 |
| Pb | pg m$^{-3}$ | 120 | 130 | 96 | −26 |
| Sb | pg m$^{-3}$ | 3.7 | 5.2 | 2.4 | −54 |
| Se | pg m$^{-3}$ | 130 | 150 | 110 | −27 |
| Zn | pg m$^{-3}$ | 170 | 210 | 120 | −43 |

[a]Values of percent change were calculated from February–June to July–January.

this mechanism for air–sea exchange. However, the validity of studies based on artificial collectors, including plates, buckets, and the like, is open to question because these surrogate surfaces clearly cannot mimic the dynamics of the ocean surface. Nevertheless, the studies that were done with a plastic plate at Enewetak did show that, in terms of dry deposition, the various classes of trace elements behaved differently.

For each dry deposition sample from Enewetak, the dry deposition velocity for each element was calculated by dividing the measured dry deposition rate for the element by the corresponding concentration of the element in bulk aerosols (i.e., dry deposition velocity is equal to the dry deposition rate divided by the concentration). The dry deposition velocities calculated in this manner were grouped by element, and a one-way analysis of variance showed that deposition velocities for the various elements were different at $p < 0.01$. Moreover, a multiple-comparison procedure based on Scheffe's method showed that the deposition velocities could be used to classify the elements into two groups. One group contained Na, Ca, Mg, and K, which are all primarily derived from the sea. The second group contained Fe, Mn, and Pb, and these elements, except Pb, were derived from crustal weathering. The dry deposition velocities were 0.41 (Fe), 0.43 (Mn), and 1.3 (Pb). The other group's dry deposition velocities were 4.6 (K), 5.1 (Ca), 5.1 (Mg), and 6.4 (Mg).

The general trend of higher deposition velocities for the sea salt elements can be ascribed to the differences in the mass-particle size distributions of the two classes of elements. Sea salt particles are large compared with those in a mineral aerosol; this size difference is due to the relative proximity of the source for sea salt and to the removal of large mineral aerosols by dry deposition during transport. As discussed later, the dry deposition velocities of particles are strongly affected by their size, and large particles tend to have high deposition velocities. Thus, the higher deposition velocities for the sea salt elements are consistent with the observed differences in mass-particle size distributions.

**RECYCLING OF TRACE ELEMENTS BETWEEN THE OCEAN AND THE ATMOSPHERE.**    Sea salt aerosols are produced by bursting bubbles, and they are enriched with many substances because of the scavenging of surface active materials by rising bubbles and the rupture of the sea surface microlayer by bursting bubbles (*14*). The production and deposition of these sea salt particles results in a recycling of trace elements between the atmosphere and oceans. Therefore, studies on air–sea transfer must take into account the recycling of substances associated with sea spray in order to determine the net or new inputs into the system. In this section, studies on Pb are used to illustrate the recycling of material between the atmosphere and oceans.

The amount of recycled Pb in the dry deposition samples (Table II) was calculated by first using measured deposition rates for sea salt that were based on the Na data (*5*). Sea salt aerosols are enriched with Pb over bulk sea water concentrations, however, and therefore the enrichments of Pb in sea salt aerosols (higher Pb-to-Na ratios in sea salt aerosols compared with the ratio in bulk sea water) also had to be taken into account. Enrichments of Pb in sea salt aerosols result from the scavenging of pollutant Pb by rising bubbles and from the fractionation of the sea surface microlayer (*3*). The magnitude of the Pb enrichments in sea salt aerosols is still uncertain, but the enrichments apparently range from 2000 to 20,000 (*14, 15*). The flux of Pb resulting from the deposition of crustal material was estimated by assuming no fractionation

Table II. Recycling of Trace Elements between the Ocean and the Atmosphere

| Element | $MMR^a$ (μm) | Amount Recycled | |
| | | Dry Deposition (%) | Wet Deposition (%) |
|---|---|---|---|
| Zn | 0.5–1 | 12 | 15 ± 13 |
| Pb | 0.3–0.6 | 60 ± 27 | 30 |
| V | 0.97 ± 2.7 | 30 | 33 ± 32 |
| Cu | 0.3–0.4 | ~70 | 48 ± 52 |
| Se | | | 17 ± 15 |
| Ag | | | 24 |

[a] MMR is mass median radius ± geometric standard deviation.

of Pb in the mineral aerosol; this crustal component was only a minor fraction of the Pb deposition, amounting to <10% of the total deposition (5). In three dry deposition samples from Enewetak, recycled Pb was estimated to be approximately 54%, 57%, and 100% of the total Pb dry deposition (5), and in a fourth sample the amount of recycled Pb could not be precisely calculated but clearly exceeded 25%. Thus, material recycled from the sea surface apparently amounted to a sizable fraction of the Pb in the dry deposition samples.

DRY DEPOSITION MODELS. Models have been used to investigate the dry deposition of particles to the natural water surfaces because of the difficulty in collecting samples from natural surfaces and in interpreting results of studies based on surrogate surfaces. None of the models yet includes all of the relevant physics, however, and the existing models are also limited because they do not take into account nonsteady-state conditions. Furthermore, the validity of the models cannot be verified through field measurements. Nevertheless, we investigated two models (16, 17) to determine how the specific processes considered in them may affect the air–sea transfer of several trace substances (18).

Both models separate the atmosphere into two layers. In the constant flux layer, atmospheric turbulence and gravitational settling govern particle transport. In the very thin deposition layer (<~1 mm) just above the air–sea interface, atmospheric turbulence is negligible, and particles grow in response to the high relative humidity there. This growth of particles in the deposition layer was taken into account in both models, but the model of Williams (17) also allowed for differences in the transfer velocities to smooth and broken water surfaces. That is, different transfer coefficients for smooth and broken water surfaces ($k_{ss}$ and $k_{bs}$, respectively) were used in his model. According to Williams (17), the deposition of particles to water surfaces broken by waves should be enhanced by two processes: the scavenging of particles by impaction and the coagulation of particles with sea spray droplets.

Comparisons of the dry deposition rates for Al that were calculated from the deposition models suggest that surface waves and the scavenging of particles by sea spray do not greatly affect the dry deposition rate of mineral aerosol (Table III). In contrast, the deposition rate for $^{210}$Pb, which is primarily found in submicrometer particles, was as much as 5 times higher when the effects of broken water surfaces were taken into account. These results suggest that the air–sea exchange of substances that are predominantly associated with small particles may be significantly enhanced by deposition to water surfaces broken by waves (*18*).

**Table III. Dry Deposition Rates Calculated from Models of Slinn and Slinn (16) and Williams (17)**

|  |  | Williams | |
| --- | --- | --- | --- |
| $u^a$ (m s$^{-1}$) | *Slinn and Slinn* | $k_{bs}{}^b = 5$ cm s$^{-1}$ | $k_{bs} = 10$ cm s$^{-1}$ |
| | | Al$^c$ | |
| 5 | 3.1 | 3.1 | 3.2 |
| 10 | 4.3 | 5.1 | 5.9 |
| 15 | 7.6 | 11 | 14 |
| | | $^{210}$Pb$^d$ | |
| 5 | $0.44 \times 10^{-9}$ | $0.46 \times 10^{-9}$ | $0.49 \times 10^{-9}$ |
| 10 | $0.68 \times 10^{-9}$ | $1.0 \times 10^{-9}$ | $1.4 \times 10^{-9}$ |
| 15 | $1.2 \times 10^{-9}$ | $2.6 \times 10^{-9}$ | $3.8 \times 10^{-9}$ |

$^a$The abbreviation $u$ denotes wind speed.
$^b$The abbreviation $k_{bs}$ denotes the broken water surface transfer coefficient.
$^c$All values for Al were based on a mass median diameter of 1.0 μm and a concentration of 20.0 ng m$^{-3}$. Deposition rates are in units of femtograms per square centimeter per second.
$^d$All values for $^{210}$Pb were based on a mass median diameter of 0.26 μm and a concentration of $8.1 \times 10^{-3}$ dpm m$^{-3}$. Deposition rates are in disintegrations per minute per square centimeter per second.

**WET DEPOSITION.** Aerosols are removed from the atmosphere by scavenging in clouds and below clouds; these processes are called "rainout" and "washout", respectively, but the two terms are being used less frequently because of the difficulty in distinguishing the two processes (*19*). One of the conceptual tools used to interpret precipitation data is the scavenging ratio, which is sometimes called a washout factor. The formula we used (*5*) for calculating a scavenging ratio (SR) is

$$SR = (C_{rain} \times \rho)/C_{air} \qquad (3)$$

where $C_{rain}$ is the concentration of an element in rain (g kg$^{-1}$), $C_{air}$ is the corresponding concentration in air (g m$^{-3}$) for the aerosol sample (approximately 9000 m$^3$ sampled in an average of 7 days) closest in time to the rain event, and $\rho$ is a density term (1.20 kg m$^{-3}$ at 20 °C and 760 mm Hg) that is used to make the scavenging ratio dimensionless. The scavenging ratios are thus calculated on a mass-weighted basis. (Other researchers calculate scavenging ratios on a volume-weighted basis.)

Analysis of the rainwater samples from Enewetak showed that scavenging ratios for the trace elements typically were between 300 and 1000; these scavenging ratios are similar to those that have been observed over the continents (20). But more interesting was the observation that the scavenging ratios for the elements apparently were not related to the mass-particle size distributions of the elements. One might expect that the sea salt elements, which are on the largest particles, would have higher scavenging ratios than the other elements because some calculations indicate that large particles are most effectively scavenged by raindrops (21). However, the scavenging ratios for the various groups of elements at Enewetak were similar despite their clearly different mass-particle size distributions.

The similarity in scavenging ratios for the various elements may be in part a result of the low concentrations of particles in the remote marine atmosphere. The concentrations of airborne particles over the oceans—around 300 particles cm$^{-3}$—are much lower than in the atmosphere over the continents (22). It has been suggested that under these conditions, virtually all particles, regardless of their size, are incorporated into rain droplets during the nucleation process (21). In contrast, over the continents, where the aerosol concentrations are higher, the chemical properties of the aerosols, particularly their hygroscopicity, also presumably affect the scavenging of particles by precipitation (19).

We note that the concentrations of trace elements in the air are often determined from samples collected close to the Earth's surface, whereas the concentrations in rain are most likely determined by processes occurring several kilometers in the atmosphere. Thus, the vertical distributions of trace elements in the atmosphere should affect the scavenging ratios of the elements. However, during storms, surface air may be transported to heights of several kilometers, where nucleation and in-cloud scavenging occur (19). Vertical motions within storms therefore may contribute to the similar scavenging ratios observed for elements that presumably have different vertical distributions in the atmosphere.

Because we were interested in distinguishing the new material entering the ocean from the material recycled from the sea surface, we also estimated the recycled component of wet deposition, again by using Na as an indicator of the deposition of sea salt and estimates of the enrichments of trace elements in sea salt aerosols (5). These calculations indicate that the percentage of recycled material in wet deposition may be smaller than in dry deposition but still significant (Table II).

The data are not extensive, but generally less than half of the wet deposition could be attributed to recycled sea spray.

**ATMOSPHERIC DEPOSITION AND THE CHEMISTRY OF THE OCEANS.** The air–sea transfer of trace elements is important from a geochemical standpoint for several reasons. First, the air–sea transfer of dust at Enewetak is similar to the rate at which sediments have been deposited on the sea floor over the past 5000–10,000 years (Figure 5) (5). The air–sea exchange rates for three crustal elements (Al, Fe, and Th) and one intermediate element (V) at Enewetak are of the same order as their rates of deposition to the sea floor at a site approximately 1500 km away. The similarity between the fluxes of Al and the other crustal elements into the oceans and the deposition of these elements in the sediments indicates that the air–sea exchange of dust supplies a significant fraction of the material that ultimately is incorporated in marine sediments. Mineralogical studies also suggest that air–sea-transfer processes can account for 75%–90% of the abiogenic sedimentation in the open ocean (23). Thus, wind-borne material from the continents eventually becomes a major component of the sediments of the oceans.

Estimates of the inputs of various elements into the atmosphere from industrial emissions and fossil fuel combustion suggest that pollution sources may enrich elements, including Cd, Cu, and Zn, in the marine atmosphere (9). These elements are particularly interesting to geochemists not only because they are enriched but also because they react with particles and organisms in the oceans. Their reactivity means

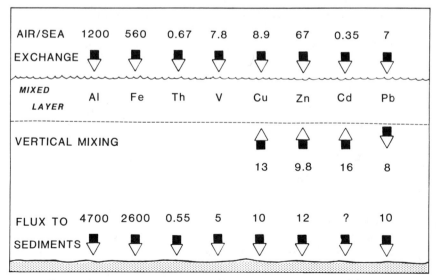

*Figure 5. Comparison of trace element fluxes in the North Pacific.*

that they will reside in the water column for hundreds to thousands of years. Results from the SEAREX experiments and related projects have shown that atmospheric deposition also affects the chemistry of these reactive trace elements in the surface waters of the open ocean.

The air–sea transfer of Cu is similar to the rate of Cu deposition in marine sediments, and evidently this is true for Zn as well, although the dry deposition rate for Zn at Enewetak was calculated from a single sample and thus may not be representative of the true flux. Simply comparing the atmospheric and sedimentation fluxes of Cu and Zn may not be meaningful. These elements are reactive in the oceans, and they are most likely deposited on the sea floor far from the place where they enter the oceans. But the fluxes of trace elements into surface waters due to air–sea exchange may be compared with the inputs from vertical mixing as estimated by Bruland (24). For Cu and Zn, the rates of air–sea transfer are similar to the inputs of these elements to surface waters due to vertical mixing (Figure 5). These results demonstrate that air–sea-transfer processes may significantly affect the chemistry of reactive trace elements in the ocean.

The air–sea exchange of pollutant Pb has resulted in a net downward flux of Pb through the water column. The eolian fluxes of Pb at Enewetak were estimated to be 7 ng cm$^{-2}$ year$^{-1}$ (Figure 5) (5), which is virtually the same as an estimate for Pb deposition at Enewetak made by Settle and Patterson (3). The air–sea transfer of pollutant Pb has changed the vertical profile of Pb concentrations in the oceans from the preindustrial steady state; Pb is now enriched in surface waters (8). Moreover, the transport of anthropogenic Pb through the water column is superimposed on the natural fluvial inputs of lead, and the full impact of pollutant Pb on the composition of bottom sediments will not be seen for some time because of the reactivity of Pb in the water.

The relative importance of wet and dry deposition as mechanisms in the air–sea transfer of trace elements remains to be determined conclusively. For the limited number of deposition samples available, wet deposition appeared to contribute more than half of trace element fluxes, but analytical uncertainties and the variability in the data do not allow us to conclude that either wet or dry deposition is the predominant air–sea transfer mechanism. Wet removal is episodic, and results from a network of islands in the North Pacific, which comprise the SEAREX Asian dust network, suggest that a major fraction of the yearly flux of dust can take place in the few weeks when precipitation occurs during periods of high atmospheric dust concentrations (25). In comparison, dry deposition is more constant, and the settling of dry particles is important because it is a mechanism for recycling certain trace elements between the atmosphere and the oceans.

The effects of the air–sea transfer of trace elements on the biology and chemistry of the oceans also remain to be determined. In particular, further studies need to be conducted on the temporal and spatial variability in the fluxes of essential trace elements such as Fe, and on the fluxes of nutrients such as nitrate and phosphate (26). Measurements of the deposition of dry particles to the ocean surface are required to test validity of the dry deposition models. The scavenging of particles by rain, the reactions that occur in clouds, and the solubilization of particles in raindrops all need to be investigated in greater detail to understand more fully the impact of atmospheric deposition on the oceans. Further studies also need to be conducted on the fates and effects of trace elements after they are deposited in the oceans and as they are transported through the water column.

### Differences in Atmospheric Deposition to Lakes and Oceans.

Differences in the deposition of trace elements to lakes and oceans may result from differences in the relative proximity of the sites to trace element sources as well as differences in meteorological conditions. A survey of 125 sites in northern France has shown that the concentrations of Cd and Pb in soils decreased with distance between the sites and the major emission sources (27). Other observations have shown that the rapid fallout of large particles by dry deposition may effectively remove a major fraction of large pollution aerosols near their sources (28). Measurements of trace metal fluxes by Eisenreich (29) have shown first that atmospheric deposition represents a significant contribution to the total trace element input to Lake Michigan, and second that urban and industrial emissions have resulted in higher atmospheric deposition rates of trace metals to the southern basin of Lake Michigan.

The deposition of trace elements to remote oceanic islands such as Enewetak is governed by the same processes that operate over the continents and the Great Lakes. However, the wet deposition of trace elements may be affected by the type of rain (i.e., cold rains versus warm rains) (19); by the vertical distributions of the substances of interest; and by differences in duration, intensity, and droplet-size spectrum of the precipitation (30). Thus, differences in the efficiencies with which particles are removed by precipitation scavenging over oceans and lakes may be expected. Indeed, the scavenging ratios for trace elements at Enewetak evidently were not related to the mass-particle size distributions of the elements, and as previously discussed, this relationship is different from what has been observed over the continents.

Dry deposition rates of trace metals to oceans and lakes will differ because of generally lower concentrations of trace metals over the oceans, but the deposition rates also may be affected by changes in the mass-particle size distribution of the aerosol as large particles are preferentially lost close to their sources. The dry deposition of trace elements such as Fe, Pb, Ti, and V to Lake Michigan may equal the inputs from wet deposition under some conditions (31), but other studies have indicated that wet deposition may account for more than half of the total deposition of $^{210}$Pb, $^{210}$Po, Fe, Pb, and other trace metals (32, 33). The relative importance of wet versus dry deposition as mechanisms for the transfer of materials between the atmosphere and freshwater lakes needs to be studied further, but dry deposition is thought to be increasingly important for substances associated with larger particles (32).

Finally, the recycling of particulate materials between lakes and the atmosphere has not been investigated, and this recycling may be an area that warrants future study. Early work by Monahan (33) showed that the whitecap coverage of lakes increased abruptly when the wind velocity increased from 7 to 8 m s$^{-1}$, but for a given wind velocity the whitecap coverage over lakes was less than that over oceans; this observation was particularly true at higher wind velocities. The composition of aerosols produced by bubbles bursting in freshwater is essentially unknown, and their role in cycling materials between the surface water of lakes and the atmosphere has not yet been thoroughly studied. Moreover, materials on the surface of the land near the shores of the lakes may be resuspended and eventually deposited in the lakes. Even though the atmospheric deposition over lakes and oceans may differ both quantitatively and qualitatively, this transfer process is clearly important for the geochemistry of both environments.

## Abbreviations and Symbols

| | |
|---|---|
| $\mu$ | wind speed |
| $\rho$ | density of air |
| $C_{air}$ | concentration in air |
| $C_{rain}$ | concentration in rain |
| EF | enrichment factor |
| $k_{bs}$ | broken water surface transfer coefficient |
| $k_{ss}$ | smooth water surface transfer coefficient |
| MMR | mass median radius |
| SR | scavenging ratio |
| Z | hypothetical element |
| (Z/Al) | concentration ratio of Z to Al |
| (Z/Na) | concentration ratio of Z to Na |

## Acknowledgments

We thank Barbara Ray, Hal Maring, Alan Hewitt, Alex Pszenny, C. K. Unni, and Patrick Harder for their participation in the SEAREX experiments. This research was supported by the National Science Foundation under Grants OCE 77-13071, OCE 77-13072, OCE 81-11895, and OCE 84-05605 as part of the SEAREX Program.

## References

1. Duce, R. A.; Arimoto, R.; Ray, B. J.; Unni, C. K.; Harder, P. J. *J. Geophys. Res.* **1983**, *88*, 5321-5342.
2. Liu, B. Y. H.; Pui, D. Y. H.; Rubow, K. L. In *Aerosols in the Mining and Industrial Work Environments;* Marple, V. A.; Liu, B. Y. H., Eds.; Ann Arbor Science: Ann Arbor, MI, 1983; Vol. 3, pp 989-1038.
3. Settle, D. M.; Patterson, C. C. *J. Geophys. Res.* **1982**, *87*, 8857-8869.
4. Duce, R.; Unni, C. K.; Ray, B. J.; Prospero, J. M.; Merrill, J. T. *Science (Washington, D.C.)* **1980**, *209*, 1424-1522.
5. Arimoto, R.; Duce, R. A.; Ray, B. J.; Unni, C. K. *J. Geophys. Res.* **1985**, *90*, 2391-2408.
6. Merrill, J. T.; Bleck, R.; Avila, L. *J. Geophys. Res.* **1985**, *90*, 12, 927-12, 936.
7. Taylor, S. R. *Geochim. Cosmochim. Acta* **1964**, *28*, 1273-1285.
8. Schaule, B.; Patterson, C. C. *Earth Planet. Sci. Lett.* **1981**, *54*, 97-116.
9. Lantzy, R. L.; Mackenzie, F. T. *Geochim. Cosmochim. Acta* **1979**, *43*, 511-515.
10. Mosher, B. W.; Duce, R. A. *J. Geophys. Res.* **1983**, *88*, 6761-6768.
11. Miyake, Y.; Tsunogai, S. *J. Geophys. Res.* **1963**, *68*, 3989-3993.
12. Rasmussen, R. A.; Khalil, M. A. K.; Gunawardena, R.; Hoyt, S. D. *J. Geophys. Res.* **1982**, *84*, 3086-3090.
13. Jaenicke, R. In *Chemistry of the Unpolluted and Polluted Troposphere*; Georgii, H. W.; Jaeschke, W., Eds.; Reidel: Dordrecht, 1982; pp 341-373.
14. Weisel, C. P.; Duce, R. A.; Fasching, J. L.; Heaton, R. W. *J. Geophys. Res.* **1984**, *89*, 11607-11618.
15. Patterson, C. C., personal communication.
16. Slinn, S. A.; Slinn, W. G. N. *Atmos. Environ.* **1980**, *14*, 1013-1016.
17. Williams, R. M. *Atmos. Environ.* **1982**, *16*, 1933-1938.
18. Arimoto, R.; Duce, R. A. *J. Geophys. Res.* **1986**, *91*, 2787-2792.
19. Scott, B. C. In *Atmospheric Pollutants in Natural Waters*; Eisenreich, S. J., Ed.; Ann Arbor Science: Ann Arbor, MI, 1981; pp 3-21.
20. Gatz, D. F. In *Precipitation Scavenging (1974)*; Semonin, R. B.; Beadle, R. W., Eds.; Technical Information Center, Energy Research and Development Administration: Springfield, VA, 1977; pp 71-87.
21. Slinn, W. G. N. In *Air-Sea Exchange of Gases and Particles*; Liss, P. S.; Slinn, W. G. N., Eds.; North Atlantic Treaty Organization, Advanced Study Institute Series No. 108; Reidel: Dordrecht, 1983; pp 299-405.
22. Blanchard, D. C.; Syzdek, L. D. *J. Phys. Oceanogr.* **1975**, *2*, 255-262.
23. Blank, M.; Leinen, M.; Prospero, J. M. *Nature* **1985**, *314*, 84-86.
24. Bruland, K. W. *Earth Planet. Sci. Lett.* **1980**, *47*, 176-198.
25. Uematsu, M.; Duce, R. A.; Prospero, J. M. *J. Atmos. Chem.* **1985**, *3*, 123-138.
26. Duce, R. A. In *The Role of Air-Sea Exchange in Geochemical Cycling*; Buat-Menard, P., Ed.; North Atlantic Treaty Organization, Advanced Study Institute Series No. 185; Reidel: Dordrecht, 1983; pp 497-529.
27. Godin, P. M.; Feinberg, M. H.; Ducauze, C. *J. Environ. Pollut.* **1985**, *10*, 97-114.
28. Davidson, C. I. *Powder Technol.* **1977**, *18*, 117-125.
29. Eisenreich, S. J. *Water, Air, Soil Pollut.* **1980**, *13*, 287-301.
30. Buat-Menard, P.; Duce, R. A. *Scope Workshop on Metal Cycling in the Environment*, in press.

31. Gatz, D. F. *Water, Air, Soil Pollut.* **1975,** 5, 239–251.
32. Sievering, H. *Water, Air, Soil Pollut.* **1976,** 5, 309–318.
33. Talbot, R. W.; Andren, A. W. *J. Geophys. Res.* **1983,** 88, 6752–6760.
34. Monahan, E. C. *J. Atmos. Sci.* **1969,** 26, 1026–1029.

RECEIVED for review May 6, 1986. ACCEPTED August 4, 1986.

# WATER COLUMN PROCESSES

# Mechanisms Controlling the Distribution of Trace Elements in Oceans and Lakes

James W. Murray

School of Oceanography, University of Washington, Seattle, WA 98195

*Many of the same mechanisms that control the distribution of trace elements in the ocean are also important in lakes. Specifically, these include nutrientlike biological recycling, sediment fluxes, oxidation–reduction cycling, and scavenging by particles. The influence of each of these mechanisms can be seen in the trace element profiles of lakes despite the fact that lakes are intrinsically much more difficult to study than oceans. This difficulty arises because (1) lakes are not at steady state, and the magnitude of the controlling mechanisms varies with time; and (2) the large sediment-to-water-volume (sediment–volume) ratios, together with rapid horizontal mixing, result in sediment fluxes that tend to mask the other processes. Nevertheless, lakes are more accessible and in most cases easier to sample than the ocean. Because of the wide variety in types of lakes, isolating variables is possible by choosing the lake with the right properties. Examples of different processes are illustrated by using new trace element data from Lake Zurich and Lake Washington.*

F EW SYSTEMATIC STUDIES HAVE BEEN CONDUCTED on trace elements in the water columns of lakes. The cycles of iron and manganese, which are not always trace elements, have been investigated in some detail, but studies of other trace elements are rare. This scarcity of data is unfortunate because the study of trace elements in the ocean has been an important source of clues for understanding the physical, biological, and chemical processes. Trace metal analyses in lakes could be equally beneficial.

0065–2393/87/0216–0153$09.00/0

In the open ocean, dissolved trace element profiles are generally considered to be at steady state, although little verification exists for this assumption. The vertical and horizontal distributions reflect a combination of biological, physical, and geochemical processes. Integrated and systematic studies have been used to understand these processes; some of the best evidence has come from the shapes of the dissolved profiles. In the interior of the ocean (away from the boundaries), the rate of horizontal mixing is fast ($K_H = 10^8$ cm$^2$ s$^{-1}$) and is controlled by large, essentially boundless eddies. One-dimensional, vertical-advection diffusion models work in the ocean, and profiles can be fit to extract chemical information (e.g., production and consumption rates).

Chemical limnology has not yet reached the same state of sophistication. Essentially the same processes act to control trace element distributions in lakes and oceans; however, the concentrations in lakes are more dynamic. The shapes of the profiles are driven by time-dependent processes varying from days to seasons.

The Navier-Stokes equations are valid in lakes, but the advective and diffusive scales are very different and highly variable. The distinction between advective and turbulent mass transport is arbitrary in that it depends on the time and length scales of interest ($1$). Energy input depends on the size of the lakes; thus, mixing is more variable ($2$). The boundaries play a much more important role in lakes. Because of these complications, extracting chemical information by applying one-dimensional models to vertical profiles in lakes is much more difficult than with oceans. One-dimensional models can be used if the sediment-to-water-volume (sediment–volume) ratio is taken into account ($3$), but in many cases two- or three-dimensional models had to be developed ($4, 5$).

The main processes expected to be important in lakes are

- nutrientlike (stoichiometric) biological cycling;
- fluxes across the sediment–water interface;
- coupling between the redox variations of Mn, Fe, S, C, and other elements;
- scavenging by adsorption onto the sinking particles or directly at the sediment–water interface; and
- anthropogenic inputs.

In this review, a specific discussion of anthropogenic effects will not be included. Most studies of this nature have focused on the sedimentary record ($6$–$9$) or on the effect of lake acidification on trace element mobility ($10, 11$). Anthropogenic inputs are commonly received by the surface of lakes from atmospheric fallout or surface runoff. Once in lakes, these inputs are distributed by the other four mechanisms listed. Understanding the basic processes controlling the distribution of all elements in lakes is necessary to understand the distribution of pollutants.

The transient nature of lake processes adds extra variability, and it is impossible to assume steady state in most lakes. However, I predict that the variability can be exploited to establish the dynamics of the controlling processes and the biogeochemical cycles. The more rapid processes in lakes may also serve as models for the analogous but slower oceanic processes. That lakes may serve as test tubes for understanding oceanographic processes has been proposed (*12*). This idea may be valid, but by the time we take into account the temporal and spatial variations, the correct analogy might be more like test tubes of annually oscillating Liesegang rings, which are due to rhythmic precipitation.

In this chapter, the important mechanisms acting to control the distribution of trace elements in sea water will be reviewed with emphasis on how these processes are reflected by the water column profiles. A review of what is known about trace elements in lakes will be presented with emphasis on the processes listed previously. When possible, data from Lake Zurich and Lake Washington will be used to illustrate certain points.

## Trace Elements in Sea Water

Owing to their low concentrations, ease of contamination, and the complex matrix, trace elements in sea water are a challenge to analyze. Data prior to about 1975 were so unreliable that the principle of oceanographic consistency (*13*) was proposed as a guide for the acceptance of results. According to these guidelines, analytical results should form smooth vertical profiles; have correlations with other elements that share the same controlling mechanism; and be consistent with known physical, biological, and geochemical processes.

Most naturally occurring elements in sea water have now been analyzed, and at least one oceanographically consistent profile has been obtained for most of them. Most of these results were published after 1975 and benefited from major advances in the sensitivity of analytical equipment; the development of preconcentration techniques; and the increased awareness of elimination and control of contamination during the sampling, storage, and analytical steps. Recent reviews by Quinby-Hunt and Turekian (*14*) and Bruland (*15*) give good summaries of sea water concentrations and include the appropriate references for specific elements. Writing such summaries for lakes is not possible at present. The equilibrium distribution of dissolved components in inorganic sea water is summarized in Turner et al. (*16*).

## Description of Water Column Profiles

Many mechanisms and processes act to control the distribution of trace elements in sea water. One of the most successful approaches for understanding these processes has been to carefully study the shape of the water column profiles. Six main types of profiles in the open ocean reflect a slightly longer list of mechanisms (Table I). Most of the reactions take place at the phase discontinuities between the atmosphere, biosphere, lithosphere, and hydrosphere. Reviewing this list is important because most of these mechanisms should also be active to a greater or lesser extent in lakes.

**Table I. Characteristic Types of Element Profiles in the Ocean and Probable Controlling Mechanisms**

| Type of Profile | Elements and Compounds | Mechanisms |
|---|---|---|
| Salinity similarity | Na, K, Mg, $SO_4$, F, Br, U, Mo, Cs | Conservative elements of low chemical reactivity |
| Sea surface enrichment | $^{210}Pb$, Mn | Atmospheric input (natural) |
| | $^{90}Sr$, Pb | Atmospheric input (bomb tests or pollution) |
| | $H_2O_2$, NO, $Fe^{2+}$, $Cu^+$ | Photochemistry |
| | Mn, Cu | Shelf-sediment input |
| | $H_2$, $N_2O$, $As(CH_3)_2$ | Biological production |
| Photic-zone depletion with deep sea enrichment | Ca, Si, $CO_2$, $NO_3$, $PO_4$, Cu, Ni, Cd | Biological uptake and regeneration |
| Middepth maxima | | |
| ~3000 m | $^3He$, Mn, $Ch_4$, $^{222}Rn$ | Hydrothermal input |
| 200–1000 m | $^3H$ | Isopycnal transport |
| | Mn, $NO_2$, Fe | Redox chemistry in the oxygen minimum |
| Bottom-water enrichment | $^{222}Rn$, $^{228}Ra$, Mn, Si | Flux out of the sediments |
| Deep ocean depletion | $^{210}Pb$, $^{230}Th$, Cu | Scavenging by settling particles |

**Conservative Elements.** Some elements have very low reactivity in sea water with respect to the mixing time of the ocean (1000–1500 years). These conservative elements have constant ratios to each other in the ocean, and the shape of their profiles is similar to that of salinity. The list of conservative elements includes ions of some of the major species (Na, K, Mg, B, Cl, Br, $SO_4$, and F) as well as Li, U, Rb, Mo, Sb, Cs, and Tl. In most cases, the weak chemical reactivity reflects the speciation of these species. Most of these elements exist as singly charged ions (e.g., $Cs^+$, $Tl^+$, $Rb^+$, and $Li^+$) or as unreactive complex species [$(UO_2CO_3)_3^{4-}$, $MoO_4^{2-}$, and $Sb(OH)_6^-$].

Elements that are conservative in sea water cannot automatically be assumed to be conservative in lakes because of fundamental differences in the background electrolytes and the relative importance of removal mechanisms.

**Sea Surface Enrichment.**    Several elements, isotopes, or chemical species have elevated concentrations in the surface-mixed layer of the ocean. Several mechanisms can lead to this type of profile. Each of these mechanisms results in an input directly into the surface water. In many cases, determining the exact origin for a given element may be difficult. For example, the source of Mn can be from several of the mechanisms listed in this section. Distributions over horizontal and vertical spatial scales are required to unravel the specific origin.

Atmospheric inputs can result in sea surface enrichment (17). The input can be from natural sources (e.g., $^{210}Pb$, Sn, and possibly Cu), from pollution (e.g., stable Pb), or from atmospheric bomb tests (e.g., $^{90}Sr$). The input of aerosols from the atmosphere may be the major source of these elements. Hodge et al. (18) have found that substantial amounts of heavy metals associated with these aerosols are readily solubilized in sea water. Their calculations suggest that the aerosol fluxes may control the surface sea water concentrations of some elements, especially Pb, Zn, Cu, and Mn.

Sunlight-induced photochemical reactions can produce a wide variety of chemical species in the photic zone (19). Nonequilibrium species such as $H_2O_2$ (20) and NO (21) can be produced and lead to maxima at the sea surface and in the upper part of the photic zone. In addition, several studies suggest that the reduced species $Fe^{2+}$, $Mn^{2+}$, and $Cu^+$ are produced by photochemical reactions. In some cases, humic or fulvic substances may be involved. The photoreduction of $MnO_2$ appears to involve organic matter (22). In the case of Fe, the highly photoreactive species $FeOH^{2+}$ may be reduced by light at the surface of iron colloids (23). The reduction of Cu(II) to Cu(I) (24) may influence copper toxicity (25).

Biological processes appear to be the origin of the surface maximum for another group of chemical species. Specific examples include the reduced gases ($H_2$, $CH_4$, and $N_2O$), reduced elements [Cr(III), I(I), and As(III)], and methylated metals [$As(CH_3)_2$]. Frequently, the high concentrations of these reduced chemical species are associated with high values of $NO_2^-$ or $NH_4^+$. These associations have been used to infer a biological origin, but in most cases the specific origin is unclear (26). Reduced forms of arsenic are a notable exception. Andreae and Klumpp (27) found that significant amounts of arsenite and methyl and dimethyl arsenite are produced in pure cultures of marine phytoplankton. The chemical similarity between phosphate and arsenate suggests that these ions should compete for the phosphate uptake system of marine algae. Phytoplankton may attempt to detoxify arsenate by transforming it to arsenite.

Rivers, continental-shelf sediments, and pollution in the coastal zone have been proposed to be important sources of elements and chemical species to the ocean surface. After their input, they can mix horizontally or be transported in large eddies into the interior of the ocean leading to surface maxima. The tracer $^{228}$Ra is released by shelf sediments and has been used to calculate horizontal transport rates (28, 29). Manganese, beryllium, and copper maxima in some regions have been suggested to have this origin (30, 31). The isotope $^{137}$Cs, input into the Irish Sea by waste emissions from the Sellafield nuclear reprocessing plant, has been traced to the Greenland Sea (32).

The relative importance of these processes in lakes has not been investigated, although there is no reason to expect that they do not all play an important role.

**Nutrientlike Profiles.**    Many elements have profiles that resemble those of the nutrients and show photic-zone depletion with deep ocean enrichment (*see* references in reference 15). Some elements such as Cd, Sr, and As show a shallow regeneration like the labile species $PO_4$ and $NO_3$ and are hypothesized to be associated with the organic matter fraction of biological debris. The profiles of other elements like Ba, $^{226}$Ra, Cr, and Ge reflect deep regeneration like the elements in the shells of organisms (e.g., Si and Ca). Other species such as Ni, $IO_3^-$, Cu, and Se have nutrientlike profiles that reflect a combination of processes.

Many elements that show nutrientlike profiles have been shown to be essential trace elements (33). In other cases, most notably Cd, Ba, and Ag, no known biological role has been shown. In almost all cases, the relationship between the biogenic carrier phases and the trace elements is uncertain. Collier and Edmond (34), for example, found that the primary carrier phase for most elements is the nonskeletal organic fraction of the particulate material. Biogenic carbonate and silicate shells have a very low trace metal content. As a result, interpretation of dissolved-element correlations such as Ba with Si is extremely difficult.

In lakes with a strong seasonal biological cycle, seasonal variations may be seen in metal–nutrient relationships, especially during the early summer months. These relationships have not been extensively investigated to date.

**Midwater Maxima.**    Several mechanisms can produce a dissolved profile with a maximum at middepth in the ocean. These mechanisms can usually be resolved by looking at the depth of the maxima and the association with other elements. For example, hydrothermal circulation at midocean ridges is the largest single source of dissolved-Mn to the world oceans (35), and as a result, dissolved Mn profiles have a maximum at the depth of the ridge crest (2500–3000 m). $^3$He, $CH_4$, and $^{222}$Rn also have profiles that reflect hydrothermal input (36).

The other depth range where midwater maxima are observed is in the range of 200–1000 m. Horizontal input along isopycnal surfaces results in a $^3$H maximum of surface origin (37). Other elements, especially gases, that equilibrate with the atmosphere at the sea surface may also be influenced, but few well-documented examples are known [*see* Gammon et al. (38) for a discussion of the chlorofluoromethanes].

The most important process in this depth range is the consumption of dissolved oxygen to produce the oxygen minimum and the interesting redox chemistry associated with this zone. The reduced forms of several species such as Mn(II), Cr(III), $CH_4$, $NO_2^-$, and $N_2O$ show maxima at the oxygen minimum (39). The oxidized pair of these redox couples, such as Cr(VI) and $NO_3^-$, sometimes exhibits a minimum at the same depth.

Hydrothermal processes generally are not important in lakes (a notable exception is Lake Ratorura in New Zealand), and isopycnal transport from the surface has also not been shown to be important. But horizontal inputs from inflowing rivers can often result in distinctive features, especially during the summer months when lakes are thermally stratified. The river input does not mix uniformly into the surface layer but forms a discrete layer at the depth where the densities of the river and lake water are equal. These densities depend primarily on temperature and concentration of suspended matter (40). A maximum in vertical eddy diffusion coefficients is occasionally observed at this depth (5). Isopycnal transport is otherwise generally important in the hypolimnion of lakes because of the large area–volume ratios (4). The redox reactions that result from oxygen depletion are extremely important in the deep water of most lakes.

**Bottom-Water Enrichment.**    In the ocean, only a few elements have profiles that can clearly be attributed to a flux out of the sediments. The tracers $^{222}$Rn and $^{228}$Ra clearly have a sediment source, and their half-lives and weak chemical reactivity have led to their use to model vertical mixing in the deep ocean (41, 42). Dissolved Mn is input into the ocean from organic-rich hemipelagic sediments (39), and Cu appears to be input from a wide range of sediment types (43, 44). Silica input from biogenic sediments can be seen in the large-scale, bottom-water silica distributions (45).

Sediment fluxes are an important factor to consider in the chemistry of lakes because of the large sediment–volume ratios that increase with depth (46).

**Deep Ocean Depletion.**   Scavenging is defined as the adsorption and removal of dissolved elements from the water column by sinking particles. Scavenging is thought to be an important process in the ocean and leads to dissolved-element profiles characterized by progressive depletion with depth. In general, metal concentrations in sea water are too low for solubility equilibrium, and this scavenging mechanism is believed to be a major process (47, 48). The effect of scavenging is more pronounced for the more reactive elements (e.g., as reflected by the strength of the hydrolysis constants).

The depletion referred to above is sometimes difficult to define. For certain radioactive isotopes (e.g., $^{210}Pb$, $^{230}Th$, $^{234}Th$, and $^{231}Pa$), secular equilibrium should exist, and as a result, depletion can be defined with reference to the parent isotopes (e.g., $^{226}Ra$, $^{234}U$, $^{238}U$, and $^{235}U$) (49, 50). For other elements, such as Cu, comparison has to be made to a hypothetical profile that would exist if no scavenging took place (43). The conclusions in this case are model-dependent. Collier and Edmond (34) have shown a strong correlation between the deep ocean scavenging rate of Cu and the primary productivity at the sea surface. This correlation implicates particulate organic matter as the scavenging agent.

Scavenging is thought to be an important process in lakes, especially because of the well-developed maxima of iron and manganese oxides that often exist. Most evidence for this mechanism has come from study of the particulate involved rather than from the dissolved phase.

**Near-Shore Profiles.**   The discussion to this point has emphasized trace element profiles in open ocean sea water. In some near-shore locations, where the circulation is restricted, dissolved oxygen is consumed more rapidly than it is replenished, and anoxic conditions exist. Reduction of sulfate results in significant concentrations of sulfide. Notable examples include the Black Sea, Cariaco Trench, Saanich Inlet, and Framvaren Fjord. These locations have been studied in detail and are excellent examples of dissolved trace element profiles across the oxygen–hydrogen sulfide interface. Two particularly significant and relevant processes for our later discussion of lakes are (1) the interface cycling occurring between metal oxides and their reduced, dissolved counterparts and (2) the precipitation of metal sulfides in the presence of dissolved sulfide.

The qualitative behavior of trace elements across the redox front can be divided into two general categories, transition and class B metals, depending on the free-ion, outer-electron shell configuration and deformability (*51*). The class B metals (Cu, Zn, and Cd) tend to form less soluble sulfides and much stronger complexes with sulfidic ligands than the transition metals (Mn, Fe, and Co). Transition metals do not form strong complexes with reduced-sulfur species and tend to have dissolved maxima just below the interface. In the deep water, the dissolved-metal concentrations decrease with increasing sulfide because of solid metal sulfide formation. For the class B metals, the situation is more complicated because these metals tend to form complexes with sulfidic ligands. Their solid metal sulfides are also less soluble. Thus, the dissolved concentration reflects the competition between sulfide complexation and precipitation reactions. In general, for class B metals the total soluble metal increases as the total sulfide decreases.

The classic profiles for Fe and Mn from the Black Sea (*52, 53*) show that both dissolved elements increase rapidly below the oxygen zero boundary. Particulate maxima for both elements are present just above this interface and result from the oxidation of the upward flux of the reduced, dissolved forms of both elements. Similar pairs of profiles have been reported for Saanich Inlet (*54*), Framvaren Fjord (*55*), and the Cariaco Trench (*56*). The best data for the trace metals Cu, Cd, Zn, Ni, and Co are from Saanich Inlet and Framvaren Fjord. Cobalt tends to be influenced by manganese. Nickel is remarkably uninfluenced by the presence of sulfide. Copper, zinc, and cadmium decrease rapidly to low concentrations presumably because of the formation of insoluble metal sulfides.

The iron and manganese cycles have been thoroughly studied in lakes, and seasonal variability is a characteristic feature. The sulfate concentration is generally much lower in lakes than in the ocean; however, insoluble sulfide formation can be an important process in some cases and cannot be automatically neglected.

## Trace Elements in Lakes

Except for Fe and Mn, which have been studied in great detail, few studies have been done of either the distribution or the controlling mechanisms of trace elements in lakes. Most of the studies that have been conducted focused on analyses of suspended matter, sediment trap material, or sediments. For most trace elements, no vertical profiles exist, let alone information on the seasonal variations. Given the recent history of trace element analyses in sea water where dissolved profiles have been used to great advantage, a comparable set of lake water data should be equally beneficial.

The processes that control the distributions of trace elements in sea water were summarized in Table I. With the exception of hydrothermal input, all of these processes are potentially important in lakes. For some of the mechanisms, such as the photochemical reactions, virtually nothing is known. A schematic representation of the main mechanisms operating in lakes is shown in Figure 1. In this section, four mechanisms will be reviewed. These mechanisms are biologically related stoichiometric metal–nutrient relationships, sediment or benthic fluxes, oxidation–reduction reactions of Mn and Fe, and scavenging by solid phases.

**Metal–Nutrient Relationships.** Some of the same biochemical processes probably occur in oceanic and lake waters although the time and depth scales should be different. Earlier, I reviewed the correlations that exist for many trace elements with nutrient elements (e.g., P and Si) in sea water. Morel and Hudson (57) proposed that the Redfield model for the biological control of algal nutrients in the sea can be extended to the trace elements in lakes by using stoichiometric formulas such as $C_{106}H_{263}O_{110}N_{16}P_1$ (Fe, Zn, and Mn)$_{0.01}$(Cu, Cd, Ni, etc.)$_{0.001}$. This extension of the Redfield model is only a hypothesis at present because some of the elements that have shown nutrient relationships have no clear biological role.

Strong evidence from several studies indicates that the removal of trace elements is strongly influenced by biological processes, specifically uptake by plankton and subsequent transport by biological debris. As part of the MELIMEX limnocorral studies in Lake Baldegg, Baccini et al. (58) used trace element, mass-balance calculations to show that phytoplankton and periphyton are the main factors responsible for metal removal. The residence times decreased according to the sequence Zn > Cu > Cd > Pb = Hg. Santschi and Schindler (59) also observed that Zn removal was more extensive than Cu removal in Lake Biel. The residence time of the elements could be predicted if the primary productivity, settling rates, and hydraulic properties were known.

Hamilton-Taylor et al. (60) found that the sequence of algae caught in sediment traps closely followed the succession in the lake. The most abundant genus, *Asterionella*, has especially high metal content. In particular, Cu appears to be taken up preferentially over other metals and not rapidly recycled. As a result, Cu is quickly removed from the surface waters early in the growing season, and dissolved Cu decreases from its winter concentration of about 100 nM to summer values of about 20 nM. Talbot and Andren (61) found that the naturally produced radioactive tracers [210]Pb and [210]Po were rapidly removed from Crystal Lake by biological activity. The residence time of [210]Pb was 0.095 years and that of [210]Po was 0.26 years, apparently because [210]Po is more readily recycled from the organic matter. Sholkovitz and Copland (62) studied the elemental composition of suspended matter in Esthwaite Water. Phytoplankton production resulted in large excess particulate concentrations of P, S, Mg, Ca, Ba, and K in the epilimnion. These

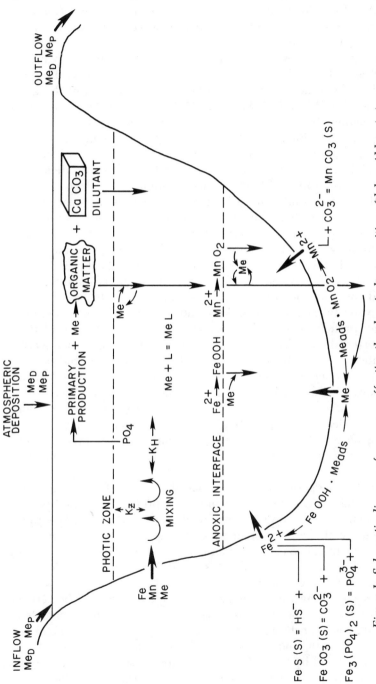

Figure 1. *Schematic diagram of processes affecting the chemical composition of lakes. Abbreviations are as follows:* $K_Z$ *and* $K_H$ *represent the vertical and horizontal eddy diffusion coefficients, respectively;* $Me_D$ *and* $Me_P$ *represent the dissolved and particulate forms of the trace metals, respectively; and* L *represents complexing ligands.*

excesses follow the concentration of particulate organic carbon and the population of the dominant summer genus, the dinoflagellate *Ceratium*. Interestingly, some elements that are conservative in sea water (e.g., K, Mg, and Ca) may be nonconservative in lakes.

Subsequent studies of the particulate Ba in Esthwaite Water by Finlay et al. (63) have added a new dimension to the biological uptake mechanism. They determined that a freshwater protozoa of the genus *Loxodes* accumulates $BaSO_4$ as discrete, ultracellular, spherical granules. The Ba accumulation by these protozoa is sufficient to account for most of the particulate Ba in the lake. What other trace metals may be involved in similar cycles is not yet known, but this example suggests that the biological influences on trace elements in lakes may depend on the species composition of the biological community.

Sediment trap studies have been one of the main approaches used to study the seasonality and stoichiometry of the metal–nutrient relationships. Giovanoli et al. (64) used sediment trap samples from Lake Zurich to show that the composition of the settling organic matter $[(CH_2O)_{114}(NH_3)_{16}(H_3PO_4)]$ is very close to the theoretical ratio for bulk plankton $[(CH_2O)_{106}(NH_3)_{16}(H_3PO_4)]$ (65). Sigg et al. (66) and Sigg (67) extended the concept of constant algal composition to heavy metals. She analyzed the major composition and trace metal contents of sediment trap particles from Lake Constance, established the correlations, and reported the results as ratios to P. The results suggest an average composition of $C_{113}N_{15}P$ $Cu_{0.008}$ $Zn_{0.06}Pb_{0.004}Cd_{0.00005}$. Further studies of the same type conducted in Lake Zurich (68) also resulted in significant correlations of Cu and Zn with P in sediment trap material. These studies suggest a stoichiometry of $C_{97}N_{16}P$ $Cu_{0.006}Zn_{0.03}$. Correlations with P were not clearly seen for Pb, Cd, or Cr. In many of these studies, $CaCO_3$ is an important component of sediment trap samples but acts almost exclusively as a diluent from the point of view of metal concentrations.

Sigg (67) suggested two ways that algal interactions could result in metal uptake with constant stoichiometric proportions. The first way is that metal ions may be taken up by algae through surface coordination reactions. If the number of sites on algal surfaces is proportional to the P content of the cells, then an increase in algal cells, reflected by P, will result in increased metal uptake, and the metal–phosphorus ratio will remain roughly constant.

The second way for the uptake is that metal ions may be actively taken up by cells either to satisfy their micronutrient requirements or because they are substituted for micronutrients. In this case, metal ions can colimit growth, and the concentration of the metal ions, in turn, could be regulated by the cells. This regulation results in constant ratios of trace metals to essential nutrients like $PO_4$. Furthermore, the metal-phosphorus ratio would most likely be a function of the solution composition and type of organisms characteristic to a lake. As a result, the metal–phosphorus ratio in algae may be more variable from lake to lake.

Following trace element uptake in the photic zone, algae sink and decompose. This decomposition releases the elements back to the water column. An implicit assumption in the constant stoichiometric approach is that trace element regeneration occurs without fractionation or in the same proportions as originally taken up.

Until now, the biological, constant-stoichiometric-proportion model has not been clearly seen in the profiles of metals in lakes. Good reasons exist to expect difficulty in seeing such clear-cut examples of metal–nutrient relations in lakes as seen in the ocean (*67*). For example:

1. Because the depth range is greatly compressed, lakes may be less sensitive to such vertical processes than the sea.

2. These regeneration effects may be blurred or obscured by other factors such as Fe and Mn recycling and atmospheric or sediment inputs.

3. Metals released by regeneration may be more efficiently scavenged in lakes than in the ocean (more about this later).

4. Most importantly, the changes due to nutrient and trace element regeneration are much more difficult to identify in lakes because compared to the ocean, lakes are nonstationary, highly dynamic systems.

Despite these reservations, we observed examples of metal–nutrient relationships in our work. We recently analyzed water samples from Lake Zurich and Lake Washington by using trace metal clean techniques. These analyses were on unfiltered samples stored at pH < 2 and analyzed by an APDC preconcentration step and atomic absorption spectrophotometry (*69, 70*). Thus, these analyses represent total dissolvable metal. The results for Cd and $PO_4$ are shown in Figure 2 for Lake Zurich (July 1983) and in Figure 3 for Lake Washington (June 1985). In these examples, both elements increase with depth below about 20 m. The resulting Cd–$PO_4$ correlations were $1.4 \times 10^{-5}$ mol Cd/mol $PO_4$ ($r$ = 0.93) for Lake Zurich and $7.0 \times 10^{-5}$ mol Cd/mol $PO_4$ for Lake Washington. By comparison, the molar ratio observed in the central North Pacific is $3.5 \times 10^{-4}$ mol Cd/mol $PO_4$ (*71*) and the ratio in marine plankton ranges from 4 to $7 \times 10^{-4}$ mol Cd/mol $PO_4$ (*34*).

Proper interpretation of these profiles and molar relations is difficult. To begin with, if these molar ratios truly correspond to a biogenic mechanism, then the resulting increase in Cd profile below 20 m is

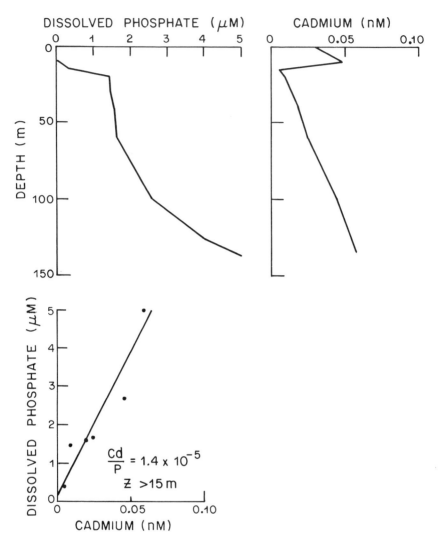

*Figure 2. Total dissolvable cadmium and dissolved-phosphate data from Lake Zurich. Samples were collected in July 1983. The relationship between Cd and PO$_4$ is shown for samples from depths greater than 15 m.*

small, only 15–70 pM/$\mu$M PO$_4$. In addition, the PO$_4$ profiles in lakes change progressively with time. Such relations could only be expected to be seen in the summer months when the PO$_4$ concentration in the hypolimnion has increased sufficiently so that the resulting increase in Cd can be detected. During the winter in most lakes, Cd and PO$_4$ are uniformly distributed by mixing. Unfortunately, in the summer the mechanism of scavenging by recycled Mn and Fe may remove Cd from

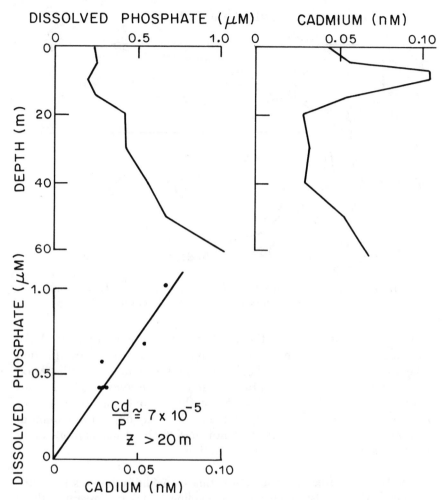

*Figure 3. Cadmium and phosphate data from Lake Washington. Samples were collected in June 1985. The relationship between Cd and PO₄ is shown for samples from depths greater than 20 m.*

the water column and mask the Cd-PO$_4$ relationship. Thus, the time window for observing such relationships in lakes may be limited. In oligotrophic Lake Washington, extensive Mn–Fe recycling does not take place in the water column, and the increase of Cd in the deep water can be seen late into the summer (Figure 4). The origin of the increase in Cd (and PO$_4$) is probably due to remineralization of biological particles at or near the sediment-water interface within the hypolimnion.

Another feature that probably has a biological origin, although the

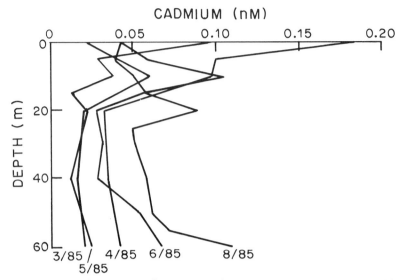

*Figure 4. Cadmium profiles from Lake Washington in 1984. Monthly profiles are shown from March to August.*

exact cause is unclear, is the shallow oxygen minimum. In Lake Washington, this oxygen minimum first appears after stratification develops; it intensifies and "shoals" through the summer months. Maxima in Mn, Fe, Ni, and possibly Cu and Cd are associated with this feature. The data for August 1985 are shown in Figure 5. Sigg (67) observed a similar maximum for Mn and Fe in the oxygen minimum at 20 m in Lake Constance. The origin of these features needs more study but may be due to zooplankton respiration.

**Sediment Fluxes.**    The return flux from the sediments is an important aspect of the trace element cycling in lakes. Several authors suggested that, in general, the concentrations of most elements in lakes are a function of the fluxes from the sediments, horizontal and vertical transport, and chemical transformations in the water. Numerical models have been constructed to demonstrate these effects (72, 73). In many cases, the concentration in any layer in a lake appears to be related to the area of sediment contact per volume of water for that layer (74). This ratio can increase with depth by as much as a factor of 10 in some lakes (73), and an increase in the importance of the absolute sediment flux results.

From sediment trap studies, biological material and iron oxides have been shown to be the main carrier phases for tracer elements in lakes (62, 68). Manganese oxide may also play an important role (67). All of these carrier phases are very reactive at the sediment–water

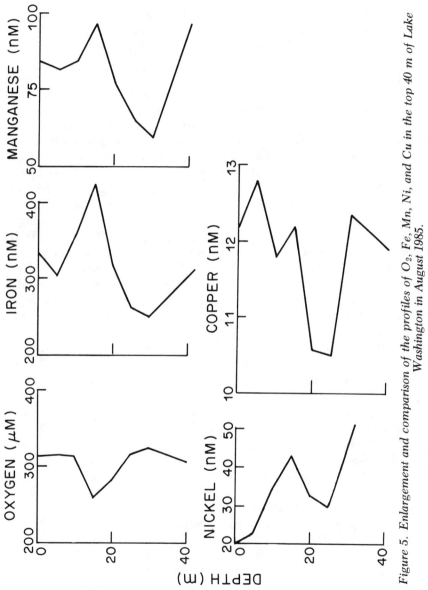

*Figure 5. Enlargement and comparison of the profiles of $O_2$, Fe, Mn, Ni, and Cu in the top 40 m of Lake Washington in August 1985.*

interface, and significant trace element recycling occurs before burial. Additional reactions occur after burial and release trace elements to the pore water. Transport from the sediments occurs mostly as molecular diffusion (75). In general, the effects of bioturbation and irrigation have not been shown to enhance the fluxes in lakes as extensively as seen in coastal marine locations (76). However, enhanced mixing of pore waters to depths as deep as 10 cm due to convective overturn may occur during the fall (75).

Sediment fluxes have been estimated for Fe and Mn by several different techniques. The simplest approach is to calculate the increase in the inventory of excess metal in the hypolimnion from a time series of profiles. The assumption is made that the metal increase in the hypolimnion is entirely due to the sediment flux. Using this approach, Davison et al. (77) calculated that under anoxic conditions, the flux of Fe from sediments in Esthwaite Water is $0.1-1.0 \times 10^{-3}$ mol m$^{-2}$ day$^{-1}$. Davison and Woof (78) calculated that the apparent sediment flux of Mn in Rostherne Mere was $1.2-2.3 \times 10^{-3}$ mol m$^{-2}$ day$^{-1}$.

A second approach is based on sediment trap analyses. Hamilton-Taylor et al. (60) calculated the ratio of the accumulation in a deep sediment trap with the accumulation rate in the underlying sediments. The ratios for Fe (1.1), Pb (1.4), and Zn (1.5) are close to 1. This closeness implies that recycling is small. In addition, Fe and Pb correlate with Al, and this correlation implies an association with detrital material. The ratios for Mn (8.6) and Cu (9.5) are much greater than 1; these values indicate large-scale recycling, probably in a flocculant layer at the sediment–water interface. Interestingly, the recycling of Mn apparently occurs in June and July, while that of Cu occurs in March and April. Apparently, this recycling is controlled by independent processes that have not been resolved. In another sediment trap study, Sigg et al. (68) found that the flux of Mn at 130 m was 50 times larger than the flux at 50 m. During this time interval (before May), no net increase in the Mn inventory occurred in the water column. Thus, the sediment trap flux of about $1 \times 10^{-3}$ mol m$^{-2}$ day$^{-1}$ should correspond to the flux from the sediments. After May, Mn begins to accumulate in the bottom water. Our data for total dissolved Mn in Lake Zurich from July to September 1983 are shown in Figure 6. Analyses were conducted by direct injection atomic absorption spectrophotometry. The integrated increase in total Mn corresponds to a sediment flux of $1.3 \times 10^{-3}$ mol m$^{-2}$ day$^{-1}$. This value is in good agreement with the estimate of Sigg et al. given previously. In the spring and early summer (ie., before bottom-water anoxia), the Mn flux from the sediments is rapidly returned to the sediments. When the anoxic boundary moves into the water column, the Mn flux from the sediments accumulates in the water column.

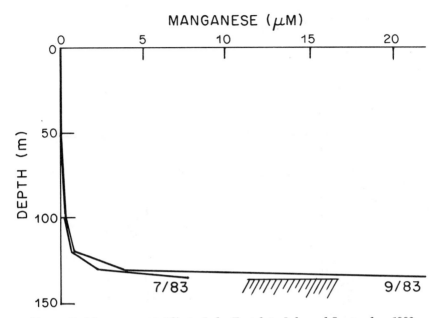

Figure 6. *Manganese profiles in Lake Zurich in July and September 1983.*

The Lake Washington Mn profiles from February to August 1985 are shown in Figure 7. The profiles are uniform and have a slight increase with depth in the winter (e.g., February and March 1985). In April, May, and especially June, Mn increases in the bottom layer below 40 m. In the midwater column (10–40 m), Mn decreases relative to the winter reference profiles. The highest concentrations reached in Lake Washington are an order of magnitude less than those in Lake Zurich.

We can use the increase in Mn in the deep layers to estimate the flux of Mn from the sediments. Using estimates of the area of the lake at different depths (J. Lehman, unpublished data), we calculated the volume of the lake at different depth intervals. From the area of a given depth horizon and the increase in Mn concentration, we calculated the apparent flux. For the water in Lake Washington below 50 m, the required flux of Mn is $1.12 \times 10^{-4}$ mol m$^{-2}$ day$^{-1}$.

We also calculated the sediment flux by two additional independent techniques. In the first technique, Mn increases in the pore water (Figure 8) by an average value of 74 $\mu$M in the top 1 cm. Assuming a porosity ($\phi$) of 0.90 and an in situ diffusion coefficient ($D$) of $3.3 \times 10^{-6}$ cm$^2$ s$^{-1}$ (K. M. Kuivila, personal communication), the diffusive flux out of the sediments ($J_{sed}$) can be calculated by using the equation

$$J_{sed} = \phi\, D\, \delta C/\delta z \qquad (1)$$

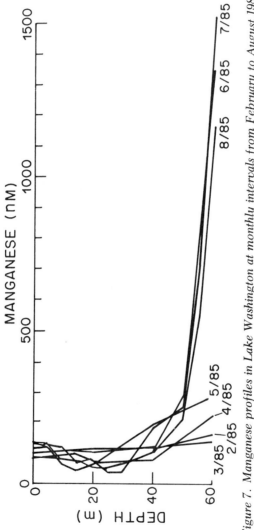

*Figure 7. Manganese profiles in Lake Washington at monthly intervals from February to August 1985.*

where $\delta C/\delta z$ is the concentration gradient. For these conditions, the flux is $1.86 \times 10^{-4}$ mol m$^{-2}$ day$^{-1}$. This value is a lower limit because the actual gradient may be larger.

For the second technique, we developed a bottom lander with two flux chambers for measuring sediment fluxes directly. During one deployment in Lake Washington in February 1982, we analyzed the samples for Mn and calculated a sediment flux of $3.7 \times 10^{-4}$ mol m$^{-2}$ day$^{-1}$ (Figure 8). This chamber flux is about twice as large as the diffusive flux. This size difference may be due to underestimation of the pore water gradient, but may also suggest an enhanced flux due to a fluff layer at the sediment–water interface or to nonmolecular transport attributable to tube worms located at the interface.

The differences between the sediment fluxes estimated by three independent techniques agree within a factor of 3. This agreement is the best expected from such a comparison. More studies are needed on this subject, however, because the sediment flux would be expected to vary over the seasons as the flux of organic carbon and the depth of the anoxic layer vary.

**Redox Coupling.**    Oxidation–reduction cycles are important components of the biogeochemistry of many lake waters and sediments. An important coupling occurs between these redox reactions and many other inorganic and organic elements and compounds. The redox cycles of Fe and Mn are probably the most important in this regard. The $SO_4$–$HS^-$ couple probably also plays an important role in most lakes, especially with regard to sedimentary diagenesis. Even in soft water lakes, the low sulfate concentrations (50–100 $\mu M$) are usually higher than the water column concentrations of Mn and Fe. All of these cycles vary seasonally, especially in lakes with annually developed anoxic hypolimnia. Extensive reviews have been published on this subject (*67, 79, 80*).

In virtually all lakes, dissolved oxygen is completely consumed in the sediments (*81, 82*). Anoxic conditions also exist on a seasonal or a permanent basis in the bottom water of lakes that have appropriate morphology and biological productivity. The percentage of lakes that have these conditions is actually quite small. The layer where dissolved oxygen approaches zero is called the anoxic interface. After oxygen is consumed, the oxidized Mn(IV) and Fe(III) oxides are reduced to soluble Mn(II) and Fe(II). The upward transport of the Mn(II) and Fe(II) result in their reoxidation to particulate forms, which then settle into the anoxic zone to repeat the cycle. As a result, these cycles have been referred to as the "ferrous wheel" (*83*) and "manganous wheel" (*84*). A schematic model of the generalized Mn and Fe cycles is shown in Figure 9. Depending on the nutrient and productivity levels and the degree and duration of stratification, the anoxic interface will be either

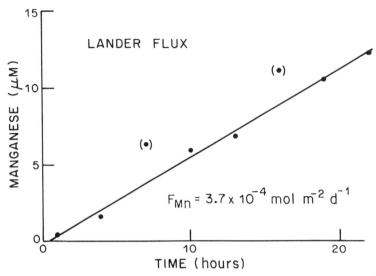

*Figure 8a. The increase in manganese with time in a chamber on the University of Washington tripod lander. The calculated flux from the sediments was $3.7 \times 10^{-4}$ mol $m^{-2}$ $day^{-1}$.*

in the sediments or in the water column. Because lakes are dynamic systems, the location of the anoxic interface can vary with time.

As a result of these cycles around the anoxic interface, the residence times of these elements can be significantly longer than the residence time of water in the lake. Campbell and Torgersen (83) found in a meromictic lake that the residence time of water was 3.5 years and that of iron was 15.3 years. As a result of iron recycling, the monimolimnion is more efficient at retaining iron.

Two important differences in the redox chemistries of Mn and Fe influence their distribution about the anoxic interface (80). The first difference is that Mn(IV) oxides can be reduced to dissolved Mn(II) at higher redox potentials than Fe(III) oxides can be reduced to dissolved Fe(II) (85). As a result, in some lakes where the anoxic interface is in the water column, soluble Mn is supplied almost exclusively from in situ reduction in the water column while soluble Fe is supplied by reduction in the sediments (77, 86).

The second difference is that the oxidation of dissolved Fe(II) to particulate Fe(III) oxides occurs much more rapidly than that of dissolved Mn(II) to Mn(III, IV) oxides. As a result, observing a vertical displacement in the particulate maxima of Fe and Mn is possible. Ferrous oxidation occurs very rapidly in the presence of dissolved oxygen (85). The ferric oxide formed from this oxidation is poorly crystalline, contains a large amount of organic carbon, and is negatively

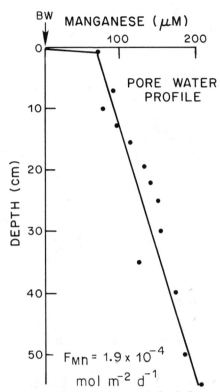

*Figure 8b. Manganese in the pore water of the sediments of Lake Washington. From the gradient across the sediment-water interface, we calculated a flux of $1.9 \times 10^{-4}$ mol $m^{-2}$ day$^{-1}$.*

charged (87). Manganous oxidation, on the other hand, is known to be very slow at the pH value of lakes in the absence of suitable catalysts (88, 89). The observed rates of oxidation in natural waters appear to be much faster; microbial catalysis is thought to be the main cause (90, 91). The controversial role of *Metallogenium* as a Mn-oxidizing organism has yet to be established (92).

As a result of these two factors, Mn and Fe exhibit different time and depth distributions in lakes. A good example of the situation in which the anoxic interface leaves the sediments and goes into the water column during the summer (Esthwaite Water) is described by Sholkovitz and Copland (62). Lake Washington is an example where the anoxic interface remains in the sediments all year. Our data for the seasonal distribution of total dissolvable Mn and Fe in the water column shows that the separation of these two elements can still be clearly seen. (Figure 10). For these elements, the bulk of the water column concentrations are in the particulate form.

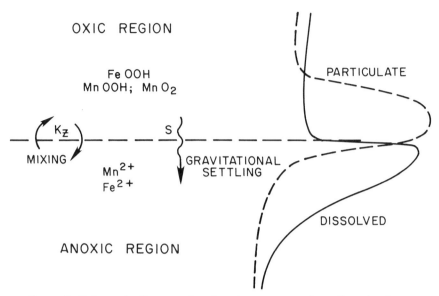

*Figure 9. Schematic diagram showing the processes that affect the distribution of the redox-sensitive elements Fe and Mn across the oxic–anoxic interface in lakes. Abbreviations are as follows: $K_Z$ represents vertical mixing, and S represents the gravitational settling rate of the particles.*

The flux and type of organic carbon and the formation of iron sulfides are additional factors that play important roles in the redox cycles. In many lakes, the Mn remobilization tends to coincide with the time of algal blooms (60, 77); this remobilization is, in fact, driven by the flux of organic carbon to the sediments. Recent results have challenged the classic picture that electron acceptors are reduced in the well-ordered sequence of $O_2$, $MnO_2$, $FeOOH$, $SO_4$, and $CO_2$. In Rostherne Mere, Davison et al. (77) observed that Mn(II) builds up to high concentrations, and no evidence has been presented for the buildup of dissolved iron despite the observation of a large flux of organic carbon to the sediments. They initially argued that when the $SO_4$ concentrations are relatively high, sulfate reduction in the sediments can result in the iron being trapped in the sediments as iron sulfide. As a result, little iron is released from the sediments, and the lake appears poised at Mn reduction. Iron sulfide can form in most lakes and has been documented (93). Pyrite apparently can form directly from the low concentrations of ferrous and sulfide ions (94). In subsequent studies, Davison et al. (95) argued that the system may be much more complicated. In Rostherne Mere, iron and sulfate are simply bypassed as electron acceptors in favor of $CO_2$ reduction to methane. A large fraction of the flux of organic carbon is made of species of dinoflagellates and cyanobacteria that are not readily decomposed in the water column or at the sediment–water interface. Apparently, organic matter decomposition is displaced to deeper layers in the sediments. The reasons why Fe and $SO_4$ are bypassed in favor of methanogenesis is still unclear.

The sulfur cycle can also be involved in photochemical reactions under certain conditions. In shallow, stratified lakes with anoxic hypolimnia [e.g., Priest Pot ($<4$ m)], light can penetrate into the hypolimnion and support photosynthetic bacteria, which oxidize sulfide to sulfate (96). Under these conditions, iron sulfide does not form in the water column, and a microbially mediated photochemical oxidation completes the sulfur cycle.

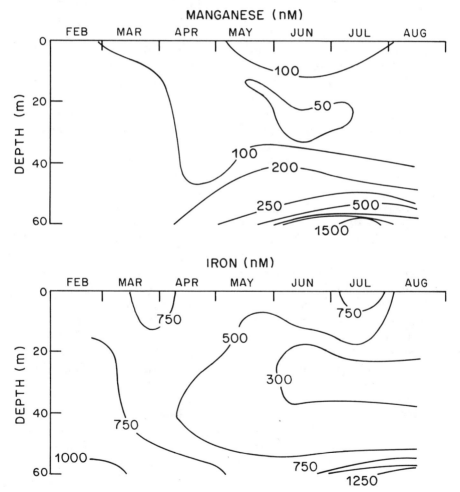

Figure 10. Variations in Mn and Fe in Lake Washington during 1985.

**Scavenging.** Adsorption and removal by particles is thought to be a significant removal process in lakes because of the high concentrations and fluxes of particles. As pointed out by Morel and Hudson (57), however, the evidence for scavenging must be viewed critically because separating the effects due to scavenging from active (stoichiometric) uptake by phytoplankton can be difficult. The ocean is deep, and the importance of scavenging can be clearly seen from the progressive enrichment in sediment traps or depletion relative to parent isotopes with depth. Most lakes are not deep enough for such trends to be clearly identifiable; thus, other approaches need to be used.

Despite these complications, the natural distributions in lakes have been used to show the importance of scavenging. Baccini and Joller (97) studied the mechanisms controlling the distribution of Cu and Zn in Lake Baldegg. They found that both elements were adsorbed by the organic matter produced in the epilimnion. Copper was adsorbed more strongly than zinc, and this result is opposite to Baccini et al. (58) MELIMEX results and Santschi and Schindler's (59) observations in Lake Biel. Some fraction of both elements was remineralized in the hypolimnion. Additional Cu was removed by iron oxides that formed above the anoxic interface. As much as 50% of the Cu transported to the hypolimnion is associated with this particulate iron. By comparison, Zn is almost entirely removed by particulate organic matter. Biogenic $CaCO_3$ is produced in large quantities in the lake but apparently is not an important scavenging phase. In addition, no correlation of particulate Cu or Zn with particulate Mn was found. This result surprisingly suggests that manganese oxides are also not significant scavengers.

Many of the same conclusions about scavenging were reached by Sholkovitz and Copland (62) and Sholkovitz (80) after a study of Esthwaite Water. They found that particulate organic carbon was the master variable for controlling the elements P, S, Mg, Ba, Ca, and K in the epilimnion. In the hypolimnion, Fe is the master variable for these elements and particulate organic carbon as well. This finding suggests two possible explanations: (1) these elements and organic carbon may adsorb directly on newly formed iron oxides, or (2) organic matter adsorbs and provides the link for adsorbing the other elements. Again, particulate Mn has very different time and depth variations and does not appear to play a major role for scavenging these elements.

A manipulative type of whole-lake experiment has been used to demonstrate that adsorption and settling particles are both important processes (98). Lake 224, in the Experimental Lakes Area in north-western Ontario, was spiked with six $\gamma$-emitting isotopes: [75]Se, [203]Hg, [134]Cs, [59]Fe, [65]Zn, and [60]Co. All of these isotopes decreased exponentially with time, as would be expected for first-order, concentration-dependent processes. The removal half times ranged from 14 days for Hg to 52 days for Se. The exact removal pathway appeared to depend on the affinity of the respective elements for suspended particulate matter. The elements that had a strong affinity for particles (e.g., Fe and Co) tended to be removed by the settling particles and thus accumulated in the sediments at the same rate everywhere. Elements such as Cs that are mostly dissolved tended to stay in the epilimnion and were apparently removed by adsorption directly onto the shallow sediments.

Although rate constants for whole-lake removal were obtained, the large-scale nature of the experiment made sufficient constraint of the removal mechanisms difficult. For this reason, a large number of additional radiotracer experiments were conducted in enclosures set up in Experimental Lakes 302S and 114 (*99*). These experiments lend further support to the argument that two main removal mechanisms are responsible for trace elements from the water column of shallow lakes.

The first removal mechanism is that particle-reactive elements, like Sn, Fe, Co, and Mn, adsorb to, or are taken up by, suspended particles (mostly plankton in these experiments) and subsequently settle to the bottom of the lakes. The second removal mechanism is that more soluble elements, like Cs and Zn, have an intermediate affinity for adsorption on biogenic particles. These elements appear to be removed by direct adsorption to sediments after molecular diffusion through the diffusive sublayer overlying the sediments. Penetration into the sediments is aided by physical mixing processes, biological mixing processes, or both processes in the surface sediments.

In addition to the field studies just reviewed, experimental and theoretical studies of scavenging from lakes have been done. Tessier et al. (*100*) used metal analyses of pore water and selective extraction techniques to calculate apparent in situ adsorption equilibrium constants for the oxidizing sediments of an oligotrophic lake. These equilibrium constants were compared with laboratory determinations of these constants on synthetic iron oxides. They concluded that the agreement between the lab and field data is good for Cd, Ni, and Zn. This agreement supports the argument that adsorption on iron oxides is the major removal process.

Sigg et al. (*101*) argued that leaching procedures yield arbitrarily defined fractions and that a sounder approach for modeling purposes is to calculate partition coefficients from total analyses of particulate matter or sediment trap samples. Sigg used this approach with some success in Lake Constance (*67*) and Lake Zurich (*68*). In both cases, trace elements were concluded to be mainly associated with organic matter. Even though the data in hand did not allow them to distinguish between binding to surface sites (adsorption) and active uptake by cells, the researchers suggested that the data could be reported in the form of a Redfield-type stoichiometry as described earlier.

In addition, Fe and Pb have relatively large distribution coefficients, and as a result their residence times are about the same as the residence time of particles. The residence times of Cd, Zn, and Cu are longer than the residence times of particles.

We have one set of data that indicates that the effect of scavenging can be seen in the water column profiles. We analyzed total dissolvable Mn, Fe, Cu, Ni, and Cd from one station in Lake Zurich in July and September 1983 (Figure 11). Between these two dates, a large increase

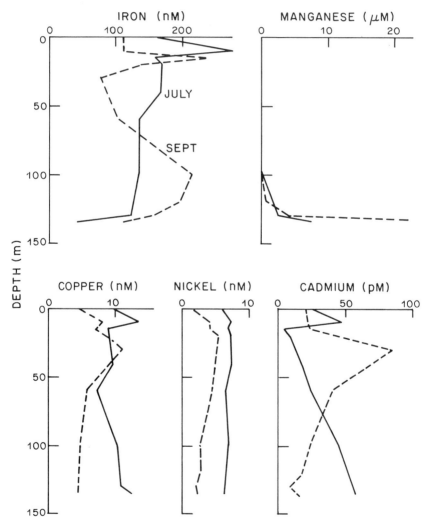

Figure 11. Profiles of Fe, Mn, Cu, Ni, and Cd in Lake Zurich in July and September 1983. The change in the profiles of Cu, Ni, and Cd from July to September can be interpreted as scavenging from the water column.

occurred in the concentrations of Fe and Mn in the hypolimnion. The profiles of Cu, Ni, and Cd increase with depth in the hypolimnion in July and then decrease with depth (<30 m) in September. From this data, we suggest the hypothesis that in lakes of this type, metal–nutrient regenerative-type profiles can be seen in the early summer. By late summer and early fall, when the anoxic interface moves into the water column, the effects of scavenging tend to dominate the profiles. This data set is very suggestive, but more detailed time series in several lakes with different hypolimnetic characteristics is clearly needed.

## Acknowledgments

This paper could not have been written without the assistance of W. Stumm and L. Sigg. They helped collect the necessary samples from Lake Zurich and provided the ancillary data used to illustrate several important points in this chapter. Valuable comments were obtained from R. F. Anderson, P. Santschi, W. Davison, L. Sigg, D. Imboden, and the American Chemical Society reviewers. B. Paul and L. S. Balistrieri performed trace metal analyses, and K. McKillop typed the manuscript. This research was supported by Natural Science Foundation Grant OCE 8511580. This manuscript is University of Washington contribution number 1684.

## References

1. Imboden, D. M.; Schwarzenbach, R. P. In *Chemical Processes in Lakes*; Stumm, W., Ed.; Wiley: New York, 1985; p 1.
2. Okubo, A. *Deep-Sea Res.* 1971, *18*, 789.
3. Imboden, D. M.; Gachter, R. *Ecol. Modell.* 1978, *4*, 77.
4. Imboden, D. M.; Emerson, S. *Limnol. Oceanogr.* 1978, *23*, 77.
5. Nyffeler, U. P.; Schindler, P. W.; Wirz, U. E.; Imboden, D. M. *Schweiz. Z. Hydrol.* 1983, *45*, 45.
6. Salomons, W.; Forstner, U. *Metals in the Hydrocycle*; Springer–Verlag: New York, 1984; p 333.
7. Evans, H. E.; Smith, P. J.; Dillon, P. J. *Can. J. Fish Aquat. Sci.* 1983, *40*, 570.
8. Nriagu, J. O.; Kemp, A. L. W.; Wong, K. T.; Harper, N. *Geochim. Cosmochim. Acta* 1979, *43*, 247.
9. Crecelius, E. A.; Piper, D. Z. *Environ. Sci. Technol.* 1973, *7*, 1053.
10. White, J. R.; Driscoll, C. T. *Environ. Sci. Technol.* 1985, *19*, 1182.
11. Johnson, N. M.; Driscoll, C. T.; Eaton, J. S.; Likens, G. E.; McDowell, W. H. *Geochim. Cosmochim. Acta* 1981, *45*, 1421.
12. Stumm, W. *Chemical Processes in Lakes*; Wiley: New York, 1985; p 435.
13. Boyle, E. A.; Edmond, J. M. *Nature (London)* 1975, *253*, 107.
14. Quinby-Hunt, M. S.; Turekian, K. K. *EOS* 1983, *64*, 130.
15. Bruland, K. W. In *Chemical Oceanography*; Riley, J. P.; Chester, R., Eds.; Academic: New York, 1983; Vol. 8, p 157.
16. Turner, D. R.; Whitfield, M.; Dickson, A. G. *Geochim. Cosmochim. Acta* 1981, *45*, 855.

17. Hardy, J. T.; Apts, C. W.; Crecelius, E. A.; Fellingham, G. W. *Limnol. Oceanogr.* **1985**, *30*, 93.
18. Hodge, V.; Johnson, S. R.; Goldberg, E. D. *Geochem. J.* **1978**, *12*, 7.
19. Zafiriou, O. C.; Joussot-Dubien, J.; Zepp, R. G.; Zika, R. G. *Environ. Sci. Technol.* **1984**, *18*, 358.
20. Zika, R. G.; Moffett, J. W.; Petasne, R. G.; Cooper, W. J.; Saltzman, E. S. *Geochim. Cosmochim. Acta* **1985**, *49*, 1173.
21. Zafiriou, O. C.; McFarland, M. *J. Geophys. Res.* **1981**, *86*, 3173.
22. Sunda, W. G.; Huntsman, S. A.; Harvey, G. R. *Nature (London)* **1983**, *301*, 234.
23. Waite, T. D. Ph.D. Thesis, Massachusetts Institute of Technology, Cambridge, 1983.
24. Moffett, J. W.; Zika, R. G. *Mar. Chem.* **1983**, *13*, 239.
25. Jones, G. J.; Waite, T. D.; Smith, J. D. *Biochem. Biophys. Res. Commun.* **1985**, *128*, 1031.
26. Wood, J. M.; Wang, H.-K. *Environ. Sci. Technol.* **1983**, *17*, 582.
27. Andreae, M. O.; Klumpp, D. *Environ. Sci. Technol.* **1979**, *13*, 738.
28. Kaufman, A.; Trier, R. M.; Broecker, W. S. *J. Geophys. Res.* **1973**, *78*, 8827.
29. Brewer, P. G.; Spencer, D. W. In *Marine Chemistry in the Coastal Environment*; Church, T. M., Ed.; ACS Symposium Series 18; American Chemical Society: Washington, DC, 1975; p 80.
30. Boyle, E. A.; Huested, S. S.; Jones, S. P. *J. Geophys. Res.* **1981**, *86*, 8048.
31. Kremling, K. *Nature (London)* **1983**, *303*, 225.
32. Livingston, H. D.; Swift, J. H.; Ostlund, H. G. *J. Geophys. Res.* **1985**, *90*, 6971.
33. Mertz, W. *Science (Washington, D.C.)* **1981**, *213*, 1332.
34. Collier, R.; Edmond, J. M. *Prog. Oceanogr.* **1984**, *13*, 113.
35. Bender, M. L.; Klinkhammer, G. P.; Spencer, D. M. *Deep-Sea Res.* **1977**, *24*, 799.
36. Lupton, J. E.; Craig, H. *Science (Washington, D.C.)* **1981**, *214*, 13.
37. Jenkins, W. J. *J. Mar. Res.* **1980**, *38*, 533.
38. Gammon, R. H.; Cline, J.; Wisegarver, D. *J. Geophys. Res.* **1982**, *87*, 9441.
39. Murray, J. M.; Spell, B.; Paul, B. In *Trace Metals in Seawater*; Wong, C. S.; Boyle, E.; Bruland, K. W.; Burton, J. D.; Goldberg, E. D., Eds.; Plenum: New York, 1983; p 643.
40. Wright, R. F.; Matter, A.; Schweingruber, M.; Siegenthaler, U. *Schweiz. Z. Hydrol.* **1980**, *42*, 101.
41. Broecker, W. S.; Cromwell, J.; Li, Y. H. *Earth Planet. Sci. Lett.* **1968**, *5*, 101.
42. Sarmiento, J. L.; Feely, H. W.; Moore, W. S.; Bainbridge, A. E.; Broecker, W. S. *Earth Planet. Sci. Lett.* **1976**, *32*, 357.
43. Boyle, E. A.; Sclater, F. R.; Edmond, J. M. *Earth Planet. Sci. Lett.* **1977**, *37*, 38.
44. Sawlan, J. J.; Murray, J. W. *Earth Planet. Sci. Lett.* **1983**, *31*, 213.
45. Edmond, J. M.; Jacobs, S. S.; Gordon, A. L.; Mantyla, A. W.; Weiss, R. F. *J. Geophys. Res.* **1979**, *84*, 7809.
46. Davison, W.; Woof, C.; Rigg, E. *Limnol. Oceanogr.* **1982**, *27*, 987.
47. Goldberg, E. D. *J. Geol.* **1954**, *62*, 249.
48. Balistrieri, L.; Brewer, P. G.; Murray, J. M. *Deep-Sea Res.* **1981**, *28*, 101.
49. Craig, H.; Krishnaswami, S.; Somayajulu, B. L. K. *Earth Planet. Sci. Lett.* **1973**, *17*, 295.
50. Bacon, M. P.; Anderson, R. F. *J. Geophys. Res.* **1982**, *87*, 2045.
51. Emerson, S.; Jacobs, L.; Tebo, B. In *Trace Metals in Seawater;* Wong, C. S.;

Boyle, E.; Bruland, K. W.; Burton, J. D.; Goldberg, E. D., Eds.; Plenum: New York, 1983; p 579.

52. Spencer, D. W.; Brewer, P. G. *J. Geophys. Res.* **1971**, *76*, 5877.
53. Spencer, D. W.; Brewer, P. G.; Sachs, P. L. *Geochim. Cosmochim. Acta* **1972**, *36*, 71.
54. Jacobs, L.; Emerson, S. *Earth Planet. Sci. Lett.* **1982**, *60*, 237.
55. Jacobs, L.; Emerson, S.; Skei, J. *Geochim. Cosmochim. Acta* **1985**, *49*, 1433.
56. Bacon, M. P.; Brewer, P. G.; Spencer, D. W.; Murray, J. W.; Goddard, J. *Deep-Sea Res.* **1980**, *27*, 119.
57. Morel, F. M. M.; Hudson, R. J. M. In *Chemical Processes in Lakes*; Stumm, W., Ed.; Wiley: New York, 1985; p 251.
58. Baccini, P.; Ruchti, J.; Wanner, O.; Grieder, E. *Schweiz. Z. Hydrol.* **1979**, *41*, 202.
59. Santschi, P. W.; Schindler, P. W. *Schweiz. Z. Hydrol.* **1977**, *39*, 182.
60. Hamilton-Taylor, J.; Willis, M.; Reynolds, C. S. *Limnol. Oceanogr.* **1984**, *29*, 695.
61. Talbot, R. W.; Andren, A. W. *Geochim. Cosmochim. Acta* **1984**, *48*, 2053.
62. Sholkovitz, E. R.; Copland, D. *Geochim. Cosmochim. Acta* **1982**, *46*, 393.
63. Finlay, B. J.; Hetherington, N. B.; Davison, W. *Geochim. Cosmochim. Acta* **1983**, *47*, 1325.
64. Giovanoli, R.; Brutsch, R.; Diem, D.; Osman-Sigg, G.; Sigg, L. *Schweiz. Z. Hydrol.* **1980**, *42*, 89.
65. Redfield, A. C.; Ketchum, B. H.; Richards, F. A. In *The Sea*; Hill, M. N., Ed.; Interscience: New York, 1963; Vol. 2, p 26.
66. Sigg, L.; Sturm, M.; Stumm, W.; Mart, L.; Nurnberg, H. W. *Naturwissenschaften* **1982**, *69*, 546.
67. Sigg, L. In *Chemical Processes in Lakes*; Stumm, W., Ed.; Wiley: New York, 1985; p 283.
68. Sigg, L.; Sturm, M.; Kistler, D. *Limnol. Oceanogr.*, in press.
69. Boyle, E. A.; Edmond, J. M. *Anal. Chim. Acta* **1977**, *81*, 189.
70. Jones, C. J.; Murray, J. W. *Limnol. Oceanogr.* **1984**, *29*, 711.
71. Bruland, K. *Earth Planet. Sci. Lett.* **1980**, *37*, 38.
72. Imboden, D. M.; Emerson, S. *Limnol. Oceanogr.* **1978**, *23*, 77.
73. Hesslein, R. H. *Can. J. Fish. Aquat. Sci.* **1980**, *37*, 552.
74. Hutchinson, G. E. *Ecol. Monogr.* **1941**, *11*, 23.
75. Hesslein, R. H. *Can. J. Fish. Aquat. Sci.* **1980**, *37*, 545.
76. Kuivila, K. M.; Murray, J. W. *Limnol. Oceanogr.* **1984**, *29*, 1218.
77. Davison, W.; Heaney, S. I.; Talling, J. F.; Rigg, E. *Schweiz. Z. Hydrol.* **1980**, *42*, 196.
78. Davison, W.; Woof, C. *Water Res.* **1984**, *18*, 727.
79. Davison, W. In *Chemical Processes in Lakes*; Stumm, W., Ed.; Wiley: New York, 1985; p 31.
80. Sholkovitz, E. R. Ibid., p 119.
81. Kelly, C. A.; Rudd, J. W. M. *Biogeochem.* **1984**, *1*, 63.
82. Kuivila, K. M.; Devol, A. H.; Murray, J. W.; Lidstrom, M. E.; Reimers, C. E. *Limnol. Oceanogr.*, in press.
83. Campbell, P.; Torgersen, T. *Can. J. Fish. Aquat. Sci.* **1980**, *37*, 1303.
84. Mayer, L. M.; Liotta, F. P.; Norton, S. A. *Water Res.* **1982**, *16*, 1189.
85. Stumm, W., Morgan, J. J. In *Aquatic Chemistry*; Stumm, W., Ed.; Wiley: New York, 1981; p 780.
86. Davison, W. *Nature (London)* **1981**, *290*, 241.
87. Tipping, E.; Woof, C.; Cooke, D. *Geochim. Cosmochim. Acta* **1981**, *45*, 1411.

88. Morgan, J. J. In *Principles and Applications of Water Chemistry*; Faust, S. D.; Hunter, J. V., Eds.; Wiley: New York, 1967; p 561.
89. Diem, D.; Stumm, W. *Geochim. Cosmochim. Acta* **1984**, *48*, 1571.
90. Emerson, S.; Kalhorn, S.; Jacobs, L.; Tebo, B. M.; Nealson, K. H.; Rosson, R. A. *Geochim. Cosmochim. Acta* **1982**, *46*, 1073.
91. Chapnick, S. D.; Moore, W. S.; Nealson, K. H. *Limnol. Oceanogr.* **1982**, *27*, 1004.
92. Jaquet, J.-M.; Nembrini, G.; Garcia, J.; Vernet, J.-P. *Hydrobiologia* **1982**, *91*, 323.
93. Doyle, R. W. *Am. J. Sci.* **1968**, *266*, 980.
94. Davison, W.; Lishman, J. P.; Hilton, J. *Geochim. Cosmochim. Acta* **1985**, *49*, 1615.
95. Davison, W.; Reynolds, C. S.; Finlay, B. J. *Water Res.* **1985**, *19*, 265.
96. Davison, W.; Finlay, B. J. *J. Ecol.* **1986**, *74*, 663.
97. Baccini, P.; Joller, T. *Schweiz. Z. Hydrol.* **1981**, *43*, 176.
98. Hesslein, R. H.; Broecker, W. S.; Schindler, D. W. *Can. J. Fish. Aquat. Sci.* **1980**, *37*, 378.
99. Santschi, P. H.; Nyffeler, U. P.; Anderson, R. F.; Schiff, S. L.; O'Hara, P.; Hesslein, R. H. *Can. J. Fish. Aquat. Sci.* **1986**, *43*, 60.
100. Tessier, A.; Rapin, F.; Carignan, R. *Geochim. Cosmochim. Acta* **1985**, *49*, 183.
101. Sigg, L.; Stumm, W.; Zinder, B. In *Complexation of Trace Metals in Natural Waters*; Kramer, C. J. M.; Duinker, J. C., Eds.; Nyhoff/Junk: 1984; p 251.

RECEIVED for review May 6, 1986. ACCEPTED October 1, 1986.

# Metal Speciation in Natural Waters: Influence of Environmental Acidification

Peter G. C. Campbell[1] and André Tessier

Université du Québec, INRS–Eau, C. P. 7500, Sainte-Foy, Québec G1V 4C7, Canada

*Changes in metal speciation in solution that may result from an increase in the acidity of natural waters over the pH range from 7 to 4 are reviewed. Attention is focused on 10 metals (Ag, Al, Cd, Co, Cu, Hg, Mn, Ni, Pb, and Zn) that exist in cationic forms and are of potential concern in the context of freshwater acidification. For those metals for which changes in the degree of complexation ($\Delta pM$ and $\Delta pM'$) as a function of pH have been studied experimentally, reasonable qualitative agreement with their predicted sensitivity to pH changes is observed. For two of the metals (Al and Cu), important changes in both inorganic and organic speciation are predicted over the pH range from 7 to 4; confirmatory experimental data are available for Cu, whereas for Al the observed changes in $\Delta pM'$ appear to be less important than predicted. In contrast, observed changes in $\Delta pM'$ for Pb in organic-rich systems are much greater than would be predicted from equilibrium calculations. Calculations for Hg suggest that its (inorganic) speciation should vary markedly over the studied pH range, but no unambiguous experimental data relating to pH-induced changes in Hg speciation could be found. In the six remaining cases, predicted to show little sensitivity to pH changes in this range, supporting experimental evidence exists for four metals (Ag, Cd, Mn, and Zn).*

METALS IN NATURAL WATERS may exist in a variety of dissolved, colloidal, or particulate forms (Table I). Because these various forms

---

[1]To whom correspondence should be addressed

0065–2393/87/0216–0185$06.75/0

Table I. Possible Physicochemical Forms of Trace Metals in Natural Waters

| Type of Speciation | Principle | Form | Description | Examples |
|---|---|---|---|---|
| Physical | Differences in dimension | Dissolved | (e.g., < 0.002 μm) | |
| | | Colloidal | | |
| | | Particulate | (e.g., > 0.1 μm) | |
| Chemical | Differences in association | Dissolved | Aquo complexes | $Cd(H_2O)_6^{2+}$ |
| | | | Inorganic complexes | $PbCO_3$ |
| | | | Organic complexes | Cu–glycinate |
| | | | Mixed ligand complexes | Fulvic acid–Fe–$PO_4$ |
| | | Colloidal | Colloidal metal hydroxides | $Cu(OH)_2$ |
| | | | Adsorbed on inorganic colloids | $M^{z+}$–colloidal $Fe(OH)_3$ |
| | | | Adsorbed on organic colloids | $M^{z+}$–colloidal humic acid |
| | | | Adsorbed on mixed colloids | $M^{z+}$–organic coatings |
| | | Particulate | Adsorbed | $Cd^{2+}$–clay |
| | | | Carbonate bound | $MnCO_3$ |
| | | | Occluded in Fe or Mn oxides | $Cu^{2+}$–$Fe_2O_3$; $Pb^{2+}$–$MnO_2$ |
| | | | Organic bound | $Cu^{2+}$–humic acid |
| | | | Sulfide bound | $FeS_2$ |
| | | | Matrix bound | Aluminosilicates |
| | Differences in oxidation state | Oxidation–reduction | | Fe(II), Fe(III) |
| | | | | Mn(II), Mn(IV) |
| | | | | Hg(0), Hg(I), Hg(II) |
| | Differences in bonding (covalent) | Organometallic compounds | | $CH_3Hg^+$, $CH_3HgCH_3$ |

often exhibit different physical and chemical properties, measurement of the total metal concentration in a given geochemical compartment will generally provide little indication of the metal's potential interactions with other abiotic or biotic components of the environment. The corollary, of course, is that knowledge of the speciation of a metal (i.e., its distribution among the various possible physicochemical forms) is necessary for understanding its geochemical behavior (mobility and transport) and biological availability.

Many speciation parameters can be defined (*1*), but from a geochemical or biological point of view, the most important parameter is surely the degree of complexation of the metal, $\Delta pM = \log [M]_T/[M^{z+}]$, where $[M]_T$ is the total dissolved-metal concentration, and $[M^{z+}]$ is the free aquo ion concentration (*2*). Factors influencing the degree of complexation of a particular metal in a natural water sample include

- the nature and concentration of the competing ligands ($L_1$, $L_2$, ... $L_x$);
- the stability of the various $M(L_x)_n$ forms;
- the concentrations of competing cations, including $Ca^{2+}$ and $Mg^{2+}$;
- the redox potential and temperature;
- the pH; and
- the degree to which equilibrium is attained (reaction kinetics) (*2-3*).

In the present review, particular emphasis is accorded to one of these factors, namely the ambient water pH, and the influence of environmental acidification from pH 7 to 4 on the speciation of dissolved metals is examined. These considerations are of obvious relevance to the current debate regarding the environmental impacts of acid precipitation. Indeed, several authors have speculated that some of the biological effects of acid deposition, both in the aquatic and terrestrial environments, may be due not only to the increase in hydrogen ion activity per se but also to changes in the concentration or speciation of certain metals (*4-6*).

The objective of the present study has thus been to review possible changes in dissolved-metal speciation that may result from an increase in the acidity of natural waters over the pH range from 7 to 4. We have not considered the changes in total dissolved-metal concentrations that may also result from environmental acidification; a review of such geochemical mobilization processes has been published (*7*). Attention has been focused on some 10 metals that exist predominantly in cationic forms and are of potential concern in the context of freshwater acidification (7–8): Ag, Al, Cd, Co, Cu, Hg, Mn, Ni, Pb, and Zn.

## Methods

Possible approaches to the determination of the speciation of metals in natural waters can be grouped into two major categories (9–10). The first category comprises thermodynamic calculations involving the use of measured total concentrations ($[M]_T$ and $[L]_T$) as well as pH, redox potential ($E_H$), temperature ($T$), and published stability-constant values. These calculations are used to compute the equilibrium contribution of the various species (11). The second category includes experimental methods involving the determination of the physical properties (e.g., size) or chemical reactivity (e.g., electrochemical lability and ion-exchange behavior) of the metal of interest.

In the first category, provided that the assumption of equilibrium is valid, a major current limitation is the absence of reliable thermodynamic data for many of the species suspected to be present in natural waters (notably those involving natural organic matter present as dissolved ligands or colloidal organic material). In addition, because those equilibrium constants that are available were often determined under conditions differing greatly from those prevailing in natural waters, semiempirical corrections for temperature and ionic strength are generally required. Results obtained from the chemical equilibrium approach will obviously be sensitive to the choice of thermodynamic constants used in the equilibrium calculations (12). In addition, the ability of these simple models to represent metal interactions with multifunctional polyelectrolytes (e.g., fulvic acids) has been questioned (13).

The second, or experimental, category is limited by the lack of experimental techniques sufficiently sensitive and selective to detect individual metal forms at the concentrations normally found in natural waters. Current techniques allow only the classification of metal forms into various operationally defined categories, according to their physical properties or chemical reactivity (Table II). Such experimental approaches to metal speciation have recently been critically reviewed (1, 9–10, 14).

The ideal speciation method would be sufficiently sensitive ($10^{-10}$–$10^{-6}$ mol/L) and selective to be used directly on a natural water sample, would involve minimal perturbation of the sample, and would furnish an analytical signal proportional to the (chemical) reactivity of the element of interest (1). Of the various possible approaches outlined in Table II, direct potentiometric determinations of free metal ion

Table II. Possible Experimental Methods for the Determination of
Metal Speciation

| Principle | Examples |
| --- | --- |
| Physical property | |
|   Size | Dialysis, ultrafiltration, gel permeation, chromatography |
|   Solubility | Solvent extraction |
|   Electronic structure | Electron-spin resonance |
| Chemical reactivity | |
|   Electrochemical determinations | Potentiometry, stripping voltammetry |
|   Ion exchange | Cation-exchange resins (e.g., Chelex) |
|   Adsorption | Macroreticular hydrophobic resins (e.g., XAD) |
|   Complexometric determinations | Chelation–extraction (e.g., oxine–MIBK), colorimetry |

activities best approach this ideal and yield the least ambiguous results. However, few of the metals of current concern (Ag, Cd, Cu, and perhaps Pb) are amenable to direct determination with ion-selective electrodes at realistic concentrations not too far removed from environmental levels. Note also that studies carried out under these conditions (i.e., close to the analytical detection limit) in low ionic strength media and in the presence of variable hydrogen ion concentrations are particularly difficult.

Electrodeposition techniques followed by stripping analysis offer considerably improved detection limits; for a given metal, changes in its degree of complexation in solution as a function of pH should affect the electrochemical response. Many of the metals of current concern (Cd, Cu, Mn, Pb, and Zn) are in fact amenable to stripping analysis. Unfortunately, however, the interpretation of changes in the analytical signal [stripping current ($i_p$) and potential ($E_p$)] with shifts in pH is not straightforward. For example, do increases in $i_p$, as frequently observed on acidification of (prefiltered) samples, reflect changes in solution chemistry (i.e., decomplexation leading to an increase in the labile metal concentration) or rather artifacts at the electrode–solution interface? Changes in the hydrogen ion concentration may, in fact, alter the polarographic reduction kinetics of various labile metal species (15) and may also influence the adsorption of dissolved organic matter (DOM) on the working electrode (16). Thus, the electrode response is affected by acidification of the sample, and changes in the analytical signal cannot be ascribed unambiguously to pH-induced changes in metal speciation in solution.

Size-separation procedures involving such techniques as equilibrium dialysis, ultrafiltration, or size-exclusion chromatography can also be used to probe metal speciation in solution as a function of pH. Provided that pH changes do not affect the physical support (membrane or chromatographic substrate) used to effect the separation, a shift of a given metal to smaller forms at lower pH values can be interpreted in terms of a dissociation of (large) complexed forms (17–18).

Finally, a variety of ion-exchange and complexometric techniques can in principle be used to follow pH-induced changes in metal ion reactivity, provided that the complexes of interest are either kinetically or thermodynamically undetectable by the chosen method in the original unacidified sample. An additional condition that must also be met is that the reaction on which the method is based should itself be insensitive to pH changes. This latter criterion can usually be satisfied in the case of ion exchange (19), but complexometric reactions often exhibit distinct pH optima. Note also that both techniques necessarily involve the perturbation of the original sample, either by the introduction of a cation-exchange resin or by the addition of a competing ligand, and often this treatment will tend to alter the sample pH, which is our master variable. Clearly, the interpretation of the results of such experiments will rarely be unambiguous.

Because the thermodynamic and experimental approaches clearly yield incomplete but complementary information, both have been considered in the present review.

## Table III. Chemical Modeling of Metal

| Case | Reference | Ca | Mg | Na | K | Cl | SO₄ | PO₄ | F | CO₃ | Alk (− log equiv/L) |
|------|-----------|----|----|----|----|----|----|----|----|----|----|
| | | | | Major Ion Concentrations[a] | | | | | | | |
| 1 | 33 | 3.00 | 3.50 | 3.50 | | 3.50 | 4.50 | 5.00 | 5.50 | 3.00 | |
| 2 | 32 | 3.40 | | 3.77 | 3.56 | 4.23 | 3.65 | 3.90 | 5.00 | 5.50 | $3.22^{b}{\rightarrow}4.7$ | |
| 3 | 29 | $1.13^{c}{\rightarrow}4.90$ | | | | 3.63 | 3.94 | 6.98 | | | $2.0{\rightarrow}4.0$ |
| 4 | 2 | 4.00 | 4.30 | | | 3.00 | 4.30 | 6.30 | | b | |
| 5 | 43 | 3.47 (7.00) | 3.77 | 3.56 | 4.23 | 3.66 | 3.93 | | 5.30 | 2.89 (3.19) | 2.85 (3.68) |
| 6 | 30 | 3.40 | 3.74 | 3.46 | 4.45 | 3.70 | 3.15 | | | | $3.36^{d}$ |
| 7 | 6, this work | 4.41 | 4.84 | 4.57 | 5.23 | 4.95 | 4.44 | 7.00 | 5.59 | 3.48 or b | |

[a]Concentrations are expressed as negative logarithms of moles per liter.
[b]Denotes variable [C]; system is assumed to be in equilibrium with atmospheric $CO_2$ ($P_{CO_2}= 10^{-3.5}$ atm, where P is pressure).
[c]Denotes variable [Ca]; the system is controlled by calcite solubility.
[d]Alkalinity at initial pH of 7.4.
[e]Protolytic equilibria of humic acid were not considered.

## Results and Discussion

**Equilibrium Calculations.** SYSTEMS STUDIED. Changes in metal speciation can be anticipated as a result of a decrease in pH due to shifts in the metal hydrolysis equilibrium (favoring the aquo metal ion, $M(H_2O)_n^{z+}$, or simply $M^{z+}$) or in complexation equilibria (competition between $M^{z+}$ and $H^+$ for the same ligands). Chemical equilibrium models can be used to identify those metals most likely to be affected by such shifts. In principle, one need only define the chemical composition of a "representative" water sample, decide whether or not the system will be allowed to equilibrate with the atmosphere ($CO_2$), impose successively lower pH values, and calculate the speciation of each metal as a function of pH. Examples of this type of approach can be found in Table III.

*Major Ions.* Defining the major ion composition of a typical natural water does not normally pose a major problem; possible sources for such data include the compilations of Livingstone (*20*) and Meybeck (*21*). Six of the seven hypothetical waters represented in Table III are indeed rather similar in inorganic composition. In the context of fresh-water acidification, however, choosing an initial ionic composition more representative of the surface waters found in those geological regions

**Complexation as a Function of pH**

| pH | Trace Metals | Organic Ligands | Comments |
|---|---|---|---|
| 9.0→5.5 | Ag, Al, Cd, Co, Cu, Hg, Mn, Ni, Pb, Zn | citrate, glycine, cysteine, NTA | |
| 8.0→6.2 | Cd, Co, Cu, Hg, Mn, Ni, Pb, Zn | citrate, aspartate, histidine, cysteine | adsorbing surfaces present: organic complexation not studied as function of pH |
| 9.0→6.5 | Cu, Mn, Zn | humate | $[Ca]_T$, $[Cu]_T$, $[Mn]_T$, $[Zn]_T$ very high |
| 8.0→6.0 | Cd, Cu, Mn, Ni, Pb, Zn | salicylate | adsorbing surface present |
| 9 (6) | Ag, Al, Cd, Co, Cu, Hg, Mn, Ni, Pb, Zn | | |
| 7.4→3.3 | Cu, Ni | humate[e] | adsorbing surface present |
| 7.0→4.0 | Ag, Al, Cd, Co, Cu, Hg, Mn, Ni, Pb, Zn | citrate, malate, phthalate, salicylate, argine, lysine, ornithine, valine | *see* Table IV |

commonly considered to be "sensitive" to acid deposition [i.e., waters with low conductivity, hardness, and alkalinity, such as those on the Canadian Precambrian Shield (22-23) or in the Adirondack region of New York (24)] would seem more appropriate. The final entry in Table III corresponds to such a choice (6).

*Organic Ligands.* To complete the hypothetical inorganic systems just described, and to account for the metal-complexing properties of the DOM present in natural waters, a number of representative organic ligands should be included. In most natural waters, the metal complexation capacity will in fact be dominated by the fulvic–humic acid fraction of the DOM (25). Conditional stability constants have been derived for a mean fulvic acid unit (26-27), but true thermodynamic constants for metal–fulvic acid and proton–fulvic acid equilibria are unfortunately not available (13, 28). Incorporation of fulvic acid(s) into a computer program is thus not straightforward, and most authors have adopted the "mixture model" approach, as defined by Sposito (27) (*see* Table III, cases 1, 2, 4, and 7). In this latter approach, a set of organic acids is selected whose reactions with trace metal cations and the hydrogen ion are well characterized, and whose functional groups are known to exist in the DOM pool. The distribution of the individual acids may be adjusted to simulate the acid–base and metal-complexing behavior of the natural DOM mixture. As pointed out by Sposito (27), the basic assumption underlying the use of a mixture model is that the measured thermodynamic stability constants for the trace metal complexes with the chosen set of organic acids will combine to approximate closely the stability constants for trace metal complexation of the natural DOM.

In the remaining two cases, attempts have been made to include a fulvic or humic component directly in the equilibrium calculations (Table III, cases 3 and 6). In the former case (29), particular attention was accorded to the competition among $H^+$, $Ca^{2+}$, and $M^{2+}$ for binding sites on the humic material, and to the strength of electrostatic interactions between sites. However, the total concentrations of dissolved Ca (and dissolved M) were not held constant during the calculations at different pH values but were assumed to be controlled by the solubilities of the appropriate solid phase ($CaCO_3$, $CuO$, $MnCO_3$, or $Zn_2SiO_4$). The resulting metal concentrations were often unrealistically high and far in excess of normal environmental levels, thus precluding transposition of the results to natural waters. In the second case (30), the protolytic equilibria of the humic acid were apparently not considered; the conditional stability constants reported for metal complexation by a freshwater humic acid at pH 8.0 (26) appear to have been used at all pH values in the pH range from 7.4 to 3.3.

*Trace Metals.* In all but one of the studies reported in Table III, total trace metal concentrations were assigned either average observed values or arbitrary but representative values gleaned from the literature (e.g., reference 31). Case 3 (*29*) constitutes an exception as mentioned earlier. Provided that the ligands are present in excess (i.e., $[L]_T \gg [M]_T$), the relative proportion of each metal form will be independent of $[M]_T$. (*See* reference 25 for a discussion on the validity of this assumption in natural waters.)

SPECIATION CHANGES AS A FUNCTION OF pH. *Inorganic Systems.* For the system most closely approximating the inorganic composition of surface waters found in regions likely to be sensitive to acid deposition [Table III, case 7 (open to the atmosphere)], four metals are predicted to show significant changes in ΔpM over the pH range from 7 to 4 (Table IV). In this pH range, metal complexation involves primarily the hydroxide, chloride, fluoride, and carbonate ligands; the pH change from 7 to 4 particularly affects those metals that undergo extensive hydrolysis (Al and Hg) and those forming strong carbonato complexes (Cu and Pb). The degree of complexation of the remaining six metals (Ag, Cd, Co, Mn, Ni, and Zn) is predicted to be pH-insensitive, with the metals existing predominantly (>99%) as their aquo complexes throughout the pH range from 7 to 4.

For the other inorganic systems represented in Table III, published metal speciation trends are qualitatively similar to those described previously for case 7. Differences among systems are most noticeable at higher pH and can be related either to the inorganic carbon concentration (the higher $[CO_3^{2-}]$, the more extensive metal complexation at high pH), or to the thermodynamic constants chosen for cation hydrolysis and carbonate complexation.

*Organic Systems.* The addition of model organic ligands to the inorganic systems described previously influences relatively few of the metals. At pH 7, and for individual ligand concentrations $<5 \times 10^{-6}$ mol/L, appreciable (>10%) metal complexation has been reported for only three of the metals of current concern: aluminum [citrate and salicylate (*6*)], copper [aspartate, citrate, and salicylate (*2, 6, 32*)], and nickel [aspartate and citrate (*32*)]. Additional organometallic interactions have been predicted for Hg, Pb, and Zn (*33*) with cysteine, but the ligand concentration ($10^{-5}$ mol/L) appears unrealistically high; the total organic sulfur concentration in natural waters is generally $<5 \times 10^{-7}$ mol/L.

On acidifying a system containing organic ligands, increased competition would be anticipated between the hydrogen ion and the metal cations for the available ligands. The published results for such acid titrations [Table III, cases 4 and 7 (*2, 6*)] do indeed show a reduced degree of complexation for Al and Cu as the pH is lowered; this

Table IV. Effect of Acidification on the Calculated Speciation of Dissolved Trace Metals in a Typical Surface Water from the Canadian Shield

| Metal | Total Concentration ($\times 10^{-7}$ M) | Percent Aquo Ion[a] $M^{z+}$ $(H_2O)_n$ | | | | Other Dissolved Species (>1%) | Solids |
|---|---|---|---|---|---|---|---|
| | | pH 7 | pH 6 | pH 5 | pH 4 | | |
| Ag | 0.10 | 99 | 99 | 99 | 99 | AgCl | |
| Al | 58.9 | <1[b] | <1[b] | 25 | 54 | $AlF^{2+}$, $AlF_2^+$, $AlOH^{2+}$, $Al(OH)_4^-$ | $Al(OH)_3$ |
| Cd | 1.00 | 98 | 99 | 99 | 99 | | |
| Co(II) | 0.10 | 99 | 99 | 99 | 99 | | |
| Cu | 1.00 | 79 | 98 | 99 | 100 | $CuOH^+$, $Cu(OH)_2$, $CuCO_3$ | |
| Hg[c] | 0.01 | <1 | <1 | <1 | <1 | $HgCl^+$, $HgCl_2$, $HgClOH$, $Hg(OH)_2$ | |
| Mn | 7.24 | 99[b] | 99[b] | 99[b] | 100[b] | | $MnO_2$[d] |
| Ni | 1.00 | 98 | 99 | 99 | 100 | | |
| Pb | 0.05 | 40 | 91 | 98 | 99 | $PbOH^+$, $PbCO_3$ | |
| Zn | 5.00 | 99 | 99 | 100 | 100 | | |

NOTE: For these results, $I = 0.001$ M, $p_\epsilon = 12$, $P_{CO_2} = 10^{-3.5}$ atm, no organic liquids were present, and the surface water was open to the atmosphere. Abbreviations are as follows: $I$ is the ionic strength, $p_\epsilon$ is the redox potential, $(E_H/2.3\ RTF^{-1})$, and $P_{CO_2}$ is the pressure of $CO_2$.
[a] The MINEQL–1 chemical equilibrium model (48) was used to calculate the theoretical speciation of each trace metal. The percentage aquo ion, $M^{z+}$ $(H_2O)_n$, was calculated by using the formula % = ([aquo ion] $\div$ [dissolved M]) $\times$ 100.
[b] Denotes the presence of a precipitated solid.
[c] Although the relative contribution of the aquo ion was <1% for Hg over the pH range, the relative proportions of the chloro and hydroxy complexes were strongly pH-dependent (e.g., $HgCl_2$ dominated at pH 4 and $Hg(OH)_2$ at pH 7).
[d] At equilibrium, Mn was calculated to be almost entirely (>99%) in particulate form.

complexation decrease is greater in the presence of the organic ligands than in the comparable inorganic systems (Table V).

**Experimental Measurements.** In view of the known sensitivity of chemical equilibrium models to the choice of stability constants used as input data and the inherent limitations of the mixture model approach, confronting the predictions of the preceding section with experimental data relating to natural waters is important. The data considered are of two types: (1) results from experimental studies, where the pH of a given water sample (synthetic solution or natural water sample) has been perturbed and changes in metal speciation followed as a function of pH, and (2) data from field surveys, where surface water samples have been collected along a natural pH gradient and analyzed; the changes in metal speciation were then ascribed to differences in sample pH. In both cases, shifts in metal speciation will be discussed in terms of the degree of complexation of the metal, $\Delta$pM. In the summary tables, values of $\Delta$pM are reported for two pH values, wherever possible for pH 7 and 5.

EXPERIMENTAL pH CHANGES.    Studies of this type generally involve pH titrations of either natural water samples or synthetic solutions containing added fulvic or humic acid as representative organic ligands. In the latter cases, the inorganic matrix is usually chosen not to approach the composition of natural waters but rather to act as a supporting electrolyte and facilitate the experimental (e.g., electrochemical) measurements. Shifts in metal speciation as a function of pH are often

Table V. Effect of Organic Ligands on the Sensitivity of Aluminum and Copper Speciation to pH Changes

| Reference | Metal | Organic Ligands | Degree of Complexation[a] | | | |
|---|---|---|---|---|---|---|
| | | | *pH 7* | *pH 6* | *pH 5* | *pH 4* |
| 2 | Cu | — | 0.12 | 0.01 | | |
| | $(10^{-7}M)$ | salicylate $(10^{-5}M)$ | 0.22 | 0.02 | | |
| this study[b] | Cu | — | 0.43 | 0.01 | 0.00 | 0.00 |
| | $(10^{-7}M)$ | mixture[c] | 0.59 | 0.13 | 0.02 | 0.00 |
| | Al | — | $5.78^d$ | $2.27^d$ | 0.60 | 0.27 |
| | $(10^{-5.23}M)$ | mixture[c] | 6.25 | 3.17 | 1.62 | 0.50 |

[a] The degree of complexation ($\Delta$pM) was calculated by using the formula $\Delta$pM = log ([total dissolved metal] $\div$ [$M^{z+}$]).
[b] The MINEQL-1 chemical equilibrium model (48) was used to calculate the theoretical speciation of each metal.
[c] *See* Table III, case 7, for mixture composition.
[d] Denotes the presence of a precipitated solid, Al (OH)$_3$(s).

followed either by electrochemical determinations (potentiometry and voltammetry) or by size-separation procedures (dialysis).

Experiments along these lines have yielded information concerning the relative complexation of different metals and the sensitivity of metal speciation to pH changes (Tables VI and VII). For a given medium and at a circumneutral pH, the observed degree of complexation differs from one metal to another and generally decreases in the order Al, Cu, Pb > Cd, Zn > Ag, Mn; experimental data for the remaining metals of interest (Co, Hg, and Ni) are lacking. This ranking corresponds qualitatively with that predicted by chemical equilibrium modeling, but the observed $\Delta pM$ (or $\Delta pM'$) values are considerably higher than those calculated with the mixture model approach (Table III, cases 1, 2, 4, and 7). For example, for the less complexed metals (e.g., Ag, Cd, and Zn), calculated values of $\Delta pM$ at pH 7 approach zero (i.e., $[M^{z+}] \rightarrow$ 100%), whereas observed values are in the $\Delta pM$ range 0.2–0.8 (see Tables VI and VII). Clearly, the mixture model approach underestimates the metal-complexing capacity of natural DOM; to achieve calculated $\Delta pM$ values close to those observed, one would have to use unrealistically high concentrations of the various monomeric ligands (amino acids and polycarboxylic acids) that have traditionally been used to approximate the metal binding components present in DOM.

As anticipated, observed values of $\Delta pM$ tend to decrease on acidification (Tables VI and VII and reference 6). A particularly convincing example of this phenomenon has been provided by Guy and Chakrabarti (17), who studied the behavior of Cu and Pb in synthetic solutions containing humic acid by using three complementary analytical techniques (potentiometry, anodic stripping voltammetry, and equilibrium dialysis; see Table VI, case 2, and Figure 1). For both metals, the proportion of free metal ion increased dramatically as the pH decreased, and the agreement among the three experimental methods was quite remarkable.

Describing these pH-induced changes in metal speciation with a view to rank the various metals with respect to their sensitivity to environmental acidification poses certain problems. Variation of $\Delta pM$ with pH can be used to this end, provided the change of $\Delta pM$ with pH is reasonably smooth or regular in the pH range of interest. For example (Table VII, case 2), Engel and co-workers (34) spiked filtered organic-rich river water with Cd or Cu (1 $\mu mol/L$) and lowered the pH from 7.7 to 5.3. For both metals, the activity of the free aquo metal ion, as determined with the appropriate ion-selective electrode, increased on acidification. Plots of $\Delta pM$ versus pH suggest a progressive decomplexation (Figure 2A), the slope being greater for Cu than for Cd. One might conclude that Cu speciation was more sensitive to pH changes than was that of Cd. Theoretical values for $\Delta pCu$, derived from equilibrium

Table VI. Experimental Evidence for pH-Induced Changes in the Speciation of Dissolved Metals in Synthetic Solutions

| Reference | Metal | Conc. ($\times 10^{-6}$ mol/L) | Sample Description | pH Range | Analytical Technique | Degree of Complexation[a] | | | |
|---|---|---|---|---|---|---|---|---|---|
| | | | | | | pH 7 | pH 6 | pH 5 | pH 4 |
| 44 | Cd | 0.9–9 | $KNO_3$ (0.002 mol/L) $NaHCO_3$ (0.0003 mol/L) HA (20 mg/L) | 6.0–8.5 | ise | 0.2 | 0.15 | | |
| 17, 40 | Cu | 0.8 | $KNO_3$ (0.001–0.1 mol/L) HA (10 mg/L) | 2.0–7.0 | ise; asv; dialysis | 1.3 | | 0.6 | |
| | Pb | 0.2 | | | | >2.0 | | 0.9 | |
| 41 | Cu | 1.0 | K acetate (0.02 mol/L) and HA (11 mg/L) | 2.0–8.0 | asv | 0.7 | | 0.4 | |
| 18 | Cd | 16 | $KNO_3$ (0.001 mol/L) $NaHCO_3$ (0.001 mol/L) HA (20 mg/L) | 3.0–8.0 | dialysis | 0.4[b] | | 0.1[b] | |
| | Cu | 16 | | | | 1.1 | | 0.75 | |
| | Pb | 16 | | | | 0.8 | | 1.1 | |
| | Zn | 16 | | | | 0.35 | | 0.0 | |
| 42 | Al | 7.7 | $Ca(NO_3)_2$ ($2.5 \times 10^{-5}$ mol/L) $MgSO_4$ ($2 \times 10^{-5}$ mol/L) FA (20 mg/L) | 3.8–5.1 | ion exchange dialysis | | | 1.0[c] | 0.4[c] 1.1[b] |

NOTE: Abbreviations are as follows: ise is ion-selective electrode, and asv is anodic stripping voltammetry.
[a] The degree of complexation ($\Delta pM$) was calculated by using the formula $\Delta pM = \log([\text{dissolved metal}] \div [M^{z+}])$ unless noted otherwise.
[b] The degree of complexation ($\Delta pM'$) was calculated by using the formula $\Delta pM' = \log([\text{dissolved Al}] \div [\text{dialyzable Al}])$.
[c] The degree of complexation ($\Delta pM'$) was calculated by using the formula $\Delta pM' = \log([\text{dissolved Al}] \div [\text{exchangeable Al}])$.

Table VII. Experimental Evidence for pH-Induced Changes in the Speciation of Dissolved Metals in Synthetic Solutions

| Reference | Metal | Conc. ($\times 10^{-6}$ mol/L) | Sample Description | pH Range | Analytical Technique | Degree of Complexation[a] | | | |
|---|---|---|---|---|---|---|---|---|---|
| | | | | | | pH 7 | pH 6 | pH 5 | pH 4 |
| 40, 18 | Cu | ~0.3 | river water and $KNO_3$ (0.1 mol/L), DOC (?)[c] | 2.0–8.0 | ise | >2.0[b] | | 1.45[b] | |
| 34 | Ag Cd Cu | 1.0 1.0 1.0 | spiked river water DOC (20 mg/L) | 5.5–7.5 | ise | 0.50 0.75 3.5 | 0.30 0.45 2.45 | | |
| 45 | Cu | ~10 | soil solution; DOC (?)[c] | 6.2–7.6 | ise | 4.7 | ~3.3 | | |
| 35 | Al | 23 | lake water and TRIS (0.001 mol/L); DOC (8.4 mg/L) | 3.3–6.9 | ion exchange | | 0.5[d] | 0.5[d] | 0.4[d] |
| 42 | Al | 15 | stream water, DOC (3.9 mg/L) | 3.4–5.6 | ion exchange | | | 0.4[d] | 0.3[d] |

NOTE: ise is ion-selective electrode.
[a] The degree of complexation ($\Delta pM$) was calculated by using the formula $\Delta pM = \log$ ([dissolved metal] $\div$ [$M^{z+}$]) unless otherwise noted.
[b] Not known whether or not the sample was filtered before acidification.
[c] (?) denotes the DOC value was not given.
[d] The degree of complexation ($\Delta pM'$) was calculated by using the formula $\Delta pM' = \log$ ([dissolved Al] $\div$ [exchangeable Al]).

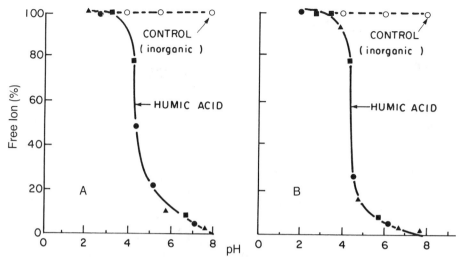

*Figure 1. Relative concentration of free metal ion as a function of pH, where A is copper–humic acid, and B is lead–humic acid. Key:* ▲*, ion-selective electrode;* ■*, anodic stripping voltammetry; and* ○*,* ●*, dialysis. (Data are from references 17 and 40.)*

calculations (Table III, case 7), also show this trend with pH, but the curve is shifted to much lower values [~2.5 log units (*see* Figure 2A)].

Not all experimental results are amenable to this treatment, however. Changes in the size distribution of metal species as a function of pH have been studied by Guy and Chakrabarti (*17–18*), who used equilibrium dialysis to differentiate between high molecular weight metal species (colloidal and pseudocolloidal forms and metals associated with fulvic or humic acids) and smaller dialyzable forms (simple inorganic species and small organic complexes). Synthetic solutions containing humic acid and one or more metals were adjusted to different pH values in the pH range from 8 to 3 and then dialyzed (Table VI, case 4). For comparative purposes, results for the same two metals (Cd and Cu) are shown in Figure 2B. Changes in Cd speciation as a function of pH are qualitatively similar to those described earlier (Figure 2A), but the observed behavior of Cu is markedly different. Values of $\Delta pCu'$ initially increase on acidification, pass through a maximum at pH ~ 6, and then decrease abruptly. Similar results were reported for Pb. Obviously, Cu speciation is highly pH-sensitive, yet a measurement of $\Delta pCu'$ at two arbitrary pH values might mask this sensitivity; comparison of $\Delta pCu'$ values at pH 7.5 and 5 would yield identical values and a $\Delta(\Delta pCu')/\Delta(pH)$ ratio of zero! Clearly, this latter criterion must be used with caution after examination of the $\Delta pM–pH$ relationship.

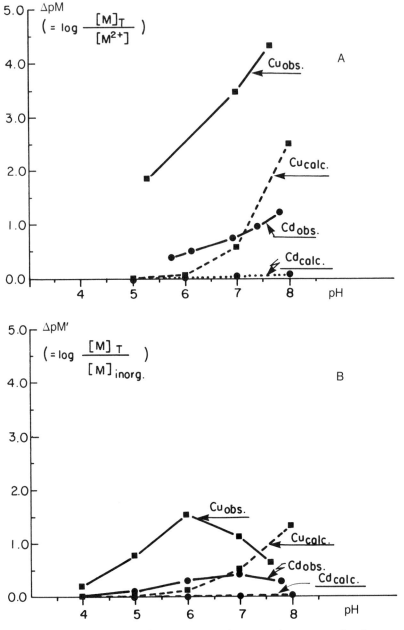

Figure 2. Variation in the degree of complexation of copper and cadmium as a function of pH, where A is spiked river water, and B is humic acid solution. The calculated values of $\Delta pM$ were derived for case 7 in Table III, where $\Delta pM = log\ ([M]_T \div [M^{z+}])$ for A, and $\Delta pM' = log\ ([M]_T \div [M]_{inorganic})$ for B. (Data are from references 18 and 34.)

Yet another type of behavior is exhibited by aluminum. Appreciably complexed at circumneutral pH (compare Cu and Pb), its apparent degree of complexation decreases only slightly over the pH range from 7 to 4 (Figures 3A and 3B). For example, Driscoll (*35*) spiked aliquots of Adirondack stream water (initial pH 4.3) with tris(hydroxymethyl)aminomethane buffer, incrementally adjusted the pH over a range of values with HCl, and allowed the resulting solutions to equilibrate for 1 week. Concentrations of nonexchangeable Al, as determined after passage of the samples through an ion-exchange column, proved relatively insensitive to variations in pH. Values of $\Delta$pAl', calculated from a ratio ([total monomeric Al] $\div$ [inorganic monomeric Al]), decreased only slightly over the studied pH range (Figure 3A). Using a batch ion-exchange technique (*36*), we obtained similar results for unbuffered stream water samples (Figure 3B). Naturally occurring DOM appears to contain components capable of binding Al even at pH values as low as 4. Theoretical values of $\Delta$pAl', derived from equilibrium calculations (Table III, case 7), show a somewhat greater pH sensitivity and pass through a distinct maximum at about pH 6 and then decrease below 0.15 on further acidification (Figure 3C).

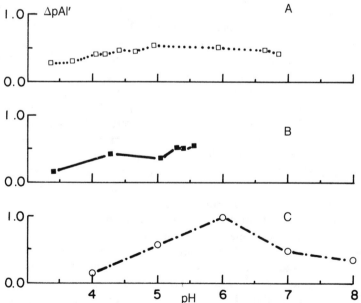

Figure 3. *Variation in the degree of complexation of aluminum as a function of pH, where A is an unnamed stream in the Adirondack Mountains; B is tributary R–5 in Lake Tantare, Quebec; and C represents calculated values derived for case 7 in Table III, where $\Delta$pAl' = log ([Al]$_T$ $\div$ [Al]$_{inorganic}$).* [*Data are from references 35 (A) and 42 (B).*]

NATURAL pH CHANGES.    In several studies, metal speciation measurements have been performed on surface water samples collected along a natural pH gradient (Table VIII). At first glance, such results would appear to offer the possibility of comparing field observations with the pH titration experiments described previously. However, other key parameters (e.g., [DOM], $[M]_T$, and [Ca + Mg]) will also tend to vary along the pH gradient, and isolating the influence of differences in sample pH is difficult.

One of the more successful attempts in this area is that of Borg and Andersson (37), who determined the concentrations of total and dialyzable metals (Al, Cd, Cu, and Pb) in 17 remote Swedish lakes. The studied lakes covered a relatively wide range in total organic carbon content (3–18 mg/L) and in pH values (4.6–6.9). In an average lake, the degree of complexation generally decreased in the order Pb ~ Al > Cu > Cd. The dialyzable portion of Cd and Al increased with decreasing pH (Figures 4A and 4B), as did that for Pb in waters with low DOM concentrations. In more humic waters, the dialyzable Pb fraction was small and was not influenced by pH (Figure 4C). The percent dialyzable Cu varied considerably (40%–95%); a negative correlation with the DOM concentration was observed, but there was no dependence on pH.

Similar field data have been collected for aluminum speciation in areas of North America subject to acid precipitation (Table VIII, cases 2 and 3). The analytical procedures used in these two studies distinguish between exchangeable and nonexchangeable forms of aluminum (35–36). In both areas studied, levels of exchangeable aluminum [$Al^{3+}$, $Al(OH)_x$, and $Al(F)_y$] were higher in the low pH surface waters; these levels reflect the pH-dependent solubility of $Al(OH)_3$ (s) (38). Concentrations of nonexchangeable Al (i.e., associated with fulvic and humic acids) were independent of pH but did show a significant positive correlation with the organic carbon concentrations. In other words, observed values of $\Delta pAl'$, which is equal to log ([dissolved Al] ÷ [exchangeable Al]), were indeed lower in the more acidic surface waters, but this difference seems to reflect a concomitant increase in dissolved and exchangeable Al more than any pH-induced dissociation of Al–fulvic or Al–humic species. A similar trend has been reported by Wright and Skogheim (39) in their study of changes in Al speciation occurring at the interface of an acid stream (pH 4.5) and a limed lake (pH 5.5). Marked decreases in concentration were noted among the inorganic forms of Al, but the fulvic or humic fraction did not change in concentration along the pH gradient.

## Conclusions

In this study of the influence of environmental acidification on metal biogeochemistry, we emphasized a single aspect, pH-related changes

**Table VIII. Field Evidence for pH-Related Changes in the Speciation of Dissolved Metals**

| Reference | Metal | Sample Description | pH Range | DOC[a] Range (mg/L) | Analytical Technique |
|---|---|---|---|---|---|
| 37 | Al, Cd, Cu, Pb | Remote Swedish lakes ($N = 17$) | 4.6–6.9 | 3–18 | In situ dialysis |
| 46 | Al | Remote Quebec lakes ($N = 50$) | 4.9–7.7 | 1–11 | Ion exchange |
| 35, 47 | Al | Remote Adirondack lakes and streams ($N = 322$) | 4.0–7.2 | 1–15 | Ion exchange |
| 39 | Al | Swedish stream (mixing zone limed lake, $N = 11$) | 4.5–5.5 | ~4 | Ion exchange |

[a]DOC denotes dissolved organic carbon.

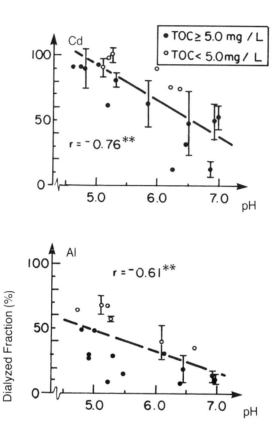

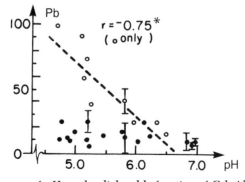

*Figure 4. Influence of pH on the dialyzable fraction of Cd, Al, and Pb in 17 Swedish lakes. The levels of significance* (p) *are as follows:* ° *denotes* p <0.05, *and* °° *denotes* p <0.01. *(Data are from reference 37.)*

in metal speciation, to the exclusion of all others. To this partial analysis should be added a consideration of other phenomena of importance in the context of acid deposition: the geochemical mobilization of metals in response to acidification, the intrinsic toxicity of the metals involved, their potential for bioaccumulation, and their interaction with the hydrogen ion at biological membranes. Only then will a clear picture of the consequences of environmental acidification emerge.

## Abbreviations and Symbols

| | |
|---|---|
| $\Delta pM$ | degree of complexation of metal M, $\Delta pM = \log([M]_T/[M^{z+}])$, and $\Delta pM' = \log([M]_T/[M]_{inorganic})$ |
| DOC | dissolved organic carbon |
| TOC | total organic carbon |
| DOM | dissolved organic matter |
| $E_H$ | redox potential (hydrogen scale) |
| $E_p$ | stripping potential |
| $i_p$ | stripping current |
| $[L]_T$ | total dissolved ligand concentration |
| M | metal (Ag, Al, Cd, Co, Cu, Hg, Mn, Ni, Pb, or Zn) |
| $[M]_T$ | total dissolved metal concentration |
| $[M^{z+}]$ | free metal ion concentration (aquo complex) |

## Acknowledgments

We wish to thank C. T. Driscoll for kindly providing some unpublished data relating to aluminum speciation. Helpful comments on an earlier version of this paper were furnished by J. Buffle and P. M. Stokes. This work was supported by the Canadian Natural Sciences and Engineering Research Council (NSERC) and by the Quebec Fonds pour la Formation de Chercheurs et l'Aide à la Recherche (FCAR).

## References

1. Buffle, J. *Trends Anal. Chem.* **1981**, *1*, 90-95.
2. Stumm, W.; Morgan, J. J. *Aquatic Chemistry: An Introduction Emphasizing Chemical Equilibria in Natural Waters*; Wiley: New York, 1981.
3. Morel, F. M. M. *Principles of Aquatic Chemistry*; Wiley: New York, 1984.
4. Dickson, W. In *Ecological Impact of Acidic Precipitation*; Drablos, D.; Tollan A., Eds.; SNSF Project: Oslo, 1980; pp 75-83.
5. Schindler, D. W.; Turner, M. A. *Water, Air, Soil Pollut.* **1982**, *18*, 259-271.
6. Campbell, P. G. C.; Stokes, P. M. *Can. J. Fish. Aquat. Sci.* **1985**, *42*, 2034-2049.
7. Campbell, P. G. C.; Stokes, P. M.; Galloway, J. N. *Acid Deposition—Effects on Geochemical Cycling and Biological Availability of Trace Elements*; U.S. National Academy of Sciences, Royal Society of Canada, Academia de la

Investigacion Cientifica of Mexico. Tri-Academy Committee on Acid Deposition, Sub-Group on Metals, National Academy of Sciences: Washington, DC, 1985.

8. Campbell, P. G. C.; Stokes, P. M.; Galloway, J. N. In *Proceedings, International Conference on Heavy Metals in the Environment, Heidelberg*; CEP Consultants: Edinburgh, United Kingdom, 1983; pp 760-763.

9. Florence, T. M. *Talanta* **1982**, *29*, 345-364.

10. Florence, T. M.; Batley, G. E. *CRC Crit. Rev. Anal. Chem.* **1980**, *9*, 219-296.

11. *Chemical Modeling in Aqueous Systems*; Jenne, E. A., Ed.; ACS Symposium Series 93; American Chemical Society: Washington, DC, 1979.

12. Nordstrom, D. K.; Plummer, L. N.; Wigley, J. M. L.; Wolery, T. J.; Ball, J. W.; Jenne, E. A.; Bassett, R. L.; Crerar, D. A.; Florence, T. M.; Fritz, B.; Hoffman, M.; Holdren, G. R.; Lafon, G. M.; Mattigod, S. V.; McDuff, R. E.; Morel, F.; Reddy, M. M.; Sposito, G.; Thrailkill, J. Ibid., pp 857-892.

13. Perdue, E. M.; Lytle, C. R. In *Aquatic and Terrestrial Humic Materials*; Christman, R. F.; Gjessing, E. T., Eds.; Ann Arbor Science: Ann Arbor, MI, 1983; pp 295-313.

14. Batley, G. E. In *Trace Element Speciation in Surface Waters and Its Ecological Implications*; Leppard, G. G., Ed.; Plenum: New York, 1983; pp 17-36.

15. Shuman, M. S.; Michael, L. C. *Anal. Chem.* **1978**, *50*, 2104-2109.

16. Brezonik, P. J.; Brauner, P. A.; Stumm, W. *Water Res.* **1976**, *10*, 605-612.

17. Guy, R. D.; Chakrabarti, C. L. *Chem. Can.* **1976**, 26-29.

18. Guy, R. D.; Chakrabarti, C. L. *Can. J. Chem.* **1976**, *54*, 2600-2611.

19. Figura, P.; McDuffie, B. *Anal. Chem.* **1977**, *49*, 1950-1953.

20. Livingstone, D. A. *Chemical Composition of Rivers and Lakes*; Paper 440G; U.S. Geological Survey. U.S. Government Printing Office: Washington, DC, 1963; pp G1-G64.

21. Meybeck, M. *Rev. Geol. Dyn. Geogr. Phys.* **1979**, *21*, 215-246.

22. Lachance, M.; Bobée, B.; Grimard, Y. *Water, Air, Soil Pollut.* **1985**, *25*, 115-132.

23. Langlois, C.; Lemay, A.; Ouzilleau, J.; Vigneault, Y. *Can. Tech. Rep. Fish. Aquat. Sci.* **1985**, *1792*.

24. Driscoll, C. T.; Newton, R. M. *Environ. Sci. Technol.* **1985**, *19*, 1018-1024.

25. Buffle, J. In *Metal Ions in Biological Systems*; Sigel, H., Ed.; Marcel Dekker: New York, 1984; Vol. 18, pp 165-221.

26. Mantoura, R. F. C.; Dickson, A.; Riley, J. P. *Estuarine Coastal Mar. Sci.* **1978**, *6*, 387-408.

27. Sposito, G. *Environ. Sci. Technol.* **1981**, *15*, 396-403.

28. Gamble, D. S.; Underdown, A. W.; Langford, C. H. *Anal. Chem.* **1980**, *52*, 1901-1908.

29. Wilson, D. E.; Kinney, P. *Limnol. Oceanogr.* **1977**, *22*, 281-289.

30. Nriagu, J. O.; Gaillard, J. F. *Adv. Environ. Sci. Technol.* **1984**, *15*, 349-374.

31. Bowen, H. J. M. *Environmental Chemistry of the Elements*; Academic: London, 1979.

32. Vuceta, J.; Morgan, J. J. *Environ. Sci. Technol.* **1978**, *12*, 1302-1309.

33. Morel, F.; McDuff, R. E.; Morgan, J. J. In *Trace Metals and Metal-Organic Interactions in Natural Waters*; Singer, P. C., Ed.; Ann Arbor Science: Ann Arbor, MI, 1973; pp 157-200.

34. Engel, D. W.; Sunda, W. G.; Fowler, B. A. In *Biological Monitoring of Marine Pollutants*; Vernberg, F. J.; Calabrese, A.; Thurberg, F. P.; Vernberg, W. B., Eds.; Academic: New York, 1981; pp 127-144.

35. Driscoll, C. T. *Intern. J. Environ. Anal. Chem.* **1984**, *16*, 267-283.

36.  Campbell, P. G. C.; Bisson, M.; Bougie, R.; Tessier, A.; Villeneuve, J. P. *Anal. Chem.* **1983,** *55,* 2246–2252.
37.  Borg, H.; Andersson, P. *Verh. Int. Ver. Theor. Angew. Limnol.* **1984,** *22,* 725–729.
38.  Johnson, N. M.; Driscoll, C. T.; Eaton, J. S.; Likens, G. E.; McDowell, W. H. *Geochim. Cosmochim. Acta* **1981,** *45,* 1421–1437.
39.  Wright, R. F.; Skogheim, O. K. *Vatten* **1983,** *39,* 301–304.
40.  Guy, R. D. Ph.D. Dissertation, Carleton University, Ottawa, Canada, 1976.
41.  Guy, R. D.; Chakrabarti, C. L.; Schramm, L. L. *Can. J. Chem.* **1975,** *53,* 661–669.
42.  Campbell, P. G. C.; Thomassin, D., INRS–Eau, Université du Québec, Sainte-Foy, Québec, Canada, unpublished data, 1985.
43.  Turner, D. R.; Whitfield, M.; Dickson, A. G. *Geochim. Cosmochim. Acta* **1981,** *45,* 855–881.
44.  Gardiner, J. *Water Res.* **1974,** *8,* 23–30.
45.  McBride, M. B.; Bouldin, D. R. *Soil Sci. Soc. Am. J.* **1984,** *48,* 56–59.
46.  Campbell, P. G. C.; Bougie, R.; Tessier, A.; Villeneuve, J. P. *Verh. Int. Ver. Theor. Angew. Limnol.* **1984,** *22,* 371–375.
47.  Driscoll, C. T.; Baker, J. P.; Bisogni, J. J.; Schofield, C. J. In *Geological Aspects of Acid Deposition*; Bricker, O. P., Ed.; Butterworth: Boston, 1984; pp 55–75.
48.  Westall, J. C.; Zachary, J. L.; Morel, F. M. M. "MINEQL, a Computer Program for the Calculation of the Chemical Equilibrium Composition of Aqueous Systems"; Technical Report No. 18; Department of Civil Engineering, Massachusetts Institute of Technology: Cambridge, MA, 1976.

RECEIVED for review May 6, 1986. ACCEPTED October 10, 1986.

# 8

# Ion Budgets in a Seepage Lake

James C. Lin[1,3], Jerald L. Schnoor[1], and Gary E. Glass[2]

[1]Department of Civil and Environmental Engineering, University of Iowa, Iowa City, IA 52242
[2]Environmental Research Laboratory—Duluth, U.S. Environmental Protection Agency, Duluth, MN 55804

*On the basis of an existing hydrologic model (trickle-down model), an ion budgets approach was applied to study the internal processes of Vandercook Lake in Wisconsin. The results indicated that alkalinity was produced by in-lake processes. Nitrogen transformations were significant, but their role in the acid–base budget was negligible because the acidifying influence of the ammonium loss was roughly balanced by the alkalizing effect of the nitrate loss. Sulfate reduction was important and accounted for 54% of the alkalinity production.*

INPUT-OUTPUT BUDGETS ARE USED IN THIS CHAPTER to estimate the internal chemical reactions that occur in lakes. As a case study, a small seepage lake in north central Wisconsin illustrates the importance of accurate hydrologic balances for dilute lakes that receive groundwater inputs. Vandercook Lake receives a small fraction of its water from groundwater inputs, but this fraction represents a significant portion of the alkalinity and cation contributions to the lake. Vandercook Lake has a low conductivity of ~15 $\mu$S/cm and Gran alkalinity of 20 $\mu$equiv/L. Thus, this lake is sensitive to inputs from acid precipitation.

Figure 1 shows several processes that modify the chemistry of lakes (*1*). These processes can occur externally or internally, in the terrestrial watershed or in the lake itself. Chemical weathering is an alkalizing process in the watershed whereby dissolved $CO_2$ or $H^+$ attacks minerals in soils, sediments, and rock and serves to increase alkalinity either by producing bicarbonate or by consuming hydrogen ion. Production of biomass, such as an aggrading canopy in the terrestrial environment or

[3]Current address: Aqua Terra Consultants, Mountain View, CA 94043

0065-2393/87/0216-0209$06.00/0

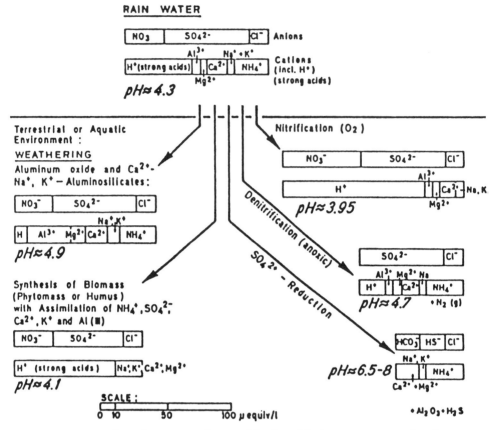

Figure 1. *Selected processes that modify the $H^+$ balance in terrestrial and aqueous environments. (Reproduced with permission from reference 1. Copyright 1985, Wiley.)*

net algal production in lakes, is often an acidifying influence due to the uptake of ammonium and basic cations ($Ca^{2+}$, $Mg^{2+}$, $Na^+$, and $K^+$) relative to acidic anions ($SO_4^{2-}$, $NO_3^-$, and $Cl^-$). The extent of acidification due to biomass synthesis is determined by the stoichiometry of the biomass and the chemical species. Nitrification and other oxidation reactions are acidifying influences. Desulfurization (sulfate reduction) and other reduction reactions are alkalizing processes. Partial decomposition of organic matter to form organic acids can cause the pH to decrease in colored, water-dystrophic systems.

The purpose of this study is to present a simple methodology, input–output mass balances or ion budgets, to estimate the important processes that modify lake chemistry. Identifying and quantifying the various watershed reactions are possible by comparing the magnitude of mass inputs and outputs (2). The ecosystem study on the Hubbard Brook Experimental Forest (3) serves as an example that illustrates the use of chemical budgets.

Seepage lakes, lakes that have no permanent inlets or outlets, are a large class of lakes sensitive to acidic deposition. The northern Midwest has one of the nation's greatest concentrations of natural lakes. The state of Wisconsin alone has about 14,000 lakes of which nearly 10,000 are classified as seepage type (4). Seepage lakes can be dominated by atmospheric inputs, and the water chemistry can be quite similar to precipitation chemistry in the most sensitive systems (5). Hydrologic flow paths through the watershed and the depth of flow through glacial till are other important factors that control the alkalinity of lakes and streams (6). Therefore, hydrologic as well as geological factors are critical with respect to lake chemistry and acidic deposition (7).

In general, seepage lakes are located at or near groundwater table "highs" in the intermediate or regional flow system. These lakes gain most of their water by precipitation and lose most of the water by evaporation and, to a lesser extent, by seepage to the groundwater aquifer.

## Site Description

Vandercook Lake (Figure 2) is located in north central Wisconsin in the vicinity of Rhinelander. It has a surface area of 40 ha, a mean depth of 3.5 m, a topographic watershed of 130 ha, and an estimated mean hydraulic residence time of 4.5 years. Vandercook Lake can be described as a soft water lake with an alkalinity in the range of 13–25 $\mu$equiv/L and is sensitive to acidic deposition. It receives acidic precipitation with a volume-weighted average pH of approximately 4.67.

Bedrock is composed of crystalline gneiss, and surficial glacial deposits are 30–60 m thick. This thickness essentially isolates the lake from bedrock. Soils are generally thin and acidic (Pence sandy loam and Rubicon sand), and glacial deposits, consisting mostly of medium-to-fine quartz sand with only traces of feldspar and clay minerals, are poor in buffering capacity. The region is forested with second-growth mixed stands of coniferous and deciduous trees.

Figure 2 indicates the locations of on-site instruments around the lake. Vandercook Lake was equipped with a bubble gauge for lake stage measurement. Twenty-two single piezometers and six piezometer nests were installed around the lake by the U.S. Geological Survey (USGS) at Madison and the Wisconsin Department of Natural Resources (WDNR) to measure the direction and magnitude of groundwater flow in the watershed.

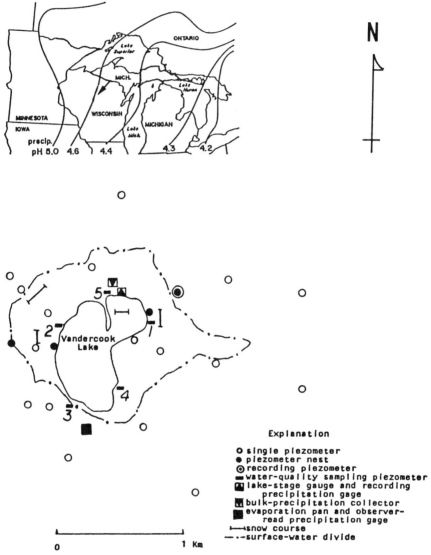

Figure 2. Location of Vandercook Lake and the positions of on-site instruments.

## Experimental Details

Lake water samples were collected with a 3.2-L nonmetallic Kemmerer sampling bottle (Wildco model 1540–C25) at a depth of 1 m and at either middepth if homothermic, or at 1 m off the bottom if stratified. Sample containers, 1-L linear polyethylene and 165-mL polypropylene (Nalgene) bottles, were rinsed twice with deionized water and twice with site water before collection of sample.

Five water quality wells for groundwater sampling were installed near shore around the lake. These wells were constructed of 38-mm polyvinyl chloride (PVC) pipe and finished at the shallow water table. The three sampling devices used were a 2.3-cm diameter PVC bailer (June–July 1983), a plastic hand pump (Black & Decker Jack Rabbit) with 4-mm silicon tubing (September 1981–February 1982), and a peristaltic pump (Horizon Ecology model 7573-80) with 4-mm silicon tubing (August 1981, March 1982, and thereafter).

Groundwater and lake water samples were filtered on-site soon after collection with a peristaltic pump (Geofilter Series 1) equipped with a pump head (Mastyerflex 7015) and silicon tubing. Water was filtered through a 0.45-$\mu$m membrane filter (Millipore) held in a 142-mm backwash stand into a prerinsed 1-L bottle. Additional aliquots were collected into 165-mL bottles for WDNR analysis. Samples were kept on ice in coolers.

Field pH and conductance measurements were made by WDNR upon return to the laboratory the same evening. Major ion and metal concentrations of a 1981 subset of the samples were determined at the U.S. Environmental Protection Agency (EPA), Environmental Research Laboratory—Duluth (ERL—D). The remaining anions and metals analyses were provided by ERL—Bellingham and ERL—Corvallis. All the measurements are in accordance with the EPA (8) and revisions to date. Precision was estimated by the analysis of field replicates collected with every 10 samples or once each day, whichever was less. The data for precipitation chemistry were from the record at nearby Trout Lake National Atmospheric Deposition Program (NADP) station.

## Computation of Ion Budgets

The concept of a mass balance is very simple: the change in mass storage equals mass inflow minus mass outflow, plus or minus reactions. Conceptually, the mass balance can be represented by

$$\Delta \text{storage} = \Sigma \text{inflow} - \Sigma \text{outflow} \pm \text{reactions} \tag{1}$$

Assuming a completely mixed lake, the concentration of ions within the lake is equal to the concentration of ions in the outflow. The mass balance equation can be expressed as

$$\frac{\Delta(C_{\text{lake}} \cdot V)}{\Delta t} = (Q_{\text{precip}} \cdot C_{\text{precip}}) + (Q_{\text{trib}} \cdot C_{\text{trib}}) +$$

$$(Q_{\text{GW}_{\text{in}}} \cdot C_{\text{GW}}) - (Q_{\text{out}} \cdot C_{\text{lake}}) -$$

$$(Q_{\text{GW}_{\text{out}}} \cdot C_{\text{lake}}) \pm \text{reactions} \tag{2}$$

where $C$ is concentration (equiv/m$^3$), $Q$ is flow rate (m$^3$/day), $V$ is lake volume (m$^3$), $t$ is the time step, GW is groundwater, in is inflow, out is outflow, precip is precipitation, and trib is tributary.

For seepage lakes, the mass inflow from tributaries and the mass outflow from a stream discharge may be quite small or even equal to zero. Therefore, this input and output can be discarded in the chemical budget computation. This smallness of inflow and outflow is especially true for Vandercook Lake. The last term in equation 2 (±reactions) can be visualized as either production or consumption of the ion via physical, chemical, or biological processes. The reaction term can be estimated by the difference of the other measured terms in equation 2. After rearrangement, equation 3 results.

$$\pm\text{reactions} = \frac{\Delta(C_{lake} \cdot V)}{\Delta t} - (Q_{precip} \cdot C_{precip}) -$$
$$(Q_{GW_{in}} \cdot C_{GW}) + (Q_{GW_{out}} \cdot C_{lake}) \quad (3)$$

Equation 3 was applied to Vandercook Lake with a monthly time step ($\Delta t$). Precipitation measurements were made on a daily basis and groundwater flow rates were calculated on a daily basis from the hydrologic model described by Lin and Schnoor (9) and Banwart (10). The chemistry concentrations of lake and groundwater were measured on a monthly basis during much of the year, but were performed approximately every 6 weeks during the winter. The data of NADP precipitation chemistry were reported weekly. The volume of the lake was estimated from the hydrologic simulation by Lin and Schnoor (9), and this volume estimate was checked with field measurements of lake stage.

Regardless of how accurate the chemical determinations are, the chemical budgets could have considerable error if the water budget has error. The importance of a good water budget is certainly without doubt. Winter (11–13) published a series of papers dealing with the possible errors in the measurement of hydrologic components. As he pointed out, among all the components, the interactions between lake and groundwater flow are the most difficult and troublesome aspects. This interaction is especially important for seepage lakes.

## Computation of Water Budget

The quantities of the hydrologic components are evaluated by the simulation of the trickle-down model (10, 14). The hydrologic model is the main driving subroutine of the trickle-down model. A schematic diagram for the hydrologic model is shown in Figure 3. A mass balance is computed for the snow, unsaturated zone, surface water (lake), and groundwater compartments. Five hydrologic equations are used to describe the dynamic nature among these compartments. Euler's method is applied to solve these equations simultaneously.

The most important hydrologic flow paths are from the atmosphere to the lake (direct precipitation), from the lake to the atmosphere (evaporation), and from the lake to the groundwater (seepage). The

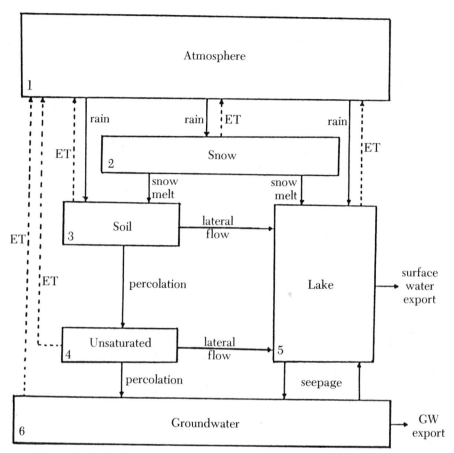

*Figure 3. Hydrologic model compartments. (Reproduced with permission from reference 9. Copyright 1986, American Society of Civil Engineers.)*

local relief is quite small, and most of the precipitation that falls in the topographic watershed does not enter the lake but percolates to groundwater. Thus, the topographic watershed does not correspond to the hydrologic watershed.

The hydrologic model was calibrated until agreement was achieved between predictions and measurements for lake stage and groundwater elevation. Simulation results for lake stage and groundwater level are shown in Figure 4 with the monthly precipitation data. Model results were within ±1 standard deviation of field data, collected at various stations and times, more than 90% of the time. A complete monthly hydrologic budget is included in Table I. Simulating the measured lake stage and groundwater level more closely was possible if time-variable lateral flow rates and percolation rates were assumed, but

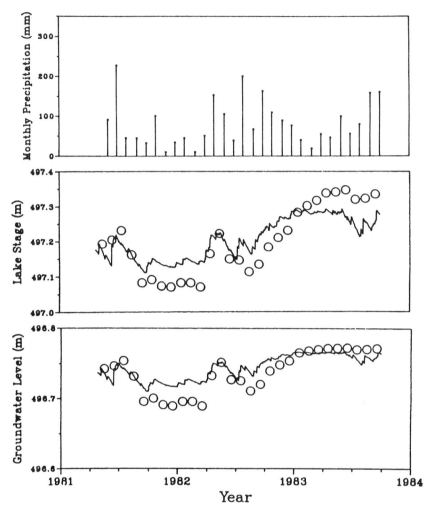

*Figure 4. Monthly precipitation data and simulation result for lake stage and groundwater level at the near-shore piezometer. Key: —— denotes simulation, and ○ denotes field data. (Reproduced with permission from reference 9. Copyright 1986, American Society of Civil Engineers.)*

for simplicity, the simulation results depicted in Figure 4 were obtained with constant hydrologic coefficients.

## Application of Bar Diagrams

Bar diagrams are a convenient method to analyze variations among lakes and their biogeochemical processes. By inspecting the bar diagram, one can surmise the reactions necessary to change precipita-

## Table I. Monthly Hydrologic Budget

| | | | Inflow | | Outflow | | |
|---|---|---|---|---|---|---|---|
| Date | Atmos. | Soil[a] | Unsat. Zone | GW | Atmos. | GW | ΔStorage |
| May 1981 | 3.343 | 0.000 | 0.001 | 0.084 | 3.177 | 1.223 | −0.972 |
| June 1981 | 8.110 | 0.000 | 0.019 | 0.083 | 3.271 | 1.171 | 3.769 |
| July 1981 | 2.850 | 0.000 | 0.012 | 0.079 | 3.862 | 1.275 | −2.196 |
| Aug. 1981 | 1.765 | 0.000 | 0.000 | 0.085 | 3.062 | 1.215 | −2.428 |
| Sept. 1981 | 0.960 | 0.000 | 0.000 | 0.089 | 2.184 | 1.115 | −2.250 |
| Oct. 1981 | 4.390 | 0.000 | 0.011 | 0.091 | 0.987 | 1.162 | 2.343 |
| Nov. 1981 | 0.195 | 0.000 | 0.001 | 0.087 | 0.263 | 1.129 | −1.109 |
| Dec. 1981 | 0.865 | 0.000 | 0.001 | 0.092 | 0.116 | 1.151 | −0.308 |
| Jan. 1982 | 2.679 | 0.000 | 0.000 | 0.091 | 0.078 | 1.162 | 1.529 |
| Feb. 1982 | 0.586 | 0.000 | 0.001 | 0.080 | 0.115 | 1.067 | −0.515 |
| March 1982 | 1.348 | 0.000 | 0.006 | 0.090 | 0.219 | 1.169 | 0.056 |
| April 1982 | 6.754 | 0.000 | 0.053 | 0.080 | 0.576 | 1.196 | 5.116 |
| May 1982 | 3.499 | 0.000 | 0.019 | 0.076 | 2.762 | 1.308 | −0.477 |
| June 1982 | 1.780 | 0.000 | 0.000 | 0.079 | 3.675 | 1.211 | −3.027 |
| July 1982 | 6.845 | 0.000 | 0.018 | 0.084 | 3.906 | 1.230 | 1.811 |
| Aug. 1982 | 2.005 | 0.000 | 0.005 | 0.082 | 2.832 | 1.242 | −1.981 |
| Sept. 1982 | 7.925 | 0.000 | 0.025 | 0.076 | 2.256 | 1.236 | 4.535 |
| Oct. 1982 | 4.655 | 0.000 | 0.023 | 0.073 | 2.143 | 1.349 | 1.258 |
| Nov. 1982 | 3.760 | 0.000 | 0.017 | 0.068 | 0.524 | 1.335 | 1.986 |
| Dec. 1982 | 2.013 | 0.000 | 0.006 | 0.067 | 0.178 | 1.414 | 0.494 |
| Jan. 1983 | 1.990 | 0.000 | 0.000 | 0.064 | 0.141 | 1.458 | 0.455 |
| Feb. 1983 | 1.254 | 0.000 | 0.005 | 0.058 | 0.205 | 1.316 | −0.204 |
| March 1983 | 1.892 | 0.000 | 0.030 | 0.065 | 0.231 | 1.448 | 0.308 |
| April 1983 | 1.972 | 0.000 | 0.014 | 0.062 | 0.469 | 1.412 | 0.167 |
| May 1983 | 3.387 | 0.000 | 0.009 | 0.064 | 2.069 | 1.451 | −0.059 |
| June 1983 | 3.450 | 0.000 | 0.008 | 0.063 | 3.682 | 1.394 | −1.554 |
| July 1983 | 2.620 | 0.000 | 0.004 | 0.070 | 4.455 | 1.376 | −3.137 |
| Aug. 1983 | 6.560 | 0.000 | 0.017 | 0.072 | 3.528 | 1.356 | 1.765 |
| Sept. 1983 | 6.534 | 0.000 | 0.020 | 0.068 | 2.338 | 1.337 | 2.946 |
| SUM | 95.986 | 0.000 | 0.325 | 2.219 | 53.303 | 36.909 | 8.318 |

SOURCE: Reproduced with permission from reference 9. Copyright 1986, American Society of Civil Engineers.
NOTE: Abbreviations are as follows: Atmos. is atmosphere and Unsat. Zone is unsaturated zone. All units are in inches of water per month per surface area of the lake (40 ha).
[a]For hydrologic simulation, snowmelt was included in the compartment calculation.

tion chemistry to lake water chemistry. Figure 5 shows the bar diagrams of average water chemistry for Vandercook Lake, groundwater chemistry for well number 1, and precipitation chemistry at Trout Lake NADP station. On a normality basis, the ratio of total cations to total anions was 0.932 for the lake water, 0.99 for the groundwater, and 0.979 for the precipitation. A good charge balance of ions was thus obtained.

The ratio of chloride concentration in lake water to chloride concentration in precipitation was approximately 3:1 (Figure 5). Assum-

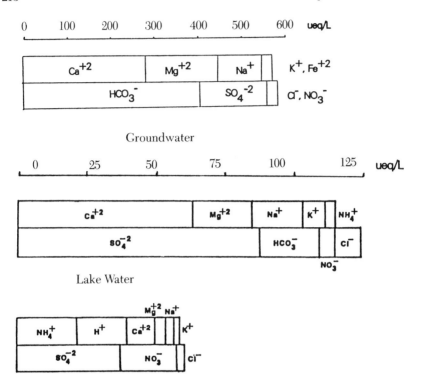

Figure 5. *Bar diagrams of water chemistry for the lake, groundwater, and precipitation.*

ing that chloride ion is conservative in the lake watershed (no reactions), the concentration ratio gives the ratio of total inflows to the lake (precipitation and groundwater) to the outflows from the lake (groundwater seepage) or the evapoconcentration factor. According to the hydrologic simulation, this ratio for Vandercook Lake was 2.7, which is in relative agreement with the chloride ion concentration ratio.

By comparison of these diagrams, the increases of basic cations ($Ca^{2+}$, $Mg^{2+}$, $Na^+$, and $K^+$) and bicarbonate anion ($HCO_3^-$) are quite apparent for lake water compared with precipitation. Besides the influence of evapoconcentration, the phenomena may be attributed to terrestrial and aquatic reactions of ion exchange and chemical weathering. However, causes for the increase of sulfate (evapoconcentration) and the decrease of nitrate (algae uptake or denitrification) were more ambiguous. Clearly, both ammonium ion and nitrate ion were removed from the lake by biological processes. Sulfate was removed from the water column by processes such as sulfate reduction and ester sulfate formation in sediments (*15*), but evaporation served to increase the concentration of sulfate in lake water compared with precipitation. The sulfate concentration in Vandercook Lake should have been ~104 $\mu$equiv/L based on a 2.7-fold evaporative concentration factor of precipitation. Instead, the average sulfate concentration in the lake, ~85 $\mu$equiv/L, indicated a biological removal of about 19 $\mu$equiv/L. Sulfate reduction appears to have occurred, but for a more quantitative estimate, ion budgets are very useful.

## Results of Chemical Budget

On the basis of a monthly water budget, together with chemical measurements of inputs and lake water, the ion budgets were computed. The time period for monthly calculations was from May 1981 to September 1983. (Results were similar for a 2-year period from May 1981 to April 1983.) Ten chemical species were investigated, including six cations ($Ca^{2+}$, $Mg^{2+}$, $Na^+$, $K^+$, $NH_4^+$, and $Fe^{3+/2+}$), three anions ($SO_4^{2-}$, $Cl^-$, and $NO_3^-$), and alkalinity. Table II demonstrates the computation and shows the monthly budgets for alkalinity. During the 29-month period, net alkalinity production in the lake was 32.2 kequiv, which is equivalent to 33.2 mequiv/m$^2$ year.

Table III summarizes the budgets for all 10 chemical species. As one might expect, there was a net production of all the base cations due to dissolution reactions and ion exchange at the sediment–water interface in the lake. There was a net loss of sulfate of 18.0 mequiv/m$^2$ year. Assuming that the net loss of sulfate was due to a reduction reaction or an ion exchange–sorption reaction, an equivalent amount of alkalinity will also be produced according to reactions 4 and 5.

$$\overset{\text{bacteria}}{SO_4^{2-} + 2CH_2O \longrightarrow H_2S + 2HCO_3^-} \tag{4}$$

$$SO_4^{2-} + \ \underset{\substack{\text{sesquoxide} \\ \text{exchanger}}}{R}\overset{\substack{\text{HO} \quad \text{OH} \\ \diagdown \quad \diagup}}{} \longrightarrow R\text{-}SO_4 + 2OH^- \tag{5}$$

Comparing the budgets for alkalinity and sulfate, 40% of total input sulfate was reduced, but this amount is not large enough to account for all the alkalinity produced. Approximately 54% of alkalinity production is caused by sulfate reduction. Two studies (*16, 17*) reported that in the Experimental Lakes Area (ELA), northwestern Ontario, sulfate reduc-

Table II. Monthly Alkalinity Budget for Vandercook Lake

| Date | RXNS[a] | GW_{out} | Precip_{in} | GW_{in} | ΔStorage |
|------|---------|----------|-------------|---------|----------|
| May 1981 | 1.03 | 0.16 | −0.43 | 0.26 | 0.69 |
| June 1981 | 1.67 | 0.16 | −1.07 | 0.25 | 0.70 |
| July 1981 | −2.94 | 0.15 | −0.97 | 0.28 | −3.81 |
| Aug. 1981 | 14.76 | 0.24 | −0.48 | 0.32 | 14.36 |
| Sept. 1981 | 3.98 | 0.25 | −0.18 | 0.32 | 3.86 |
| Oct. 1981 | −4.58 | 0.22 | −1.16 | 0.35 | −5.61 |
| Nov. 1981 | −1.69 | 0.20 | −0.14 | 0.34 | −1.69 |
| Dec. 1981 | 10.14 | 0.28 | −0.28 | 0.38 | 9.97 |
| Jan. 1982 | 10.68 | 0.35 | −0.72 | 0.38 | 9.99 |
| Feb. 1982 | 13.94 | 0.41 | −0.11 | 0.35 | 13.77 |
| March 1982 | 5.36 | 0.48 | −0.54 | 0.41 | 4.75 |
| April 1982 | −8.65 | 0.42 | −1.27 | 0.52 | −9.82 |
| May 1982 | −9.08 | 0.39 | −0.70 | 0.37 | −9.80 |
| June 1982 | −21.87 | 0.20 | −0.15 | 0.32 | −21.91 |
| July 1982 | −11.85 | 0.11 | −1.00 | 0.34 | −12.63 |
| Aug. 1982 | 15.03 | 0.22 | −0.29 | 0.33 | 14.85 |
| Sept. 1982 | −5.34 | 0.18 | −1.19 | 0.32 | −6.30 |
| Oct. 1982 | −0.46 | 0.18 | −1.31 | 0.36 | −1.59 |
| Nov. 1982 | 6.19 | 0.22 | −0.47 | 0.38 | 5.87 |
| Dec. 1982 | 6.27 | 0.28 | −0.40 | 0.33 | 5.92 |
| Jan. 1983 | −7.07 | 0.23 | −0.36 | 0.29 | −7.38 |
| Feb. 1983 | −0.44 | 0.20 | −0.32 | 0.24 | −0.72 |
| March 1983 | −2.59 | 0.20 | −0.36 | 0.27 | −2.88 |
| April 1983 | −0.29 | 0.19 | −1.05 | 0.26 | −1.26 |
| May 1983 | −0.51 | 0.18 | −1.02 | 0.27 | −1.44 |
| June 1983 | −4.83 | 0.13 | −1.04 | 0.27 | −5.73 |
| July 1983 | 5.72 | 0.17 | −0.84 | 0.29 | 5.01 |
| Aug. 1983 | 18.81 | 0.30 | −1.58 | 0.32 | 17.26 |
| Sept. 1983 | 0.77 | 0.29 | −0.68 | 0.29 | 0.09 |
| Total | 32.22 | 7.00 | −20.99 | 9.39 | 14.52 |

NOTE: All units are kiloequivalents. [a]RXNS = GW_{out} − Precip_{in} − GW_{in} + Δstorage.

Table III. Summary of Chemical Budgets in Vandercook Lake

| Ions | RXNS | GW_{out} | Precip_{in} | GW_{in} | ΔStorage |
|------|------|----------|-------------|---------|----------|
| Alkalai | 33.2 | 7.2 | −20.7 | 9.7 | 15.0 |
| $SO_4^{2-}$ | −18.0 | 34.3 | 41.9 | 3.6 | −6.8 |
| $NO_3^-$ | −25.8 | 2.0 | 20.4 | 0.1 | −7.3 |
| $Ca^{2+}$ | 27.1 | 26.6 | 11.2 | 6.9 | 39.6 |
| $Mg^{2+}$ | 20.0 | 11.4 | 3.7 | 4.0 | 25.7 |
| $Na^+$ | 1.2 | 6.6 | 2.3 | 2.1 | −1.0 |
| $K^+$ | 3.5 | 3.7 | 0.7 | 0.3 | 2.7 |
| $Cl^-$ | 2.7 | 3.1 | 2.6 | 0.3 | 2.5 |
| $NH_4^+$ | −20.6 | 1.0 | 18.8 | | −2.8 |
| $Fe^{3+/2+}$ | 2.2 | 0.5 | | 0.5 | 2.2 |

NOTE: All units are in milliequivalents per square meter per year. All measurements were calculated from May 1981 to September 1983.

tion accounted for 85% of alkalinity production in Lake 223 and 53% in Lake 239. If dry deposition of sulfate is considered, where dry deposition is estimated to be ~20% of wet deposition, then the amount of alkalinity produced by sulfate reduction is even greater but is counterbalanced by concomitant dry deposition of acidity that serves to increase the magnitude of the alkalinity RXNS term in Table III as well. However, the large terms for $\Delta$storage and the large reaction terms for $Ca^{2+}$ and $Mg^{2+}$ in Table III may indicate some problem with these budgets. Magnesium was less than detectable in the early portion of the study, and this fact may account for the large $\Delta$storage and RXNS terms. The failure to include bulk or dry deposition of calcium can account for a significant portion of the RXNS term for that ion.

Cation production and neutralization reactions, such as mineral dissolution and ion exchange, likely account for the remaining alkalinity generation in Table III. These reactions are indicated by reactions 6 and 7.

$$CaAl_2Si_2O_8 + 3H_2O \longrightarrow Ca^{2+} + Al_2Si_2O_5(OH)_4 + 2OH^- \quad (6)$$

$$\text{anorthite} \qquad\qquad\qquad \text{kaolinite}$$

$$Ca\text{-}X + 2H^+ \longrightarrow H_2X + Ca^{2+} \quad\quad (7)$$

$$\text{cation}$$
$$\text{exchanger}$$

Another important result from Table III is that the biological consumption of nitrate (an alkalizing process) is roughly balanced by the consumption of ammonium ion (an acidifying process). The possible reaction of nitrate may be denitrification, $NO_3^-$ assimilation, or both of these processes. In any case, the net effect is generation of alkalinity or consumption of $H^+$ in reactions 8 and 9.

$$\overset{\text{bacteria}}{5CH_2O + 4NO_3^- + 4H^+ \longrightarrow 5CO_2 + 2N_2 + 7H_2O} \quad (8)$$

$$\text{organic compounds}$$

$$NO_3^- + 5CO_2 + 4H_2O \xrightarrow[\text{sun}]{\text{algae}} C_5H_7NO_2 + 7O_2 + OH^- \quad (9)$$

$$\text{biomass}$$

Conversely, nitrification of ammonium or $NH_4^+$ assimilation by algae will consume alkalinity as in reactions 10 and 11.

$$NH_4^+ + 2O_2 \xrightarrow{\text{bacteria}} NO_3^- + H_2O + 2H^+ \qquad (10)$$

$$NH_4^+ + 4CO_2 + HCO_3^- + H_2O \xrightarrow[\text{sun}]{\text{algae}} C_5H_7NO_2 + 5O_2 \quad (11)$$

Figure 6 shows the laboratory-measured alkalinity values for precipitation, groundwater, and lake water. By inspection of Figure 6, the approximate 1:10 volumetric mixture of groundwater and precipitation necessary to obtain the measured alkalinity concentration of lake water can be determined. Such a calculation neglects internal reactions that may produce alkalinity, but it does illustrate that the groundwater contribution to Vandercook Lake is small. By using a completely mixed reactor model for the lake, ion budgets quantified the chemical inputs, outputs, and internal reactions. Precipitation and groundwater were treated as input variables, and water in the center of the lake at a depth of 1 m was taken to represent the outflow concentration. Thus, the flow component from groundwater should be about 2%–5% of the total volumetric inputs to the lake. Calculations indicate that a larger groundwater contribution would result in a surplus of calcium and other cations in the lake (Table IV). Sulfate concentrations for precipitation, groundwater, and lake water are shown in Figure 7. Lake water concentration of sulfate is usually higher than the precipitation concentration. The ratio of lake water sulfate to precipitation sulfate concentration was approximately 2.5. This value is in relatively good agreement with the evapoconcentration factor (2.7), which was computed from the hydrologic model. The difference (2.7 − 2.5) is due to sulfate reduction.

## Discussion

The considerable uncertainty in the budget for magnesium is due to concentrations falling below the limits of detection in the early part of the study before analytical methods were changed. Also, uncertainty in both the calcium and the magnesium ion budgets is due to the large $\Delta$storage term, which was caused by an increase in concentration as well as lake stage during the 29-month period of record.

The role of the nitrogen cycle in the acid–base balance for Vandercook Lake was negligible because the acidifying influence of the ammonium loss seems to be roughly balanced by the alkalizing effect of the nitrate loss. The small net production of $Cl^-$ is quite possibly a measure of the error in the $\Delta$storage term.

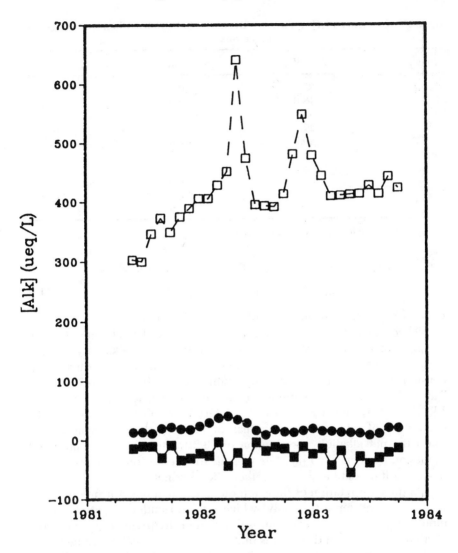

*Figure 6. Alkalinity values for precipitation, groundwater, and lake water. Key:* ■, *precipitation;* □, *groundwater; and* ●, *lake water.*

Alkalinity can be defined as an acid-neutralizing capacity (ANC) or alternatively as a charge balance (1). One way to check on the overall reactions of the ion budgets is to compare the magnitude of alkalinity reactions with the base cations minus the strong acid anion reactions.

$$[Alk] = ANC = [HCO_3^-] + 2[CO_3^{2-}] + [OH^-] - [H^+] + [A^-] \quad (12)$$

Table IV. Influence of Groundwater Recharge Rate on the Chemical Budget of
Vandercook Lake

| Ions | 2.3% $GW_{in}$ | 5.0% $GW_{in}$ | 15.0% $GW_{in}$ |
|------|------|------|------|
| Alkalai | 33.2 | 20.1 | −48.3 |
| $SO_4^{2-}$ | −18.0 | −22.5 | −31.5 |
| $NO_3^-$ | −25.8 | −25.7 | −17.7 |
| $Ca^{2+}$ | 27.1 | 18.4 | −58.1 |
| $Mg^{2+}$ | 20.0 | 15.0 | −32.7 |
| $Na^+$ | 1.2 | −1.5 | −11.7 |
| $K^+$ | 3.5 | 3.0 | 2.1 |
| $Cl^-$ | 2.7 | 2.3 | −1.3 |
| $NH_4^+$ | −20.6 | −20.6 | −20.6 |
| $Fe^{3+/2+}$ | 2.2 | 1.6 | −4.9 |

NOTE: All units are in milliequivalents per square meter per year.

$$[Alk] = [Na^+] + [K^+] + [NH_4^+] + 2[Ca^{2+}] + 2[Mg^{2+}]$$
$$+ 3[Fe^{3+}] - [Cl^-] - [NO_3^-] - 2[SO_4^{2-}] \qquad (13)$$

where $[A^-]$ is the concentration of organic acid anions.

Theoretically, the sum of the reaction terms of well-computed ion budgets should equal the reaction term from the alkalinity budget because each ion is tied to an equivalent alkalinity production or consumption (reactions 4–11). By summing all the reaction terms in Table III for the base cations minus the strong acid anions (equation 13), the net overall reaction can be determined to be equivalent to the generation of 74.5 mequiv/m² year of alkalinity, which is about 2.2 times the net alkalinity production of 33.2 mequiv/m² year computed in the ion budget. Two possible reasons for the discrepancy are lack of dry deposition data and error in the cation budgets.

In the computation of these ion budgets, the inputs for precipitation data were wet deposition only, which tends to underestimate the mass inputs, especially for calcium, sulfate, and hydrogen ion (negative alkalinity). The second reason is that the error in the cation budgets and particularly the Δstorage term in Table III is too large to allow a meaningful summation of reactions. This size problem is particularly true for magnesium. If the actual RXNS term for magnesium is zero or very small, much of the discrepancy between the alkalinity RXNS term and the reactions from equation 13 is resolved. Although each individual ion budget is probably accurate to within ±50%, the uncertainty may be too great to make the summation of reaction terms meaningful.

With respect to the quality of precipitation data, two additional factors need to be considered. The first factor is the difference in

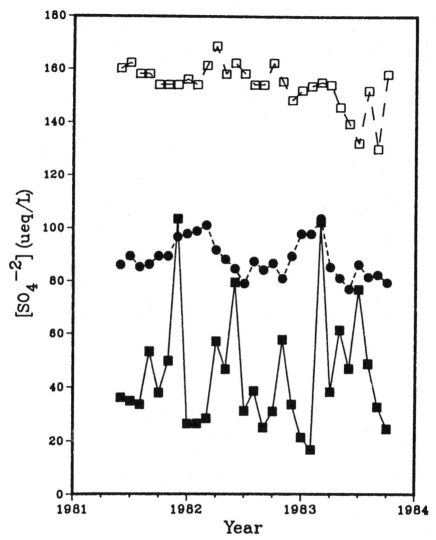

*Figure 7. Sulfate concentrations for precipitation, groundwater, and lake water. Key: same as Figure 6.*

volume of precipitation at Vandercook Lake (95.99 in) compared with Trout Lake (85.04 in) during the same period of record. There are probably some problems in extrapolating Trout Lake data to Vandercook Lake even though the lakes are close (3 km). Considering the sampling effort, 114 weekly precipitation samples were collected, and among these, 94 chemical analyses were reported. This sample number probably gives an uncertainty of about 20% in the precipitation data.

## Abbreviations and Symbols

| | |
|---|---|
| $[A^-]$ | concentration of organic acid anions |
| $C$ | concentration |
| GW | groundwater |
| in | inflow |
| out | outflow |
| $Q$ | flow rate |
| $t$ | time |
| trib | tributaries |
| $\Delta V$ | change in lake volume |

## Acknowledgments

We thank M. D. Johnson (WDNR) and D. A. Wentz (USGS) for providing support and field data, and J. M. Eilers (EPA), S. A. Banwart, and J. E. Carleton for creative input to this project. K. E. Webster (WDNR) provided helpful comments and data for the manuscript. This research was supported by the U.S. EPA, Environmental Research Laboratories at Duluth and Cornvallis (CR-810395-03). Jerald Schnoor is grateful to Werner Stumm and the Swiss Federal Institute for Water Resources and Water Pollution Control (EAWAG) for additional encouragement and support.

## References

1. Schnoor, J. L.; Stumm, W. In *Chemical Processes in Lakes*; Stumm, W., Ed., Wiley: New York, 1985; Chapter 14.
2. Dillon, P. J.; Jeffries, D. S.: Scheider, W. A.; Yan, N. D. In *Ecological Impact of Acid Precipitation*; Drablos, D.; Tollan, A., Ed.; SNSF Project: Oslo, Norway, 1980; pp 212-213.
3. Likens, G. E.; Bormann, F. H.; Pierce, R. S.; Eaton, J. S.; Johnson, N. M. *Biogeochemistry of a Forest Ecosystem*; Springer-Verlag: New York, 1977; pp 146.
4. Eilers, J. M.; Glass, G. E.; Webster, K. E.; Rogalla, J. A. *Can. J. Fish. Aquat. Sci.* **1983**, *40*, 1896-1904.
5. Eilers, J. E. "Distribution and Inventory of Softwater Lakes in Wisconsin"; ERL-D Report CR809484; U.S. Environmental Protection Agency: Duluth, MN, 1984; p 20.
6. Chen, C. W.; Gherini, S. A.; Peters, N. E.; Murdock, P. S.; Newton, R. N.; Goldstein, R. A. *Water Resour. Res.* **1984**, *20(12)*, 1875-1882.
7. Schnoor, J. L.; Nikolaidis, N. P.; Glass, G. E. *J. Water Pollut. Control Fed.* **1986**, *38(2)*, 139-148.
8. *Methods for Chemical Analyses of Water and Waste*; U.S. Environmental Protection Agency. Environmental Monitoring and Support Laboratory, Office of Research and Development: Cincinnati, OH, 1979; EPA-600/4-79-020.
9. Lin, J. C.; Schnoor, J. L. *J. Environ. Eng.* **1986**, *112(4)*, 677-694.
10. Banwart, S. A. M.S. Thesis, University of Iowa, Iowa City, 1983.

11. Winter, T. C. *Verh. Int. Ver. Theor. Angew. Limnol.* **1978,** *20,* 438–444.
12. Winter, T. C. *Water Resour. Bull.* **1981,** *17(1),* 82–115.
13. Winter, T. C. In *Modeling of Total Acid Precipitation Impacts;* Schnoor, J. L., Ed.; Butterworth: Boston, 1984; Chapter 5.
14. Lin, J. C. Ph.D. Thesis, University of Iowa, Iowa City, 1985.
15. Baker, L. A.; Brezonik, P. L.; Edgerton, E. S.; Ogburn, R. W., III. *Water, Air, Soil Pollut.* **1985,** *25,* 215–230.
16. Cook, R. B.; Kelly, C. A.; Schindler, D. W.; Turner, M. A. *Limnol. Oceanogr.* **1986,** *31(1),* 134–148.
17. Schindler, D. W.; Turner, M. A.; Stainton, M. P.; Linsey, G. A. *Science (Washington, D.C.)* **1986,** *232,* 844–847.

RECEIVED for review May 6, 1986. ACCEPTED July 29, 1986.

# Mechanisms of Alkalinity Generation in Acid-Sensitive Soft Water Lakes

Patrick L. Brezonik, Lawrence A. Baker, and Todd E. Perry

Department of Civil and Mineral Engineering, University of Minnesota, Minneapolis, MN 55455

*In-lake alkalinity generation (IAG) is important in regulating alkalinity in lakes that have large water residence times ($t_w$), particularly seepage lakes, which receive most of their water from direct precipitation and thus receive little alkalinity from watershed weathering processes. Major IAG processes include nitrate assimilation, sulfate reduction, and cation production; conversely, ammonium assimilation consumes alkalinity. Nitrate and ammonium are both efficiently retained by lakes with $t_w >$ 1 year, and net alkalinity generation by immobilization of these ions depends upon their relative inputs. Ion budgets show that in-lake cation production occurs in some lakes, but the relative importance of possible mechanisms (e.g., cation exchange, mineral weathering, or biological recycling) is not well understood. Sulfate reduction is the dominant IAG process in most seepage lakes, and sulfate is the major acid anion in precipitation. Sulfate reduction does not require an anoxic water column but occurs by diffusion into anoxic sediments. Sulfate retention in acid-sensitive lakes can be predicted by a simple input–output model with a first-order rate constant ($k_{SO_4} \sim$ 0.5 m/year). Further research is needed to assess the relative importance of sestonic sulfur deposition and dissimilatory reduction, to determine environmental factors that control reduction rates, and to determine the extent of recycling. Techniques used to measure IAG processes are described, and limitations of each method are discussed.*

$C$ONCERN ABOUT POTENTIAL EFFECTS OF ACIDIC PRECIPITATION has stimulated much scientific interest in the factors affecting the pH and alkalinity of soft water lakes. Until recently, the primary focus of scientists interested in the processes involved in neutralizing acidic

0065-2393/87/0216-0229$09.00/0

deposition was on reactions occurring in the terrestrial portion of watersheds: weathering of primary and secondary minerals, sulfate adsorption by metal oxides in soils, and biological assimilation of nitrate. Several process-oriented models have been developed to predict both short- and long-term responses of aquatic systems to acid deposition onto watersheds (1–3), and these models account for alkalinity produced by weathering mechanisms in soils.

Within the past few years, however, scientists have become aware that in-lake alkalinity generation (IAG) can be important in neutralizing acid inputs (4–11). Schindler et al. (12) concluded that in-lake alkalinity generation is more important than terrestrial alkalinity production in lakes with watershed–lake ratios < 8:1. In-lake alkalinity production is particularly important in neutralizing H⁺ inputs to precipitation-dominated seepage lakes that are widely distributed throughout the upper Great Lakes region (NE Minnesota, northern Wisconsin, and northern Michigan) and considered particularly susceptible to acidification (12). Seepage lakes receive all or most of their water as direct precipitation onto their surface, and inputs of alkalinity from weathering reactions in their watersheds are thus small or nonexistent. Nonetheless, many seepage lakes are not acidic despite appreciable inputs of acidity from atmospheric deposition. This fact suggests that in-lake processes are important in neutralizing acid inputs and regulating alkalinity in these lakes and perhaps in others.

This chapter describes the mechanisms whereby alkalinity can be generated within lakes and summarizes information on their importance in acid-sensitive lakes. Factors affecting rates of the most important alkalinity-generating processes are reviewed in the context of developing predictive models. Methods used to measure internal alkalinity generation range from laboratory incubations (microscale) to experiments on in situ enclosures (mesoscale) and to whole-lake mass balances (macroscale). Advantages and limitations of these methods are discussed at the end of the chapter.

## Alkalinity-Generating and -Consuming Reactions

*Alkalinity* is defined as the sum of the bases titrated by strong acid to a defined end point. In natural waters, alkalinity is associated primarily with the carbonate system and often is defined in terms of concentrations of the major carbonate species:

$$\text{alkalinity} = [\text{HCO}_3^-] + 2[\text{CO}_3^{2-}] + [\text{OH}^-] + [\text{H}^+] \qquad (1)$$

Alkalinity also can be defined in terms of the electroneutrality condition (i.e., charge balance) for the major ions not included in equation 1 as the sum of the equivalents of cations minus the sum of equivalents of anions.

$$\text{alkalinity} = [Na^+] + [K^+] + 2[Ca^{2+}] + 2[Mg^{2+}] + [NH_4^+]$$
$$- [Cl^-] - 2[SO_4^{2-}] - [NO_3^-] - [A^-] \qquad (2)$$

where $A^-$ denotes organic anions. Equation 2 illustrates an important point: any process that, on a net basis, increases the concentrations of base cations without increasing concentrations of the acid anions increases alkalinity; conversely, any process that increases the acid anions without increasing base cations decreases alkalinity.

The box below lists the major processes involved in generation

---

### Major Classes of Acid Consumption or Production Processes in Lakes

| Process | Product |
|---|---|
| Nitrogen Cycle | |
|   Ammonium assimilation | Acidity |
|   Nitrate assimilation | Alkalinity |
|   Denitrification | Alkalinity |
|   Nitrification | Acidity |
| Sulfur Cycle | |
|   Dissimilatory sulfate reduction | Alkalinity |
|   Sulfate assimilation | |
|     Ester sulfate formation | Alkalinity |
|     Reduction to organic sulfides | Alkalinity |
|   Sulfide oxidation | Acidity |
| Carbon Cycle | |
|   Partial decomposition of organic matter to form organic acids | Acidity |
|   Protonation and subsequent precipitation of organic acids | Alkalinity |
| Iron Cycle | |
|   Fe(II) oxidation | Acidity |
|   Fe(III) reduction | Alkalinity |
| Mineral Weathering | |
|   Congruent dissolution of oxides, hydroxides, and carbonates | Alkalinity |
|   Incongruent dissolution of aluminosilicate minerals | Alkalinity |
|   Sulfate adsorption onto metal oxide surfaces | Alkalinity |
|   Cation exchange with anionic sites on mineral and organic sediments | Acidity or alkalinity |

or consumption of acidity within lakes. Some of the processes occur exclusively or primarily in the water column, but most occur in surficial sediments. The first four groups of processes belong to elemental (biogeochemical) cycles and are mediated by microorganisms. In general, the oxidation reactions for these elements produce acidity, and reduction to a lower oxidation state produces alkalinity. Largely because sulfuric acid is the major anthropogenic source of acidity in lakes, biological sulfate reduction has received the most attention as a mechanism for internal alkalinity generation. A variety of studies reviewed in this chapter indicates that this process is widespread in acid-sensitive lakes and can account for much of the internally produced alkalinity (Table I). Mineral weathering and cation exchange processes are abiotic reactions and are the principal alkalinity-generating mechanisms in terrestrial systems. These processes also can occur in lake sediments, however, and evidence presented in this chapter indicates that they produce a significant fraction of the IAG in some lakes.

## Nitrogen Cycle Reactions

The major nitrogen cycle reactions producing alkalinity are assimilatory reduction of nitrate and organic nitrogen, dissimilatory reduction of nitrate to $N_2$ (denitrification), and production of ammonium from decomposition (*see* box on page 234, reactions 2, 4, and the reverse of reaction 1). Conversely, microbial assimilation of ammonium consumes alkalinity (reaction 1). Oxidation of ammonium to nitrate (nitrification) converts a weak acid ($NH_4^+$) to a strong acid ($HNO_3$) and thus consumes alkalinity (reaction 3). Nitrification is mediated by various aerobic bacteria, and the process is considered to be quite pH-sensitive, with an optimum pH $> 7$ (*19*). Nonetheless, evidence exists for nitrification in acidic soils (*20*) and lakes at pH $< 5$ (*21, 22*). Denitrification also is pH-sensitive (*23*) but occurs in anoxic (generally circumneutral) pore waters of acidic lakes (*24*).

Ammonium and nitrate, the major inorganic forms of nitrogen, exhibit highly nonconservative behavior in lakes (i.e., they are lost rapidly by assimilation into organic nitrogen, which accumulates in sediments or is exported). Nitrate also can be lost by denitrification; the resulting $N_2$ is lost to the atmosphere. Therefore, ammonium inputs have an acidifying influence (*see* box on page 234, reaction 1), and nitrate inputs have a neutralizing influence (*see* box on page 234, reactions 2 and 4). Because all, or nearly all, the nitrate produced by nitrification in lakes is reduced back to the $N(-III)$ oxidation state via nitrate assimilation by phytoplankton or aquatic macrophytes, nitrification can be ignored in the acid–base balance of soft water lakes. Similarly, ammonium is regenerated from sediments, but most of it is reassimilated into organic matter, and the net direction of reaction 1 is to the right.

Table I. Ion Budgets for Soft Water Lakes

| Lake and Location | Sulfate Reduction | | Cation Production | | $NO_3^-$ Reduction | | $NH_4^+$ Retention | | Total IAG[a] | |
|---|---|---|---|---|---|---|---|---|---|---|
| | A | B | A | B | A | B | A | B | Calculated[b] A | Measured A |
| **Drainage Systems** | | | | | | | | | | |
| Langtern, Norway | 57 | 5.5 | 38 | 3 | 17 | 36.4 | 13 | 76.2 | 112 | 117 |
| Lake 239, Ontario | 39 | 21.2 | 18 | 4.7 | 19 | 86.9 | 30 | 90.9 | 76 | 117 |
| Harp, Ontario | 72 | 8.8 | 5 | 0.4 | 52 | 58.4 | 22 | 81.5 | 207[c] | 188 |
| Plastic, Ontario | 70 | 18.8 | -1 | 0.3 | 33 | 80.5 | | | 148[c] | 166 |
| **Seepage Systems** | | | | | | | | | | |
| McCloud, FL | 26-35 | 43-46 | | | 17-20 | 95 | 10-14 | 91-93 | | 29-36 |
| Lowery, FL | 44 | 75 | | | 25 | 108 | 15 | 89 | | 46 |
| Magnolia, FL[d] | 48 | 81.7 | | | 26 | 114 | 16 | 95.3 | | 44 |
| Vandercook, WI | 36 | | | | 31 | | 31 | | | |
| **Experimentally Acidified Lakes** | | | | | | | | | | |
| Lake 223, Ontario[e] | 225 | 36 | 30 | 10 | | | | | 255 | 308-385 |
| Lake 302S, Ontario | 168 | | 14 | | 21 | | | | 182 | 248 |
| Lake 302N, Ontario | 81 | | 11 | | 149 | | | | 260 | 244 |
| Little Rock, WI[f] | 15 | 47 | | | | | | | | |

NOTE: A denotes milliequivalents per square meter per year. B denotes percent of input.
[a] IAG denotes in-lake alkalinity generation.
[b] Calculated alkalinity = $SO_4^{2-}$ loss + cation production + $NO_3^-$ loss − $NH_4^+$ loss.
[c] Calculated IAG includes organic anion sink.
[d] Cation budgets uncertain because of errors in seepage inputs.
[e] Budgets based on 7-year mean in which lake was acidified to pH 5.0.
[f] Estimate based on preliminary 1985 budget (preacidification conditions).

---

### Nitrogen Cycle Reactions That Produce or Consume Alkalinity

Ammonium Assimilation:

$$106CO_2 + 106H_2O + 16NH_4^+ \longrightarrow C_{106}H_{260}O_{106}N_{16}$$
$$+ 16H^+ + 106O_2 \qquad (1)$$

Nitrate Assimilation:

$$106CO_2 + 138H_2O + 16NO_3^- \longrightarrow C_{106}H_{260}O_{106}N_{16}$$
$$+ 16OH^- + 138O_2 \qquad (2)$$

Nitrification:

$$NH_4^+ + 2O_2 \longrightarrow NO_3^- + H_2O + 2H^+ \qquad (3)$$

Denitrification:

$$5CH_2O + 4NO_3^- + 4H^+ \longrightarrow 5CO_2 + 2N_2 + 7H_2O \qquad (4)$$

---

Nearly all of the ammonium and nitrate inputs are retained by lakes that have ($t_w > 1$ year). The net effect on alkalinity budgets of seepage lakes (generally, $t_w > 5$ years) can thus be estimated from the ratio of ammonium to nitrate inputs (Table I). For example, in McCloud Lake, FL (7), nitrate sinks contributed 17–20 mequiv/m$^2$ year of internal alkalinity production, whereas ammonium sinks consumed 10–14 mequiv/m$^2$ year. Nitrogen cycle reactions thus contributed a net of 6–7 mequiv/m$^2$ year out of a total IAG of 29–36 mequiv/m$^2$ year. In contrast, nitrate and ammonium retention exactly balanced each other in Lake Vandercook, WI (11), and no net alkalinity was produced from retention of nitrogen inputs.

### Mineral Reactions

Only a few of the major categories of mineral-related reactions listed in the box on page 235 are likely to be important in the alkalinity budgets of acid-sensitive lakes. Carbonate minerals (e.g., limestone and dolomite) generally are absent or rare in the watersheds of acid-sensitive lakes, and because the water in such lakes by definition is soft and dilute, autochthonous $CaCO_3$ is not present in their sediments. A small quantity of $CaCO_3$ may enter such lakes from wind-blown dust, especially in acid-sensitive lake regions near prairies and agricultural activity. Such dust should dissolve rapidly under the highly under-

**Mineral Reactions That Produce or Consume Alkalinity**

Congruent Dissolution:

$$CaCO_3 + H^+ \longrightarrow Ca^{2+} + HCO_3^-$$ (5)

$$Me(OH)_3 + 3H^+ \longrightarrow Me^{3+} + 3H_2O \quad (Me = Al, Fe)$$ (6)

Sulfate Adsorption:

$$\begin{matrix} Al \\ \diagdown \\ OH \\ \quad + 2H^+ + SO_4^{2-} \\ OH \\ \diagup \\ Al \end{matrix} \longrightarrow \begin{matrix} Al \\ \diagdown \\ O \quad O \\ S \\ O \quad O \\ \diagup \\ Al \end{matrix} + 2H_2O$$ (7)

Incongruent Dissolution (aluminosilicate mineral weathering):

$$2NaAlSi_3O_8 + 2H^+ + 9H_2O \longrightarrow 2Na^+$$
$$+ Al_2Si_2O_5(OH)_4 + 4Si(OH)_4$$ (8)

Cation Exchange:

$$H^+ + X^-Me^+ \longrightarrow Me^+ + X^-H^+$$ (9)

where $X^-$ is the negatively charged mineral or organic particle

Precipitation or Dissolution of Organic Acids (fulvic and humic acids):

$$A^- + H^+ \longrightarrow HA(s)$$ (10)

where $A^-$ is the organic anion

saturated conditions of low-alkalinity lakes. Estimates of the importance of this alkalinity source can be obtained by comparing measurements of Ca concentrations in bulk and wet-only deposition (25) or by directly measuring particulate Ca fluxes in dry-fall collectors. Sulfate adsorption (*see* box above, reaction 7) occurs in soils rich in iron and aluminum oxides. We found no evidence that the reaction occurs in sediments of soft water lakes in Florida (6, 8, 9), but the reaction could contribute to alkalinity generation in some acid-sensitive lakes.

Dissolution of ferric hydroxide, which can be a significant component in surficial lake sediments, occurs at pH levels too low to account for significant alkalinity generation in lakes, and there is little evidence from lake surveys that the iron content of lakes increases during acidification. Downing (26) found no desorption or solubilization of Fe from lake sediments acidified to pH 3. On the other hand, dissolution of aluminum hydroxide (and other Al minerals) can contribute to the neutralization of acidic inputs in lake waters. Because dissolved Al is toxic to fish (perhaps more than $H^+$), this process cannot be considered mitigative, however. Substantial dissolution of Al does not occur until a lake is rather acidic (pH < 5.0).

Dissolution of aluminosilicate minerals (see box on page 235, reaction 8) is a major mechanism for alkalinity generation in terrestrial environments, and to the extent that such minerals occur in lakes, such reactions also may contribute to IAG. The products of aluminosilicate weathering reactions are dissolved base cations, bicarbonate, silica, and secondary aluminosilicate minerals. The generated silica is subject to uptake by diatoms, and thus an increase in dissolved $SiO_2$ may not be detected even when mineral dissolution is occurring. Thus, the principal evidence for such processes in lakes is the internal generation of 'base cations, as measured by whole-lake ion budgets (Table I), and internal production of more alkalinity than can be accounted for by other mechanisms, such as sulfate reduction and nitrate assimilation.

Cation production is not unequivocal evidence for mineral weathering, however. Simple cation exchange (see box on page 235, reaction 9) also can increase cation concentrations in lake water and consume $H^+$. In fact, differentiation between the two processes based on whole-lake ion budget measurements cannot be accomplished, although the two processes may be distinguishable on kinetic grounds in laboratory experiments with lake sediments. Dissolution of complex aluminosilicates is a relatively slow process (days to years), whereas ion exchange is rapid (minutes to hours), especially in well-stirred systems, where transport limitation is minimized. In early experiments on sediment buffering of lake pH (6), we made sequential additions of sulfuric acid to sediment slurries in well-mixed flasks. Aqueous concentrations of base cations (primarily Ca and Mg) increased as more acid was added, and most of the added $H^+$ was consumed (Figure 1). In theory, either mineral dissolution or ion exchange could have caused these results, but the finding that pH stabilized and release of cations was completed within 24 h of acid addition suggests that ion exchange was responsible.

A variety of evidence has been accumulated to support the hypothesis that cation production by weathering or ion exchange contributes to IAG in acid-sensitive lakes. Measurement approaches have ranged from laboratory acidification experiments on well-mixed sediment slurries (Figures 1 and 2) and intact cores (8, 9) to ion balance measurements on whole-lake systems (Table I). To date, studies on IAG

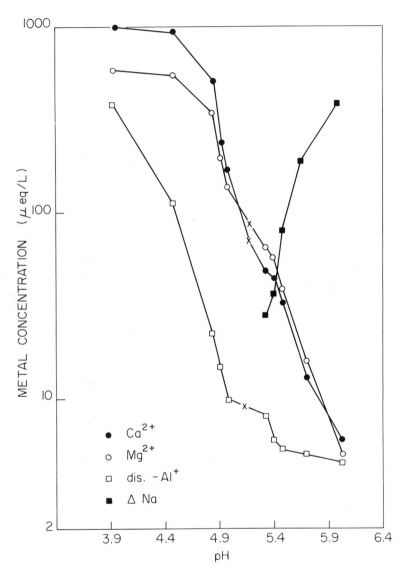

*Figure 1. Batch titration of sediment slurry from McCloud Lake, FL. This figure shows release of base cations with additions of strong acid ($H_2SO_4$) and adsorption of cations with additions of strong base. Increase in $Na^+$ at pH > 5.2 reflects addition of NaOH. (Reproduced with permission from reference 6. Copyright 1985, Reidel.)*

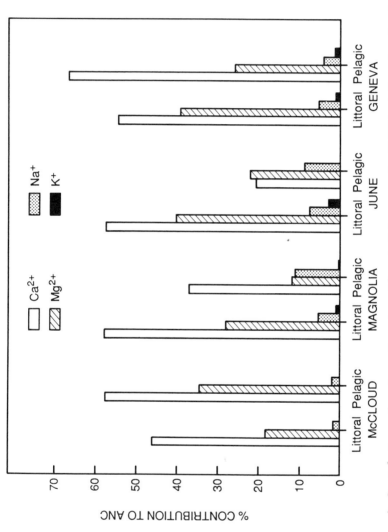

*Figure 2. Contributions of major cations to acid-neutralizing capacity (ANC) of littoral and pelagic sediments from four Florida lakes. (Reproduced with permission from reference 9. Copyright 1986, American Society for Testing and Materials.)*

by cation production processes have been primarily descriptive. There have not been enough studies to develop generalizations, and there has been little effort to place results in a theoretical framework. How cation production changes with acidification is uncertain. Cook et al. (5) found no enhanced production of cations in experimentally acidified Lake 223; if anything, cation production may have declined during acidification. In contrast, experiments at the microcosm scale (6, 8, 9, 11) and mesocosm scale (in-lake enclosures) (27, 28) have shown that experimental acidification resulted in increased cation production at least for a few months. If cation production is caused primarily by ion exchange, rates may decline over longer time periods as the pool of exchangeable cations becomes depleted.

Total concentrations of exchangeable bases in sediments of acid-sensitive soft water lakes tend to be low. Baker et al. (6) reported that the top 10 cm of sediment in McCloud Lake, FL, could neutralize about 23 times the current annual $H^+$ loading in north Florida wet deposition (40 mequiv/m² year); in so doing, sediment (pore water) pH would decline from 5.5 to 4.5. Oliver and Kelso (29) reported that total exchangeable Ca and Mg varied from 0.7 to 77 mequiv/100 g sediment in a group of Ontario lakes, and levels were correlated with lake pH and alkalinity. The sum of exchangeable base cations in sediments from Little Rock Lake, WI, ranged from 0.9 to 8.9 mequiv/100 g, and values were correlated with sediment organic content (26). The order of abundance of exchangeable cations was Ca > Mg > K > Na; divalent cations accounted for more than 80% of the base cations in all samples. This value agrees with observed cation production from acidified sediments. In all cases, Ca and Mg account for most of the cation production (6, 8, 9, 26–28). Little information is available on the total quantities of weatherable minerals in lake sediments.

Development of a predictive model for internal alkalinity generation by cation production is hampered by the fact that we still are insure of the relative importance of the two main mechanisms of cation production (ion exchange and mineral weathering). Although their net effects are the same, the mechanisms are quite different and require different types of models. Weathering, a slow process, must be described by kinetic models; ion exchange, a rapid process relative to the time scale of lake acidification, is best described by equilibrium models.

The ion exchange problem essentially is a multiple-metal, multi-ligand (exchange-site) type. As such, ion exchange models similar to those developed to quantify metal exchange with humic materials (30) may be applicable. Models developed to describe trace metal adsorption onto sediments (31) also may apply. In either case, practical models will involve conditional (apparent) selectivity constants rather than thermodynamic constants so that the complications of electrostatic effects and variable binding energies of exchange sites can be subsumed into the constant rather than treated explicitly. Because Ca and Mg are the only important base cations involved in exchange with $H^+$, only three selectivity constants ($K_{Ca,Mg}$, $K_{H,Ca}$, and $K_{H, Mg}$) need to be determined for each sediment.

Weathering of aluminosilicate minerals is a complicated and as yet poorly understood process, although it has been the subject of many studies (33–36). Weathering rates in soils depend on many factors including soil mineralogy, grain size of soil particles, hydraulic flow rates through soils, temperature, pH, and concentrations of $CO_2$ and various ligands in the soil solution (36). The same factors should be important for mineral weathering in sediments; in addition, organic coatings on mineral particles, including rocks and stones, probably affect weathering rates in aquatic systems. Many early studies (see reference 33 for a review) found mineral dissolution linear in $t^{1/2}$ (i.e., arithmetic plots of amount of dissolution versus time are parabolic). Such plots are generally assumed to imply that the overall process is transport-limited (diffusion-limited). More recent studies (37, 38) have demonstrated that parabolic kinetics are an artifact of mineral sample preparation. Grinding of samples in an effort to obtain uniform particle size disrupts grain surfaces and produces ultrafine particles that dissolve rapidly. The net effect is a decreasing rate of dissolution with time, and a parabolic kinetic plot. When these artifacts are taken into account, dissolution of mineral particles is linear in time. The processes occurring at the mineral surface can be described as congruent dissolution, and the rate-controlling step is viewed as surface reaction controlled. Dissolution occurs preferentially at surface sites with excess free energy (e.g., along crystal defects, screw dislocations, etch pits, and edges).

Dissolution of mineral surfaces does not occur by spontaneous breakdown of lattice constituents to dissolved species but is thought to involve surface coordination reactions in which hydrogen ions or ligands from solution "attack" and protonate surface–OH groups. This reaction tends to weaken M–O bonds between the central metal ion and the oxide surface and promotes detachment of an "activated" coordination complex, which renews the surface for further dissolution. The mechanism can be summarized as a two-step reaction (36):

$H^+$ attack:

$$\equiv|-OH + H^+ \xrightarrow{\text{fast}} \equiv|-OH_2^+ \xrightarrow{\text{slow}} \longrightarrow M(H_2O)_x^{z+} + \equiv| - \qquad (3)$$

Ligand attack:

$$\equiv|-OH + HA^- \xrightarrow{\text{fast}} \equiv|-A^- + H_2O \xrightarrow{\text{slow}} \longrightarrow M(A)^{(z-2)+} \equiv| - \qquad (4)$$

where $\equiv|-OH$ is a hydrous metal (Al, Si, or Fe) oxide with a functional OH group, $H_2A$ is a diprotic ligand, and M is a central metal ion of valence $z$.

In the first mechanism, dissolution rate is proportional to the degree of surface protonation to the $z$ power: rate $\propto (\equiv|-OH_2^+)^z$. The degree of surface protonation depends in a nonlinear fashion on bulk solution pH. Surface protonation can be viewed as an adsorption process. At low bulk solution pH (high $[H^+]$), the surface becomes saturated with $H^+$, and the rate is independent of pH. At intermediate pH values, surface protonation can be described by the Freundlich isotherm:

$$\log [\overset{=}{\equiv}|-OH_2^+] = n \log [H^+] + \log K_{ads} \qquad (5)$$

where $n$ is the slope of the isotherm and $K_{ads}$, the $y$-intercept, is the adsorption equilibrium constant. Over the pH range where equation 5 is valid, the dissolution rate thus depends on bulk solution $[H^+]$: rate $[H^+]^m$, where $m = nz$. For example, Stumm et al. (35) showed that $m = 0.4$, $n = 0.13$, and $z = 3.1$ for dissolution of aluminum oxide. Hydrogen ion attack thus leads to fractional-order dependence on bulk solution $[H^+]$. The second mechanism, ligand attack, leads to dissolution rates proportional to the bulk solution concentration of ligand. If surface binding by the ligand also follows an adsorption isotherm, dissolution rate may be a fractional order of $[HA^-]$. Moreover, if the $pK_a$ of the ligand is in the pH range of interest for dissolution, $[HA^-]$ will be pH-dependent at a fixed total ligand concentration.

Overall, the two mechanisms lead to "mixed" kinetics (20).

$$\text{rate} = k_{H^+}[H^+]^m + k_{HA}[HA^-]^{m'} \qquad (6)$$

At high pH, the first term in equation 6 may become small compared to the ligand attack term, and the dissolution rate will become pH-independent (if $[HA^-]$ is not pH-dependent).

This discussion is a considerable simplification of a complicated subject but raises several key questions regarding pH effects on mineral dissolution rates in acid-sensitive systems including the following: (i) At what pH do surface hydroxides on sediment minerals become fully protonated? (ii) At what pH does surface protonation cease to be important compared to ligand attack? (iii) How variable are the pH limits defined in (i) and (ii) among individual minerals and among mixed-mineral sediment deposits in acid-sensitive lakes? Few data are available to address these questions. Aluminum oxide surfaces were found (35) to be fully protonated below pH 3.5, and Schnoor and Stumm (36) stated that $H^+$ attack contributes to silicate mineral dissolution rates below pH 5. Dissolution of potassium feldspar was reported to be $H^+$-dependent below pH 5 and pH-independent in the pH range

5-10 (33), but other workers have found that dissolution rates of primary minerals (various feldspars) are independent of pH to much lower values ($<\sim$1-2) (P. Bloom, personal communication, 1986). We are not aware of any data on the pH-dependence of weathering rates for mixtures of minerals in soft water lake sediments.

The overall importance of cation production in internal alkalinity generation is not well understood. In part, this lack of understanding is due to the potential importance of these reactions being recognized only recently. Questions that need to be addressed include the following: What is the relative importance of cation exchange and mineral weathering in cation production, and how do changes in pH affect these processes? Are inputs of weatherable minerals sufficient to sustain long-term internal production of cations? What is the role of plankton in removing cations from the water column, and to what extent is this material recycled?

## Sulfate Reduction

A wide variety of sulfur cycle reactions can affect the alkalinity balance of lakes (*see* box, page 243). Probably the most important IAG process in acid-sensitive lakes is the loss of sulfate from the water column. This loss can occur in two ways: assimilation by plankton, which involves reduction to sulfide and formation of organic sulfide compounds (as well as formation of ester sulfates), followed by removal of the organic sulfur from the lake by sedimentation or export in outflows (*see* box on page 243, reaction 11), and dissimilatory reduction of sulfate to sulfide by ánaerobic bacteria (*see* box on page 243, reaction 12). The latter reduction occurs in the water column if it stratifies and the hypolimnion becomes anoxic, but the process is not limited to such lakes. There is considerable evidence that sulfate diffuses into sediments, which are nearly always anoxic within a few centimeters from the surface, and is reduced at significant rates in soft water lakes even when the overlying water is oxic (*5, 7, 24, 28, 39, 40*).

Sulfide is unstable under oxic conditions, and the reverse reaction of sulfate reduction, sulfide oxidation (*see* box on page 243, reaction 13) completes a null cycle. Thus, assimilatory reduction is a net producer of alkalinity only to the extent that organic (sestonic) sulfur is incorporated permanently in the sediment. Net alkalinity production by dissimilatory reduction requires that the reduced sulfur must either be incorporated into stable end products in the sediment [e.g., FeS (*see* box on page 243, reaction 14) or carbon-bonded sulfur compounds (*see* box on page 243, reaction 17)] or must diffuse out of the sediment, through the water column, and into the atmosphere before being reoxidized. Although rates of $H_2S$ oxidation by $O_2$ are rapid at and

## Sulfur Cycle Reactions That Produce or Consume Alkalinity

Assimilatory Reduction[a]:

$$106CO_2 + 16NO_3^- + HPO_4^{2-} + 122H_2O + 19H^+$$
$$+ 0.5SO_4^{2-} \longrightarrow C_{106}H_{264}O_{110}N_{16}P_1S_{0.5} + 139O_2 \quad (11)$$

Sulfate Reduction ($H_2S$ production):

$$SO_4^{2-} + 2CH_2O \longrightarrow H_2S + 2HCO_3^- \quad (12)$$

Oxidation of $H_2S$:

$$H_2S + 2O_2 \longrightarrow SO_4^{2-} + 2H^+ \quad (13)$$

Formation of FeS:

$$8H^+ + 4SO_4^{2-} + 4Fe(OH)_3 + 9CH_2O \longrightarrow 4FeS(s)$$
$$+ 19H_2O + 9CO_2 \quad (14)$$

Formation of Elemental S:

$$SO_4^{2-} + 2H^+ + 3H_2S \longrightarrow 4S(s) + 4H_2O \quad (15)$$

Formation of Pyrite:

$$FeS(s) + S(s) \longrightarrow FeS_2(s) \quad (16)$$

Formation of Carbon-Bonded Sulfur[b]:

$$H_2S + R-C-OH \longrightarrow R-C-SH + H_2O \quad (17)$$

Formation of Ester Sulfur:

$$ROH + SO_4^{2-} + H^+ \longrightarrow ROSO_3^- + H_2O \quad (18)$$

[a]The stoichiometry of algae used in this equation is a modification of the Redfield formula that includes sulfur (0.5% g S/g algae).
[b]There are numerous other reactions of $H_2S$ with organic compounds.

above pH 7, they decline rapidly at lower pH (*41*). At pH < 6, the rate constant is $<10^{-2}$ M/h (pseudo-second-order). For $O_2$-saturated conditions (0.3 mM) and a total sulfide concentration of 1 mg/L ($3 \times 10^{-5}$ M), the rate of sulfide oxidation would be only about $10^{-5}$ $\mu$M/h. This calculation suggests that $H_2S$ is sufficiently stable under mildly acidic conditions to allow substantial transfer to the atmosphere before reoxidation. Other volatile sulfur compounds that are potential products of sulfate reduction (e.g., dimethyl sulfide and carbonyl sulfide) are likely to be at least as stable to oxidation as $H_2S$, but quantitative information is lacking on their significance and stability in freshwater.

The first evidence that sulfate reduction could represent a significant internal source of alkalinity was found in the experimental acidification of Lake 223 in the Canadian Experimental Lakes Area (*4, 5, 39, 42, 45*). Because of sulfate reduction, much larger quantities of $H_2SO_4$ were needed to achieve and maintain target pH values in the lake than were predicted from titrations of water column alkalinity. Although sulfate reduction occurred in the anoxic hypolimnion of Lake 223, most (66%) of the loss occurred by diffusion of sulfate into anoxic sediments from overlying oxygenated water. A variety of methods have been used in recent studies to detect sulfate reduction, including sulfate mass balances on lakes (*see* box below), mass balances on experimentally manipulated enclosures in lakes (*27, 28*), laboratory studies on sediment–water microcosms (Figure 3) (*6, 8, 9, 40, 43, 44*), and measurement of sulfate profiles in sediment pore water (*7, 24, 28*,

---

### Methods of Measuring Internal Alkalinity Generation

Microscale
  Laboratory
    Ion budgets in sediment–water microcosms
    $^{35}$S tracer incorporation into sediment
  Field
    Benthic chambers to measure flux of ions across the sediment–water interface
  Interface
    Pore water equilibrators for ion profiles and Fick's law calculation of diffusional fluxes
Mesoscale
  Ion mass balances on enclosures (limnotubes) isolating a portion of lake and associated sediment
Macroscale (ecosystem level)
  Ion mass balances on whole lake
  Sulfur accumulation rates in sediment cores

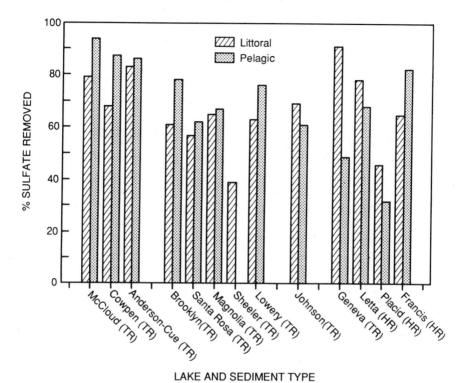

LAKE AND SEDIMENT TYPE

*Figure 3. Loss of sulfate in sediment-water microcosms (undisturbed cores and overlying water) from 13 Florida lakes. Overlying water was maintained at pH 4.0 with biweekly additions of H₂SO₄ (as necessary) over a 4-month incubation period. (Reproduced with permission from reference 9. Copyright 1986, American Society for Testing and Materials.)*

*38–40, 45*). Alkalinity generation by sulfate reduction has been measured in a variety of lakes in several geographic regions (*see* Table I).

Previous studies on sulfate reduction have been primarily site-specific. Available data are not yet sufficient for confident generalizations about mechanisms of reduction, physical and chemical factors affecting reduction rates, and the nature and stability of the reduced products. A first step in developing a predictive model was described by Baker et al. (*45*), who applied simple continuously stirred-tank reactor (CSTR) principles to sulfate mass balances for seepage lakes. This approach yielded the following equation to describe the change in $[SO_4^{2-}]$ with time in such lakes.

$$-d[SO_4^{2-}]/dt = (1/V)\{J_{SO_4} - [SO_4^{2-}](S_o + k_{SO_4}A)\} \qquad (7)$$

where $V$ is lake volume (m³), $A$ is lake surface area (m²), $J_{SO_4}$ is the total annual loading of sulfate into a lake (mequiv/year), $S_o$ is seepage outflow (m³/year), and $k_{SO_4}$ is a first-order loss coefficient for sulfate (m/year). Sulfate loss is assumed to be proportional to sediment surface area. The steady-state solution of equation 7 is

$$[SO_4^{2-}]_{ss} = \frac{J_{SO_4}}{(S_o + k_{SO_4}A)} = \frac{L_A}{[(z/t_w) + k_{SO_4}]} \tag{8}$$

where $z$ is lake mean depth, $t_w$ is water residence time, and $L_A$ is the areal loading of sulfate (mequiv/m² year). If $k_{SO_4}$ is reasonably consistent among soft water lakes, the simple model could be used for regional predictions. If $k_{SO_4}$ is not constant, the model could be used on a case-by-case basis to describe sulfate mass balances but could not be used for predictions unless further work was done to $k_{SO_4}$ (i.e., determine the physical and chemical factors affecting it). Calibration of this model using ion budget data for 14 soft water lakes shows that mean $k_{SO_4} = 0.52 \pm 0.30$ m/year (45). We hypothesize that values of $k_{SO_4}$ are reasonably consistent among lakes because sulfate reduction rates are controlled by a common physical limitation, diffusion into sediments (39, 45).

Equation 8 can be used to illustrate the effect of $t_w$ on the extent of sulfate retention (magnitude of internal sinks) in lakes (Figure 4). Sulfate appears to behave almost conservatively in lakes with small $t_w$ values; only 10% of the input is retained when $z = 5$ m and $t_w = 1$ year. For a lake with the same mean depth but with $t_w$ more typical of seepage lakes (5–10 years), about 30%–50% of the input sulfate would be lost internally.

The simple model presented assumes that sulfate reduction is first order in $[SO_4^{2-}]$, and several studies have found this to be the case over the concentration range of interest in acid-sensitive lakes (4, 28, 39, 40, 44). This finding leads to the interesting conclusion that sulfate reduction is a homeostatic, or buffering, mechanism. Accordingly, an increase in $H_2SO_4$ loading to a lake results in increased buffering by sulfate reduction, and a decrease in loading is accompanied by a decrease in buffering by sulfate reduction. The fact that some lakes have become acidic because of acid deposition indicates that the process is not completely effective as a buffering mechanism; nonetheless, many more lakes probably would be acidified if the process did not occur.

For further progress in developing relationships to assess the importance of sulfate reduction, several major questions need to be addressed. These questions include the following:

- What is the effect of temperature, pH, sediment porosity, organic content, and iron content on $k_{SO_4}$?

- What is the relative importance of various sulfur species as initial, intermediate, and long-term storage products for sulfate loss to sediments?

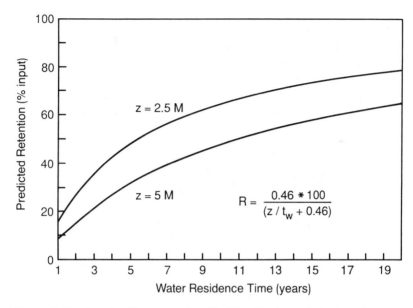

*Figure 4. Predicted sulfate retention (% of loading) as a function of water residence time ($t_w$) for lakes having mean depths (z) of 2.5 and 5.0 m.*

- Is the long-term production of organic substrates usable by sulfate reducers adequate to maintain the process?

- If iron sulfides are the major long-term reservoir for products of sulfate reduction, are rates of iron input to acid-sensitive lakes sufficient to maintain the process on a steady-state basis?

- How important is sulfate assimilation and seston deposition as a sulfate retention (alkalinity-generating) mechanism? Also, how much of the sestonic sulfur deposition is contributed by allochthonous organic sulfur inputs?

- To what extent do short-term measurements of sulfate losses (by $^{35}$S experiments, flux estimates from pore water profiles, or analysis of seston deposition in sediment traps) reflect long-term losses (i.e., how much of the sedimented and reduced sulfur is recycled to the water column)?

Sulfate reduction is inhibited at low pH (<5) (*39, 47*); this fact suggests that the process may be inhibited in acidic lakes. The sediment environment, however, is relatively well-buffered, largely by the activity of sulfate-reducing bacteria, and pore water profiles in several acidic lakes generally show circumneutral pH within a few centimeters of the surface (*24, 28, 39, 40, 48*). Thus, sulfate reduction rates apparently do not decline with acidification in most lakes. Depressed pore water pH has been observed in a few lakes (*24*), and certain sediment types may be susceptible to acidification and thus lead to reduced sulfate reduction activity.

Identification of the end products of sulfate reduction is important in understanding rate-limiting factors for the process. For example, if iron-containing compounds are the major end products, iron supply rate is a potential limiting factor. Sulfate reduction often is written as a chemical reaction resulting in formation of FeS (*see* box on page 243, reaction 12). However, acid-volatile sulfide (AVS), which generally is assumed to be FeS and closely related compounds, is a small part of the total sediment sulfur in most soft water lakes that have been studied (*17, 49–52*). For example, AVS is <1% of the total sulfur in sediment cores from Little Rock Lake, WI (Figure 5). The fraction of sediment sulfur occurring as AVS may depend on the extent of sulfate pollution.

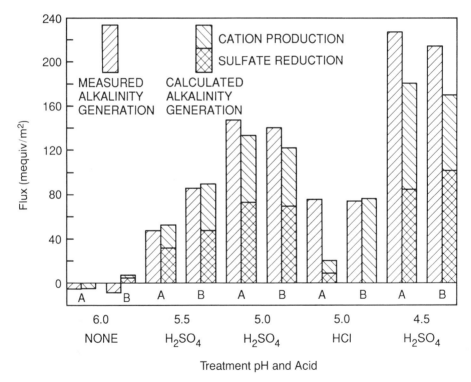

*Figure 5. Alkalinity-generating mechanisms in enclosures in Little Rock Lake. Enclosures (4-m diameter × 1-m deep, open to the sandy sediment) were maintained at treatment pH levels for 3 months with frequent additions of acid ($H_2SO_4$ or HCl as noted). (Reproduced with permission from reference 40. Copyright 1987, Todd E. Perry.)*

Only small amounts of AVS were found in sediments from two unpolluted lakes in Ontario and in a moderately contaminated lake in the Sudbury area, but AVS comprised 10%–25% of the total sediment sulfur in a highly contaminated lake ($[SO_4^{2-}] = 480$ mg/L) (50). There is some evidence (17, 39, 49) that AVS initially formed during sulfate reduction is later reoxidized or converted to more stable end products. However, even in short-term laboratory incubations using soft water lake sediments, AVS is a minor end product (39, 49).

Analyses of sediment cores and isotope tracer studies suggest that pyrite and elemental sulfur are more important than AVS. Rudd et al. (49) found that chromium-reducible sulfur (CRS), which includes pyrite and elemental sulfur, was 5–10 times AVS levels in eight lakes they studied, and $^{35}S$ tracer experiments showed that CRS was 9%–46% of the end products. According to Nriagu and Soon (50), CRS may be more important in acidified lakes than in circumneutral systems.

There is evidence that much of the reduced sulfur formed during sulfate reduction is incorporated into organic compounds. Organic sulfur often is a dominant component of total sulfur in soft water lake sediments (17, 49–52). Short-term $^{35}S$ experiments with sediment cores from six North American lakes (49) showed that organic sulfur was a major end product (28%–85% of total sulfur), but was a minor product (<15%) in two acidic Norwegian lakes. Several reactions and organic products are possible, including thiols (RSH), which recently were suggested to be important diagenetic products of sulfate reduction in marine sediments (53). The major organic products in lake sediments are not yet known, but incorporation of some of the reduced sulfur into the high molecular weight humic fraction seems likely.

The relative importance of seston deposition and dissimilatory reduction as sinks for sulfate is still uncertain. David et al. (54) found that seston deposition was more important than sulfate reduction as a sink for sulfate in South Lake in the Adirondacks; preliminary data suggest that both mechanisms are significant sulfate sinks in Little Rock Lake (51). Nriagu and Soon (50) presented several arguments in favor of the hypothesis that sedimentary organic sulfur in sulfur-contaminated lakes is a product of sediment diagenesis rather than deposition of seston. Stable isotope ($^{34}S$ or $^{32}S$) measurements showed that sediment organic sulfur was higher in $^{32}S$ than was sestonic sulfur, and the isotope ratio of the former was similar to that of reduced inorganic sulfur in the sediment, suggesting a common origin. In addition, C/S ratios in the sediment were lower than those in seston, suggesting postdepositional enrichment of sulfur. In contrast, isotopic ratios of organic S and C/S ratios in sediments were similar to those found in seston in two uncontaminated lakes, a result suggesting that sediment organic sulfur in these lakes was derived from seston deposition.

In terms of alkalinity generated, it makes no difference whether the sulfur accumulated in sediments is derived by seston deposition or by dissimilatory sulfate reduction in the sediment. The mechanism of accumulation is of considerable interest, however, from the perspective of predicting future accumulation rates for changing environmental conditions. If seston deposition is more important, the process should be controlled by the overall productivity status of the lake and be relatively independent of water column sulfate concentration. If dissimilatory reduction is more important, the process should be proportional to water column sulfate because the driving force for sulfate diffusion into sediments will increase with $[SO_4^{2-}]$. Although evidence is limited, both seston deposition and dissimilatory reduction appear to be important as sulfur sinks in pristine lakes; at higher sulfate levels characteristic of acidified lakes, dissimilatory reduction becomes the predominant sink.

## Measurement of Internal Alkalinity Generation

A wide variety of methods are available to measure rates of internal alkalinity generation and to determine the importance of various mechanisms in lakes and sediments (*see* box on page 244). The scales of measurement range from very small (e.g., laboratory studies in flasks or intact sediment cores) to whole ecosystems (e.g., ion budgets on lakes). In general, larger scale methods are more integrative in time and space but less subject to experimental control. Microscale methods such as the measurement of sulfate loss or cation release in intact sediment cores allow control of experimental conditions and thus are useful to study factors affecting process rates. On the other hand, extrapolation from small-scale measurements, especially those made in the laboratory, to whole ecosystem rates can be inaccurate. Some methods are suited for measurement of alkalinity generated by sulfate reduction but not by cation production; others (e.g., mass balances) are broadly applicable to all the mechanisms listed in the box on page 244. Selection of a method to be used in alkalinity-generation studies thus depends on the objectives of the study. Because all the methods used to estimate whole ecosystem rates have some shortcomings, approaching the problem by several methods is desirable. There is not sufficient space in this chapter for detailed discussion of all the methods contained in the box on page 244, but results from the most widely used methods are described here to illustrate their advantages and limitations.

**Total Sulfur Accumulation in Sediment Cores.** This method requires collection of representative sediment cores from a lake, accurate dating of the profile (e.g., by [210]Pb), analysis of total sulfur concentrations as a function of depth, and calculation of accumulation rates of total sulfur. Because sediment focusing and other processes lead to inhomogeneous rates of sulfur deposition over the area of a lake, collection of a single core, which usually is taken from the deepest part of the water column, is not sufficient to estimate basinwide accumulation rates. The inference that all the sulfur deposited in the sediment is derived from assimilatory uptake or dissimilatory reduction of sulfate within the lake–sediment system, which represents an internal sulfate sink, depends upon the assumption that allochthonous inputs of particulate sulfur in organic or mineral particles are negligible. Sediment sulfur derived from allochthonous inputs of particulate organic or mineral sulfur does not contribute to alkalinity generation. Profiles of total sulfur in three [210]Pb-dated cores from Little Rock Lake, WI (Figure 6), show that accumulation rates vary as a function of water column depth and that concentrations are relatively constant and low below 15 cm. An increase in sulfur content occurred around 1900, and the largest increase occurred at the 9-m site, near the deepest part of the lakes. Based on total sulfur content and sediment deposition rates for each core, we calculated areal rates of sulfur accumulation of 15, 17, and 30 mmol/m$^2$ year. Basinwide deposition was estimated to be 17 mmol/m$^2$ year (*51*).

**Whole-Lake and Mesocosm Ion Balances.** Mass balance or input–output (I–O) analysis is commonly used in chemical engineering as well as in geochemical studies to determine the fate of a reactive substance (i.e., its extent of retention or transformation) in a given system. The approach is attractive for several reasons: (i) the method is holistic; (ii) calculation of mass balances is conceptually simple; (iii) the method is based on one of the most fundamental precepts of science (conservation of mass); and (iv) mass balance calculations, at least for some systems, rely on routine monitoring data (e.g., streamflows or ion concentrations) that either are readily available or easily obtained. Whole-lake mass balances can indicate whether an ion is conservative (I = O) or whether the lake acts as a source (I < O) or sink (I > O). However, mass balances do not provide any information on mechanisms for loss or gain. For example, in-lake losses of sulfate could be caused by diffusion into anoxic sediments and subsequent reduction to forms that are retained in the sediment or by planktonic uptake, conversion to organic sulfur, and seston sedimentation. As noted above, knowing the relative importance of each mechanism is important in order to develop predictive models, but this knowledge can only be obtained by in-lake studies (e.g., sediment trap analysis or pore water analysis).

Accurate mass balances are difficult to obtain on seepage lakes, which lack channelized surface inflows and outflows. Groundwater inseepage and outseepage and atmospheric dry deposition are impor-

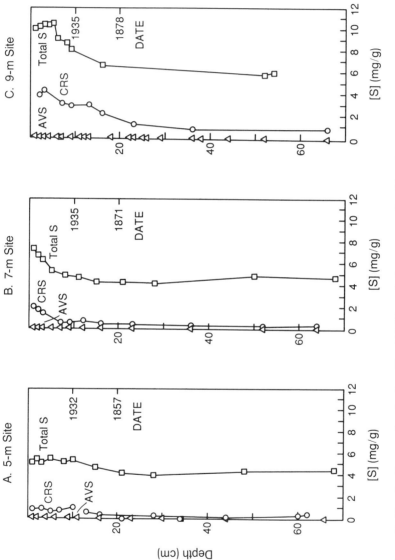

*Figure 6. Total sulfur, acid-volatile sulfur (AVS), and chromium-reducible sulfur (CRS) in three cores from the north basin of Little Rock Lake. Dating was by $^{210}Pb$.*

tant components of ion budgets for seepage lakes, but satisfactory methods are not available to measure them accurately. Groundwater inflows calculated from observed hydraulic gradients in the lake and the surrounding watershed are highly uncertain. Although groundwater may be a small part of the total water input, it may be a significant source of major ions. In Vandercook Lake, for example, Lin et al. (*11*) found that groundwater was only 2% of the total water input but contributed 38% of the total $Ca^{2+}$ input, 52% of the total $Mg^{2+}$ input, and neutralized 47% of the atmospheric $H^+$ input. Methods to estimate aerosol and gas deposition also are unreliable. Most studies use wet and dry collectors or bulk precipitation collectors to estimate total atmospheric input (a few studies ignore dry-fall inputs altogether), but these methods are unlikely to collect aerosols with the same efficiency as the lake surface and may not collect certain gas species at all. Dry fall is an important component of the ion budgets of seepage lakes that cannot be ignored; Baker et al. (*7*) reported that dry deposition contributed >30% of total atmospheric deposition for all major ions to a Florida lake. Finally, seepage lakes tend to have long water residence times ($t_w > 5$ years). Consequently, annual inputs of major ions are small compared with amounts stored within a lake, and relatively small inaccuracies for calculated changes in storage over a short period (mass balances often are based on only 1 or 2 years of data) tend to produce large inaccuracies in calculated sources and sinks.

Ion mass balances also have been used to evaluate internal sources of alkalinity in large enclosures (mesocosms) installed in lakes for experimental acidification experiments (*27, 28, 44*). Results demonstrate that both sulfate reduction and cation production are involved (Figure 5). Such experiments are less expensive to conduct than whole-lake manipulations, and the scale allows some degree of replication and multiple levels of acidification. Mesocosms avoid some of the problems of spatial heterogeneity inherent in microcosm studies, and because such experiments are done under in situ conditions, they avoid problems of trying to simulate environmental conditions in the laboratory. On the other hand, problems with ice limit most in situ enclosure experiments to one field season in northern climates, and sunlight tends to degrade most plastic sheeting used to construct enclosures in 1 year or less. Wall effects can be a serious problem in enclosure experiments to study nutrient or trace metal dynamics, but they probably are not important in acidification- or alkalinity-generation experiments. A major concern in such experiments, however, is the integrity of the enclosure; improper sealing of enclosures at the sediment–water interface or small holes in the enclosure walls cause spurious gains or losses of ions by exchange of lake water with that in the enclosure. The extent of such exchange can be determined by adding a small quantity of a conservative tracer such as NaCl to each enclosure.

**Pore Water Profiles (Diffusional Flux Calculations).** Concentration–depth profiles of ions involved in alkalinity generation or consumption can be obtained in the region of the sediment–water interface by use of pore water equilibrators (55). These devices consist of rectangular Plexiglas bars with machined chambers at 1-cm intervals. These chambers are filled with distilled water and covered with a sheet of membrane filter material held in place by a top plate. The samplers are inserted vertically into the sediment and left in place until the solution in each chamber reaches equilibrium with the pore water surrounding it. Subsequent analysis of water in the chambers leads to depth profiles of the ionic composition of the pore water (Figures 7 and 8).

Pore water profiles reflect the combined effects of several transport and reaction terms: diffusion of ions into or out of the sediment zone (depending on the concentration gradient), chemical or biological reactions (sources or sinks) in the sediment, vertical advective movement of water through the sediment (outseepage or inseepage), mixing by activity of zoobenthos (bioturbation), mixing by rising gas bubbles formed biologically in the sediment (gas ebullition), compaction of sediment with depth, and accretion of new sediment by settling particles. In many cases, only a few of the processes are important. For example, accretion and compaction usually are small compared to the diffusion and reaction terms in lake sediments. Diagenetic models have been developed to describe concentration profiles and calculate fluxes to or from sediments in terms of all these processes (56–58), but the resulting equations are quite complicated and include coefficients that are difficult to evaluate.

A simple diffusion reaction model for sulfate reduction in lake sediments will be described. The model yields equations that quantitatively describe vertical profiles of $[SO_4^{2-}]$ in the pore water and predict fluxes of $SO_4^{2-}$ across the sediment–water interface. Five assumptions are made to derive the model:

1. The oxic–anoxic interface occurs at the sediment surface; no reduction occurs in the water column.

2. Profiles result solely from diffusion, accretion, and reaction (dissimilatory reduction). Compaction and advective flows are not considered.

3. Reduction is first order with respect to $[SO_4^{2-}]$.

4. Rate dependency on other factors (pH, temperature, and availability of organic substrate) is subsumed into a site-specific rate constant that is independent of depth within the sediment.

5. The system is at steady state.

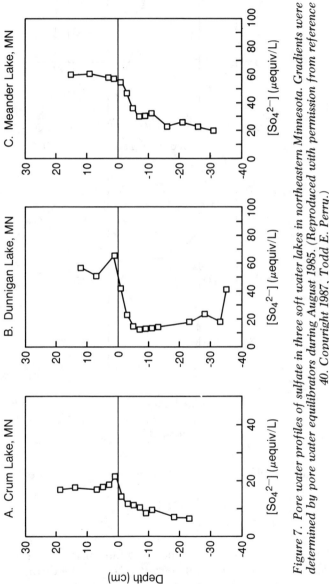

*Figure 7. Pore water profiles of sulfate in three soft water lakes in northeastern Minnesota. Gradients were determined by pore water equilibrators during August 1985. (Reproduced with permission from reference 40. Copyright 1987, Todd E. Perry.)*

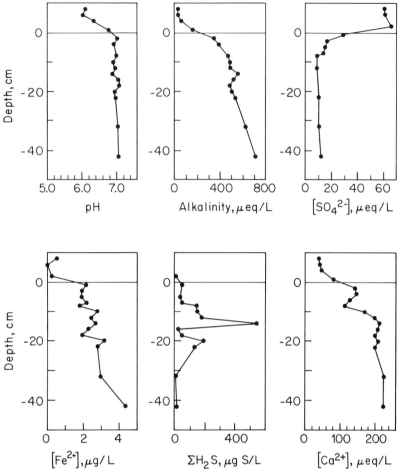

*Figure 8. Pore water profiles from the south basin of Little Rock Lake, August 1984.*

The fundamental diffusion reaction equation for assumption 5 is

$$\frac{\partial[SO_4^{2-}]}{\partial t} = \frac{D_s \, \partial^2[SO_4^{2-}]}{\partial z^2} - \frac{r_s \, \partial[SO_4^{2-}]}{\partial z} + R \qquad (9)$$

where $D_s$ is the diffusion coefficient of sulfate in the sediment and is equal to $\phi D_{H_2O}$; $\phi$ is the sediment porosity; $D_{H_2O}$ is the diffusion coefficient of sulfate in water; $r_s$ is the sedimentation rate; $R$ is the sulfate reaction rate, which is $-k_{red}[SO_4^{2-}]$; and $z$ is the depth in sediment. Assumed boundary conditions are as follows: $\partial[SO_4^{2-}]/\partial t = 0$, $[SO_4^{2-}] = [SO_4^{2-}]_o$ at

$z = 0$, and $[SO_4^{2-}] = 0$ at $z = \infty$. Integration of equation 9 for these conditions gives an equation that defines the sulfate profile (58).

$$[SO_4^{2-}] = [SO_4^{2-}]_o \exp\left[\frac{r_s}{2D_s} - \left(\frac{r_s^2}{4D_s^2} + \frac{k_{red}}{D_s}\right)^{1/2}\right] z \qquad (10)$$

If $r_s/4D_s \ll k_{red}$ (i.e., accretion is small compared to effects of diffusion and reaction), equation 10 simplifies to

$$[SO_4^{2-}] = [SO_4^{2-}]_o \exp[-(k_{red}/D_s)^{1/2}z] \qquad (11)$$

Sulfate fluxes ($F_{SO_4}$) into the sediment are calculated by Fick's first law:

$$F_{SO_4} = -D_s(\partial[SO_4^{2-}]/\partial z)_{z=0} \qquad (12)$$

The sulfate concentration gradient at the interface can be estimated by drawing a tangent to the measured profile at $z = 0$. Alternatively, a mathematical expression can be derived by differentiating equation 11 with respect to $z$ and evaluating the derivative at $z = 0$.

$$d[SO_4^{2-}]/dz \big|_{z=0} = -[SO_4^{2-}]_o(k_{red}/D_s)^{1/2} \qquad (13)$$

Substitution of equation 13 into equation 12 yields

$$F_{SO_4} = [SO_4^{2-}]_o(k_{red}D_s)^{1/2} \qquad (14)$$

The simple model predicts that sulfate concentrations decrease exponentially with depth in the sediment (equation 11). Comparison with measured sulfate profiles (Figure 7) shows that the decrease with depth is rapid and quasi-exponential. However, concentrations do not fall to zero (a boundary condition of the model) but instead decline to a constant (low) concentration, typically about 5-10 $\mu$equiv/L. The residual sulfate may reflect exhaustion of available organic substrate at depth or a threshold requirement for sulfate by microbial sulfate reducers.

Sulfate profile data can be fit to equation 11 by nonlinear regression programs, and if the fit is good, it provides a measure of $k_{red}$, the site-specific rate constant for sulfate reduction. The variability of $k_{red}$ among sediments of different characteristics (e.g., organic content) could provide useful information on factors affecting sulfate reduction in lake sediments. The occurrence of a peak in $[SO_4^{2-}]$ just above the sediment–water interface, as shown for Dunnigan and Crum Lakes in Figure 5, is a common, albeit not universal, feature of sulfate profiles that we have measured. This peak is strong evidence for recycling processes near the interface; the most reasonable explanation is upward diffusion of $H_2S$ or other mobile reduced sulfur into the oxic zone at the interface, where reduced sulfur is reoxidized. Further studies are needed to quantify this process.

Pore water profile data for other ionic species (Figure 8) affecting alkalinity provide strong evidence that several mechanisms of alkalinity generation occur in surficial lake sediments. Both alkalinity and pH increase rapidly with depth in sediment pore water; gradients for $Ca^{2+}$ and $Mg^{2+}$ indicate that these cations diffuse from the sediment into the overlying water. Development of quantitative models for IAG based on pore water profiles and sediment diagenesis modeling approaches should be pursued.

## Abbreviations and Symbols

| | |
|---|---|
| $A$ | lake surface area |
| $A^-$ | organic anions |
| AVS | acid-volatile sulfide |
| CRS | chromium-reducible sulfur |
| $D_{H_2O}$ | diffusion coefficient of sulfate in water |
| $D_S$ | diffusion coefficient of sulfate in sediment |
| $F_{SO_4}$ | sulfate flux |
| IAG | in-lake alkalinity generation |
| $J_{SO_4}$ | total annual loading of sulfate into a lake |
| $k_{SO_4}$ | first-order loss coefficient for sulfate |
| $K$ | selectivity constant |
| $K_a$ | acid dissociation constant |
| $K_{ads}$ | adsorption equilibrium constant |
| $L_A$ | areal loading |
| $\equiv|\text{-OH}$ | hydrous metal oxide having a functional –OH group |
| $r_s$ | sedimentation rate |
| $R$ | sulfate reaction rate |
| $S_o$ | seepage outflow |
| $t_w$ | water retention time |
| $V$ | lake volume |
| $z$ | lake mean depth |

## Acknowledgments

Work reported in this chapter was supported by cooperative agreements with the U.S. Environmental Protection Agency—Duluth (J. Eaton, project officer) and —Corvallis (R. Church, project officer) by a grant from the Minnesota Water Resources Research Center and by a contract with Environmental Science and Engineering, Inc., Gainesville,

FL. Lead-210 dating of Little Rock Lake cores was done by Daniel Engstrom, Limnological Research Center, University of Minnesota. We appreciate the fruitful discussions we had on this subject with Robert Cook, Jerald Schnoor, Dave Schindler, Carol Kelly, John Rudd, and Leslie Sherman.

## References

1. Gherini, S.; Mok, L.; Hudson, R. J. M.; Davis, G. F. *Water, Air, Soil Pollut.* **1985**, *26*, 425–459.
2. Schnoor, J. L.; Carmichiel, G. R.; Van Schepen, F. A. In *Energy and Environmental Chemistry*; Keith, L. H., Ed.; Ann Arbor Science: Ann Arbor, MI, 1982; Vol. 2.
3. Christopherson, N.; Wright, R. F. *Water Resour. Res.* **1981**, *17*, 377–389.
4. Cook, R. B.; Schindler, D. W. *Environ. Biogeochem., Ecol. Bull. (Stockholm)* **1983**, *35*, 115–127.
5. Cook, R. B.; Kelly, C. A.; Schindler, D. W.; Turner, M. A. *Limnol. Oceanogr.* **1986**, *31*, 134–148.
6. Baker, L. A.; Brezonik, P. L.; Edgerton, E. S.; Ogburn, W. O., III. *Water, Air, Soil Pollut.* **1985a**, *25*, 215–230.
7. Baker, L. A.; Brezonik, P. L.; Edgerton, E. S. *Water Resour. Res.* **1986a**, *22*, 715–722.
8. Baker, L. A.; Perry, T. E.; Brezonik, P. L. In *Lake and Reservoir Management: Practical Applications*; Taggart, J., Ed.; North American Lake Management Society: Merrifield, VA, 1985b; pp 356–360.
9. Perry, T. E.; Brezonik, P. L.; Pollman, C. D. In *Impact of Acid Rain and Deposition on Aquatic Biological Systems*; Special Technical Publication 928; Isom, B. G.; Dennis, S. D.; Bates, J. M., Eds.; American Society for Testing and Materials: Philadelphia, PA, 1986; p 114.
10. Schindler, D. W.; Turner, M. A.; Stainton, M. P.; Linsey, G. A. *Science (Washington, D.C.)* **1986a**, *232*, 844–847.
11. Lin, J. C.; Schnoor, J. L.; Glass, G. L., in this book.
12. Schindler, D. W.; Rudd, J. W. M.; Kelly, C. A.; Turner, M. A. Presented at the International Symposium on Acidic Precipitation, Muskoka, Ontario, Canada, 1986b; paper B-2 (07).
13. Eilers, J. M.; Glass, G. E.; Webster, K. E.; Rogalla, J. A. *Can. J. Fish. Aquat. Sci.* **1983**, *40*, 1896–1904.
14. Wright, R. *Hydrobiologia* **1983**, *101*, 1–12.
15. Hultberg, H. *Ecol. Bull.* **1985**, *37*, 133–157.
16. Baker, L. A.; Pollman, C. D.; Brezonik, P. L. "Florida Acid Deposition Study, Phase IV"; ESE 83-152-060200120; Environmental Sciences and Engineering: Gainesville, FL, 1985c.
17. Mitchell, M. J.; David, M. B.; Uutala, A. J. *Hydrobiologia* **1985**, *21*, 121–127.
18. Dillon, P. J., Ontario Ministry of the Environment, personal communication, 1986.
19. "Nitrates: An Environmental Assessment"; report prepared by Panel on Nitrates, Coordinating Committee for Scientific and Technical Assessments of Environmental Pollutants; National Academy of Sciences: Washington, DC, 1978.
20. Lee, J. R.; Stewart, G. R. *Adv. Bot. Res.* **1978**, *6*, 1–43.
21. Schindler, D. W.; Turner, M. A.; Hesslein, R. H. *Biogeochem.* **1985**, *1*, 117–133.

22. Baker, L. A. Ph.D. Thesis, University of Florida, Gainesville, 1984.
23. Focht, D. D.; Verstraete, W. *Adv. Microb. Ecol.* **1977**, *1*, 135–214.
24. Rudd, J. W. M.; Kelly, C. A.; St. Louis, V.; Hesslein, R. H.; Furutani, A.; Holoka, M. H. *Limnol. Oceanogr.* **1986a**, *31*, 1267–1280.
25. Brezonik, P. L.; Hendry, C. D., Jr.; Edgerton, E. S.; Schulze, R. L.; Crisman, J. L. *Acidity, Nutrients, and Minerals in Atmospheric Precipitation over Florida: Deposition Patterns, Mechanisms, Ecological Effects*; U.S. Environmental Protection Agency: Corvallis, OR, 1983; EPA-600/3-83-004.
26. Downing, G. M.S. Thesis, University of Minnesota, 1986.
27. Schiff, S. L.; Anderson, R. F. *Can. J. Fish. Aquat. Sci.* **1986**, *43*, in press.
28. Baker, L. A.; Perry, T. E.; Brezonik, P. L., submitted for publication in *Biogeochem.*
29. Oliver, B. G.; Kelso, J. R. M. *Water, Air, Soil Pollut.* **1983**, *20*, 379–389.
30. Gamble, D. S.; Schnitzer, M. *Geochim. Cosmochim. Acta* **1982**, *46*, 983.
31. Benjamin, M. M.; Leckie, J. O. *Environ. Sci. Technol.* **1981**, *15*, 1050–1057.
32. Balastrieri, L.; Murray, J. *Geochim. Cosmochim. Acta* **1983**, *47*, 1091–1098.
33. Aagaard, P.; Helgeson, H. C. *Am. J. Sci.* **1982**, *282*, 237–285.
34. Holdren, G. R., Jr.; Berner, R. A. *Geochim. Cosmochim. Acta* **1979**, *43*, 1161–1171.
35. Stumm, W.; Farrer, G.; Kunz, G. *Croat. Chem. Acta* **1983**, *56*, 593–611.
36. Schnoor, J. L.; Stumm, W. *Environ. Sci. Technol.* **1986**, *20*, in press.
37. Berner, R. A.; Holdren, G. R., Jr. *Geology* **1977**, *5*, 369.
38. Berner, R. A.; Sjoberg, E. L.; Velbel, M. A.; Krom, M. D. *Science (Washington, D.C.)* **1980**, *207*, 1205–1206.
39. Kelly, C. A.; Rudd, J. W. M. *Biogeochem.* **1984**, *1*, 73–77.
40. Perry, T. E. M.S. Thesis, University of Minnesota, 1987.
41. Chen, K. Y.; Morris, J. C. *Environ. Sci. Technol.* **1972**, *6*, 529–537.
42. Schindler, D. W.; Wagemann, R.; Cook, R. B. *Can. J. Fish. Aquat. Sci.* **1980**, *37*, 342–354.
43. Hongve, D. *Verh. Int. Ver. Theor. Angew. Limnol.* **1978**, *20*, 743–748.
44. Perry, T. E.; Baker, L. A.; Brezonik, P. L. In *Lake and Reservoir Management*; Redfield, G.; Taggart, J. F.; Moore, L. M., Eds.; North American Lake Management Society: Merrifield, VA, 1986; Vol. 2, pp 309–312.
45. Kelly, C. A.; Rudd, J. W. M.; Cook, R. B.; Schindler, D. W. *Limnol. Oceanogr.* **1983**, *27*, 868–882.
46. Baker, L. A.; Brezonik, P. L.; Pollman, C. D. *Water, Air, Soil Pollut.* **1987**, *30*, 89–94.
47. Zinder, S. H.; Brock, T. D. In *Sulfur in the Environment, Part II: Ecological Impacts*; Nriagu, J. O., Ed.; Wiley–Interscience: New York, 1978; pp 445–466.
48. Herlihy, A. T.; Mills, A. L. *Biogeochem.* **1985**, *2*, 95–99.
49. Rudd, J. W. M.; Kelly, C. A.; Furatani, A. *Limnol. Oceanogr.* **1986b**, *31*, 1281–1291.
50. Nriagu, J.; Soon, Y. K. *Geochim. Cosmochim. Acta* **1985**, *49*, 823–834.
51. Baker, L.; Brezonik, P. L.; Engstrom, D. A., in preparation.
52. Cook, R. B. Ph.D. Thesis, Columbia University, New York, 1981.
53. Luther, G. W., III; Church, T. M.; Scudlark, J. R.; Cosman, M. *Science (Washington, D.C.)* **1986**, *232*, 746–749.
54. David, M. B.; Mitchell, M. J. *Limnol. Oceanogr.* **1986**, *30*, 1196–1207.
55. Hesslein, R. H. *Limnol. Oceanogr.* **1976**, *21*, 912–914.
56. Lerman, A. *Geochemical Processes*; Wiley–Interscience: New York, 1979.
57. Berner, R. A. *Early Diagenesis*; Princeton University: Princeton, 1980.
58. Vanderborght, J.-P.; Wollast, R.; Billen, G. *Limnol. Oceanogr.* **1977**, *22*, 794–803.

RECEIVED for review May 6, 1986. ACCEPTED October 13, 1986.

# WATER–SEDIMENT PROCESSES

# Hydrophobic Organic Compounds on Sediments: Equilibria and Kinetics of Sorption

Alan W. Elzerman and John T. Coates

Environmental Systems Engineering, Clemson University, Clemson, SC
29634–0919

*Modeling the fate and distribution of hydrophobic compounds in aquatic environmental systems has received considerable attention. Hydrophobic organic compounds are relatively insoluble, nonpolar compounds that have large octanol–water partition coefficients (log $K_{ow} > 2$). A number of pollutants of interest are hydrophobic compounds such as polychlorinated biphenyls (PCBs), polycyclic aromatic hydrocarbons (PAHs), and the DDT group of pesticides. Hydrophobic compounds tend to have large partition coefficients onto suspended and bottom sediments in aquatic systems, a factor resulting in significant effects on their distribution, fate, and effects. The purpose of this chapter is to examine in a tutorial manner the empirical, conceptual, and mechanistic knowledge that has developed concerning sorption of hydrophobic compounds onto sediments. Thermodynamic and kinetic models are examined in an attempt to understand the basis for current practical approaches and theoretical concepts. Estimation procedures and the development of generalized evaluative approaches are emphasized.*

$\mathbf{M}$ANY ORGANIC COMPOUNDS OF INTEREST IN ENVIRONMENTAL systems are hydrophobic, for example, 1,1,1-trichloro-2,2-bis($p$-chlorophenyl)ethane (DDT) and related pesticides, polychlorinated biphenyls (PCBs), and polycyclic aromatic hydrocarbons (PAHs). Hydrophobic compounds derive their classification from being nonpolar, insoluble in water, and having a relatively large octanol–water partition coefficient

0065–2393/87/0216–0263$21.00/0

($K_{ow}$). Typically, compounds that have water solubilities $<100$ mg/L and $K_{ow} > 10^2$ are considered hydrophobic.

Table I presents some examples of hydrophobic compounds, their molecular weights, aqueous solubilities at 25 °C, and log $K_{ow}$ values. The compounds are the hydrophobic examples from a list of "benchmark" compounds of varying characteristics used by Neely and Blau (1) for comparisons of estimation and modeling techniques. Imboden and Schwarzenbach (2) used a similar approach with a group of "model" compounds. Selection of such a list of test compounds for the purposes of evaluating models and estimation techniques is often very useful, particularly in conjunction with an overall approach of generalized, or "generic", modeling. Generic modeling is not site-specific or tied to one or a few specific situations but rather seeks general relationships related to processes that are usually expected to operate and predict the major aspects of what is generally expected to occur. In short, generic models provide evaluative assessments for general cases and provide a place to start for looking at specific cases. Generalized modeling requires, therefore, knowledge of relationships and processes that have widespread applicability as tools to apply in the analysis of different scenarios or classes of cases. The development of one set of generic modeling tools is the subject of this chapter.

**Table I. Physicochemical Parameters for Selected Hydrophobic Compounds**

| Compound | MW (g/mol) | Solubility (mg/L, 25 °C) | log $K_{ow}$ |
|---|---|---|---|
| 1,2,4-Trichlorobenzene | 181.4 | 48.8 | 4.10 |
| 2-Dichlorobiphenyl | 188.7 | 5.0 | 4.54 |
| 2,4,2',4'-Tetrachlorobiphenyl | 291.9 | 0.068 | 6.31 |
| Chlorpyrifos | 350.6 | 0.87 | 4.90 |
| DDT | 354.5 | 0.0033 | 6.36 |
| Bis(2-ethylhexyl) phthalate | 390.6 | 2.49 | 5.3 |

SOURCE: Data are from reference 1.

Because of environmental concerns related to hydrophobic compounds, more and more pesticides and industrial chemicals are being designed to be not only more biodegradable, but also more hydrophilic. Investigation of the analysis, sources, distribution, fate, and effects of hydrophilic organic compounds, therefore, is also requisite, and hydrophilic organic compounds are the object of considerable current research. Perhaps more challenging is to successfully understand the factors controlling the fate and effects of molecules that can behave as though they are hydrophilic or hydrophobic, depending on the sorbent and ambient conditions [e.g., organic molecules that can ionize as a function of pH or that contain functional groups that can participate in hydrogen bonding or other specific bonding interactions (3-5)]. However, the characteristics that initially made hydrophobic compounds of concern—their potential toxic effects, tendency to be recalcitrant and persistent, and observed accumulation in sediments and organisms—will maintain the significance of hydrophobic pollutants for some time.

To understand factors that determine the fate of a chemical in the environment (controlling factors) and to predict the concentration of a chemical in an environmental system, the source term(s) must be known, and then the effects of transport and transformations in the system must be included to make a complete analysis. Typically, the system is divided into parts that individually are more simple, and a compartmental model develops. Simple divisions such as the water column and the sediment as the only two compartments are sometimes sufficient. More detailed divisions improve capabilities but also increase the complexity of the model. Physical, chemical, and biological processes that influence fluxes at the boundaries of the compartments and internal transformations within the compartments must then be quantified.

In the simplest cases when equilibrium exists within and between the compartments with respect to the reference compound, equilibrium models can be used to calculate expected distributions. For example, models based on a balanced fugacity between environmental compartments are currently popular (6, 7). Kinetic considerations, when necessary, must be built into the model with time-dependent mathematical expressions for concentrations. A common approach is to combine sources, sinks, and transformations into a model based on conservation of mass (a mass balance model). Mass balance models relate accumulation in each compartment to the net accumulation of all inputs, outputs, and internal transformations associated with that compartment (2, 8-12). Often, the analysis is further simplified by assuming the compartments behave as idealized "reactors", for example, a batch, plug-flow, or continuously stirred tank reactor (CSTR) (2, 13). Mass balance models can be more sophisticated and adaptable than equilibrium models but also generally require more data and effort and do not always provide significantly better results.

The processes usually considered, when applicable, include advective and dispersive transport; settling; chemical, photochemical, and biological degradation or transformation; hydrolysis; volatilization; and sorption. Sorption is a convenient term that refers collectively to both adsorption onto an interface and absorption beyond the interface into a compartment defined by an interface. Definition of interfaces can be difficult and even arbitrary, so distinction between adsorption and absorption can be difficult, hence the utility of the term sorption to encompass both. Desorption is the process that is the reverse of sorption. The term sorbate refers to the material being removed from solution to the surface, and the term sorbent refers to the sorbing solid.

Sorption onto solids is important because it results in removal from the dissolved phase and incorporation into a solid phase. When sorbed, the chemical is in a different environment and possibly a different form that may interact and react differently or at a modified rate (14). Transport of the chemical will also be affected because the chemical, in the sorbed state, is tied to the movement of the solid. As discussed below, sorption can also result in removal of the chemical from the system by bottom sediment or settling particles.

Significant progress has been made in utilizing modeling approaches to identify the types of data required (15) and the processes that are likely to be important (2) for predicting the fate and distribution of chemicals in the environment for different classes of generic cases. Sorption is often a particularly important process affecting hydrophobic compounds. As the term implies, hydrophobic compounds are not very soluble in water, and so they are more likely to associate with surfaces in aquatic systems than more soluble compounds. Even under ambient conditions where the majority of the hydrophobic material remains in the dissolved state (see later discussion), the overall impact of processes affecting the fraction that is sorbed can be significant. Reaction rates for some processes may be greater in the sorbed state, and some reactions may be unique to the sorbed species. Transport to a lake system can occur predominantly in a sorbed state. Once in a lake system, the fate, distribution, and effects of hydrophobic materials, whether of natural or anthropogenic origin, are affected or can even be predominantly controlled by their association with the particulate phase. Therefore, sorption is an important phenomenon in aquatic systems.

Unfortunately, our knowledge of sorption phenomena is far from complete or even sufficient for some practical applications and problems. Some areas have received less attention than others. For example, rates have received less attention than extents of sorption, and desorption has been studied less than sorption. In some areas, state-of-the-art knowledge is still more empirical than theoretical or mechanistic. Consequently, investigation of sorption processes has been and remains an active area of research.

For many practical purposes, finding mathematical relationships between the appropriate variables and putting them into empirical models suffices. However, to understand sorption of hydrophobic compounds in lake systems, scientific approaches to fundamental and mechanistic understanding must also be invoked. For processes having a characteristic rate of reaction much faster than the time scale of interest in the system, thermodynamic (equilibrium) approaches are usually sufficient. However, if the characteristic rate of the process is slow compared to the time scale of interest, kinetic approaches must also be used (*see* references 2, 10, and 11 for further discussion).

This chapter focuses on the physical and chemical processes of sorption and desorption and the approaches used by environmental chemists and others to investigate, model, and try to understand sorption and desorption of hydrophobic compounds. Because both the rate and extent of sorption and desorption are of interest, the applications of thermodynamic and kinetic approaches to investigating and modeling hydrophobic compounds in aquatic systems are discussed. Biological processes are not emphasized in the scope of this chapter but are certainly important to sorption considerations. Biodegradation or biotransformation can be very significant to the fate of a chemical. However, other than possible biodegradation or biotransformation, biological uptake is usually more important to the organisms involved because of possible toxic effects than to the fate and distribution of the hydrophobic compound. The total mass of the compound in all organisms combined is generally insignificant compared to the total mass of the compound in the overall aquatic system due to the relative magnitude of the masses of biomass and water. It should be noted, however, that many of the approaches to quantifying sorption of hydrophobic materials discussed in this chapter have also been used to predict concentrations in organisms, for example, in estimating bioconcentration factors (*16–18*).

The objective of this chapter is to serve as an introduction and to be tutorial rather than exhaustive, focusing on knowledge and its development rather than on a complete historical review or the latest research results. Accordingly, emphasis is placed on explaining relevant terms, reviewing the basis of current understanding and concepts, and putting sorption of hydrophobic compounds in perspective relative to current environmental concerns and research. An attempt is made to sort out what we know, what we think we know, and what we think we need to know, and show some applications of our knowledge using practical "tools" that have been developed along with fundamental understanding. Finally, some research needs are identified to bring the reader to the threshold of current research and encourage further involvement.

## Framework

Current knowledge of sorption processes is substantial and valuable but is still developing. Understanding and advancement of knowledge are often aided when placed within the framework of history, foundations, and the approaches from which the knowledge is obtained. An understanding of the impetus for current efforts is also appropriate. A more complete understanding can be fostered from consideration of the subtle interplay of practical and fundamental knowledge as they develop together. Practical approaches, such as empirical regressions of data that are used "because they work", are often available before fundamental understanding of "why they work" is achieved; therefore, scientific endeavors are stimulated to develop consistent theoretical explanations.

Sorption of hydrophobic compounds in aquatic systems provides an excellent case for consideration of interrelated development of empirical, practical, and theoretical approaches. Consequently, a central theme throughout this chapter is progression from observations to more and more detailed and fundamental knowledge. At some points, compromises between the practical considerations of meeting immediate needs and the time and monetary expenses of questing for more fundamental knowledge become evident.

Typically, observations lead to questions, which in turn lead to attempts to obtain information useful in answering the questions and in beginning the process of hypothesis formulation, testing, reformulation, and retesting, which is characteristic of what is generally called the scientific method. In the course of this process, simplifications are necessary (i.e., development of models, whether empirical, physical, conceptual, mathematical, or theoretical) to aid our understanding and make the information manageable. Thus, a progression from empirically derived relationships, through various models, to theoretical hypotheses, and finally to detailed mechanistic understanding is normal, although occasional leaps in advancement do occur.

For example, an observation of higher PCB concentrations in the sediments of some parts of a lake compared to other parts of the same lake logically leads to questions about what factors control such distributions. Collection of data on sediment concentrations and other possibly related parameters, such as organic carbon content of the sorbent, follows. Eventually, models are developed based on observed relationships between the data, for example, models relating sorption of hydrophobic compounds to the organic carbon content of sediments.

A progression of phases may thus be defined in the development of understanding. The progression demonstrates use of both inductive and deductive reasoning. Note that the phases are artificial constructs (models) that aid our understanding, are not mutually exclusive, and in many cases progress concurrently and sometimes discontinuously. One possible delineation of the phases follows:

### Define System

   delineate boundaries
   define inputs and outputs (fluxes)
   determine internal transport characteristics
   define internal transformations

### Empirical Phase

make observations and ask questions
collect data and look for generalizations
develop empirical models

### Conceptual Phase

hypothesize interactions, processes, etc.
examine possible surrogate parameters
infer theoretical concepts and causal relationships
identify criteria and constraints to test concepts
develop and use models (all types)
verify and validate models
perform sensitivity and uncertainty analyses on models

### Fundamental Mechanistic Phase

consider all available information
deduce possible explanations
develop fundamental theoretical models
test models in the lab and in the field
refine interpretations
attempt mechanistic explanations
retest and look for internal consistency

Although certainly interesting and worthwhile, further discussion of the procedural and philosophical aspects of these phases is beyond the scope of this chapter. An excellent introduction and demonstration of the application of a similar approach to understanding and managing lake systems can be found in references 10 and 11. The above delineation will, however, be used to organize the discussion that follows. Definition of the system will not be directly addressed. The focus is on several types of generic models that are intended for general evaluations rather than site-specific analyses. The process of developing the approaches has been effectively described by Karickhoff (*19*).

*Natural sorbents may be biotic and abiotic, may be organic, inorganic, or chemical composites thereof, and may range in size from macromolecules to gravel. Organic pollutants vary in size (single carbon to macromolecular compounds) and in water solubility from complete miscibility to virtual insolubility. This high degree of variability and complexity in sorbent–sorbate composition, chemical character, and potential sorptive interactions seems to preclude the development of simplistic sorption models with any degree of theoretical legitimacy. The key to such model development involves dis-*

*criminating initially the essential phenomenological behavior
(trends) while neglecting what could be termed "second-
order" effects or incidental details. These "idealized" models
can then be made more general by the addition of "corrective
terms" that hopefully can be causally related to those details
neglected in the idealization process. . . . It should be stated at
the outset that the prescribed usage precludes the utilization
of regression equations developed and calibrated for site-
specific circumstances.*

## Empirical Phase

Large amounts of data are available on sorption of hydrophobic
compounds and provide a substantial base for consideration. A significant portion of the data is limited by the means by which it was obtained
or by the specific system from which it was derived. Nonetheless, much
of what we know emanates from interpretation of empirical data, and
this data form the base for development of many useful empirical models. This chapter focuses on types of data more than on actual values of
parameters, amounts sorbed, or concentrations in example systems. Specific data are available in many of the references cited (20–27). Table II
presents some data on PCBs in sediments and indicates concentration
levels encountered and the degree of variability between samples. Variability between samples is often significant for hydrophobic compounds
because of inherent variability from site to site caused by specific inputs
and ambient conditions, and because of analytical uncertainties resulting
from the low concentrations of analytes and complex matrices characteristic of environmental samples.

**Table II. Selected Data on Polychlorinated Biphenyls in Sediments**

| | | Concentrations[a] | | |
|------|------------------------|------|---------|------|
| Year | Location | Mean | Maximum | Ref. |
| 1969 | Escambia Bay, FL | 0.69 | 486 | 28 |
| 1974 | Western Gulf of Mexico | 0.03 | 0.33 | 29 |
| 1974 | Puerto Rico | 0.16 | 0.64 | 29 |
| 1975 | Puget Sound, WA | 0.12 | 0.64 | 30 |
| 1976 | Baltimore Harbor, MD | 0.61 | 84.2 | 31 |
| 1977 | Raritan Bay, NJ | 110 | 2035 | 32 |
| 1978 | Sheboygan River, WI | 12 | 190 | 33 |
| 1978 | Severn Estuary, UK | 1.1 | 1.69 | 34 |
| 1979 | Lake Superior | 0.13 | 0.39 | 35 |
| 1980 | Hudson River, NY | 11.4 | 140 | 36 |
| 1982 | Southern Lake Michigan | 48.7 | 201 | 37 |
| 1985 | Lake Hartwell, SC | 12 | 88 | 38 |

[a]Concentration is in units of milligrams of total PCBs per kilogram of dry sediment.

The presentation of data is organized much as an investigation might be designed to begin assessment of the factors that affect, or control, sorption of hydrophobic compounds in a lake, including a "scoping" or "screening" study followed by a series of improvements and refinements in approach and understanding. A scoping study usually is neither comprehensive nor complete but is a way to start and acquire the base needed for planning more detailed studies. After some initial probing and observation, designing more and more sophisticated investigations is usually possible.

**Making Observations and Asking Questions.** One of the most important steps in reaching useful answers is to ask questions at the right time, about the right subjects, and to ask the right questions. Scientific progress, like our personal lives, is a series of tests of reality; we directly benefit from the tests when we are aware of their occurrences and outcomes. Whether the experience is only evaluated after the fact, or planned in advance as an experiment, the better the question asked, the more useful the answer (which makes asking good questions part of the art of the science).

For example, observing that some industrially produced chemicals must be released to the environment leads to questions about the subsequent distribution, fate, and effects of the released material. Knowing that different chemical compounds have different physical and chemical properties should lead to questions concerning how the different properties might lead to different fates and effects in the environment, including possible uneven distributions and localized accumulations. For example, recognizing that many of the hydrophobic compounds produced resist degradation, are more soluble in lipid materials than water, and are potentially toxic can lead to concern about possible bioaccumulation before the reality is observed. Stimulated by the potential consequences of bioaccumulation, the scientist should design measurement techniques, collect data, determine variables, and develop evaluative and predictive models.

Sometimes a collection of data that was obtained for a variety of purposes can also raise questions. For example, consider again the data in Table II on PCBs in sediments of various aquatic systems. Only the mean and maximum values from each reference are presented. These data do not allow for a complete evaluation of each system but do provide an indication of typical values obtained, maximum concentrations encountered, and the variability that can exist in concentrations within and between systems. Although the limitations of the data [e.g., errors related to the notorious difficulties in analytical determinations of PCBs (39)] cannot be obtained from the table, some inferences can be made. For example, PCBs appear to accumulate in sediments, and a number of additional questions can be generated from this inference.

Simple observations of concentrations in the water column can also lead to useful questions. For example, consider the difference between concentration (an intensity factor), total mass loading or inventory (the total mass of a compound in the system, which is a capacity factor), and ultimate fate (which might be removal by physical, chemical, or biological processes, or, especially for persistent compounds, might be accumulation in one or more compartments as a sink). Also, consider the importance of distinguishing between different forms and associations of the compound, for example, particulate versus dissolved forms.

Not surprisingly, the concentration of hydrophobic materials in aquatic systems is usually extremely low, on the order of parts per billion ($\mu$g/L), parts per trillion (ng/L), or less because, by definition, their solubilities are low and therefore they would not tend to remain in aqueous systems. For the same reason, hydrophobic compounds would be expected to accumulate at interfaces. In fact, concentrations in the suspended particulate matter (mass of hydrophobic compound per mass of particulate matter) are generally 1–3 and sometimes 4–5 orders of magnitude greater than the surrounding dissolved-phase concentration (*see* Table III and the related discussion). However, the total mass of a hydrophobic compound in the particulate phase of the total system will also be a function of the suspended solids concentration (*see* Figure 1, Table III, and related discussion). Because suspended solids concentrations are also generally low (e.g., 1–10 mg/L), the total mass of the hydrophobic chemical in the solution phase is usually greater than in the particulate phase (*35, 42*).

However, if some process concentrates the hydrophobic materials in the particulate phase, transport and reactivity consequences can result. Considering ultimate fate, even though the "standing crop" of a hydrophobic compound in the water column has more mass in the dissolved phase than in the particulate phase, settling of the particles to the bottom can result in the bottom sediment being the most important sink of the hydrophobic compound over long time scales (*see* references 43–45 on PCBs in sediments and references 46–48 for other compounds).

Hydrophobic compounds can also leave the aqueous phase to enter the atmosphere (*see* reference 7 and Chapters 1, 3, and 13). Although many hydrophobic compounds have low vapor pressures, their low solubilities result in significant Henry's law constants and volatility from aqueous systems (*26, 49*). In some systems, volatilization of hydrophobic compounds may be a significant ultimate sink (*50, 51*). Volatilization can also be affected by sorption (*52*). Ultimate fate, therefore, is a relative term that depends on the time scale of reference, and additional fates may have to be considered if longer time scales are used.

**Determining Variable Parameters.**   Determining variable parameters requires identification of factors that need to be quantified, development of measurement techniques or estimation procedures through relationships to other measured or calculated values, reducing data to useful forms, and constant judging of the quality of the numbers obtained. Determining parameters is a necessary step for developing and understanding models; unfortunately, defining parameters and simply making measurements are often also considered sufficient steps and lead to the production of many numbers that serve little or no purpose for advancing understanding. Numbers are just numbers unless they can be used to develop information. Continual adjustment and improvement of parameters is also important, and this activity continues through all stages.

In the case of sorption of hydrophobic compounds on sediments, parameters are required to evaluate the extent of sorption and the rate. Thermodynamic considerations of the extent of sorption will be discussed first. Knowledge of sorption reaction rates generally lags behind that of sorption extent and will appropriately be discussed later. Similarly, discussion of desorption will not be emphasized initially. In most cases, sorption in environmental systems has been considered to be both fast and reversible. Current research indicates that these assumptions are not always valid; therefore, they will receive attention later in the chapter.

Parameters describing the distribution of a compound between a solid and a liquid phase have been defined much the same as for other multiphase systems, by defining a distribution coefficient $(K_d)$ or a partition coefficient $(K_p)$ (7, 53, 54). Distinctions between $K_d$ and $K_p$ can be made, but the terms are often used interchangeably. However, distinguishing between true equilibrium partition coefficients and observed distributions is important. One notation proposed to make the distinction clear is to call the observed or empirical distributions ratios; make no assumption regarding equilibrium, reversibility, or sorption mechanism; and give these ratios a symbol such as $R_d$ (55).

The partition coefficient $(K_p)$ is the term of choice for this discussion and is defined as

$$K_p = \frac{C_s}{C_w} \tag{1}$$

where $C_s$ is the concentration in the solid phase due to sorption, and $C_w$ is the concentration dissolved in the water phase at equilibrium.

If $C_s$ is in units of micrograms of sorbate per gram of sorbent, and $C_w$ is in units of micrograms per milliliter, then $K_p$ will have units of milliliters per gram. Values of $K_p$ are usually reported for equilibrium conditions, but equilibrium may not exist or be confirmed, in which case values should be reported as observed partition coefficients as discussed previously.

By using equation 1, the relationship between $K_p$, suspended solids concentration, and the fraction of sorbate that will be sorbed at equilibrium can be quantified. Many authors (e.g., references 56 and 57) have derived (or referenced the derivation) and used equations similar to equation 2 for quantifying distribution between two phases.

$$F = \frac{K_p[\text{SPM}]}{1 + K_p[\text{SPM}]} \tag{2}$$

where $F$ is the fraction of total mass of sorbate sorbed, $K_p$ is the partition coefficient (mL/g), and [SPM] is the concentration of suspended particulate matter in suspension (g/mL).

Equation 2 shows that as the product $K_p[\text{SPM}]$ increases, the fraction sorbed increases and eventually approaches 1. By using equation 2, plots of the relationship can also be constructed. For example, Figure 1 shows the fraction sorbed as a function of the suspended solids concentration for different $K_p$ values, assuming that the $K_p$ values remain constant over the range of suspended particulate matter concentration. Such plots help visualize the relationship and are tools for quick assessments of when sorption may be important. Figure 1 indicates that only small fractions of sorbate will be sorbed when $K_p < 10^3$ and [SPM] $< 100$ mg/L. In contrast, if $K_p > 10^5$, a significant fraction of sorbate will be in the particulate phase even at relatively low suspended particulate matter concentrations. Figure 2 shows a similar plot with a different shape due to the log scale for suspended particulate matter concentration and includes some laboratory and field data on sorption of PCBs for comparison.

Equations 1 and 2 can also be used to generate a table of values showing the quantitative relationships between $K_p$; [SPM]; and the dissolved, particulate-phase, and suspension concentrations of a sorbate (Table III). A table of values such as Table III can be used for many purposes. Only a few generalizations will be made here. Not surprisingly, increasing $K_p$ values are accompanied by decreasing values of fraction dissolved, although at a suspended particulate matter concentration of 100 mg/L, $K_p$ values greater than $10^4$ are required to attain more sorbate in the particulate phase than the dissolved phase. As pointed out by Mills et al., the concentration of sorbate in the particles ($C_s$, or mass of sorbate per mass of dry sorbent) remains relatively high over almost the entire range of $K_p$ and [SPM] values (40).

The next task in developing parameters is to collect $K_p$ data, empirically determine the factors that affect $K_p$, and then develop conceptual and theoretical explanations for the results.

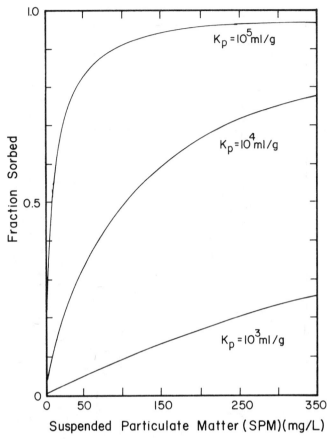

*Figure 1. Theoretical fraction sorbed as a function of suspended particulate matter concentration for selected $K_p$ values. (Reproduced with permission from reference 41. Copyright 1985, J. A. Vandeven.)*

**Collecting Data and Looking for Generalizations.** Collecting data might include actually collecting samples and analyzing them or gathering information collected by others. In either case, the dependence on and limitations imposed by the availability of analytical techniques and quality of available data are obvious. As more specific and detailed information is sought, the dependence on analytical techniques to help generate and verify concepts becomes greater. One of the biggest challenges (another part of the art of the science) is to make the most of available data without surpassing its applicability and quality. Ignoring data because it is not perfect is as senseless as making unsupported interpretations of data; the question is what can the data tell you.

Partition coefficient data for hydrophobic compounds exhibit signif-

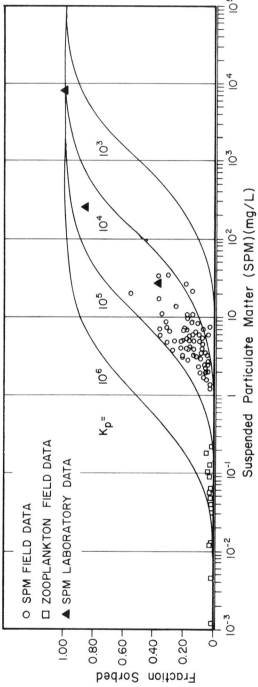

*Figure 2. Theoretical fraction sorbed versus log of suspended particulate matter concentration for selected $K_p$ values (solid lines) with example data for sorption of total PCBs on SPM and zooplankton from Puget Sound and for laboratory uptake studies with SPM (points). (Reproduced from reference 30. Copyright 1979, American Chemical Society.)*

Table III. Relationships between the Partition Coefficient, $K_p$, Suspended Particulate Matter Concentration, and Dissolved- and Particulate-Phase Concentrations of Sorbate at Equilibrium

| $K_p{}^a$ $(C_s/C_w)$ | $[SPM]^b$ $(\mu g/mL)$ | Fraction Dissolved $(C_w/C_T)^c$ | If $C_T = 100$ $\mu g/mL$ | | |
|---|---|---|---|---|---|
| | | | $C_w$ | $C_s$ | $C_{SPM}{}^d$ |
| $10^0$ | 1 | 1.0 | 100 | 100 | 0.0 |
| | 10 | 1.0 | 100 | 100 | 0.0 |
| | 100 | 1.0 | 100 | 100 | 0.0 |
| | 1000 | 1.0 | 100 | 100 | 0.0 |
| | 10000 | 1.0 | 99 | 99 | 0.1 |
| $10^1$ | 1 | 1.0 | 100 | $1 \times 10^3$ | 0.0 |
| | 10 | 1.0 | 100 | $1 \times 10^3$ | 0.0 |
| | 100 | 1.0 | 99.9 | 999 | 0.1 |
| | 1000 | 1.0 | 99.0 | 990 | 1.0 |
| | 10000 | 0.9 | 90.9 | 909 | 9.1 |
| $10^2$ | 1 | 1.0 | 100 | $1 \times 10^4$ | 0.0 |
| | 10 | 1.0 | 99.9 | $1 \times 10^4$ | 0.1 |
| | 100 | 1.0 | 99.0 | $9.9 \times 10^3$ | 1.0 |
| | 1000 | 0.9 | 90.9 | $9.1 \times 10^3$ | 9.1 |
| | 10000 | 0.5 | 50.0 | $5.0 \times 10^3$ | 50.0 |
| $10^3$ | 1 | 1.0 | 99.9 | $1.0 \times 10^5$ | 0.1 |
| | 10 | 1.0 | 99.0 | $9.9 \times 10^4$ | 1.0 |
| | 100 | 0.9 | 90.9 | $9.1 \times 10^4$ | 9.1 |
| | 1000 | 0.5 | 50.0 | $5.0 \times 10^4$ | 50.0 |
| | 10000 | 0.1 | 9.1 | $9.0 \times 10^3$ | 90.9 |
| $10^4$ | 1 | 1.0 | 99.0 | $9.9 \times 10^5$ | 1.0 |
| | 10 | 0.9 | 90.9 | $9.1 \times 10^5$ | 9.1 |
| | 100 | 0.5 | 50.0 | $5.0 \times 10^5$ | 50.0 |
| | 1000 | 0.1 | 9.1 | $9.1 \times 10^4$ | 90.9 |
| | 10000 | 0.0 | 1.0 | $9.9 \times 10^3$ | 99.0 |

SOURCE: Data are from reference 40.
$^a K_p$ is the equilibrium partition coefficient (mL/g). $^b[SPM]$ is the suspended particulate matter concentration. $^c C_T$ is the total concentration of adsorbate ($\mu g/mL$), and $C_T = C_w + C_{SPM}$. $^d C_{SPM}$ is the concentration of sorbed adsorbate in suspension ($\mu g/mL$), and $C_{SPM} = 10^{-6} (C_s) [SPM]$.

icant variability for different sorbate–sorbent pairs and in reported values for a given pair and system. Some of the variability results from analytical difficulties in making the measurements, especially for larger $K_p$ values because a large $K_p$ value can result in very low solution-phase concentrations that can be difficult to determine accurately. McCall et al. (58), Baes and Sharp (59), and Lyman et al. (24) provide discussions of some of the problems associated with $K_p$ measurements (*see* Figures 11 and 12 and the related discussion). However, variability in $K_p$ data also indicates that additional factors probably affect $K_p$ values for specific cases and that further investigation of the controlling factors should be undertaken.

A common approach at this juncture is to define parameters for and collect data on other characteristics of the samples and systems and look for relationships between parameters. Application of previous knowledge and experience can improve the efficiency of this step (more art in the science). For example, the following list of factors related to characteristics of the system under investigation, the sorbent, and the sorbate might be generated for consideration (*see also* reference 60 for a useful table of chemical properties that are important for questions such as "where will it go?" and "how fast will it get there?").

### System Characteristics

Macroenvironment (bulk and system characteristics)
    all microenvironment factors
    inputs (amount, composition, timing, location, and distribution)
    physical transport of water
    physical transport of sorbing particles
    competing outputs (e.g., volatilization or degradation)
    postdepositional relocation of sediments
    postdepositional transformations (diagenesis)
Microenvironment (close to sorbing surface)
    concentration (activity) gradient of sorbate
    temperature
    pH
    ionic strength
    competing sorbents and sorbates
    species modifiers (e.g., ligands)

### Sorbent Characteristics

Composition
    bulk
    surface
Surface state
    charge
    energy
    reactive groups
    previous sorption history
Morphology
    size of particles
    shape and pore structure (specific surface area and pore size)

### Sorbate Characteristics

Molecular structure
    reactive groups

> charge (in conditions of microenvironment)
> polarity
> size
> shape
> flexibility
> Association with other components of the system

Considering this list, some factors can be discounted in terms of their importance to sorption of hydrophobic compounds, and other factors can be emphasized on the basis of previous knowledge or known chemical concepts. For example, ionic strength would be expected to be more important to sorption of ions, where interactions between ions in solution and at surface sorption sites could occur, than to sorption of uncharged hydrophobic compounds. In fact, sorption of hydrophobic compounds by sediments has been observed to be relatively independent of ionic strength and usually sorption increases by no more than a factor of 1.2 with increasing ionic strength (*61*).

The relative independence of sorption of hydrophobic compounds from effects of ionic strength implies that some other mechanism relatively unaffected by ionic strength is important. A reasonable hypothesis is a mechanism that minimizes the importance of interactions of the hydrophobic molecule with sites affected by ionic strength on the surface of the sorbent, such as a mechanism of partitioning, or "solubilization", of the hydrophobic material into a relatively nonpolar phase in the solid matrix. Hydrophobic compounds partitioning into sorbent-phase organic matter is the key to much of the discussion that follows.

By using these considerations to design lab and field studies and interpreting the data collected, several factors related to accumulation of persistent hydrophobic compounds in sediments have been identified. These factors are usually considered first in questions of sediment accumulation of hydrophobic compounds. Within a given system, a reasonable prediction of sediment concentrations of a given hydrophobic chemical can usually be made from these factors using the modeling techniques discussed in this chapter in conjunction with modeling of sediment dynamics in the system. These "first-order" factors for sorption of persistent hydrophobic compounds on sediment are generally

- inputs to the system;
- dissolved-phase concentration, which determines the concentration gradient of sorbate;
- organic carbon content of the sediment;
- degree of hydrophobicity of the compound;
- particle size of the sediment;

• sediment transport in the system; and
• sediment reworking and resuspension.

Many references are available that discuss these factors, including most of the references at the end of this chapter and other chapters in this book. For example, Armstrong and Swackhamer (62) and Weininger et al. (37) discuss physical processes that affect concentrations in the sediment at any given location and at different depths in a sediment core. Increasing hydrophobic compound concentration in sediments with increasing organic carbon content, and increasing hydrophobic compound concentrations with decreasing particle size are discussed by Hiraizumi et al. (63), Steen et al. (45), Choi and Chen (64), and Karickhoff et al. (65). Pavlou has presented a model of the effect of decreasing particle size on increasing surface area–mass ratios and $K_p$ (66). As discussed later, Wu and Gschwend extend particle size considerations to sorption kinetics as well as to thermodynamics (67). Smaller particles also often have higher organic carbon contents than larger particles (see discussion later in the chapter for importance of this factor). Smaller particles also tend to accumulate in the more quiescent deposition areas of aquatic systems because of sediment settling and redistribution processes. These factors combine to support the common observation that the highest sediment concentrations of hydrophobic compounds (and other materials that tend to be sorbed) usually occur in the more quiescent deposition areas associated with smaller particles.

As with most observations and generalizations, exceptions are well known. Particle size and organic carbon content are not always clearly related to sediment concentration of hydrophobic compounds (68). Expected correlations may not be observed, especially when dealing with materials that are actually mixtures of compounds such as the PCB mixtures Aroclor 1016 and 1242 (45). Postdepositional transport and transformation may also destroy expected correlations.

One method of factoring out variability in field and lab data to make comparisons between samples and develop predictive relationships is to seek some related parameter that shows a correlation to the parameter of concern and normalize all the data to that parameter (i.e., for each sample, divide the parameter of interest by the correlated parameter). Organic carbon content of the sorbent has proven to be a useful parameter for this purpose when considering the extent of sorption of hydrophobic compounds. A normalized parameter, the organic carbon partition coefficient ($K_{oc}$), from which the apparent concentration of sorbed material in the organic carbon phase of the sorbent can be calculated, is often determined as follows (56):

$$C_{s,oc} = K_{oc} C_w = \frac{K_p}{OC} C_w \qquad (3)$$

where $C_{s,oc}$ is the concentration of sorbate in the organic phase of sorbent at equilibrium, $C_w$ is the concentration of sorbate in the aqueous phase at equilibrium, $K_{oc}$ is the organic carbon referenced partition coefficient $(K_p/OC)$, and OC is the fraction of organic carbon in the sorbent and equals the percent organic carbon in the sorbent divided by 100.

The use of $K_{oc}$ values for uncharged hydrophobic compounds is now widespread, and the term "carbon referenced sorption" is commonly applied to this empirical approach. Karickhoff (*61*), Lyman et al. (*24*), and Lyman (*60*) list some of the many collections of $K_{oc}$ data that are available. The lists include values for soils and sediments for compounds such as triazines, PCBs, PAHs, and some pesticides. Some examples of $K_{oc}$ values for the same compounds listed in Table I are as follows: 1,2,4-trichlorobenzene, 3.69; 2-dichlorobiphenyl, 4.32; 2,4,2′,4′-tetrachlorobiphenyl, 4.78; chlorpyrifos, 4.24; 1,1,1-trichloro-2,2-bis(*p*-chlorophenyl)ethane (DDT), 5.39; and bis(2-ethylhexyl) phthalate, 5.00 (*1*). These $K_{oc}$ values were estimated by using Karickhoff's method (*19*), which is similar to equation 6. This empirically based approach has led to improvements in our understanding and conceptualization and has provided practical tools. Karickhoff (*61*) concluded:

*Carbon referenced sorption has been tested for a wide variety of both compounds and sediment and soils. . . . From these findings combined with data compilations . . . it can be safely generalized that for uncharged organic compounds of limited water solubility ($<10^{-3}$ M) that are not susceptible to speciation changes or other special complex formation in the sediment suspension of interest, sorption is typically 'controlled' by organic carbon and is amenable to the $\mathrm{K}_{oc}$ format of quantification.*

**Developing Empirical Models.** Empirical models for predicting sediment concentrations of compounds that can be sorbed are most often related to sediment characteristics (e.g., particle size and organic carbon content of the sorbent), which in turn are related to sources of particles, particle distribution processes, and to physicochemical characteristics and concentration of the sorbate in the solution. These parameters take into account the most important considerations. Further discussion can conveniently be divided into two areas, namely (1) factors that influence the sources, nature, transport, and accumulation of sorbent particles, and (2) isotherms and factors that influence the extent that a given sorbate sorbs on a given sorbent (including concentrations and factors that influence the equilibrium $K_p$ for different sorbates and sorbents). These two areas are addressed separately.

**SOURCES, NATURE, TRANSPORT, AND ACCUMULATION OF SEDIMENT.** This first group of factors is related to particle sources (inputs and internal generation), the sorbent characteristics of the particles that are present, and to the many physical, chemical, and biological processes in aquatic systems that affect particle fate and distribution. A detailed discussion of these topics is beyond the scope of this chapter. Appropriate discussions are available (*10, 11, 69, 70,* and Chapters 11 and 13). As discussed in these references, dynamic physical forces in most aquatic systems and the physical and biological processes that cause sediment mixing and resuspension frequently lead to complexities in predicting the fate of particles. Consequently, models involving transport of particles are often more complex than the empirical models and must include conceptual and theoretical considerations that remove these transport models from the list of empirical models.

However, particle size is an important factor in determining the extent (and the rate) of sorption as well as affecting how the particles are transported. Surface area per unit mass is inversely related to particle size. Total available surface area is related to the porosity of the particle structure, which also can vary with particle size (*67*). Finally, because of variations in sources of particles [e.g., living organisms compared to dead biogenic, mineral, or coagulation-produced particles (*69* and Chapters 11 and 13)] and variations in composition, sorbent characteristics of particles can be related to particle size.

**ISOTHERMS AND RELATED FACTORS.** Factors that affect the extent of sorption of a given sorbate on a given sorbent include temperature, pH, ionic strength, and solution-phase concentration. As discussed above, sorption of nonionizable hydrophobic compounds is not greatly affected by ionic strength over the ranges generally encountered in environmental systems. Temperature is an important factor in sorption, and the effect is related to the mechanism of sorption. Generally, increasing temperature results in decreasing extent of sorption for physical mechanisms (e.g., when the attraction at the surface results from relatively weak van der Waals forces), but the opposite can be true for chemical sorption mechanisms (*71*). As discussed later, sorption of hydrophobic compounds is primarily a result of a physical mechanism and inversely related to the solubility of the sorbate in water; therefore, decreasing sorption with increasing temperature is expected (*61*). However, very hydrophobic compounds can show increasing sorption with increasing temperature, and this relationship may actually be an artifact of faster rates of sorption and therefore the appearance of a greater extent of sorption in a given time frame of observation (*see* reference 61 for more discussion of the effect of temperature on sorption of hydrophobic compounds). The effect of pH on sorption of hydrophobic compounds is greatly influenced by whether they ionize and, to a lesser extent, on whether pH changes affect the sorbent surface structure or composition.

The relationship between extent of sorption and solution-phase concentration of sorbate occupies a special place in any discussion of sorption because, at a stated constant temperature, plots of this relationship, called isotherms, are a common way of presenting sorption results, reducing large quantities of data to manageable forms, and investigating sorption processes. Discussion of sorption isotherms is treated in more detail elsewhere (71-74).

The two most common treatments of isotherms are the Langmuir and the Freundlich models, although numerous modifications and several other models are available. The Langmuir isotherm model actually derives from a theoretical treatment of sorption of gases onto solids, although when applied to sorption of solutes onto solids in aqueous systems, its application becomes semiempirical at best. Modifications of the Langmuir isotherm model to account for additional factors and effects have occurred and result in extensions to the basic equation, which is equation 4 (74-77). The Freundlich isotherm model has an empirical foundation from fitting a curve to data, although a theoretical foundation has been developed to explain why it works (73). Even if experimental data for sorption onto soil or sediment fit either of these models (i.e., the data can be adequately described by the model), this apparent consistency is not sufficient to conclude that the assumptions of the model have been met. The equations do not give mechanistic information (78).

Both models are intended to describe equilibrium conditions only. Isotherms are usually generated experimentally by determination of a series of points from a series of batch experiments in which varying amounts of sorbent and sorbate are allowed to come to equilibrium. Ascertaining that equilibrium has been attained can be more difficult than it would first seem. This situation is discussed later in relation to kinetics.

The characteristic shapes of both isotherms can be generated from mathematical relationships that have been developed and widely used. The mathematical formulations of the Langmuir and Freundlich models are given by equations 4 and 5, respectively, and generic plots of calculated isotherms are presented in Figure 3.

$$C_s = \frac{K_1 K_2 C_w}{1 + K_1 C_w} \tag{4}$$

$$C_s = K_F C_w^{1/n} \tag{5}$$

where $C_s$ is the equilibrium concentration of sorbate in sorbed phase, $C_w$ is the equilibrium concentration of sorbate in solution, $K_1$ is a constant related to energy of sorption in the theoretical model but is simply a fitting constant in empirical use, $K_2$ is a constant related to capacity of

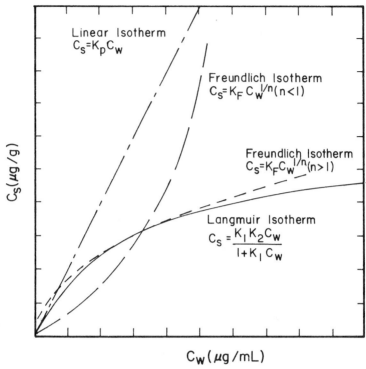

*Figure 3. Generalized calculated sorption isotherms. (Modified from reference 40.)*

sorption (maximum amount of sorption), $K_F$ is the Freundlich "capacity" constant, and $n$ is the Freundlich "intensity" constant.

Many transformations of these models are available that allow determination of the constants from linear plots. However, as Harter (75) points out, care must be exercised in using the linearized forms of the Langmuir equation because these forms can remove some of the variability and obscure whether the data actually fit the Langmuir model. Harter (75) states

> In summary, the [simple form of the] Langmuir equation remains useful for empirical description of sorption data. However, it is apparent that soil sorption data frequently does not fit the Langmuir model, and the linearity of the 'Langmuir plot' is an inadequate test of fit because plotting a function of concentration vs. concentration decreases data variability.

Harter also presents the "family" of curves, similar to Figure 3, that can be generated by using the Langmuir model and different values for the ratio $K_2/K_1$ and suggests that data should be tested against the basic Langmuir equation (equation 4) for fit. Similar inaccuracy problems can also be encountered if extrapolations from a few data points are made with the model to obtain values for the maximum amount of sorption $(K_2)$ (75).

Isotherm models have been used extensively and significant amounts of data have been generated, so only features most salient to sorption of hydrophobic compounds in aquatic systems are mentioned here. One of the most significant differences between the Langmuir and Freundlich models is their predictions at high sorbate concentrations. The Langmuir model predicts saturation of the sorbent and a leveling off of the amount sorbed at high solution concentrations. This observation is shown in Figure 3 and predicted in equation 4 when $K_1C_w \gg 1$. The Freundlich model predicts continually increasing amounts sorbed with increasing solution concentrations and therefore may not be applicable at high concentrations because a maximum amount of sorption cannot be predicted. If uptake of sorbate occurs continuously with increasing sorbate concentration, a mechanism other than sorption, such as precipitation, may be operating. Values of $n > 1$ can be used in a Freundlich model to obtain a plot that appears to plateau, similar to the Langmuir plot, as shown in Figure 3. Also, because the Freundlich equation is generally not applicable at low sorbate concentrations unless $n = 1$, the Freundlich equation finds its greatest applicability at intermediate concentrations of sorbate. Changing values of $n$ in one model may also be used to fit data over a range of sorbate concentrations.

Perhaps the point most relevant to a discussion of sorption in environmental systems is the similarity of the two models under some conditions. At low concentrations of sorbate in solution, the Freundlich $n$ is typically equal to 1, $K_1C_w$ in the Langmuir equation becomes insignificant relative to 1, and the two models become essentially equivalent. Furthermore, the models become equivalent to the linear sorption model, which is represented by the definition of $K_p$ (equation 1), and a plot of the amount sorbed versus solution-phase concentration at equilibrium is linear and has a slope equal to $K_p$. Solution-phase concentrations less than approximately one-half of the compound solubility are usually considered in the safe range to assume "linear sorption" (61) when availability of sorbent is not limiting.

In summary, at low sorbate concentrations the extent of sorption for a given temperature with sorbent present in excess generally increases linearly with solution-phase sorbate concentration. At high solution-phase sorbate concentrations, the sorption capacity of the sorbent may be exceeded and saturation of the sorbent may occur. At concentrations above the solubility of the sorbate, precipitation may occur. Because concentrations of hydrophobic compounds encountered in the environment are generally low compared to their solubilities and availability of sorbent, the linear model is frequently and conveniently applicable.

In practice, the Freundlich model is often used for modeling sorption in aquatic systems, especially in water and waste water treatment, when the simple linear model is insufficient. The Freundlich model often produces a better description of the data than the Langmuir model. Values for the coefficient $n$ are typically 0.5–1.2 (60). At high sorbate concentrations, the ever-increasing solid-phase concentration of sorbate with increasing solution-phase concentrations predicted by the Freundlich model is also observed in many cases. Note, however, that a change from a sorption mechanism to a precipitation mechanism cannot be distinguished in an isotherm.

Variations in sorption model parameters determined for particular sets of data can sometimes yield useful insights. Although resulting from an empirical model, variations in coefficients such as $n$ in the Freundlich model must have an explanation. The explanation may be the presence of experimental artifact or error, but variations can also indicate changes in sorption mechanisms or energies of interaction (72, 73). Additionally, the models presented deal with only a single sorbent and single sorbate, and the equilibrium is independent of the path leading to equilibrium (i.e., the coefficients should depend only on the final equilibrium state of the system). Significant information can be obtained by looking at results for mixtures and sequences of perturbations of the system, although mechanistic explanations of the results are not always easily derived. In mixtures of sorbents and sorbates the extent and rate of sorption can be related to not only the compounds present, but also the history of the system (i.e., timing, sequencing, and intermediate treatments) and the interactions between the sorbents and the sorbates (76, 77). The results of such interactions are collectively referred to as "competitive effects" and can have a significant effect on the outcome of simultaneous equilibria and kinetics.

Competitive effects are the focus of significant research efforts, although detailed discussion is beyond the scope of this chapter. Competitive effects are more likely to be significant when sorbate concentrations on the sorbent are higher, for example, in carbon sorption treatment of drinking water (77), than in natural water systems with relatively low concentrations of hydrophobic compounds. However, apparent competitive effects are sometimes observed even in simple mixtures at environmentally relevant concentration levels (79).

Observing relative invariability of empirical coefficients can also lead to worthwhile information. The fact that measured Freundlich $n$ values remained constant over a wide range of sorbate solution-phase concentrations for sorption experiments conducted with hydrophobic compounds and soils led Chiou et al. (80, 81) to hypothesize that sorption of hydrophobic compounds results from a mechanism involving physical partitioning into an organic phase rather than a mechanism involving surface attraction. As discussed later in relation to estimating $K_p$, this conceptualization of the process has led to useful semiempirical quantitation and prediction techniques and has provided a basis for developing fundamental understanding of sorption of hydrophobic compounds on soils and sediments.

## Conceptual Phase

As expected and already noted, the empirical phase has stimulated consideration of many concepts. In fact, observed deviations from empirical models have stimulated current research on detailed mechanistic processes to obtain explanations. The choice of topics for this section is, therefore, somewhat arbitrary, but one of the exciting aspects of the conceptual phase is that it bridges observation and understanding. The major focus is to show how some of the observations in the empirical phase can be reconciled with accepted concepts of thermodynamics and to begin to apply known chemical concepts to make advancements in understanding.

**Estimating $K_p$.** This section discusses models based on organic carbon content of the sorbent and physicochemical characteristics of the sorbate molecules, which in combination have been found to be good predictors of $K_p$. In this process, more and more use of theoretical concepts has resulted in getting more into the conceptual phase, although the realm of empiricism has not been left entirely. In the words used in the title of one of the classic papers in this area (82), these approaches are actually semiempirical. Semiempirical approaches serve to stimulate development and testing of concepts. In this approach, theory is used to guide formulation of mathematical relationships, and empirical calibration is used to obtain values of coefficients applicable to specific cases.

Physical and chemical characteristics of the sorbent and the sorbate are of primary importance in controlling sorption at given concentration levels. Many different physical and chemical mechanisms can be responsible for sorption (71–73). The common requirement of these mechanisms is that sufficient attraction between the sorbate and the sorbent must be established to maintain the presence of some sorbate on the surface of the sorbent. The stronger the attraction between the surface and the sorbate and the weaker the attraction between the sorbate and the solvent, the greater the extent of sorption. However, in any specific case, sorption may be the result of a number of interactions, and the sorbate–sorbent surface interactions may not even be the dominant factor. Overall, the energies associated with changes in solvent–solvent, solvent–sorbate, and sorbate–sorbate interactions as sorption occurs must also be taken into account. Simple models will only be possible when some of these factors are insignificant compared to others.

In the case of sorption of hydrophobic compounds, a mechanism involving physical partitioning into organic matter in or sorbed onto the particles may actually be a better description than a surface attraction mechanism as introduced previously (80, 81). In this concept, sorption can be viewed as preferential solubilization of the hydrophobic compound in the sediment organic matter over the aqueous phase. Whether or not such a process is called sorption or simply partitioning quickly becomes a discussion of semantics and definitions of the location of the interface, although such discussions provide stimulus to conceptualization. The fact that the equilibrium $K_{oc}$ tends to be independent of the specific sediment or soil (56, 61) suggests the common presence in many sediments and soils of organic phases of sufficiently similar characteristics that sorption of hydrophobic materials is similar for the different samples. A mechanism of the hydrophobic compounds essentially dissolving into the organic phase can thus be supported, and predictive relationships based on the solubility characteristics of the sorbate in water and in organic phases can be pursued. Application of chemical principles of solubility can then help advance the theory and use of the models. Estimation of $K_p$ values is, therefore, possible by estimating $K_{oc}$ values and converting to $K_p$ using equation 3 (or equation 7 or another appropriate formulation).

A significant advance made in this area was the realization that estimates of $K_{oc}$ can be made from physicochemical characteristics of the sorbate. Furthermore, because the solubilities of hydrophobic compounds are more similar in nonpolar organic solvents than in water, the approach of using a single organic solvent in a model to represent the organic matter responsible for sorption of hydrophobic compounds was recognized as potentially useful. As discussed later, these approaches have resulted in some very practical tools (see reference 61 for a more detailed discussion of the thermodynamic basis).

At equilibrium, the chemical potential, or fugacity, of a single sorbate is equal in all phases, so the ratio of the concentrations multiplied by the appropriate activity coefficients in any two phases will be constant (assuming constant activity coefficients independent of the sorbate concentration). The concentration-based $K_p$, as defined in equation 1, will also be constant as long as the activity coefficients are constant and will be a function of the ratio of the activity coefficients of the compound in the two phases. Therefore, differences in $K_p$ for two compounds in the same two phases at the same temperature are related to differences in the ratio of the activity coefficients in the two phases for each compound.

If the two phases are water and an organic phase, $K_p$ is related to the activity coefficient of the compound in water and in the organic phase. Thus, if the activity coefficients for hydrophobic compounds are similar in nonpolar organic solvents and in sediment and soil organic matter, partitioning of a hydrophobic compound into them is primarily a function of their individual activity coefficients in the water phase. Hence, obtaining initial predictions of sorption into the sorbent organic phase is possible by looking at the behavior of the hydrophobic compound in just one representative organic phase, chosen to be convenient to work with in the laboratory. The $K_p$ value for partitioning of the hydrophobic compound into the representative organic phase becomes a "surrogate" parameter for the $K_{oc}$ of the compound partitioning into sorbent organic matter. Furthermore, because the aqueous solubility of a compound is a function of its aqueous-phase activity coefficient, exploiting this relationship and predicting partitioning into the organic phase from the aqueous solubilities of the hydrophobic compounds should also be possible.

The model solvent usually chosen is octanol, and the surrogate parameter (surrogate $K_p$ in this case) used is the octanol–water partition coefficient ($K_{ow}$) (*see* references 24, 65, and 83 for discussion of experimental and mechanistic causes of variations in $K_{ow}$ values). $K_{oc}$ is then related to $K_{ow}$ through a model such as the following (i.e., a line is fitted to the experimental points and the line can be mathematically described by an equation of this form):

$$\log K_{oc} = a \log K_{ow} + B \qquad (6)$$

where $K_{oc}$ is the organic carbon referenced partition coefficient; $K_{ow}$ is the octanol–water partition coefficient; and $a$ and $B$ are empirically determined coefficients that are not unique but are specific to the compounds and sorbents tested.

Dzombak and Luthy (*84*), Karickhoff (*19, 61*), Lyman et al. (*24*), and Lyman (*60*) have reviewed specific $K_{oc}$ results that have been obtained by estimation from $K_{ow}$. A summary plot is presented in Figure 4 that indicates the range of $K_{ow}$ over which the estimation is most applicable. Karickhoff (*61*) also included some indications of possible reasons for failure of the model outside of the linear range, as indicated in Figure 4. Note that a model can be useful for predictions when it works, and for giving us clues to new frontiers of understanding when it does not work completely.

$K_{oc}$ and $K_{ow}$ values can also be estimated from other physicochemical characteristics of hydrophobic sorbate molecules, including their aqueous solubilities (Figure 5). Because aqueous solubilities of hydrophobic compounds are related to their aqueous activity coefficients, the

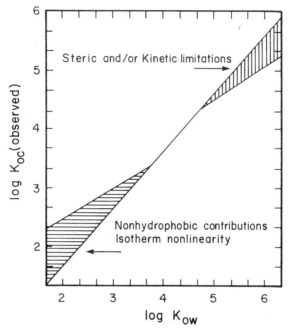

*Figure 4. Log $K_{oc}$ versus log $K_{ow}$. (Modified with permission from reference 61. Copyright 1984, American Society of Civil Engineers.)*

relationship between partition coefficients and activity coefficients should carry through to aqueous solubilities, and partition coefficients of hydrophobic compounds into the organic phase should be inversely related to aqueous solubilities. Limitations of the solubility estimator approach have been observed for high melting point solids and very hydrophobic compounds but may be compensated by additions to the equation to account for second-order factors (*24, 85, 86*).

Models based on molecular fragments of the adsorbate molecule and based on solubility parameters have also been advanced (*19, 24, 61, 84, 87, 88*). Modifications to the extent of sorption caused by the presence of organic solvents in the liquid phase (cosolvent effects) are now also being predicted (*91, 92*). Correlations between $K_{ow}$ and chromatographic behavior of hydrophobic compounds on reverse-phase high-performance liquid chromatography (HPLC) columns have also been shown to be useful for obtaining $K_{ow}$ data (*93, 94*).

An important consequence of the partitioning model concerns what determines the maximum amount of sorption, or the sorption capacity, of the sorbent. By using a partitioning mechanism, the volume or mass of organic phase in the sorbent available for the sorbate to partition into can be found to be the direct determinant of the sorption capacity of the

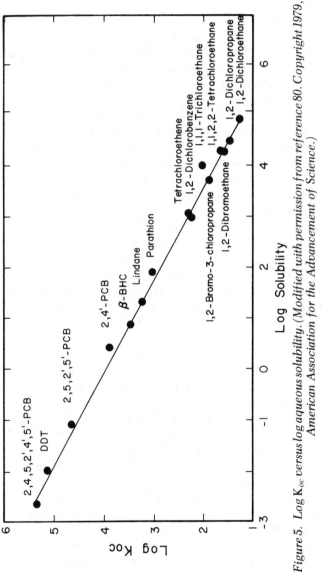

*Figure 5.  Log* K$_{oc}$ *versus log aqueous solubility. (Modified with permission from reference 80. Copyright 1979, American Association for the Advancement of Science.)*

sorbent. For other mechanisms, the total mass of sorbent, total available surface area, or total number of reactive sites on the sorbent are factors that can be important to sorption capacity. Obviously, designing experiments to investigate sorption mechanisms exploiting these differences is a possibility.

Of course, when the fraction of organic carbon in the sorbent is very small (the critical value will depend on the physicochemical characteristics of the specific sorbate as well as the surface properties of the sorbent), mixed sorption mechanisms are more likely, and a more sophisticated multicomponent model that sums the fractions sorbed by the different phases is required. For example, Schwarzenbach and Westall (27) found that a model similar to equation 6 was applicable to modeling sediment sorption of hydrophobic hydrocarbons and chlorinated hydrocarbons when the sediment organic carbon content exceeded 0.1%. Presumably, other mechanisms became significant at lower organic carbon contents. [Note the consequences to groundwater modeling because aquifers may contain less than 0.1% organic carbon (see reference 95 for an introduction).] Similarly, Kahn et al. (96) found that the amount of expandable clays affected the sorption of acetophenone on low organic carbon content sediments and soils. This effect may also be related to the more polar nature of acetophenone compared to other hydrophobic compounds. Means et al. (97) found the ratio of organic carbon to montmorillonite clay content affected the precision of estimates of the sorption of substituted PAHs. An example formulation of a multicomponent model is provided by Karickhoff (61):

$$K_p = K_m CM f_A^m + K_{oc} OC f_A^{oc} \qquad (7)$$

where $K_p$, $K_m$, and $K_{oc}$ are the partition coefficients of the mixture, mineral component, and organic component, respectively; CM and OC are the fractional masses of the mineral and organic components, respectively; and $f_A^m$ and $f_A^{oc}$ are the active (available) fractions of the mineral and organic compounds, respectively.

Note the use of the factor "available" organic phase in equation 7, which recognizes that not all organic matter in the particles will necessarily be accessible to the sorbate. In fact, accessibility of sorbent to the sorbate and reversibility of sorption can be a function of sample pretreatment (38, 98, 99) and time allowed for equilibration. Accessibility is also a function of the morphology of the particles and the size, shape, and rigidity of the sorbate (see the kinetics discussion and reference 67). Finally, whenever mechanisms other than partitioning of hydrophobic compounds into the organic phase of the sorbent may be significant, alterations to the model may be required (4, 5, 19, 61). Because these areas are in the focus of current research, they will be discussed in detail later.

Each of the models has its own advantages and disadvantages and each helps increase understanding when the conditions under which it seems to be or not be a good predictor are analyzed. The process of proposing mechanisms and testing them for consistency with observations is continual. Estimation procedures have gained in popularity as the need for the results has increased in environmental planning and assessment. Particularly because of the need to evaluate the many organic chemicals that are produced and developed each year, modeling techniques that yield generic results and that can use parameters that are easily measured or estimated have received increasing attention. Two terms used to describe the relationships that are sought summarize the goal of the generic modeling approach. These terms are the property-reactivity correlations (PRCs) and the structure–activity relationships (SARs) (*100*). Current efforts to determine quantitative relationships are reflected in the now popular term "quantitative structure–activity relationships" (QSARs).

Certainly, nothing can replace actual validation in the field (e.g., a field measurement of $K_p$). In some cases field measurement could be faster and easier, especially if data required for an estimation are not available. In all cases, field verification should be sought. Model testing (verification of the formulation and validation in actual application to the system) is an important process that has many interrelated aspects (*101*). However, surrogate parameter modeling fills a need for estimations and making inquiries into the consequences of actions and decisions. Similarly, mathematical models, formulated on whatever basis, are useful for performing sensitivity analyses and asking "what if" questions.

For evaluative generic modeling, availability of data is a prime consideration. Compilations of data are becoming increasingly available (*24, 25*), and computer data bases are increasingly popular. Collections of methods of measuring and estimating physicochemical parameters are also available (*24, 60, 102, 103*). Increased availability of data and models helps in applications to practical problems and stimulates more people to make further developments. Consequently, models based on the most readily available data are desired.

**Kinetic Models.**  Simple observation of the rates of sorption and desorption is an empirical operation. Likewise, application of mathematical descriptions of the results, such as a first-order kinetic model, can also be done on a purely empirical basis (i.e., "if it works, use it"). Many examples of useful applications of different empirical approaches exist. However, the nature of the study of kinetics also includes consideration of the actual mechanisms by which the reactions occur and a look at the system on a three-dimensional molecular size scale with variations over time. Kinetic investigations are very dependent on underlying concepts that are assumed (actively or passively), and results can be very sensitive to experimental protocol. Therefore, although discussion of kinetics could have begun with empirical approaches, this discussion is begun in this conceptual-phase section and continues into the fundamental mechanistic section.

Defining objectives carefully in a kinetic study is very important because many of the results obtained are operationally defined (i.e., dependent on the experimental protocol). For example, Figure 6 presents data from Weber et al. (99) obtained for sorption of the PCB mixture Aroclor 1254 on river sediment. The rate of sorption appeared to be relatively fast, and equilibrium appeared to be approached in a few minutes and essentially attained for practical purposes in a few hours. Particle size affected not only the extent of sorption, as discussed previously, but also perhaps the rate, and sorption by the natural composite seems to have been similar to sorption by the smaller class of particles tested. When 24-h equilibrations with fresh aliquots of river water were used in a step-wise desorption experiment, the process appeared to be reversible (Figure 7). In this system, PCB sorption appeared fast and reversible, and these conclusions are valid to the extent that the experiment is applicable to a question and system being considered. To conclude that these data would be relevant to another system or experiment, or that they indicate sorption occurred by a simple, fast, reversible physical mechanism on the surface could be erroneous, however. Similarly, the data in Figures 6 and 7 do not prove equilibrium was attained. A clue to the need for a more in-depth look is the relationship of particle size to sorption, which demands an explanation that accounts for faster sorption by smaller particles.

The central question concerning the above kinetic, or any, study, concerns the general applicability of the results. Many references to irreversible sorption and "nonsingular sorption isotherms" (i.e., the paths of sorption and desorption appear different, sorption and desorption isotherms are not coincident, or hysteresis occurs) are available (4, 104–106). Reversibility must be judged in relation to the time allowed for it to occur. For practical purposes, reversibility must be determined in relation to a chosen time frame relevant to the system under consideration. The question of true, thermodynamic, or "ultimate" reversibility, on the other hand, is another matter. Attainment of true equilibrium might be just academic in many practical applications but is important to the development of theoretical understanding.

The interrelationship of kinetics and observed thermodynamics is complicated but also very important. Traditionally, equilibrium is assumed to have been reached in sorption experiments when the solution-phase concentration stops changing with time, as seen in Figure 6. However, sensitivities of analytical techniques often limit detection of small changes in solution-phase concentrations, hampering recognition

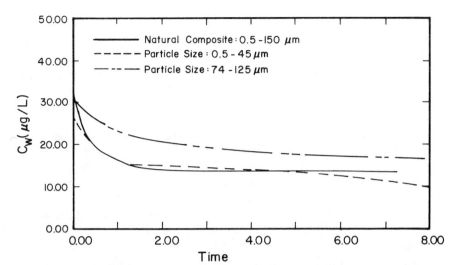

Figure 6. *Sorption of Aroclor 1254 as a function of time on Rouge River basin solids of different particle sizes, shown as decreasing water-phase concentration. (Modified with permission from reference 99. Copyright 1980, Ann Arbor Science.)*

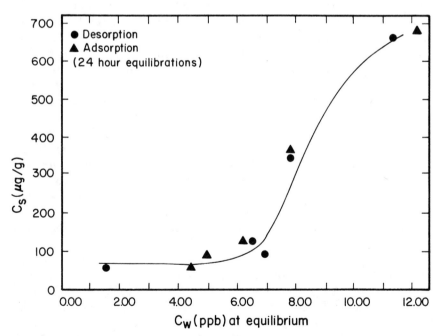

Figure 7. *Experimental adsorption and desorption data for Aroclor 1254 on Saginaw River sediments. (Modified with permission from reference 99. Copyright 1980, Ann Arbor Science.)*

of continuing changes. In any event, if an open system is being considered, a constant solution-phase concentration of sorbate is a necessary, but not sufficient condition of equilibrium.

Similarly, reversibility can be difficult to determine. Reversibility is assumed complete if all of the sorbate can be removed by putting the sorbent into a series of aliquots of fresh solvent containing no sorbate. However, desorption may occur at much slower rates than sorption (4, 79, 107). Indeed, a $K_p > 1$ indicates that sorption occurs faster than desorption if a dynamic equilibrium exists. Consequently, determining when sufficient time has been allowed to attain desorption equilibrium can be difficult, and determining this time would be a necessary condition for assessing true irreversibility. Also, if more than one step or mechanism is involved in sorption and these steps occur at significantly different rates, equilibrium can mistakenly be assumed to have been attained at the conclusion of the first fast step. In such cases when true equilibrium has not been attained, the rate of desorption becomes dependent on the incubation time allowed for the preceding sorption step to occur. Figure 8 shows the rate of desorption of hexachlorobenzene from sediment as a function of prior sorption equilibratici or incubation time. The data were collected by using a gas purge technique to collect the analyte as a function of time as the analyte was desorbed from the sediment and removed from the solution by the gas purge. Because the rate of removal by the purge is much faster than the desorption rate, the overall removal rate reflects essentially the desorption rate.

Schweich and Sardin (108) have reviewed the relationship of kinetics and reversibility of sorption to the variety of observations that can be obtained in batch and column sorption experiments. They showed that results can depend on the experimental protocol employed. Whether or not equilibrium has been attained in a system also affects the application of equilibrium models to data obtained from the system. For example, Rao et al. (109) and Goltz and Roberts (110) discussed the consequences of nonequilibrium to the patterns that will be observed for migration of solutes in groundwater and some of the changes needed in modeling approaches.

Not surprisingly, rates of desorption of hydrophobic compounds from sediments and soils have received increasing attention because desorption and sorption may occur at rates much slower than previously thought (107). For example, Coates and Elzerman (79) reported PCB sorption and desorption rates, and Oliver (111) reported desorption rates for 20 chlorinated organic compounds that have first-order half-lives in the range of days to months. Both of these investigations used modifications of the gas purge technique of Karickhoff and Morris (107) described previously in relation to Figure 8. The gas purge procedure provides an alternative and superior method for evaluating the attain-

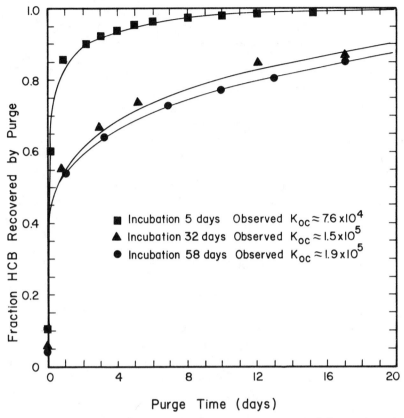

*Figure 8. Effect of adsorption incubation time on rate of desorption of nonlabile fraction of hexachlorobenzene. (Modified with permission from reference 61. Copyright 1984, American Society of Civil Engineers.)*

ment of sorption equilibrium based on a criterion of observing desorption rates that are independent of sorption incubation time. This observation indicates balanced fugacity had been attained in all phases before the desorption step was commenced.

Several concepts have evolved to explain kinetic observations of the sorption of hydrophobic compounds (*112*). In some cases, a simple first-order model is applicable, where the rate of sorption is simply proportional to the difference between the equilibrium concentration and the current concentration in the sorbed phase. One approach to quantifying results when a simple equilibrium is not reached is to consider levels of partial attainment of equilibrium in stated time frames, as done, for example, in the "quasi-equilibrium" model of DiToro and Horzempa (*104, 113*).

Another approach is to divide the overall process into a number of steps. Many authors have reported observing a rapid (on the order of minutes) initial stage of sorption or desorption followed by a slower (days to months) second stage (*56, 61, 79, 107*). A distinction has been made, which introduces the nature of the concept, identifying a "nonlabile" fraction that sorbs or desorbs slowly, in contrast to a "labile" fraction that sorbs or desorbs more quickly. The nonlabile fraction, which may appear irreversibly sorbed in desorption experiments that are too short, is distinguished from a thermodynamically defined irreversible fraction, which should never desorb. Again, defining the relevant time frame is important because it will affect the appearance of being irreversible versus nonlabile (*see* reference 108 for further discussion and references on the relationship between slow kinetics and the appearance of irreversibility). Another term that has been used for the fraction that may include both the nonlabile and the truly irreversible components when experimental details do not allow distinction between them is the "resistant" fraction (*104, 113*).

In the formulation proposed by Karickhoff (*56*) and others, the labile fraction is composed of the material that undergoes sorption or desorption within approximately 1 h, while the fraction requiring longer time to attain equilibrium is called the nonlabile fraction. The concept underlying this model is that sorbate can exchange quickly on the surface of the sorbent and then some can more slowly diffuse into the organic phase of the sorbent. A "two-box" model of sorption logically follows; the first box is conceptualized as the surface and the second box as the interior of the organic phase of the particle, where sorbate can diffuse inward. The total amount sorbed is the sum of the two fractions. Consideration of the validity of this conceptualization requires a molecular level view of the process and will be discussed further in the fundamental mechanistic-phase section.

The two-box model of kinetics of sorption of hydrophobic compounds is a simplification that often works well. Note, however, that the model is incapable of mechanistically defining the boxes or even of distinguishing the actual number of boxes. In reality, because sediments and soils are complex mixtures, numerous fractions may occur that have a continuum of applicable sorption–desorption rates, which when blended appear as a group of slow and another group of fast processes. Similarly, the existence of different rates does not confirm the actual existence of different fractions. For example, the second box that has slow kinetics could actually be a result of a slow surface transformation.

The mathematical formulation of Karickhoff's model (*56*) is based on two steps involving the two boxes defined previously, and results in a "two-box serial model". A similar "two-box parallel model" can also be formulated with the reactions occurring concurrently (*114*). The first step, which defines the labile fraction, is described by an equilibrium distribution because it is controlled by relatively fast processes. The second step, which defines the nonlabile fraction, is controlled by kinetics as the surface material diffuses into the organic phase of the sorbent, as follows:

$$C \underset{}{\overset{k_1}{\longleftrightarrow}} S_l \underset{}{\overset{k_d}{\longleftrightarrow}} S_{nl} \tag{8}$$

where $C$ is the sorbate in solution; $S_l$ and $S_{nl}$ are the sorbates in the labile and nonlabile fractions, respectively; $k_1$ is the equilibrium constant for first-stage distribution; and $k_d$ is the rate constant for movement of sorbate from the labile to the nonlabile fraction (usually assumed to be first order).

The two-box model presented above reasonably describes much of the data obtained for kinetics of sorption of hydrophobic compounds but still has limitations. As pointed out by Karickhoff (*61*), one of the limitations of this model is that isotropic distribution of the sorbate is assumed to occur in the solution, sorbed labile, and sorbed nonlabile phases (i.e., within each phase, the distribution is homogeneous and no fugacity gradients appear). On short time scales compared to the rate of movement of sorbate in the different phases, changes in the concentration of sorbate in one phase may not be accurately reflected by the model in other phases. Of course, an advantage of this limitation is that deviations in results from the model provide the basis for further investigations, although it is also easy to be fooled (consider the indications of Figure 8, and *see* reference 79 for further discussion of the observation that desorption can occur more quickly after short sorption incubation times than long incubation times).

A consequence of the formulation of the model is that the total amount sorbed by a given amount of sediment at a given solution-phase concentration will be controlled by $k_d$ and the incubation time until equilibrium is attained. The amount of sorbate in the labile phase responds to the concentration (actually the activity) of adsorbate in solution over short time periods. However, the amount of sorbate in the nonlabile phase depends on the time and the rate at which it moves from the labile phase to the nonlabile phase. Hence, if the change from the labile to the nonlabile phase results from diffusion in the organic phase or pores in the sorbent, intraparticle mass transfer will be the overall limiting step in this model (*see* reference 73 for more discussion of intraparticle transfer).

Again, the potential interplay of kinetics and observed thermodynamics is evident. The extent of sorption observed is dependent on the time allowed, and slow kinetics may obscure the detection of true equilibrium. Consider again the previously mentioned effect of temperature

on sorption of hydrophobic compounds, where the observed increasing
sorption with increasing temperature for very hydrophobic compounds
may be a kinetics artifact (*61*). Also, consider the common observation
(*see* reference 107 for examples) that solvent extraction of hydrophobic
compounds added to sediments is often inversely related to the incubation time allowed between inoculation and extraction (i.e., the longer
time allowed for the sorbate to enter the nonlabile phase, the more difficult extraction will be). Finally, the amount and distribution of the
organic phase in the sorbent will affect diffusion distances and therefore,
the kinetics of sorption of the nonlabile phase; thus, a possible mechanism is suggested to explain particle size and particle aggregation
effects. The two-box model, therefore, does not give a complete physical picture but leads to logical extensions into more mechanistic pictures
represented by the gel permeation and intraparticle diffusion models
that are discussed later in this chapter.

Considerable variability in results obtained for different sorbents
and sorbates has also been observed with the two-box model. Experimental results for given sets of hydrophobic sorbates and sediments
indicate the fraction of sorbate in the labile fraction varied from 20%
to 60%, and that $k_d$ for a given sorbate could vary by 3 orders of magnitude
for different sediments (*61*). Because of the serial transfer nature of the
two-box model, the fraction of labile sorbate will continuously change
with time until true equilibrium is attained, and variability in some
results can thus be related to the previous history of the sample.

Karickhoff (*61*) also observed that the rate of desorption of the nonlabile phase was inversely related to the final extent of sorption when he
said, "as a 'rule-of-thumb', the characteristic time $(1/k_d)$ for diffusive
release from the nonlabile state approximates numerically the $K_{oc}$ (in
minutes)". The predictive capability of a related generalization is
represented in Figure 9. This capability can also be stated as follows
(*61*):

$$1/k_d \approx 0.3 \, K_p \qquad r^2 = 0.87 \qquad (9)$$

where $1/k_d$ is expressed in hours.

Further consideration indicates the time to reach equilibrium (in
hours) will approach $K_p$, or be more than 1 year for $K_p > 10^5$ [for a
first-order process, $t_{1/2} = 0.693/k$, therefore, $t_{1/2}$ (h) $\approx 0.2 \, K_p$ from equation 9]. This relationship appears to indicate that compounds with greater
hydrophobicity will diffuse more slowly in the organic phase of the sorbent, or, perhaps, have access to more organic phase, but a mechanistic
explanation must await further research. Again, the observation of
increasing rates of sorption of very hydrophobic materials with increasing temperature is relevant. Isaacson and Frink (*4*) point out that models

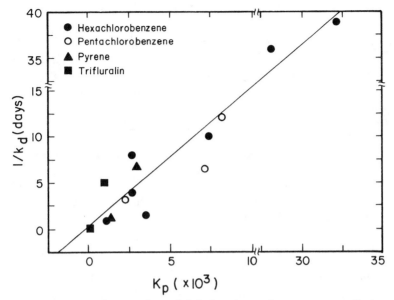

*Figure 9. Dependence of nonlabile-fraction release constant* ($k_d$) *on* $K_p$. *(Modified with permission from reference 107. Copyright 1985, Pergamon Press.)*

that use a "corrected organic fraction" or "active organic fraction" for hydrophobic sorption to obtain better results actually reflect the need to account for nonuniform penetration of the sorbates in the sorbents when the sorption mechanism is not simply a partitioning into the organic phase.

Other model formulations are possible, such as a surface-site reaction model. However, surface partitioning of hydrophobic sorbates rather than surface reaction mechanisms seem to make a better accounting of sorption of hydrophobic materials when organic carbon is present in the sorbent (*115*). When other mechanisms of attraction between sorbate and sorbent become significant, the simple generalizations of the hydrophobic partitioning model may not apply. For example, additional important factors have been identified in the following cases: sorption of phenolic compounds by lake sediment that has been treated to partially remove the organic matter so sorbent–sorbate hydrogen bonding is increased (*4*), sorption of (ionic) phenolate (*3*), sediment sorption of triazine and dinitroaniline compounds that have log $K_{ow}$ values close to 2 (*116*), and sorption of ionizable quinoline as a function of pH (*5*).

## Fundamental Mechanistic Phase

Sorption of hydrophobic compounds in aqueous systems is just one of the processes that must be considered in an overall evaluation of their fate and interactions. Taken together, "hydrophobic interactions", or all of the processes peculiar to hydrophobic compounds in aqueous systems, are not well understood, and their investigation is an infant science (*117*). Fundamental mechanistic understanding is still a distant goal, impeded by the complexities of the systems, the expense of time and money required to make progress, and sometimes the lack of interest in pursuing an idea beyond immediate needs. Perhaps a complete understanding is not possible if today's certainty always yields to tomorrow's refinement and the next day's major reassessment. Certainly, the goal of a more thorough understanding is always possible, and in the process many direct and indirect beneficial outcomes can accrue.

The difference between the conceptual phase and the mechanistic phase is more in degree than in kind. Again, the perspective of Karickhoff (*19*) yields valuable insight:

> *A fundamental prerequisite to model development is definition of purpose. Model structure (i.e., mathematical format and degree of abstraction, level of temporal and spatial detail, and requirement for supporting physical rationale) is intimately a function of intended use. The sorption models described herein are designed for what might be loosely characterized as engineering applications and can serve the following purposes:*
>
> *1. A priori estimation of pollutant sorption behavior using commonly measured (or computed) physical properties of pollutant and sorbent and incorporating commonly measured environmental variables*
>
> *2. Guidance for ordering, interpreting and, in a nonrigorous sense, "understanding" existing data*
>
> *3. Evaluation or screening of existing data based on "expected" behavior*
>
> *4. Extrapolation of sorption measurements (field or laboratory) outside the domain of measurement.*
>
> *For the most part, model validity derives more from long-term accretion of phenomenological evidence than from mechanistic rigor. Physical rationale is offered, however, in support of mathematical functionality and is used to define (or extend) appropriate models in instances where the current state of knowledge or available data are insufficient for model definition.*

Research in this area raises questions at least as much as it answers them. However, there are significant results to present and current research to discuss. To be comprehensive here would not be possible or desirable; therefore, a particular sequence of information consistent with the previous discussion will be pursued as an example. A concept of hydrophobic compounds participating in a sorption mechanism that may be described as partitioning into an organic phase in the sorbent; the relationships of practical tools to traditional thermodynamic principles; the development of empirical models for quantitative assessments; and the use of surrogate parameters to aid in predictions, extrapolations, and faster initial assessments have all been discussed. However, in many cases the models do not seem to work as well as for others, and the actual mechanisms involved must be addressed more directly on a three-dimensional, molecular scale of the sorbent and sorbate (e.g., such factors as surface and pore characteristics of the sorbent and molecular size, topography, and rigidity of the sorbate). Current research is addressing these areas.

More complex models of hydrophobic partitioning mechanisms are possible and forthcoming. For example, models based on radial diffusion of sorbate into spherical sorbent particles are under development (*67, 114*). These models modify the assumption of infinite depth and planar source diffusion incorporated into Karickhoff's two-box model. Radial diffusion models take into account variations with particle size, and this inclusion gives the model the potential to explain the particle size effect on sorption kinetics indicated in Figure 6. Unfortunately, radial diffusion models also increase the data requirements for modeling. Other empirical evidence also suggests sorption models should take into account particle size. Karickhoff et al. (*65*), for example, reported a bell-shaped dependence of $K_p$ on particle size; the highest values were obtained for the medium or fine silt, and lower values were obtained increasing in size to the sand or decreasing in size to the clay fractions. For the whole sediment, they proposed the overall $K_p$ value can be modeled as the sum of the $K_p$ values for each particle size fraction weighted by the mass of each fraction in the sediment, as follows:

$$K_p^* = \sum_i K_{pi} f_i \tag{10}$$

where $K_{pi}$ is the $K_p$ of the $i$th fraction, and $f_i$ is the fraction of the total mass represented by the fraction $i$.

The point is that detailed models of sorption of hydrophobic compounds must take into account particle size. Initial predictions have been shown to work well just on the basis of organic carbon content of the sorbent. These predictions led to the conclusion that the organic matter involved is similar regardless of its source. Of course, this conclusion is only partially true, and the generalization, in fact, works in part because it is assumed to be true when all the organic matter fractions and sources are averaged together (*4*). The composition of the organic matter in sediments actually varies from sample to sample and is not necessarily homogeneous within a sample. In fact, composition variations with different particle size classes are common.

More detailed models must also account for the structure of the sorbent particles, in particular any pores that might exist. Models that are based on a concept of partitioning into an organic gel (gel partitioning models) have been extended to include pores of different sizes in the gel, which allows for variations in the penetrability and availability of the organic phase to hydrophobic compounds (61). Variations in the pores with ambient conditions, such as pH, must also be handled. An example of a modified gel permeation model incorporating the swelling characteristics of sediment organic matter that result in changes in pore sizes is offered by Freeman and Cheung (118).

More complex models may offer more sophisticated treatments but usually at the expense of ease of use and understanding of the related factors and mathematical formulations. A radial diffusion or gel permeation model may be closer to a representation of the actual sorption mechanism than more simple models, but they are still incomplete representations that are not fundamental mechanistic descriptions. Particle size is still a gross general parameter when compared to the molecular size level, three-dimensional understanding actually required. A recent advancement looks at intraparticle diffusion in particles of different sizes and with pore structures (67). Further application of already available treatments of macropore and micropore transport control of kinetics (73) will probably be beneficial in this area.

Another active area of research addresses an apparent failure of available models and established chemical principles and focuses on mechanisms on a molecular scale. At the same time, this area of research is an extension of what has been discussed previously in this chapter. Results from numerous sorption experiments have indicated that $K_p$ for both organic and inorganic sorbates can or can appear to vary with the concentration of sorbent in the suspension (i.e., $K_p$ decreasing with increasing suspended particulate matter concentrations). For example, Figure 10 presents the data on hydrophobic compounds from the now-classic paper of O'Connor and Conolly (119). (The original figure also presented similar data for inorganic sorbates.)

Other evidence indicates the effect is not the same over all ranges of sorbent concentrations. For both organic and inorganic sorbates, decreasing $K_p$ values appear above some critical range of suspended particulate matter concentration. A quick consideration indicates the relationship cannot be extrapolated over all concentration ranges. If the relationship did operate over all concentration ranges, in the limit of low suspended particulate matter concentrations, $K_p$ values would get very large; in the extreme, one sorbent particle would sorb all or most of the sorbate! The critical range of suspended particulate matter concentra-

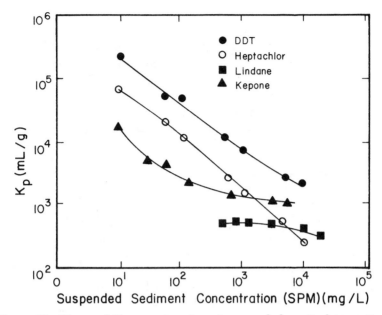

*Figure 10. Observed* $K_p$ *as a function of suspended particulate matter concentration. (Modified with permission from reference 119. Copyright 1980, Pergamon Press.)*

tion varies with the sorbate and sorbent, but for reference, Coates (*120*) found the range to be between 100 and 1000 mg/L for sorption of PCBs on river sediments.

This observation is contrary to traditional concepts of an equilibrium $K_p$ from thermodynamics. In fact, the observation appears to undermine basic tenets of equilibrium approaches, which should stimulate significant questioning, research, and possibly reformulation of concepts or inclusion of more factors. For a given sorbate–sorbent pair at equilibrium at a given temperature in a given solution, the concentration of sorbate in the sorbed phase should be related only to the concentration of sorbate in the solution phase and not directly to the amounts of material present, hence resulting in a constant $K_p$ value. The $K_p$ value should be independent of the masses of sorbent and sorbate involved for a given equilibrium solution-phase concentration and therefore independent of the suspended particulate matter concentration. The total mass of sorbate in the sorbed phase should increase with sorbent concentration in the suspension (Figure 1 and Table III), but the ratio of the equilibrium concentrations in the two phases should be constant and result from a balancing of the fugacity in the two phases.

If $K_p$ is a function of suspended particulate matter concentrations, many practical ramifications result. Because $K_p$ is an integral parameter in many models of fate and distribution in aquatic systems, at least a functional understanding of the factors that affect it is required. For example, the common practice for determining $K_p$ is to make measurements in the lab at concentrations significantly higher than would apply in the field systems to be modeled, hence inappropriate values could be produced. In addition, models have generally been constructed with a constant value of $K_p$ for each sorbent and sorbate pair at a given temperature. A variable $K_p$ value not only increases data requirements, but also makes the model more complicated and probably more site-specific. Different values of $K_p$ would be required for sediments, aquifers, and the water column, and even for different times in the water column if resuspension, storm events, or other disturbances of the system were significant.

Actual fate and distribution could also be affected. For example, a decrease in $K_p$ when suspended sorbent is incorporated into the bottom sediment (a change in sorbent concentration from perhaps 5 to 500,000 mg/L) could result in loss of sorbate back to the solution phase. Chapra and Reckhow (11) present results of model predictions that show a significant effect of changes in $K_p$ related to suspended particulate matter concentration on water column and sediment concentrations and the residence time of sorbable materials in lake systems. Although DiToro et al. (121) did not find evidence of this release upon increases in solids concentration in sediments in data on diffusion and partitioning of hexachlorobiphenyl, investigations of this question are just beginning.

Because researchers are faced with abandoning classical thermodynamic concepts, discovering additional thermodynamic principles, or finding other explanations, numerous papers are appearing in the literature on this subject. Indeed, several alternative explanations are developing. Two types of considerations currently provide the most promising explanations. Questions concerning some previously overlooked important component(s) of the system and experimental artifacts have been raised. Not surprisingly, no simple or single answer has emerged, and examples resulting from both of these possibilities and combinations of the two have been documented. In the process, significant fundamental knowledge and directives for further research have resulted.

A number of factors appear to be relevant to the subject. (1) Hydrophobic compounds may become associated with other materials in solution, for example, humic acids (52, 122–126). If the hydrophobic compound is part of another soluble species, a model with a single sorbate containing the compound is no longer appropriate, but appropriate models can be formulated (55, 127, 128). (2) Competitive sorption mechanisms can also make a single sorbate and single sorbent model inapplicable if an unrecognized sorbate that is a function of the solids concentration competes with the sorbate of interest. (3) Other sorbents may be present but not properly accounted for in a two-phase model. For example, a colloidal phase acting as a sorbent that is not separated from the dissolved phase in analytical procedures will increase the apparent dissolved-phase concentration and decrease the apparent $K_p$. If the concentration of the nonseparated phase increases with the solids concentration, the effect will be to decrease the apparent $K_p$ as solids concentration increases (*129*). Gschwend and Wu (*129*) and Brownawell and Farrington (*130*), among others, have shown that data from some systems can be adequately represented by a three-phase model that adds sorption onto a colloidal phase. However, other analytical artifacts can also result in similar observations. For example, loss of analyte during filtration or centrifugation may not be recognized if a complete mass balance is not performed (e.g., if sorbed-phase concentration is obtained only by difference by looking at the loss from the dissolved phase). A loss of a constant amount to the apparatus or filter, for example, will constitute a varying fraction of the total amount apparently sorbed at different sorbate concentration levels. Depending on the magnitude of the loss compared to the amount sorbed at different sorbate concentration levels, the result can appear as $K_p$ being a function of solids concentration (*41*) (Figures 11 and 12). (4) Effects of some as yet undetermined particle interactions that cause desorptions have also been hypothesized to explain the observations (*131*). (5) Analytical artifacts associated with premature identification of equilibrium can also result in $K_p$ being an apparent function of suspended solids concentration if the rate of attainment of equilibrium is dependent on the suspended solids concentration. Aggregation of particles may be a factor if the aggregation affects the accessibility of the sorbent to the sorbate, for example, increasing required intraparticle diffusion distances. Because aggregation can be a function of solids concentration, the result can be an apparent effect of solids concentration on the observed $K_p$. Different sediments do show different rates of desorption of hydrophobic compounds (*79, 107*). Sediments that tend to form aggregates tend to exhibit the slowest kinetics (*79, 120*). Intraparticle and interparticle bridging mechanisms induced by other sorbates have also been identified (*132*), and these mechanisms may also prove to be significant. Identification of a "resistant" fraction that does not desorb in the course of the timeframe of the experiment (*133, 134*) provides a similar explanation for observations. As previously discussed, the greater the tendency for a compound to be sorbed (the higher its $K_{oc}$ value), generally the slower will be the attainment of equilibrium (*107*) and the more likely that equilibrium will not be properly identified.

At this point, calculations based on equations 1 and 2 are useful tools to guide understanding. An error in determination of $K_p$, whether resulting from separation of dissolved and particulate phases, determination

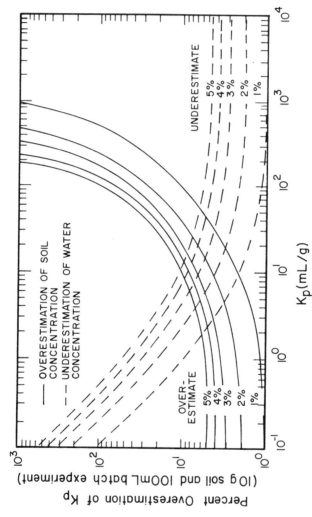

*Figure 11. Errors in log $K_p$ resulting from overestimation of sorbent concentration or underestimation of dissolved-phase concentration in a 100-mL batch experiment with 10 g of soil. (Reproduced with permission from reference 59. Copyright 1983, American Society of Agronomy.)*

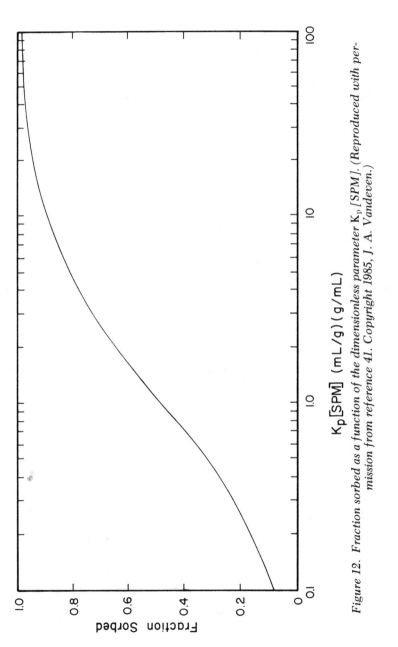

*Figure 12. Fraction sorbed as a function of the dimensionless parameter $K_p[SPM]$. (Reproduced with permission from reference 41. Copyright 1985, J. A. Vandeven.)*

of concentrations, loss of analyte, or kinetic effects, will not have the same consequences in all cases. For example, Figure 11 shows the effect on measured $K_p$ values of relatively small (1%-5%) errors in determining dissolved-phase and particulate-phase concentrations in a representative batch experiment. The magnitude of the effect is heavily dependent on the $K_p$ value of the sorbate–sorbent pair. Figure 12 presents another perspective and generalizes the relationship between fraction sorbed and the product $K_p$ times the suspended particulate matter concentration, or $K_p$[SPM]. At the high and low ends of the range of $K_p$[SPM] plotted, the fraction of sorbate sorbed is less sensitive to $K_p$[SPM] than at intermediate values. Reversing the approach, the determination of $K_p$ (assuming a constant [SPM]) is more sensitive to analytical errors in determination of fraction sorbed at the high and low end of the range of $K_p$[SPM] in Figure 12 than at intermediate values.

Karickhoff and Morris (*107*) also showed that the fraction of sorbate in the labile (fast-sorbing) fraction of the previously discussed two-box kinetic model can be a function of suspended solids concentration and $K_p$. They observed that the fraction in the labile phase decreases with increasing values of the parameter $K_p$[SPM] above a critical value of this parameter. In this case, a mechanism involving increasing aggregation of particles with increasing [SPM] and slower intraparticle diffusion of the more hydrophobic compounds with higher values of $K_p$ might be hypothesized to explain an increasing fraction in the nonlabile fraction with increases in $K_p$[SPM]. Whatever the mechanism, this relationship might explain some reported observations of suspended particulate matter concentrations affecting apparent $K_p$ values.

Apparently, there are many potential reasons why $K_p$ could appear to vary with the concentration of sorbent. Lioux (*135*) has presented arguments that, in some cases, there may be a thermodynamic explanation related to entropic effects that does not require invoking an analytical error or misrepresentation of the components of the system. One obvious need is for further investigation of associations of hydrophobic compounds with other soluble materials, competitive sorption effects, and mechanisms of sorption and desorption. Also, concepts of the actual structure of sorbents and associated organic phases must be refined until a fundamental, molecular scale, three-dimensional understanding is accomplished.

Progress is possible and observable. As a final example, consider the current status of understanding of the nature of the organic phase responsible for partitioning of hydrophobic compounds. Initially, a simple normalization of $K_p$ data by the total organic carbon content of the sorbent was used. The approach was modified to consider an active fraction of the organic carbon, then the concept of a gel with particular dimensions was introduced followed by a gel or particles of different sizes that could have varying sizes and lengths of pores. Interactions between particles and the organic phase and between different particles that come together to form aggregates have been recognized as affecting particle size and geometry, the availability of the sorbent to the sorbate, and sorption and desorption kinetics. Humic and fulvic acids have been identified as important common components of the organic sorbent phase of sediments and soils (*136*). Finally, actual "pictures" can be proposed of how organic molecules might align on particle surfaces to create a sorbent matrix (*137*), how the organic phase might change under different conditions and alter sorption kinetics, and how molecular characteristics of sorbate molecules might relate to interactions at the surface.

## Acknowledgments

We thank Sam Karickhoff for his inspiration and invaluable help, Brian Looney for very helpful comments on the manuscript, and the National Science Foundation for partial support of our research and the preparation of this manuscript through Grant ISP–8011451 to Environmental Systems Engineering.

## Abbreviations and Symbols

| | |
|---|---|
| $a$ | empirically determined coefficient |
| $B$ | empirically determined coefficient |
| $C_s$ | concentration in the solid phase due to sorption |
| $C_{s,oc}$ | concentration of sorbate in the organic phase of the sorbent |
| $C_w$ | concentration in the dissolved phase |
| $f_A^m$ | active fraction of the mineral component |
| $f_A^{oc}$ | active fraction of the organic component |
| $f_i$ | fraction of total mass represented by the fraction $i$ |
| $F$ | fraction of total mass of sorbate sorbed |
| $k_1$ | equilibrium constant for first-stage (labile-phase) distribution |
| $k_d$ | rate constant for movement of sorbate from the labile into and out of the nonlabile fraction |
| $K_1$ | constant related to the energy of sorption in the Langmuir model |
| $K_2$ | constant related to the capacity of sorption in the Langmuir model |
| $K_F$ | Freundlich "capacity" constant |
| $K_m$ | partition coefficient of the mixture |
| $K_{oc}$ | organic carbon referenced partition coefficient |
| $K_{ow}$ | octanol–water partition coefficient |
| $K_p$ | partition coefficient |
| $K_{pi}$ | The $K_p$ of the $i$th fraction |

OC    fraction of organic carbon in the sorbent
$n$    Freundlich "intensity" constant
$S_l$    sorbate in the labile fraction
$S_{nl}$    sorbate in the nonlabile fraction
$t_{1/2}$    half-life

## References

1. Neely, W. B.; Blau, G. E. *Environmental Exposure from Chemicals;* CRC Press: Boca Raton, FL, 1985; Vol. 1.
2. Imboden, D. M.; Schwarzenbach, R. P. In *Chemical Processes in Lakes;* Stumm, W., Ed.; Wiley–Interscience: New York, 1985; pp 1–30.
3. Schellenbarg, K.; Leuenberger, C.; Schwarzenbach, R. P. *Environ. Sci. Technol.* 1984, *18(9)*, 652–657.
4. Isaacson, P. J.; Frink, C. R. *Environ. Sci. Technol.* 1984, *18(1)*, 43–48.
5. Zachara, J. M.; Ainsworth, C. C.; Felice, L. J.; Resch, C. *Environ. Sci. Technol.* 1986, *20(6)*, 620–627.
6. Mackay, D.; Patterson, S. *Environ. Sci. Technol.* 1982, *16*, 654A–660A.
7. Mackay, D.; Patterson, S.; Joy, M. In *Fate of Chemicals in the Environment;* Swann, R. L.; Eschenroeder, A., Eds.; ACS Symposium Series 225; American Chemical Society: Washington, DC, 1983; pp 175–196.
8. Baughman, G. L.; Burns, L. A. In *The Handbook of Environmental Chemistry;* Hutzinger, O., Ed.; Springer–Verlag: New York, 1980; Vol. 2, Part A, pp 1–18.
9. Haque, R. *Dynamics, Exposure, and Hazard Assessment of Toxic Chemicals;* Ann Arbor Science: Ann Arbor, MI, 1980.
10. Reckhow, K. H.; Chapra, S. C. *Engineering Approaches for Lake Management;* Ann Arbor Science (Butterworth Group): Boston, 1983; Vol. 1.
11. Chapra, S.; Reckhow, K. H. *Engineering Approaches for Lake Management;* Ann Arbor Science (Butterworth Group): Boston, 1983; Vol. 2.
12. Swann, R. L.; Eschenroeder, A. *Fate of Chemicals in the Environment;* ACS Symposium Series 225; American Chemical Society: Washington, DC, 1983.
13. Imboden, D. M. In *Lakes: Chemistry, Geology, Physics;* Lerman, A., Ed.; Springer–Verlag: New York, 1978; pp 341–356.
14. Steen, W. C.; Paris, D. F.; Baughman, G. L. In *Contaminants and Sediments;* Baker, R. A., Ed.; Ann Arbor Science: Ann Arbor, MI, 1980; Vol. 1, pp 477–482.
15. Mill, T. In *Dynamics, Exposure, and Hazard Assessment of Toxic Chemicals;* Haque, R., Ed.; Ann Arbor Science: Ann Arbor, MI, 1980; pp 297–322.
16. Mackay, D. *Environ. Sci. Technol.* 1982, *16(5)*, 274–278.
17. Bruggeman, W. A. In *The Handbook of Environmental Chemistry;* Hutzinger, O., Ed.; Springer–Verlag: New York, 1982; Vol. 2, Part B, pp 83–102.
18. Veith, G. D.; Kosian, P. In *Physical Behavior of PCBs in the Great Lakes;* Mackay, D.; Patterson, S.; Eisenreich, S. J.; Simmons, M. S., Eds.; Ann Arbor Science: Ann Arbor, MI, 1983; pp 269–282.
19. Karickhoff, S. W. In *Environmental Exposure from Chemicals;* Neely, W. B.; Blau, G. E., Eds.; CRC Press: Boca Raton, FL, 1985; Vol. 1, pp 49–64.
20. Callahan, M. A.; Slimak, Michael W.; Gabel, Norman W.; May, Ira P.; Fowler, Charles F.; Freed, J. Randall; Jennings, Patricia; Durfee, Robert L.;

Whitmore, Frank C.; Maestri, Bruno; Mabey, William R.; Holt, Buford R.; Gould, Constance. *Water-Related Fate of 129 Priority Pollutants;* U.S. Environmental Protection Agency: Washington, DC, 1979; EPA–440/ 4–79–029(a+b).

21. Baker, R. A. *Contaminants and Sediments*, 2 vols.; Ann Arbor Science: Ann Arbor, MI, 1980.
22. Reinbold, K. A.; Hassett, J. J.; Means, J. C.; Banwart, W. L. *Adsorption of Energy-Related Organic Pollutants: A Literature Review;* U.S. Environmental Protection Agency: Washington, DC, 1979; EPA–600/3–79–086.
23. Hassett, J. J.; Means, J. C.; Banwart, W. L.; Wood, S. G. *Sorption Properties of Sediments and Energy-Related Pollutants;* U.S. Environmental Protection Agency: Washington, DC, 1980; EPA–600/3–80–041.
24. Lyman, W. J.; Reehl, W. F.; Rosenblatt, D. H. *Handbook of Chemical Property Estimation Methods;* McGraw–Hill: New York, 1982.
25. Verschureren, K. *Handbook of Environmental Data on Organic Chemicals;* Van Nostrand Reinhold: New York, 1983.
26. Mackay, D.; Shiu, W. Y.; Billington, J.; Huang, G. L. In *Physical Behavior of PCBs in the Great Lakes;* Mackay, D.; Patterson, S.; Eisenreich, S. J.; Simmons, M. S., Eds.; Ann Arbor Science: Ann Arbor, MI, 1983; pp 59–70.
27. Schwarzenbach, R. P.; Westall, J. *Environ. Sci. Technol.* 1981, *15*, 1360–1367.
28. Duke, T. W.; Lowe, J. I.; Wilson, A. J. *Bull. Environ. Contam. Toxicol.* 1970, *5*, 171–180.
29. Dennis, D. S. In *National Conference on PCBs;* U. S. Environmental Protection Agency: Washington, DC, 1976; EPA–560/6–75–004, pp 183–194.
30. Pavlou, S. P.; Dexter, R. N. *Environ. Sci. Technol.* 1979, *13(1)*, 65–71.
31. Morgan, R. P.; Sommer, S. E. *Bull. Environ. Contam. Toxicol.* 1979, *22*, 413–419.
32. Stainken, D.; Rollwagen, J. *Bull. Environ. Contam. Toxicol.* 1979, *23*, 690–697.
33. Bremmer, K. E. In *Management of Bottom Sediments Containing Toxic Substances;* U.S. Environmental Protection Agency: Washington, DC, 1979; EPA–600/3–79–102, pp 261–287.
34. Cooke, M.; Nickless, G.; Povey, A.; Roberts, D. J. *Sci. Total Environ.* 1979, *13*, 17–26.
35. Eisenreich, S. J.; Hollod, G. J.; Johnson, T. C.; Evans, J. In *Contaminants and Sediments;* Baker, R. A., Ed.; Ann Arbor Science: Ann Arbor, MI, 1980; Vol. 1, pp 67–94.
36. Bopp, R. F.; Simpson, H. J.; Olsen, C. R.; Kostyk, N. *Environ. Sci. Technol.* 1981, *15*, 210–216.
37. Weininger, D.; Armstrong, D. A.; Swackhammer, D. L. In *Physical Behavior of PCBs in the Great Lakes;* Mackay, D.; Patterson, S.; Eisenreich, S. J.; Simmons, M. S., Eds.; Ann Arbor Science (Butterworth Group): Ann Arbor, MI, 1983; pp 423–439.
38. Dunnivant, F. M. M.S. Thesis, Clemson University, Clemson, SC, 1985.
39. Erickson, M. D. *Analytical Chemistry of PCBs;* Butterworth: Boston, 1986.
40. Mills, W. B.; Dean, J. D.; Porcella, D. B.; Gherini, S. A.; Hudson, R. J. M.; Frick, W. E.; Rupp, G. L.; Bowie, G. L. *Water Quality Assessment: A Screening Procedure for Toxic and Conventional Pollutants in Surface and Groundwater;* Revised ed.; U.S. Environmental Protection Agency: Washington, DC, 1985; EPA–600/6–85/002.
41. Vandeven, J. A. M.S. Thesis, Clemson University, Clemson, SC, 1985.
42. Eisenreich, S. J.; Capel, P. D.; Looney, B. B. In *Physical Behavior of PCBs*

*in the Great Lakes;* Mackay, D.; Patterson, S.; Eisenreich, S. J.; Simmons, M. S., Eds.; Ann Arbor Science: Ann Arbor, MI, 1983; pp 181–212.

43. Fisher, J. B.; Petty, R. L.; Lick, W. *Environ. Pollut. Ser. B.* **1983**, *5*, 121–132.

44. Haque, R.; Schmedding, D. W. *J. Environ. Sci. Health* **1976**, *11*, 129–137.

45. Steen, W. C.; Paris, D. F.; Baughman, G. L. *Water Res.* **1978**, *12*, 655–657.

46. Barnes, M. A.; Barnes, W. C. In *Lakes: Chemistry, Geology, Physics;* Lerman, A., Ed.; Springer–Verlag: New York, 1978; pp 127–152.

47. Hites, R. A.; Lopez-Avila, V. In *Contaminants and Sediments;* Baker, R. A., Ed.; Ann Arbor Science, Ann Arbor, MI, 1980; Vol. 1, pp 53–66.

48. Wakeham, S. G.; Farrington, J. W. Ibid., pp 3–32.

49. Mackay, D. In *Environmental Exposure from Chemicals;* Neely, W. B.; Blau, G. E., Eds.; CRC Press: Boca Raton, FL, 1985; Vol. 1, pp 91–108.

50. Neely, W. B. *Sci. Total Environ.* **1977**, *7*, 117–129.

51. Murphy, T. J.; Pokojowczyc, J. C.; Mullin, M. D. In *Physical Behavior of PCBs in the Great Lakes;* Mackay, D.; Patterson, S.; Eisenreich, S. J.; Simmons, M. S., Eds.; Ann Arbor Science: Ann Arbor, MI, 1983; pp 49–58.

52. Hassett, J. P.; Milicic, E. *Environ. Sci. Technol.* **1985**, *19(7)*, 638–643.

53. McCall, P. J.; Swann, R. L.; Laskowski, D. A. In *Fate of Chemicals in the Environment;* Swann, R. L.; Eschenroeder, A., Eds.; ACS Symposium Series 225; American Chemical Society: Washington, DC, 1983; pp 105–124.

54. Patterson, S. In *Environmental Exposure from Chemicals;* Neely, W. B.; Blau, G. E., Eds.; CRC Press: Boca Raton, FL, 1985; Vol. 1, pp 217–232.

55. Higgo, J. J. W.; Rees, L. V. C. *Environ. Sci. Technol.* **1986**, *20(5)*, 483–490.

56. Karickhoff, S. W. In *Contaminants and Sediments;* Baker, R. A., Ed.; Ann Arbor Science: Ann Arbor, MI, 1980; Vol. 2, pp 193–206.

57. Bierman, V. J.; Swain, W. R. *Environ. Sci. Technol.* **1982**, *16(9)*, 572–579.

58. McCall, P. J.; Laskowski, D. A.; Swann, R. L.; Dishburger, H. J. In *Test Protocols for Environmental Fate and Movement of Toxicants; Proceedings of the Association of Official Analytical Chemists Symposium, Washington, DC, October, 1980;* Zweig, G.; Beroza, M., Eds.; Association of Official Analytical Chemists: Arlington, VA, 1981; pp 89–109.

59. Baes, C. F.; Sharp, R. D. *J. Environ. Qual.* **1983**, *12(1)*, 17–26.

60. Lyman, W. J. In *Environmental Exposure from Chemicals;* Neely, W. B.; Blau, G. E., Eds.; CRC Press: Boca Raton, FL, 1985; Vol. 2, pp 13–48.

61. Karickhoff, S. W. *J. Hydraul. Eng.* **1984**, *110(6)*, 707–735.

62. Armstrong, D. E.; Swackhamer, D. L. In *Physical Behavior of PCBs in the Great Lakes;* Mackay, D.; Patterson, S.; Eisenreich, S. J.; Simmons, M. S., Eds.; Ann Arbor Science: Ann Arbor, MI, 1983; pp 229–244.

63. Hiraizumi, Y.; Takahashi, M.; Nishimura, H. *Environ. Sci. Technol.* **1979**, *13*, 580–584.

64. Choi, W. W.; Chen, K. Y. *Environ. Sci. Technol.* **1976**, *10*, 782–786.

65. Karickhoff, S. W.; Brown, D. S.; Scott, T. A. *Water Res.* **1979**, *13*, 241–248.

66. Pavlou, S. P. In *Contaminants and Sediments;* Baker, R. A., Ed.; Ann Arbor Science: Ann Arbor, MI, 1980; Vol. 2, pp 323–332.

67. Wu, S.; Gschwend, P. M. *Environ. Sci. Technol.* **1986**, *20(7)*, 717–725.

68. Glooschenko, W. A.; Strachan, W. M.; Sampson, R. C. *Pestic. Monit. J.* **1976**, *10(2)*, 61–67.

69. O'Melia, C. R. In *Chemical Processes in Lakes;* Stumm, W., Ed.; Wiley-Interscience: New York, 1985; pp 207–224.

70. Hilton, J. *Limnol. Oceanogr.* **1985**, *30(6)*, 1131–1143.

71. Stumm, W.; Morgan, J. J. *Aquatic Chemistry,* 2nd ed.; Wiley-Interscience: New York, 1981.

72. Adamson, A. W. *Physical Chemistry of Surfaces*, 3rd ed.; Wiley: New York, 1976.
73. Ruthven, D. M. *Principles of Adsorption and Adsorption Processes;* Wiley: New York, 1984.
74. Laub, R. J. In *Chromatography and Separation Chemistry Advances and Developments;* Ahuja, S., Ed.; ACS Symposium Series 297; American Chemical Society: Washington, DC, 1986; pp 1–33.
75. Harter, R. D. *Soil Sci. Soc. Am. J.* **1984**, *48*, 749–752.
76. Crittenden, J. C.; Luft, P.; Hand, D. W.; Oravitz, J. L.; Loper, S. W.; Art, M. *Environ. Sci. Technol.* **1985**, *19(11)*, 1037–1041.
77. Yonge, D. R.; Keinath, T. M. *J. Water Pollut. Control Fed.* **1986**, *58(1)*, 77–81.
78. Veith, J. A.; Sposito, G. *Soil Sci. Soc. Am. J.* **1977**, *41*, 697–702.
79. Coates, J. T.; Elzerman, A. W. *J. Contam. Hydrol.* **1986**, *1(1/2)*, 191–210.
80. Chiou, C. T.; Peters, L. J.; Freed, V. H. *Science (Washington, D.C.)* **1979**, *206*, 831–832.
81. Chiou, C. T.; Porter, P. E.; Schmedding, D. W. *Environ. Sci. Technol.* **1983**, *17(4)*, 227–231.
82. Karickhoff, S. *Chemosphere* **1981**, *10(8)*, 833–846.
83. Chiou, C. T.; Schmedding, D. W. In *Test Protocols for Environmental Fate and Movement of Toxicants; Proceedings of the Association of Official Analytical Chemists Symposium, Washington, DC, October, 1980;* Zweig, G.; Beroza, M., Eds.; Association of Official Analytical Chemists: Arlington, VA, 1981; pp 28–42.
84. Dzombak, D. A.; Luthy, R. G. *Soil Sci.* **1984**, *137(5)*, 292–308.
85. Banerjee, S.; Yalkowsky, S. H.; Valvani, S. C. *Environ. Sci. Technol.* **1980**, *14(10)*, 1227–1229.
86. Miller, M. M.; Wasik, S. P.; Huang, G.-L.; Shiu, W.-Y.; Mackay, D. *Environ. Sci. Technol.* **1985**, *19*, 522–529.
87. Arbuckle, W. B. *Environ. Sci. Technol.* **1983**, *17(9)*, 537–542.
88. Campbell, J. R.; Luthy, R. G. *Environ. Sci. Technol.* **1985**, *19(10)*, 980–985.
89. Banerjee, S. *Environ. Sci. Technol.* **1984**, *18(8)*, 587–591.
90. Banerjee, S. *Environ. Sci. Technol.* **1985**, *19(4)*, 369–370.
91. Nkedi-Kizza, P.; Rao, P. S. C.; Hornsby, A. G. *Environ. Sci. Technol.* **1985**, *19(10)*, 975–979.
92. Fu, J.; Luthy, R..G. *J. Environ. Eng.* **1986**, *112(2)*, 346–366.
93. Rapaport, R. A.; Eisenreich, S. J. *Environ. Sci. Technol.* **1984**, *18(3)*, 163–170.
94. Woodburn, K. B.; Doucette, W. J.; Andren, A. W. *Environ. Sci. Technol.* **1984**, *18(6)*, 457–459.
95. Mackay, D. M.; Roberts, P. V.; Cherry, J. A. *Environ. Sci. Technol.* **1985**, *19(5)*, 384–392.
96. Kahn, A.; Hassett, J. J.; Banwart, W. L.; Means, J. C.; Wood, S. G. *Soil Sci.* **1979**, *128(5)*, 297–302.
97. Means, J. C.; Wood, S. G.; Hassett, J. J.; Banwart, W. L. *Environ. Sci. Technol.* **1982**, *16*, 93–98.
98. Banwart, W. L.; Kahn, A.; Hassett, J. J. *J. Environ. Sci. Health B* **1980**, *15*, 165–179.
99. Weber, W. J.; Sherrill, J. D.; Pirbazari, M.; Uchrin, C. G.; Lo, T. Y. In *Dynamics, Exposure, and Hazard Assessment of Toxic Chemicals;* Haque, R., Ed.; Ann Arbor Science: Ann Arbor, MI, 1980, pp 191–213.
100. Mabey, W. R.; Podoll, R. T. *Estimation Methods for Process Constants and*

*Properties Used in Fate Assessments;* U.S. Environmental Protection Agency: Washington, DC, 1984; EPA–600/s3–84–035.

101. Donigian, A. S. In *Fate of Chemicals in the Environment;* Swann, R. L.; Eschenroeder, A., Eds.; ACS Symposium Series 225; American Chemical Society: Washington, DC, 1983; pp 151–171.

102. *Test Protocols for Environmental Fate and Movement of Toxicants; Proceedings of the Association of Official Analytical Chemists Symposium, Washington, DC, October, 1980;* Zweig, G.; Beroza, M., Eds.; Association of Official Analytical Chemists: Arlington, VA, 1981.

103. "Guidelines for Testing of Chemicals"; Organization for Economic Cooperation and Development: Paris, France, 1981.

104. DiToro, D. M.; Horzempa, L. M. *Environ. Sci. Technol.* 1982, *16(9),* 594–602.

105. Horzempa, L. M.; DiToro, D. M. *Water Res.* 1983a, *17(8),* 851–859.

106. Corwin, D. L.; Farmer, W. J. *Environ. Sci. Technol.* 1984, *18(7),* 514–517.

107. Karickhoff, S. W.; Morris, K. R. *Environ. Toxicol. Chem.* 1985, *4,* 469–479.

108. Schweich, D.; Sardin, M. *J. Hydrol. (Amsterdam)* 1981, *50,* 1–33.

109. Rao, P. S. C.; Ralston, D. E.; Jessup, R. E.; Davidson, J. M. *Soil Sci. Soc. Am. J.* 1980, *44,* 1139–1146.

110. Goltz, M. N.; Roberts, P. V. *J. Contam. Hydrol.* 1986, *1(1/2),* 77–93.

111. Oliver, B. G. *Chemosphere* 1985, *14(8),* 1087–1106.

112. Miller, C. T.; Weber, W. J. *Ground Water* 1984, *22(5),* 584–592.

113. DiToro, D. M.; Horzempa, L. M. In *Physical Behavior of PCBs in the Great Lakes;* Mackay, D.; Patterson, S.; Eisenreich, S. J.; Simmons, M. S., Eds.; Ann Arbor Science: Ann Arbor, MI, 1983; pp 89–113.

114. Miller, C. T.; Weber, W. J. *J. Contam. Hydrol.* 1986, *1(1/2),* 243–261.

115. Rao, P. S. C.; Davidson, J. M.; Jessup, R. E.; Selim, H. M. *Soil Sci. Soc. Am. J.* 1979, *43,* 22–28.

116. Brown, D. S.; Flagg, E. W. *J. Environ. Qual.* 1981, *10,* 382–386.

117. Ben-Naim, A. *Hydrophobic Interactions;* Plenum: New York, 1980.

118. Freeman, D. H.; Cheung, L. S. *Science (Washington, D.C.)* 1981, *214,* 790–792.

119. O'Connor, D. J.; Conolly, J. P. *Water Res.* 1980, *14,* 1517–1523.

120. Coates, J. T. Ph.D. Thesis, Clemson University, Clemson, SC, 1984.

121. DiToro, D. M.; Jerls, J. S.; Clarcia, D. *Environ. Sci. Technol.* 1985, *19(12),* 1169–1176.

122. Hassett, J. P.; Anderson, M. A. *Environ. Sci. Technol.* 1979, *13(12),* 1526–1529.

123. Chiou, C. T.; Malcolm, R. L.; Brinton, T. I.; Kile, D. E. *Environ. Sci. Technol.* 1986, *20(5),* 502–508.

124. Means, J. C.; Wijayaratne, R. *Science (Washington, D.C.)* 1982, *215,* 968–970.

125. Carter, C. W.; Suffet, I. H. In *Fate of Chemicals in the Environment;* Swann, R. L.; Eschenroeder, A., Eds.; ACS Symposium Series 225; American Chemical Society: Washington, DC, 1983; pp 215–230.

126. McCarthy, J. F.; Jiminez, B. D. *Environ. Sci. Technol.* 1985, *19(11),* 1072–1076.

127. Curl, R. L.; Keoleian, G. A. *Environ. Sci. Technol.* 1984, *18(12),* 916–922.

128. Voice, T. C.; Weber, W. J. *Environ. Sci. Technol.* 1985, *19(9),* 789–796.

129. Gschwend, P. M.; Wu, S. *Environ. Sci. Technol.* 1984, *19(1),* 90–96.

130. Brownawell, B. J.; Farrington, J. W. *Geochim. Cosmochim. Acta* 1986, *50,* 157–169.

131. DiToro, D. M. *Chemosphere* 1985, *14(10),* 1503–1538.

132. Anderson, M. A.; Tejedore-Tejedore, I.; Stanforth, R. R. *Environ. Sci. Technol.* **1985**, *19(7)*, 632–637.
133. DiToro, D. M.; Horzempa, L. M.; Casey, M. M.; Richardson, W. *J. Great Lakes Res.* **1982**, *8(2)*, 336–349.
134. Horzempa, L. M.; DiToro, D. M. *J. Environ. Qual.* **1983b**, *12(3)*, 373–380.
135. Lioux, N. Ph.D. Thesis, University of Wisconsin, Madison, 1985.
136. Thurman, E. M. *Organic Geochemistry of Natural Waters;* Nijhoff/Junk: Boston, 1985.
137. Wershaw, R. L. *J. Contam. Hydrol.* **1986**, *1(1/2)*, 29–45.

RECEIVED for review June 10, 1986. ACCEPTED October 15, 1986.

# 11

# The Role of Particulate Matter in the Movement of Contaminants in the Great Lakes

Brian J. Eadie and John A. Robbins

Great Lakes Environmental Research Laboratory, National Oceanic and Atmospheric Administration, 2205 Commonwealth Boulevard, Ann Arbor, MI 48104

*Particle-contaminant interactions and subsequent behavior of the particulate matter control the long-term concentration of many compounds in aquatic systems. Even in deep systems such as the Great Lakes, particle settling times from the water column are less than 1 year. After reaching the bottom, contaminant-laden particles are redistributed by episodic cycles of resuspension and redeposition, resulting in focusing, which is the spatially inhomogeneous distribution of contaminants in sediments. Bioturbation, coupled with focusing, provides source material to the resuspendible pool. The combination of these processes mediates both the composition and long-term behavior of contaminants in these lakes.*

IN THE GREAT LAKES, AS IN MANY AQUATIC SYSTEMS, the movement of particulate matter plays a crucial role in determining the levels and fate of most contaminants. Persistent hydrophobic contaminants have a strong affinity for particles. As a result, the process of sorption and settling into the sediments is generally a major removal mechanism. This mechanism is very important in large systems such as the Great Lakes, where hydraulic residence times are long (3–190 years), and particle residence times in the water column are generally much less than 1 year. Thus, contaminants with a high degree of affinity for settling particles can be very efficiently scavenged from the system. However, studies of the long-term behavior of certain fallout radionuclides and stable contaminants (*1, 2*) have shown that higher levels persist in the lakes than expected if scavenging were the sole transport mechanism. This persistence is thought to be due primarily to the combination of bioturbation and resuspension (3), in which contaminants initially transferred to sediments are brought back up into the water column. At the sediment surface, freshly deposited matter is typically mixed with older material as a result of the movement and feeding activities of organisms inhabiting the upper layers of sediment (1–10 cm thick). As a result of mixing, materials that would have been buried are reintroduced into the resuspendible pool.

The processes that characterize the behavior of hydrophobic contaminants are shown in their simplest form in Figure 1. The processes of particle–contaminant transfer, settling and resuspension, and bioturbation and burial are intimately interconnected and control the phase distribution, bioavailability, and long-term behavior of most trace contaminants in aquatic systems. This chapter provides a description of these processes as they relate to the well-studied Great Lakes system and provides access to the appropriate literature. Emphasis is placed on the behavior of organic contaminants, selected radionuclides, and lead, which is an example of a major inorganic contaminant.

## Lake Particulate Matter

Particle–contaminant affinities are related to the chemical composition of the substrate (4). The composition of the pool of particles in the Great Lakes varies considerably because of combinations of physical, chemical, and biological processes (Figure 2). In winter and early spring, the lake is isothermal and storm events easily resuspend vast quantities of fine sedimentary matter, which has undergone considerable alteration and has relatively low organic carbon and high clay mineral content. During the isothermal period, the composition of the water column is virtually the same from top to bottom. As spring progresses, the lake warms and stratifies. Primary production increases especially within the warm near-surface waters or *epilimnion* (about 30-m maximum depth), and the phytoplankton and zooplankton then dominate the particle pool. During late spring and early summer, phytoplankton concentration is reduced primarily because of particle sinking and zooplankton grazing. These processes generate large quantities of nonliving organic particles that descend through the epilimnion. During the stratified period, the epilimnion is largely decoupled from underlying colder waters (*hypolimnion*) so that particle transport is one way, down. Little of the particulate material residing near the bottom of the deep open lakes can be resuspended into the epilimnion during the stratified period. In the nearshore region, regular upwellings that have periods of a few days reinject resuspended shelf and slope sediments as well as materials from the ben-

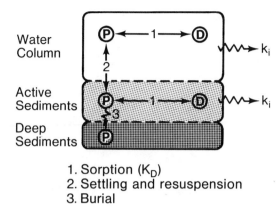

1. Sorption (K$_D$)
2. Settling and resuspension
3. Burial

*Figure 1. Schematic diagram of the minimum fundamental processes required to account for the long-term behavior of particle-associated contaminants in the Great Lakes. Abbreviations are as follows: P represents particle-associated contaminant, and D is the dissolved fraction. The major processes discussed in this manuscript are shown as (1) partitioning, (2) settling and resuspension, and (3) bioturbation and burial. The rate constants (k$_i$, i = 1 . . . n) represent losses from the system from processes such as outflow, radioactive decay, evaporation, and chemical or biological degradation.*

thic nepheloid layer (BNL) into surface waters (5). During late summer and early fall, CaCO$_3$ precipitates in the lower Great Lakes (6), and these particles constitute 20%–90% of the particle pool. Following this period, the lake cools, stratification breaks down, and fine sedimentary materials are again resuspended throughout the waters of the lake.

These cyclic alterations in the composition of particulate matter are reflected in the spatial and temporal variations in amount of total suspended matter (TSM) in the water column. The Great Lakes vary in their average open lake TSM concentrations, from Superior (0.2–1.0 mg/L), Huron and Michigan (0.5–2.0 mg/L), and Ontario (1.0–4.0 mg/L) to Erie (2.0–6.0 mg/L). In Lake Michigan, the TSM in near-surface, open lake waters is approximately 0.7 mg/L during the isothermal period and decreases during the months of June and July, presumably as a result of zooplankton grazing and rapid settling of detritus. With the onset of calcite formation in late July through September, the TSM increases by roughly a factor of 3. With the breakdown of stratification, the concentration of TSM decreases again.

Profiles of light transparency reveal seasonal changes in the vertical structure of TSM at a 75-m-deep station, 10 miles offshore in southeastern Lake Michigan (Figure 3). In spring, when the lake is isothermal, the transparency is constant from the surface down to approximately 15 m

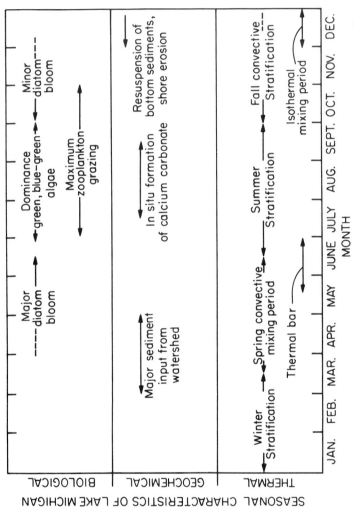

*Figure 2. Seasonal processes in the Great Lakes that affect the particle pool. The seasonally varying thermal, geochemical, and biological processes combine to create a particle inventory whose amount and composition change greatly during the course of the year.*

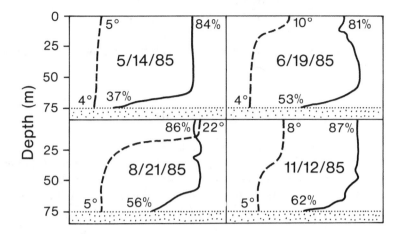

*Figure 3. Seasonal variations in the vertical distribution of temperature and light transmission at an open water site in Lake Michigan. An isothermal temperature profile in April is gradually replaced by a lens of warm surface water whose depth increases throughout the summer and early fall. In November, the lake has begun to cool significantly and remix. Profiles of transmissivity show that a region of high particle concentration (BNL) persists near the bottom (20 m) throughout the season.*

above the bottom where it decreases rapidly. As thermal stratification begins to develop, a somewhat more structured transparency profile becomes established with a pronounced minimum near the thermocline (a usually sharp boundary between the epilimnion and the hypolimnion). This minimum is attributed to an increase in biota as indicated by increased concentrations of chlorophyll in the vicinity of the thermocline. Later in the summer, when the thermocline is well developed, several inflections can be observed in the transparency profile. The general shape of the transparency profile remains constant as the lake cools and the thermocline deepens. In all cases, the near-bottom BNL is well developed, and TSM concentrations at 5 m above bottom range from 2.0–4.0 mg/L throughout the year at this site.

**Contaminant–Particle Interactions.**   The degree of partitioning of a contaminant between dissolved and particulate phases ($K_d$) is a complex function of the compound's intrinsic properties and the composition of the substrate. Because a detailed discussion of the factors influencing contaminant–particle affinities is included in this volume (7), the following comments are specific to conditions in the Great Lakes.

The affinity of a contaminant for particles is defined in terms of the partition coefficient, $K_d$, as

$$K_d = \frac{\text{concentration in particulate phase (mg/g)}}{\text{concentration in dissolved phase (mg/mL)}} \qquad (1)$$

and the fraction of contaminant associated with the suspended particulate phase $(f_p)$ is

$$f_p = \frac{K_d \, (\text{TSM})}{1 + K_d \, (\text{TSM})} \qquad (2)$$

where TSM is in units of milligrams per liter.

Under ideal conditions, in a well-mixed lake where the partitioning of a contaminant between phases represents a state of equilibrium in a process of reversible sorption, the apparent residence time with respect to removal by particle settling, $T_a$, is related to the value of $f_p$ by

$$T_a = T_s/f_p \qquad (3)$$

where $T_s$ is the mean lifetime of a particle settling through the water column and is equal to the average depth of the lake divided by the characteristic particle settling rate (approximately 1 m/day). Thus, at least under ideal conditions, a simple relationship exists between the fraction of the contaminant associated with the particle phase and the characteristic time of its removal from the system through the sedimentation process. For sufficiently large values of $K_d$, $f_p = 1$, and the rate of contaminant removal to sediments is characterized by the rate at which particles settle out. As $K_d \to 0$, $f_p = 0$, and the rate of removal through the sedimentation process becomes insignificant. The effect of $K_d$ values on the relative importance of outflow versus sedimentation as a removal pathway is illustrated in Table I.

Table I. Effect of $K_d$ on the Relative Importance of Outflow versus Sedimentation as a Removal Pathway

| Lake | Hydraulic Residence Time (years) | $T_s$ (years) | $T_a$ (years) | | | |
|------|------|------|------|------|------|------|
| | | | $K_d = 10^4$ | $K_d = 10^5$ | $K_d = 10^6$ | $K_d = 10^7$ |
| Superior | 190 | 0.41 | 82.0 | 8.6 | 1.2 | 0.49 |
| Michigan | 100 | 0.23 | 46.0 | 4.8 | 0.69 | 0.28 |
| Huron | 30 | 0.21 | 42.0 | 4.4 | 0.63 | 0.25 |
| Erie | 3 | 0.05 | 1.0 | 0.14 | 0.06 | 0.05 |
| Ontario | 8 | 0.24 | 24.0 | 2.6 | 0.48 | 0.26 |

In every lake, mean particle settling times are far shorter than hydraulic residence times. Thus, provided $K_d$ values are sufficiently high, sedimentation is the most important removal mechanism for persistent contaminants. When the apparent settling time is less than 10% of the hydraulic residence time, then the value of $K_d$ corresponds roughly to the limit below which a contaminant is not considered to be a particle-associated entity in the Great Lakes.

ORGANIC COMPOUNDS. The extent of partitioning of organic contaminants onto particles is well correlated with the compound's solubility and the characteristics of the substrate (8-11). In a comprehensive review, Kenaga and Goering (8) derived the following relationship from published data:

$$\log K_{oc} = 3.64 - 0.55 \log WS \qquad n = 106 \qquad r = -0.84 \qquad (4)$$

where WS is the solubility in water (ppm), $r$ is the correlation coefficient, and $K_{oc} = (100 \times K_d)/\%$ organic carbon of the substrate.

The use of $K_{oc}$, correlating the extent of association to the organic carbon fraction of the particulate matter, has been shown to reduce the variability in the relationship between partitioning and solubility. Solubilities have been reported (12) for the organic compounds recently identified by the International Joint Commission (13) as being of most concern in the Great Lakes (Table II).

Table II. Solubilities of Organic Compounds of Concern in the Great Lakes

| Compound | Log Solubility (ppm) |
|---|---|
| Total PCBs | −1.4−−0.7 |
| Mirex | −0.7 |
| Hexachlorobenzene | −2.3 |
| Dieldrin | −1.0 |
| DDT | −2.5 |
| Benzo[a]pyrene | −3.3 |

The calculated equilibrium distribution of contaminants between the particulate and dissolved phases is shown in Figure 4 for compounds that have the range of solubilities just presented. The range of TSM represented between the two lines in the figure (0.5-5.0 ppm) covers most Great Lakes situations. In the open lakes, the majority of these trace contaminants should be in the dissolved phase. This point has been confirmed by a limited number of measurements in the waters of the lakes (14-16).

The magnitude of $K_d$ is inversely proportional to the concentration of substrate (17, 18). This phenomenon has been described in a number

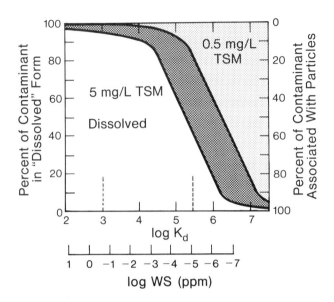

*Figure 4. Relation between fraction dissolved and the aqueous distribution coefficient ($K_d$) for contaminants. The two curves represent the range of TSM values encountered in the Great Lakes. The scale for $K_{oc}$ is established on the assumption that 10% of the substrate is organic carbon. The $K_{oc}$ term is useful for characterizing organic contaminants in the system. The relationship between $K_{oc}$ and solubility (WS) is based on the expression of Kenaga and Goering (8).*

of conceptual and numerical models (*19*). Several of the models can be made to approximate the data, but no consensus has evolved on the process responsible for the observed behavior. The consequences are most important in sediment pore waters, where measured (operationally defined) "dissolved" concentrations (*18, 20*) are several orders of magnitude higher than overlying waters, and $K_d$ values for hydrophobic organic compounds (HOCs) generally decrease from $10^5$–$10^6$ to a value of about $10^3$. Most of the increase in the dissolved phase is probably associated with elevated concentrations of very fine particles, colloidal material, or dissolved organic matter (*21*).

Although solubility is a reliable indicator of the average affinity of persistent organic compounds for particles, other more complex interactions affect both the distribution and speciation of such contaminants. For example, particle-bound PCBs photodecompose significantly more rapidly than their dissolved counterparts, and substances associated with particles will be less susceptible to removal through the air–water interface but will be exposed to zooplankton grazing. The size spectrum of the particle pool influences the rate and extent of grazing and biological repackaging by the zooplankton (*18*). Calculation of the effects of zoo-plankton (*22*) indicates that the filter-feeding animals in the epilimnion of Lake Ontario consume approximately 10% of the particulate material per day. Much of the material consumed (60%–70%) is subsequently replaced, but in physically and chemically altered form. Zooplankton grazing results in particle breakage through inefficient feeding and in the formation of both diffuse and pelletized fecal material (*23*). Both processes alter the spectrum and characteristics of suspended matter and associated contaminants. In Lake Michigan, most of the material collected in offshore sediment traps was composed of fecal pellets or degraded fecal material (*24*). Similar biological mediation of particle flux has been described for both coastal and offshore marine environments (*25–27*).

INORGANIC COMPOUNDS.  Although the phase distribution of organic contaminants can be reasonably approximated from their solubilities, the distribution of inorganic species is controlled by the geochemistry of the compound and its matrix. The most studied cases of contaminant partitioning between water and particulate matter in the Great Lakes are those of the fallout radionuclides cesium-137 and the plutonium isotopes. Radiocesium ($t_{1/2} = 30$ years) is thought to have a comparatively simple interaction with particles. The radionuclide is probably irreversibly bound by lattice substitution in certain clay minerals present in waters of the Great Lakes and certain other freshwater systems (*28*).

The behavior of plutonium is considerably more complex. Plutonium appears to be conveyed from the water column in association with amorphous silicon (diatoms) and possibly calcium carbonate. The mechanism for removal of plutonium from the lakes is further complicated by the fact that it can exist in several oxidation states that possess greatly differing aqueous solubilities (*29*). Dissolved organic carbon (DOC) strongly influences the partitioning of plutonium onto suspended particles in natural waters from a selection of systems including the Great Lakes (*30*). An order-of-magnitude increase in DOC concentrations was associated with a hundredfold decrease in the partition coefficient for plutonium. Evidently, the organic carbon in solution or in a colloidal phase can effectively compete with particulate matter for the dissolved species. Laboratory studies with tracer-spiked suspensions of sediment and DOC-enriched natural lake water confirmed the importance of DOC in regulating plutonium $K_d$ values and led to the formulation of a model for the complexation of plutonium with multiple organic ligands (*31*). In sediments, plutonium is associated predominantly with hydrous oxides of iron and manganese (*23*).

Wahlgren and Nelson (*33*) found that despite known chemical differences between radiocesium and plutonium, their field-determined partition coefficients were comparable, and that average values of $K_d$ for plutonium were not significantly different in each of the four lower Great Lakes ($2.5 \pm 0.5 \times 10^5$). They also showed that at a site in southern Lake Michigan, plutonium $K_d$ values changed little throughout the water column during the season. The general constancy of the value of $K_d$ for plutonium in both suspended-particulate fraction and in trap-collected materials indicates that a relatively limited set of chemical variables controls its distribution. Wahlgren and Nelson suggested that the uptake of plutonium is dominated by a surface coating process. A subsequent study by Alberts and Wahlgren (34) showed that plutonium exists as a simple anionic complex or is associated with particles smaller than 30 Å and negatively charged. They also determined that the proportion of both radiocesium and plutonium in the dissolved state was consistent with previous values of the distribution coefficient. Although strong variations in plutonium $K_d$ values are seen from lake to lake as a result of varying amounts of DOC, their constancy within the Great Lakes, both temporally and spatially, indicates that either the amount of DOC present is relatively constant or that changing DOC concentrations are offset by a varying ligand complexation capacity. Very little is yet known about the concentration and composition of colloidal organic matter in the Great Lakes.

In contrast with the radiocesium and plutonium, another fallout radionuclide, strontium-90 ($t_{1/2} = 28$ years), having essentially the same loading time dependence to the lakes, has a value of $K_d$ more than 2000 times smaller [$1.2 \times 10^3$ (35–37)]. The $K_d$ value for strontium-90 is so low that it cannot be considered a particle-associated nuclide. However, including strontium-90 here as an example illustrates the importance of geochemical controls on the extent of partitioning and the long-term behavior of contaminants in the lakes. The low $K_d$ for strontium-90 is due to its chemical similarity to calcium, which is the major dissolved ion in the lakes. Strontium-90 must compete with calcium in relatively nonspecific ion exchange for sites on particles in the water column and sediments. Because of its low $K_d$, losses of strontium-90 from Lake Michigan through particle settling have a characteristic time of about 400 years. Thus, removal of strontium-90 from the system is dominated by radioactive decay and outflow.

Studies of the partitioning of nonradioactive inorganic contaminants in the Great Lakes have been generally hindered by the lack of reliable analytical results (38). One exception is lead, which is the principal metal contaminant in the Great Lakes system and has a radioactive analogue, lead-210. Recent studies of this isotope in Lake Michigan have shown that lead partitioning is strongly related to the annual cycle of calcite formation in the lower lakes (39). This relationship is presumably due to the fact that lead carbonate is a very insoluble compound.

**Phase Distribution Measurements.** An important distinction must be considered between field and laboratory measurements of distribution coefficients. Most field and laboratory studies related to the Great Lakes have arbitrarily taken the 0.45-$\mu$m-diameter filter pore size as the operational distinction between particulate and solution phases. Field measurements determine the degree of partitioning at a given instant in time without regard for whether the system is in equilibrium or not. Laboratory studies usually involve the addition of a tracer or labelled compound to a multiphase system. Equilibration times are usually very short in comparison with exposure times in the natural environment. The effect of these differences is to produce values of distribution coefficients that are not necessarily comparable and may not be indicative of a reversible sorption process. Major questions relating to the spatial and temporal consistency of contaminant $K_d$ in the lakes remain. The results of recent field and laboratory experiments addressing some of these questions are described in this chapter.

LEAD.   During the period from April through October 1983, water samples were collected over a range of depths at a 147-m-deep site in the central basin of southern Lake Michigan. Samples were filtered aboard ship through a 0.45-$\mu$m filter, and the dissolved and suspended fractions were subsequently analyzed by conventional alpha spectroscopy for lead-210. In addition, the amount of suspended matter was determined from separate filtered water samples collected from the same depths. During the same period, trap strings were deployed that collected sinking particulate matter for roughly monthly intervals. Surface trap (15-m) samples have been analyzed for inorganic carbon. Average values of TSM, $f_p$, and $K_d$ for the region of the water column above the nepheloid layer and individual values within the nepheloid layer are shown in Figure 5. Vertical error bars in the figure are standard deviations from the average and indicate the extent of variability of values above the nepheloid layer. Horizontal bars in the figure indicate the exposure period of the traps.

In water above the nepheloid region, average values of TSM decrease from 0.8 mg/L in May to a minimum of 0.3 mg/L in mid-August. The decrease is even more pronounced within the nepheloid layer. The reduction in near-bottom turbulence that accompanies stratification of the lake suppresses the amount of materials resuspended off the bottom. After August, TSM increases in both regions of the water column. This increase is accompanied by a pronounced increase in the amount of inorganic carbon in surface-trap samples. During the spring and early summer, inorganic carbon is low and shows some tendency to decrease. In the late summer and fall months, trap samples are abundant in inorganic carbon; calcite accounts for as much as 44% of the total sample weight.

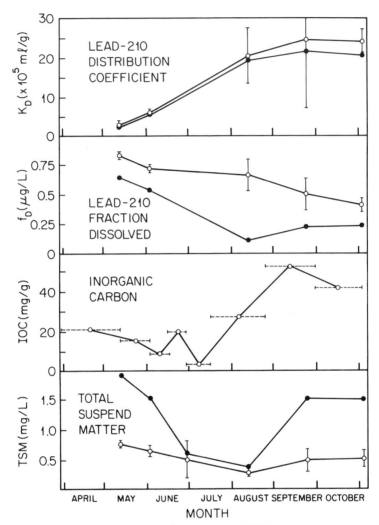

*Figure 5. Seasonal variations in the amount of TSM, fraction of dissolved lead-210 ($f_d$), and distribution coefficient for lead-210 ($K_d$) in water within the BNL and average values above it. Also shown are values of inorganic carbon concentrations for sediment-trap-collected materials from near-surface waters (15 m).*

Above the nepheloid layer, the fraction of dissolved lead-210 decreases progressively throughout the season. During the spring and early summer, average values are consistent with that for stable lead ($f_d$ = 0.75 ± 0.3) determined by Rossmann (38). By November, less than half of the lead-210 is found in the dissolved phase. Within the nepheloid region, the fraction dissolved is consistently much less than in the rest of the water column. This difference is probably due to increased amounts of TSM in the nepheloid layer because the distribution coefficients for the two parts of the water column are very close to each other throughout the observation period (Figure 5, top panel). Thus, at a given time, $K_d$ appears to be largely depth-independent for lead-210. However, $K_d$ shows a large systematic variation with season. The average $K_d$ increases from 2.7 × 10⁵ in April to about 5.8 × 10⁵ in June. This increase is accompanied by an increase in the amount of particulate organic material in the water. This increase suggests that the affinity of lead for particulate matter is affected by the presence of organic substrates. However, the major increase in $K_d$ occurs between June and August. In August and throughout much of the fall, values of $K_d$ average 22 × 10⁵. These high values of $K_d$ coincide with high-trap inorganic carbon content. Thus, for lead, $K_d$ is not constant but shows about an order-of-magnitude variation throughout a season. During the period of calcite formation, the contaminant metal can be expected to be efficiently scavenged from the lake, but its ultimate fate depends on the phase transfer it undergoes as calcite dissolves. Very little of the precipitated calcium carbonate is ultimately stored in sediments. Dissolution of calcite as it settles through the water column, enters the nepheloid layer, and is either resuspended or buried provides an opportunity for lead to redissolve and move in the dissolved phase or to reassociate with stable residents of the particle pool. Such processes combine to produce a concentration of lead that should show strong seasonal variations.

ORGANIC COMPOUNDS.  The seasonal variability of $K_d$ for organic species is illustrated by a study of equilibrium partitioning onto trap materials. Trap samples, representing nine time intervals, were collected from a depth of 15 m from the same 147-m-deep station in the center of the southern basin of Lake Michigan and then used as substrate material for laboratory measurements of equilibrium partition coefficients. Radiolabeled organic compounds covering a wide range of solubilities were added to suspensions of trap material in glass-fiber-filtered lake water at a substrate concentration of 10 mg/L. The samples were equilibrated for 3 days, separated by filtration, and analyzed for radioactivity. Real seasonal differences were observed for most of the organic compounds examined [benzo[a]pyrene, pyrene, phenanthrene (shown in Figure 6), as well as benzo[a]anthracene and beryllium-7 (40)], although the temporal responses were different. In this case, normalizing organic carbon by substrate resulted in an increase rather than the expected decrease in variability (Figure 7). No seasonal variation was observed for the partitioning of 2,4,2',4'-tetrachlorobiphenyl (TCB); its mean value of $K_d$ was 3450 (Table III). The differences in seasonal behavior among the compounds indicate that there may not be a simple normalization factor

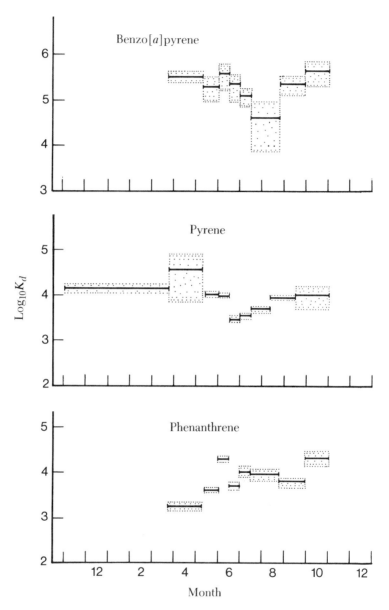

*Figure 6. Seasonal* $K_d$ *on undried trap material for benzo[a]pyrene (0.2 ppb), pyrene (150 ppb), and phenanthrene (1002 ppb). Values in parentheses are water solubilities. Average values of a triplicate set are indicated as the horizontal line. The gray area represents 1 standard deviation. Systematic, order-of-magnitude differences are seen throughout the season for these compounds.*

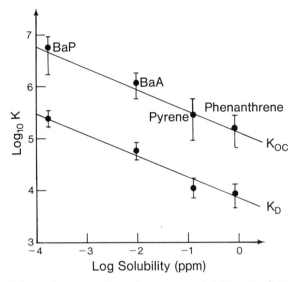

*Figure 7. Relation between $K_d$ and aqueous solubility of selected organic compounds based on equilibrium partitioning of labeled materials in standard water from Lake Michigan (adjusted to 10 mg/L TSM). Although solubility is an excellent predictor of $K_d$, the relationship is not improved by normalization to organic carbon ($K_{oc}$), as the increased scatter shows.*

(e.g., $K_{oc}$) as the substrates change throughout the year. Some intrinsic properties of the compounds (e.g., dipole moment and shape) in conjunction with substrate properties may be required to explain the details of their partitioning behavior. Another important factor to remember is that trap-collected samples, although more representative of the particle pool than sediment, are still temporally integrated. If changes in substrate composition affect equilibrium $K_d$ in an as yet unpredictable way, then we would expect to see larger temporal differences in the more variable raw lake water.

In addition to the seasonal epilimnetic samples from the station in the center of the southern basin, epilimnetic and BNL trap samples were collected for the complete stratified and unstratified periods from stations throughout the open lake in order to examine spatial homogeneity of laboratory-measured $K_d$. The measured $K_d$ values for the epilimnetic samples from these stations were not significantly different when compared to the seasonally calculated $K_d$ values from station 4. The implication is that the surface waters of the lake are well mixed spatially. The $K_d$ values measured by using BNL-collected substrate are not significantly different from the near-surface samples. This similarity supports the previously described concepts of long-term homogenization of the resuspendible pool. Interestingly, the variances for the $K_d$ values mea-

Table III. Equilibrium Partition Coefficients for Trap-Collected Material

| Period | TCB ($\times 10^3$) | Phenanthrene ($\times 10^3$) | Pyrene ($\times 10^4$) | Benzo[a]anthracene ($\times 10^4$) | Benzo[a]pyrene ($\times 10^5$) |
|---|---|---|---|---|---|
| Stratified (15 m) | 2.6 (39) | 10.0 (53) | 0.74 (58) | 3.7 (26) | 2.2 (67) |
| Unstratified (15 m) | 4.4 (41) | 7.7 (76) | 1.2 (24) | 7.4 (49) | 2.7 (40) |
| Stratified (5 m above bottom) | | 8.9 (29) | 1.3 (30) | 5.6 (29) | 2.7 (19) |
| Unstratified (5 m above bottom) | | 7.6 (26) | 1.3 (32) | 5.0 (28) | 2.2 (44) |
| Total mean | 3.4 | 7.3 | 1.1 | 5.6 | 2.5 |
| Standard deviation | 1.3 | 3.9 | 0.36 | 1.9 | 0.86 |
| Number of samples | 21 | 63 | 66 | 39 | 63 |

NOTE: The trap-collected material had a concentration of 10 mg/L. Values in parentheses are coefficients of variation.

sured on the BNL substrate are lower than those using near-surface materials (Table III). This difference may imply that the BNL particulate matter is more homogenized throughout the lake than surface particulate matter.

**Mass and Contaminant Fluxes.** An important distinction exists between the mobile particulate matter in the lake collected as the total suspended matter (TSM) associated with a water sample and that collected as settling matter in a sediment trap. The former represents the complete population of particles in the water if the sample is large enough to capture the rare, large particles, and the latter collects a biased population of settling particles: those that, because of their size or density, have sufficiently high settling velocities to be captured. This biased population, however, is responsible for removing contaminants from the water column and reintroducing them through resuspension. Studies of mass and contaminant fluxes have illustrated the importance of the resuspendible pool in particle transport within the lakes (*1, 3, 41*).

The vertical distribution of gross downward particle fluxes recorded in Great Lakes sediment-trap studies reveals that fluxes increase exponentially with a scale depth of approximately 5 m near the bottom (*3, 41, 42*) throughout the year. This near-bottom increase is consistent with TSM profiles and supports the concept of an extensive BNL. A comparison among offshore stations situated in regions of modern sediment accumulation in three of the Great Lakes at approximately the same depth (Figure 8) shows a considerable difference in the near-bottom profiles. Near-bottom trap-collected material from the slope and deep

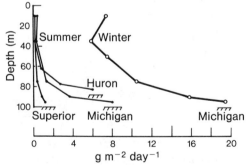

*Figure 8. Mass fluxes at open lake sites in three Great Lakes. Striking increases within the bottom 20 m (BNL) are seen in every lake but differ considerably in magnitude. Summer fluxes above the BNL are small and similar for each lake. The winter flux in Lake Michigan is much greater and reflects the movement of suspended materials up into the entire water column.*

basin of southern Lake Michigan have chemical characteristics indistinguishable from fine-grained surficial sediment collected in the basin region (3). Thus, the relative contribution from these two sources cannot be discriminated at this time.

Station depth, distance from shore, and slope all play a role in mediating local mass fluxes. Figure 9 illustrates mass fluxes measured in Lake Michigan at seven stations with depths ranging from 48 to 165 m (43). The fluxes observed near the surface (above the thermocline) during the stratified period support earlier conclusions that these traps are relatively uncontaminated with resuspended sediments, and that near-bottom fluxes during this period are amazingly consistent (5 ± 1 g/m²/day) in depths ranging from 70 to 161 m. Fluxes at station 11 in the northern basin are significantly lower than the other central lake stations 4 and 9. Stations 2 and 6 in the southern basin are situated where sediments are not presently accumulating, and yet the flux profile of station 6 is very similar to station 8 where present sediments are being deposited. This similarity indicates that nondepositional areas are receiving similar amounts of BNL materials, but for some reason (probably

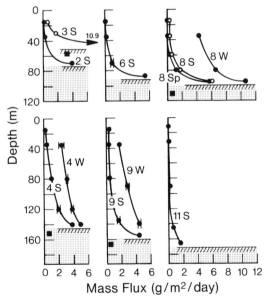

*Figure 9. Summer (S) and winter (W) mass fluxes at selected sites in Lake Michigan. The BNL is seen in each of the profiles; in summer the calculated half depth of the BNL appears nearly independent of location and approximately equal to 10 m. The magnitude of the summer BNL fluxes tends to decrease with increasing water depth. The solid squares in the sediment represent radiometrically determined local sediment accumulation rates. (Data are from reference 53.)*

related to local energy) are not retaining them for long periods of time. Fluxes are significantly larger during the unstratified period, W, from late November through May, than for the stratified period, S, from June through October, for the three offshore stations (4, 8, and 9) in the southern basin. The local sediment accumulation rates for stations 3, 4, 8, and 9 are all 5–10 times less than trap accumulation at 5 m above the bottom. Because these are depositional areas, a net import rather than export of particulate matter is presumed. To account for the large amount of extra mass in the near-bottom traps, a significant amount of local resuspension must occur; a refluxing of the unconsolidated fine-grained surficial sediments a few meters into the water results in this material being collected in the traps. During the winter, this resuspension is more intense, and the effect is extended from a few meters above bottom to throughout the water column.

A systematic development of the particle flux profile occurs throughout the year (Figure 10). Note that the mass flux scale is logarithmic and spans more than 2 orders of magnitude. The dashed line in the figure corresponds to a flux of 10 $g/m^2$ day, which is close to the average BNL flux. As was shown in the previous figure, the flux profile is nearly constant during the isothermal period. After thermal stratification develops, the surface waters become isolated from the resuspendible pool, and the greatly reduced fluxes near the surface are primarily the result of in situ biological processes and the loss of the resuspended component. The fluxes into the near-bottom traps remain high throughout the year as a result of the stability and regular recharging of the BNL.

Of particular interest is the behavior of cesium-137 in this same set of trap materials. Because negligible amounts of the radionuclide are presently being added to the lake, any amount detected in trap samples must have come from the resuspendible pool. The behavior of this signature of bottom-derived materials is shown in Figure 11. During the isothermal period, concentrations of the radionuclide are the same throughout the water column and are characteristic of concentrations in near-surface sediments from a wide area of the lake bottom. As the season progresses, concentrations within the BNL remain at about 20–30 dpm/g, which is the same activity as the local fine-grained sediments. However, above the BNL, concentrations drop drastically. By mid-summer (July 23–August 17) cesium-137 is no longer collected in the traps. This situation persists through the fall until the lake destratifies in late November, and bottom-derived materials once again appear throughout the water column. The characteristics of cesium-137 removal contrast with the calcite precipitation-mediated removal of lead discussed previously.

The assumption of offshore surface-trap isolation from resuspended material during the stratified period allows these traps to be used to

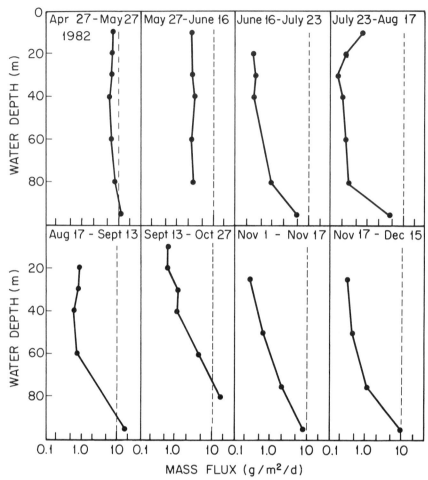

Figure 10. *Seasonal variation in the mass flux profile (log scale) at station 8 (see Figure 9) in southern Lake Michigan during 1982.*

estimate the input of particle-associated materials to the lake (3). If the input of these materials has a large atmospheric component, the use of traps may provide the best method to estimate such loads.

## Local Integration Processes

In two distinct regions adjacent to the sediment–water interface, the particulate matter loses its source identity through spatial and temporal integration processes: resuspension–deposition and bioturbation. At a depth of approximately 10–20 m above the bottom, particles enter the BNL. The BNL is a regular feature in all of the Great Lakes and appears

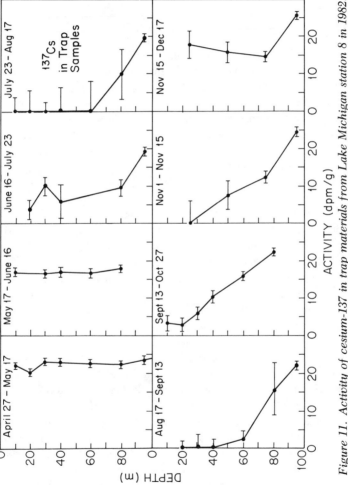

*Figure 11. Activity of cesium-137 in trap materials from Lake Michigan station 8 in 1982.*

to be composed primarily of resuspended sediments (42, 44). Materials enter the BNL from a variety of sources in addition to the particles settling down from above. In the shallow waters of the shelf and slope, surface and internal waves and occasional strong currents resuspend sediments, sorting the particles (29) and transporting them horizontally as well as vertically. Additionally, during the long period when the lakes are not thermally stratified, local resuspension appears even in deeper regions. The cycle of resuspension and redeposition has the effect of producing a resuspended pool composition that is relatively uniform throughout major basins of the lakes. The long-term consequence of this process is the eventual accumulation of particle-associated contaminants in least turbulent areas of the lake bottom. This latter process, termed focusing, operates strongly in the Great Lakes system.

Integration occurs not only in the water column but also in the upper layer of sediments by the process called bioturbation. Upon entering sediments, particles become part of the loose matrix that houses and serves as a food supply to a multitude of benthic organisms. In the Great Lakes, the amphipod *Pontoporeia hoyi* is the predominant organism (45). *P. hoyi* lives at the interface, although it does burrow down a few centimeters and also swims upward a few meters into the water column. This behavior tends to homogenize the upper few centimeters of sediment. Oligochaete worms are another major contributor to sediment mixing in the lakes. These animals characteristically burrow several centimeters into the sediments and spend a significant amount of their time feeding in a head-down position while excreting from tails protruding through the sediment–water interface. This behavior, described as a conveyor-belt process, has the effect of bringing buried materials back, in a particle-selective way, into the resuspendible pool and ultimately homogenizing near-surface deposits.

The process of sediment mixing by these and other zoobenthos has been elucidated in the Great Lakes by measurement of fallout and naturally occurring radionuclides. An early study (46) indicated that sedimentary profiles of lead-210 and cesium-137 included a region over which activities were constant. This mixed-layer zone was subsequently shown (47) to be directly related to the vertical distribution and number of organisms that can mix sediments at a given site. Several studies now show that (1) in general, organisms occur in sufficient numbers throughout the Great Lakes to homogenize near-surface sediments on time scales of years to decades (47–49), and that (2) their range of penetration into sediments is consistent with the depth of the radiometrically determined mixed zone (Figure 12). Sediment-mixed depth is expressed in terms of the cumulative dry weight of sediment. The depth of mixing varies systematically over the lake bottom, tending to be highest in the central regions of depositional basins. This trend is illustrated in Figure

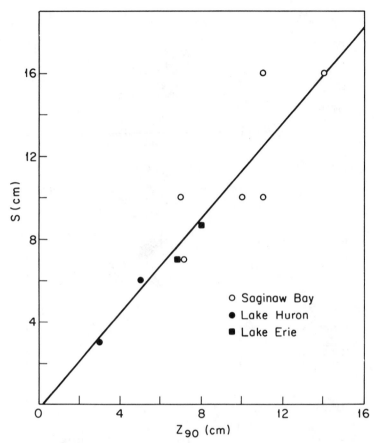

*Figure 12. Relationship between the depth of sediment mixing determined radiometrically (S) and the depth above in which 90% of the benthic macroinvertebrates occur ($Z_{90}$). This relationship implicates zoobenthos as the active agents in the mixing process.*

13, which shows the mixed depth based on cesium-137 for the southern part of Lake Huron (50). This part of the lake possesses two depositional basins separated by a north–south midlake escarpment that is nondepositional. Depositional characteristics such as sedimentation rate (cm/year) and total storage of anthropogenic elements as well as mixed depth tend to be congruent with bathymetric features. In southern Lake Michigan, the mixed depth ranges from 0 to about 6 cm. The depth of mixing correlates well with the flux of particulate organic carbon as shown in Figure 14. Oligochaete abundance is strongly related to this flux and to the mass sedimentation rate (51).

The effects of mixing on the distribution of cesium-137 are illustrated in the three principal depositional basins of Lake Erie (Figure

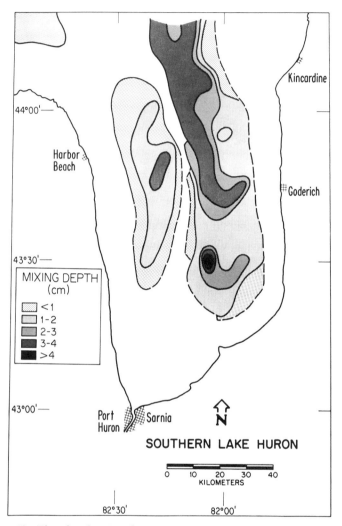

*Figure 13. The depth of sediment mixing in southern Lake Huron as determined from distributions of cesium-137. Deepest mixing tends to occur toward the centers of depositional basins in Lake Huron and the other Great Lakes.*

15). The loading of cesium-137 to the lakes is strongly peaked with most of the fallout occurring during the 7-year period from 1958 through 1965. Cesium-137 distribution in undisturbed sediments should be a narrow pulse. Inspection of Figure 15 shows that no pulse occurs at all in the western basin of Lake Erie (site F4). Radiocesium is more or less uniformly distributed down to 29 cm. This shallow part of the lake is sub-

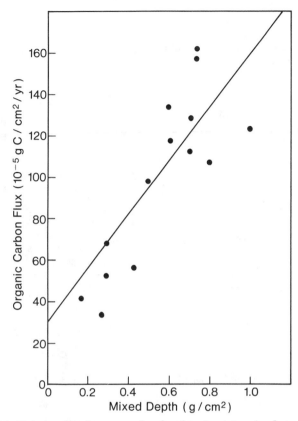

*Figure 14. Relationship between the depths of mixing (g dry sediment/ cm²) and the organic carbon flux.*

ject to strong resuspension and associated physical mixing of sediments. In addition, organisms are abundant, including large Unionid clams, which can mix sediments as deep as 20 cm (52) and together with other organisms, completely obliterate any time sequence information. In contrast, sediments from the central basin (site G16) possess a well-defined cesium-137 peak. However, far more activity of cesium-137 occurs in surface sediments than expected from current levels of fallout. Mixing in this case does not obliterate the distribution but elevates surface concentrations by roughly an order of magnitude. This distribution results from a partial mixing of sediments. The third profile in Figure 15 is a core (MS) collected from the eastern basin of Lake Erie having the highest sedimentation rate found in all the open Great Lakes. This profile clearly illustrates that when sedimentation rates are high, the effects of mixing are reduced, and the correspondence between loading and sedimentary profiles is well preserved.

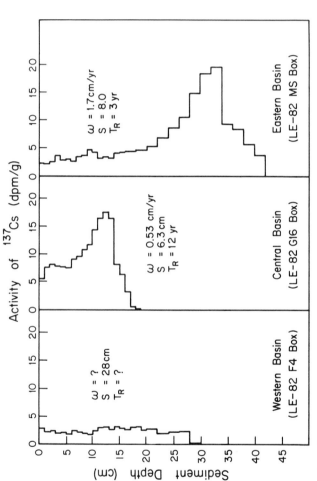

Figure 15. Distributions of cesium-137 in representative cores from the three principal depositional basins of Lake Erie. In the western basin, sediments in shallow water are so thoroughly mixed that the history of deposition of the nuclide is totally obliterated. Comparison of the central and eastern basin profiles indicates the effects of more restricted mixing. The combination of sedimentation rate (ω) and depth of mixing (S) predicts a time resolution ($T_R = S/\omega$) of 12 and 3 years in these cores. Most historical detail is preserved when the time resolution is small.

**Modeling Postdepositional Sediment Mixing.** To account for such distributions, a simple, rapid, steady-state mixing (RSSM) model was developed (*46*). Bioturbation was assumed to homogenize sediments over a zone of fixed depth. As sedimentation continues at a constant rate, the zone moves upward and leaves behind a distribution that is smeared but not subject to further change. This model has been generally useful in describing individual radionuclide and stable element (*53*) distributions, although it represents an extreme simplification of the processes and effects of mixing. Applying the RSSM model to the cesium-137 distribution at site G16 is shown in Figure 16 as the solid line, which closely follows the data. The least-squares model calculates a mixed depth of 6 cm and an average sedimentation rate of about 0.5 cm/year. The shaded portion of the figure is the distribution of cesium-137 expected at this site in the absence of mixing. Comparison of the two model distributions shows that the effect of intense mixing is to displace the peak downward, smooth out the distribution, and elevate surface concentrations. An important additional effect is that such mixing restricts the time resolution with which the contaminant history of the lakes can be inferred

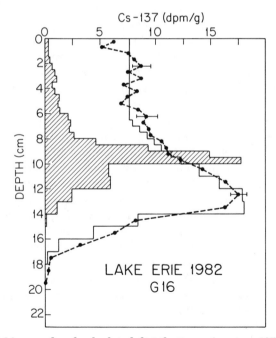

*Figure 16. Measured and calculated distributions of cesium-137 in the central basin core. The rapid steady-state mixing model (RSSM) predicts a distribution (dashed line) that closely follows the data. The unnormalized distribution expected in the absence of mixing is shown as the shaded profile.*

from sedimentary records. For example, events separated by a decade that leave markers in sediments would not be distinguishable when the time resolution is much greater than 10 years. The time resolution is expressed approximately as the ratio of mixed depth to sedimentation rate (49). At site G16, this ratio is about 12 years. Thus, events spaced 10 years apart would be poorly resolved there. In contrast, the eastern basin site (MS) has 3-year time resolution and is a desirable place to obtain accurate historical records. Studies (49, 50) have shown that the time resolution varies systematically within depositional basins and tends to be greatest toward basin margins. If the resuspendible pool is composed primarily of materials transferred from the mixed zone, then the time resolution averaged over the entire lake bottom measures the average time that particle-associated contaminants reside in the pool. Then, for substances whose primary removal pathway is through sedimentation and burial, the ultimate rate of removal from the system may be governed by this average system time resolution.

Extensive literature has been presented on the modeling of post-depositional transport of sediments (54, 55). The RSSM model outlined previously, although generally successful, cannot account for the cases of incomplete mixing. This failure is illustrated in Figure 17, which shows the distribution of selected radionuclides in a core from Lake Michigan. At this site, the mixed depth is roughly 6 cm, and the sedimentation rate is about 0.15 cm/year, so that the RSSM time resolution is about 40 years. Thus, the profile of cesium-137 is dominated by the effects of mixing and the peak is barely evident in the distribution. Distributions of two particle-associated radionuclides with short half-lives, beryllium-7 ($t_{1/2} = 53.5$ days) and cerium-144 ($t_{1/2} = 284$ days), are not uniform but decrease more or less exponentially through the mixed zone. Thus, mixing rates are sufficient to homogenize cesium-137 on time scales of decades but not to homogenize transient tracers on a seasonal or an annual basis. Greater realism in modeling bioturbation necessitates introducing finite mixing rates and accurate representations of zoobenthic interactions with the sediment matrix. One important interaction is that of particle-selective transport. Numerous studies have shown that conveyor-belt deposit feeders such as those found in the Great Lakes preferentially transport fine, organic-rich materials toward the sediment surface (55–57). In principle, this action enhances the transfer of particle-associated contaminants back into the resuspendible pool, but its significance for the overall, long-term lake response is as yet unknown.

**Sediment Focusing and Accumulation.** The focusing of sediment-associated contaminants in the Great Lakes is important and must be considered in interpreting contaminant records, estimating average removal times, and inferring mass balances from sedimentary data. Focus-

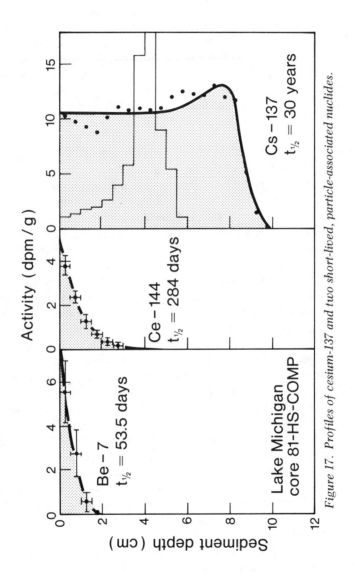

*Figure 17. Profiles of cesium-137 and two short-lived, particle-associated nuclides.*

ing and net sedimentation are the end results of continuous as well as episodic cycles of resuspension and redeposition. In near-shore and shallow waters, wind-driven waves are primarily responsible for resuspending sediments. These materials are then transported some distance, depending on local currents and settling velocities, before they again enter the sediments. In deeper waters, internal waves or currents are the agents that supply sufficient energy to resuspend and further transport the sediments. The results of this process are particle sorting (58, 59) and the accumulation of fine-grained, organic-rich sediments in confined, relatively low-energy, depositional basins within the lake. For particle sorting, the residual sediment is sufficiently large in grain size, high in density, or sufficiently cohesive so that it cannot be moved within the local energy regime.

The extent of focusing is illustrated for cesium-137 in Lake Huron (Figure 18). The delivery of the radionuclide to the lake has been essentially uniform over its surface. The total amount of the radionuclide stored on the lake bottom ($pCi/cm^2$), however, is strongly focused. At certain sites, more than 5 times as much cesium-137 is stored than could be delivered by direct atmospheric fallout. The ratio of amount stored at a given site to the average storage in the lake is termed the focusing factor. Focusing factors for cesium-137 in other lakes can also be as high (46). At sites G16 and MS (see Figure 15) in Lake Erie, focusing factors are 1.5 and 5, respectively. Where comparison has been possible, the focusing factor for one constituent at a given site is comparable to that of another provided they are particle-associated regardless of their other chemical characteristics. This situation is clearly shown (Figure 19) in maps of the distribution of sediment concentrations of various organic contaminants (60–63) in which the primarily atmospherically derived materials (64) are eventually focused into depositional zones that represent only a fraction of the total lake bottom. This situation arises because particle-bound contaminants tend to associate with suites of particles having a range of geochemical characteristics but comparable hydrodynamic properties. Because the similarity of focusing factors is an expression of the fact that contaminants are generally codeposited within depositional basins, the possibility arises of using focusing factors to estimate sedimentary mass balances from a relatively limited set of sediment cores. The extent of variability of focusing factors has not yet been well studied.

Patterns of storage or concentration such as exemplified previously appear to be established in a relatively short time. The map of storage of cesium-137 in southern Lake Michigan (65) that shows dramatic focusing effects was based on cores collected in 1971 and 1972 only 8 years after the period of peak fallout. A study of the same sites a decade later (66) has revealed a small but continuing process of focusing in which storage

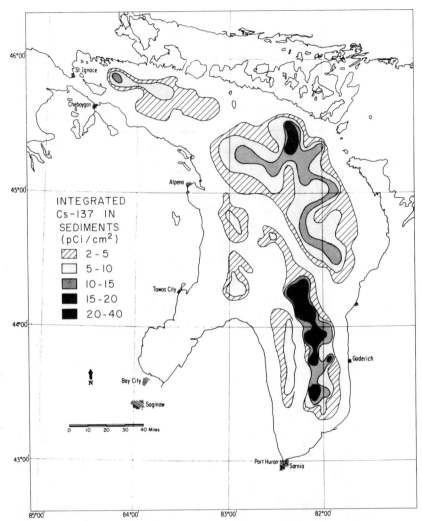

*Figure 18. Storage of cesium-137 in sediments of southern Lake Huron.*

of cesium-137 in the vicinity of the centers of depositional basins is enhanced at the expense of storage in the margins. In contrast, short-lived beryllium-7, introduced continuously from the atmosphere during the year, is not focused in Lake Erie during the summer. Most of the annual input is moved into depositional regions during the period of isothermal mixing. The rapidity of this process indicates that at most sites in the Great Lakes, focusing times will not affect the time resolution in cores. However, in areas of high sedimentation, continuing migration of particles from one site to another (straggling) may result in a measurable smearing of sedimentary distributions.

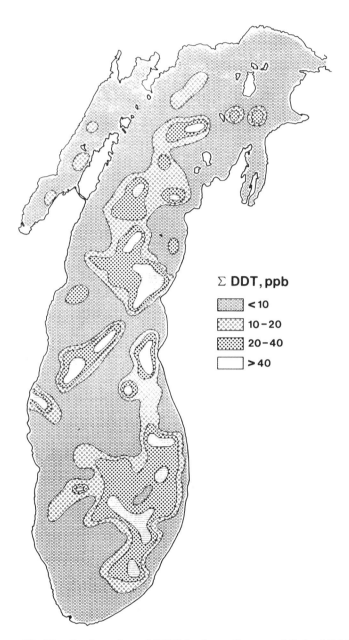

*Figure 19. Distribution of total DDT in the sediments of Lake Michigan. The use of this particle-associated contaminant was stopped nearly 10 years ago. The asymmetric distribution of DDT in surface sediments is a result of focusing. (Data are from reference 60.)*

## Geochronology in Sediment Cores

One of the most important uses of radioisotopes in environmental sciences has been for the dating of sediments. The lead-210 method (67), now established as one of the major geochronological tools (68), is routinely applied in Great Lakes studies (66). The rate of supply of lead-210 to the lakes is essentially constant from year to year and even over a season. Because of its high affinity for particles, lead-210 has a short residence time in the water. In Lake Michigan, the maximum residence time based on the maximum value of fraction dissolved (0.75) is slightly less than 1 year. Thus, lead-210 is rapidly transferred to sediments where its decay ($t_{1/2} = 22.3$ years) on burial produces an activity profile that, under ideal conditions, decreases exponentially with sediment depth. Examples of such profiles are given in Figure 20 for the sites in the central (G16) and eastern (HS) basins of Lake Erie. The data are plotted on a log scale so that, in the case of a uniform sedimentation rate, the activity would decrease linearly with depth. The line through the eastern basin profile is the least-squares fit for a mass sedimentation rate of 0.47 $g/cm^2/year$ (about 1.1 cm/year). On the average, no major change in the rate of sedimentation at this site has occurred over the time period for which the method is valid (about 100 years). In general, radiochronological studies show that sedimentation rates in these large lakes have probably not changed much, if at all, since human alteration of watershed characteristics. However, the eastern basin profile shows significant departures from linearity. Such features as the dip at 50 cm are reproducible from year to year in cores collected from the same site. Departures from linearity are even more clearly seen in the distribution of lead-210 at site G16. Here the least-squares line is omitted, and in its place, line segments are drawn to illustrate that the profile is composed of a series of zones of constant activity interspersed with sections of constant slope. Such features are common in cores and suggest that periods of constant sedimentation are interrupted by events that either mix an interval of sediment or import slugs of sediment to the site. Several previous studies (68, 69) indicate that the flat zones correspond to the occurrence of major storms on the lakes. Thus, like focusing, net sedimentation is apparently the result of both continuous and episodic processes. From the figure, lead-210 and cesium-137 at site G16 clearly share a common interval at the sediment–water interface over which their activity is constant. Both nuclides appear to be mixed to the same depth in this core. However, in some cores (46, 47, 59) depths of mixing for the two nuclides are significantly different, probably because incomplete mixing affects the distribution of steady-state and transient tracers differently.

Distributions of particle-associated contaminants and tracers in lead-210-dated cores serve at least two useful purposes: (1) they can verify

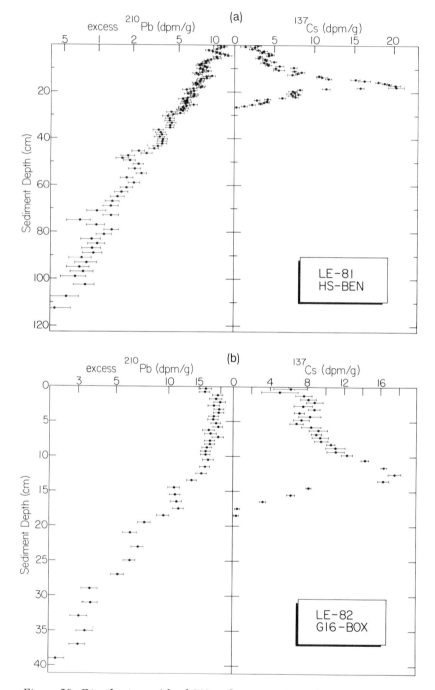

*Figure 20. Distributions of lead-210 and cesium-137 in the eastern (a) and central (b) basin cores from Lake Erie.*

models of sediment transport and (2) provide estimates of the history of contaminant loading to the lakes. Distributions of cesium-137 in each of the Lake Erie cores illustrate this first point. Independent application of the RSSM model to lead-210 and cesium-137 distributions yields self-consistent values for sedimentation and mixing parameters. The second point is illustrated by Figure 21, which shows the distribution of stable lead in the eastern basin core. The solid line that closely follows the data

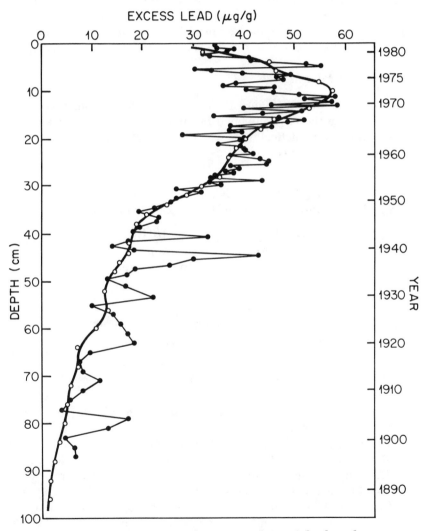

Figure 21. Concentration of anthropogenic (excess) lead in the eastern basin core (LE–81–HS). The dates indicated are based on lead-210 (see Figure 20).

points is the distribution of lead expected based on regional lead use (70) and the sedimentation and mixing characteristics based on lead-210 and cesium-137 measurements for this core. The model and data show the decline of lead in the sediments following the increasing use of unleaded gasoline in the region. Because the focusing factor for radiolead is known for this site, and the loading of radiolead to the lake is well known, the absolute loading of lead to the lake can be inferred from this core and compared (favorably) with similar estimates based on cores collected elsewhere in the lake. Because stable lead is comparatively easy to measure in sediments but not in water, the sediments provide the best measure of the validity of models representing the time dependence of concentrations in the lake.

## Contaminant Fate Model

The conceptual model (Figure 1), representing the combination of processes affecting the environmental behavior of trace contaminants, can be expressed mathematically for water column and mixed sediments by equations 5 and 6, respectively.

$$\frac{V\,dC_T}{dt} = \underset{\text{(inputs)}}{\Sigma Q_i C_i + I} \;-\; \underset{\text{(outflows)}}{\Sigma Q_o C_T} \;-\; \underset{\text{(decomposition)}}{V\Sigma k_j C_T}$$

$$-\,\underset{\text{(sedimentation)}}{\frac{AC_T F_P\,(S+R)}{\text{TSM}\cdot\rho}} + \underset{\text{(resuspension)}}{\frac{ARC_{\text{TS}}}{\rho}} \tag{5}$$

$$ZA\,\frac{dC_{\text{TS}}}{dt} = \underset{\text{(sedimentation)}}{\frac{A(S+R)F_p C_T}{\rho\cdot\text{TSM}}} - \underset{\text{(resuspension)}}{\frac{ARC_{\text{TS}}}{\rho}}$$

$$-\,\underset{\text{(decomposition)}}{ZA\Sigma K_j C_{\text{TS}}} - \underset{\text{(burial)}}{\frac{ASC_{\text{TS}}}{\rho}} \tag{6}$$

The terms for equations 5 and 6 are defined in the glossary of terms at the end of this chapter.

This model has been calibrated by using the 30-year monthly fluxes of fallout radionuclides (71) and available Great Lakes water column data (72), an exercise that allows us to select appropriate average values for Lake Michigan from within the observed range for sediment accumulation rate (0.8 mg/cm$^2$/month), sediment mixed depth (1 cm), and rate of resuspension (0.9 and 9.0 mg/cm$^2$/month in summer and winter, respectively). Other models of varying complexity that account for the behavior of fallout radionuclides in the Great Lakes also use radionuclides as a calibration tool (2, 36, 73, 74).

Application of the calibrated model to the concentration of strontium-90 in Lake Michigan yields the results shown in Figure 22. In the case of strontium-90, $K_d$ is taken to be $10^3$. Because virtually no removal by particle settling occurs, concentrations have simply built up over time, and the primary removal mechanism is radioactive decay. In the last 20 years, the amount of the radionuclide in the water has decreased very little. In contrast, concentrations of total cesium-137 in the lake (Figure 23) have decreased by about an order of magnitude since cessation of atmospheric testing. The data shown in the figure include all concentrations associated with a well-mixed lake and normalized values of plutonium concentrations and cesium-137:potassium ratios in alewife fish, which track very well with model predictions. The curve through the data points includes the effects of resuspension. If resuspension were negligible, and cesium-137 had an infinite $K_d$, the model would predict an order of magnitude lower than measured concentration in the lake by 1980. The recent data (post-1975) from the lakes show that resuspension or other mechanisms (70) involving transport from the bottom must be invoked to account for the persistence of fall-

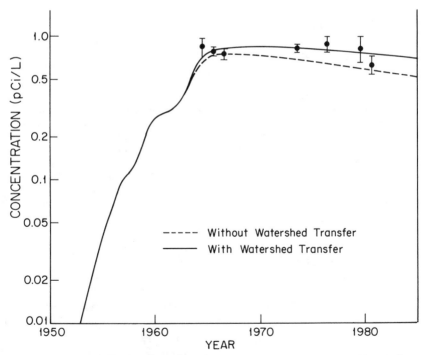

*Figure 22. Measured and predicted concentrations of strontium-90 in Lake Michigan. Because of the low affinity of this fallout radionuclide for particulate matter, its concentration in the water has simply built up in time and is being reduced primarily by radioactive decay.*

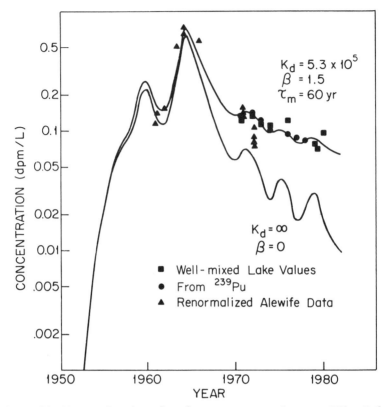

*Figure 23. Measured and predicted concentrations of cesium-137 in Lake Michigan. This radionuclide has essentially the same loading to the lake as strontium-90, but because of its affinity for particles has been largely transferred to sediments. The solid curve closely following the data is the calibrated model fit that also appears to reproduce the trend in cesium-137:potassium ratios in alewife, a planktivorous fish. The curve that undershoots the data, starting especially in the 1970s, is the prediction if all radiocesium were on particles and there was no resuspension. In the last decade, remobilization from sediments via resuspension has been the primary source of this radionuclide in the water column.*

out radionuclides. The model also predicts concentrations in the resuspendible pool (Figure 24); cesium-137 in traps sampling the BNL are shown together with renormalized cesium-137:potassium values for sculpin, which track well with the model values. Because the resuspendible pool residence time is 60 years in this simulation, decreases of cesium-137 activity are dominated by radioactive decay.

Application of the model to a stable element contaminant contrasts markedly with the fallout radionuclide case. The model concentration of lead is shown in Figure 25 along with all known measurements of lead in

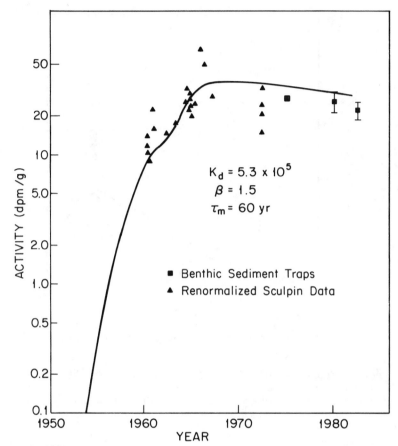

*Figure 24. Measured and predicted concentrations of cesium-137 in the resuspendible pool. Because of the long residence time of radiocesium in the pool, concentrations decrease slowly with time.*

water of Lake Michigan (*38*). The model result is displayed as an envelope that corresponds to the range of measured $K_d$ values and approximates the uncertainty in estimating total lead concentrations. System model parameters are fixed by the radionuclide calibration. The absolute loading of lead was determined by applying focusing factors (discussed previously) to sediment inventories of lead. The time dependence of lead inputs was based on regional lead use and confirmed by comparison with profiles in dated cores. The comparison of model results with observation is hopeless. First, the model calculation predicts the increase in lead due to anthropogenic loading but does not include presumably small natural background concentrations. Second, measurements of lead in Great Lakes water began with a few dubious mea-

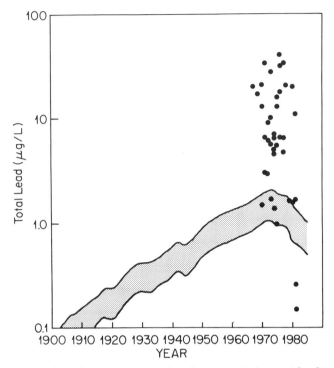

*Figure 25. Predicted concentration of anthropogenic (excess) lead in Lake Michigan. Although natural levels of lead contribute a small but uncertain additional amount to the lake, large sources of difficulty in comparison with measurements appear to be analytical and strategic.*

surements in the 1960s and continued in earnest only after 1970. Hence, most of the long history of increase of this contaminant in the system is undocumented except for the sedimentary record. Third, concentration measurements are tremendously scattered. Rossmann (38) summarized existing total lead measurements in water and found a range of values from 0.00048 to 8600 μg/L. This range is an impressive $10^7$-fold variation in values. The majority of this variability can be attributed to analytical error and sample contamination, but some is due to sampling at relatively arbitrary and inconsistent sites and depths and generally at the convenience of the experimenter (summer) without regard for the representatives of the sampling. Studies of radiolead and other tracers have shown that large natural fluctuations in levels occur as a result of seasonally dependent geochemical cycles and the changing composition of particulate matter in the lakes. Thus, reported values for lead cannot be suitably interpreted.

For organic contaminants, the only significant long-term data, other than sediment cores, are concentrations in fish. Figure 26 illustrates con-

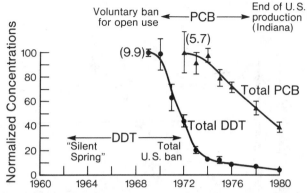

*Figure 26. The normalized concentrations of total DDT and PCBs in Lake Michigan bloater chubs. The actual maximum values (ppm) are shown in parentheses.*

centrations of two hydrophobic organic compounds, PCBs and DDT in Lake Michigan chubs. The insecticide DDT was manufactured and used in the United States until 1972. The maximum concentration (9.9 ppm) in bloater chub (midtrophic level, planktivore) declined rapidly ($t_{1/2} \sim$ 2.2 years) and has now leveled off at approximately 1 ppm (75). A few years later, a similar, but less-rapid, decline ($t_{1/2} \sim$ 5.5 years) began for PCBs in these fish. Most recent values (1982) are not statistically different from 1980 values. The long tail in the observed decline in concentration is a result of the sediment–water coupling processes discussed previously. Models of these processes are necessary both to interpret observed data and to simulate long-term lakewide responses to alternative actions.

A major value of numerical models is the ability to examine system responses to alterations in the modeled processes and sensitivity to changes in process rates. Our radionuclide-calibrated model was run to examine time-dependent responses of the lake to surrogate contaminants with a wide range of partition coefficients (Figure 27). The figure represents a simulation of Lake Michigan's response to spiked loads of persistent contaminants introduced for only one time step. For a compound with a $K_d$ of $10^3$ (approximate solubility 0.1–1 mg/L), the time to reach one-half of the initial concentration is approximately 54 years. For a more hydrophobic compound ($K_d = 10^5$; approximate solubility 1–10 $\mu$g/L), the calculated half-time is only 3 years, which is similar to the DDT and PCBs just mentioned and some polycyclic aromatic hydrocarbon data (76). The lower part of the figure illustrates the time-dependent behavior of these model materials within the mixed layer of sediment. Unfortunately, insufficient data exist for hydrophobic organic compounds in the waters of the Great Lakes to compare with these

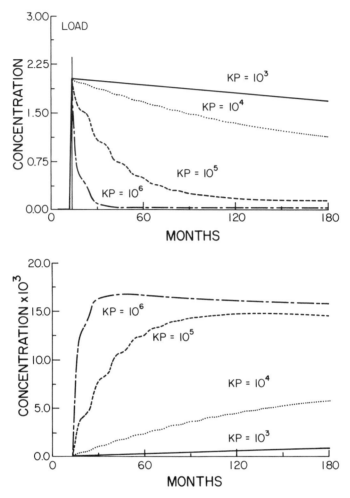

*Figure 27. The calibrated Lake Michigan model response to a spiked load of trace contaminants with varying values of $K_d$. The upper panel represents the water column; the lower panel represents the mixed sediments. Contaminants with $K_d > 10^5$ are rapidly removed from the water and build up in the resuspendible pool. Long-term changes in concentration are extremely sensitive to $K_d$ over the range that includes nearly all contaminants of interest. Strong seasonal variability in values of $K_d$ makes the choice of a representative $K_d$ for use in long-term fate models difficult.*

model results. The model is an attempt to represent our current understanding of the combination of processes mediating the behavior of hydrophobic compounds. Examination of such simulations gives some feeling for the response time of the system and for the relative importance of various processes. Well-conceived research and monitoring efforts will eventually provide the necessary data to validate and improve our predictive capability.

## Abbreviations and Symbols

| | |
|---|---|
| $\phi$ | porosity |
| $\rho$ | dry sediment density |
| $A$ | area |
| $C_D$ | dissolved contaminant concentration |
| $C_i$ | total contaminant concentration in inflow |
| $C_p$ | particle-associated contaminant |
| $C_{pw}$ | contaminant concentration in pore water |
| $C_s$ | contaminant concentration in sediment solids |
| $C_T$ | total (dissolved and particulate) contaminant concentration |
| $C_{TS}$ | total contaminant concentration in sediment |
| $f_p$ | fraction of contaminant associated with particulate matter |
| $f_{ps}$ | fraction of contaminant associated with sediment solids |
| $I$ | sum of other inputs (nonpoint and atmospheric sources) |
| $k$ | first-order decomposition rate |
| $K_d$ | degree of partitioning of a contaminant between dissolved and particulate phases |
| $K_{oc}$ | correlation coefficient of extent of association to the organic carbon fraction of the particulate matter |
| $Q_i$ | rate of inflow |
| $Q_o$ | rate of outflow |
| $R$ | resuspension rate |
| $S$ | net sedimentation rate |
| $t_{1/2}$ | half-life |
| $T_a$ | apparent residence time with respect to removal by particle settling |
| $T_s$ | mean lifetime of a particle settling through the water column |
| TSM | total suspended matter |
| $V$ | volume |
| $Z$ | thickness of sediment-mixed layer |

## Acknowledgments

We thank Kjell Johansen for his help in developing a regional source function for lead and for assistance in measurement of cesium-137 activities in Lake Huron sediments. Thanks are due to Jerry Bell for his help with our sediment traps, to N. Hawley for his measurement of TSM and transparency, and to E. Hugel for her analysis of lead-210 in water samples from Lake Michigan. We are grateful to R. Bourbonnier, A. Mudroch, and others on the staff of the Canada Centre for Inland

Waters for the use of the *CSS Limnos* to collect sediment cores from Lake Erie. We acknowledge the help of R. Rossman and E. Meriweather in determination of lead-210 and stable lead in sediment samples from Lake Erie. This manuscript was substantially improved by the thoughtful reviews and comments of Peter Sly, Peter Landrum, and Wayne Gardner. This chapter was GLERL contribution number 478.

## References

1. *Proceedings of the 2nd International Symposium on the Interactions between Sediments and Freshwater;* Sly, P. G., Ed.; Dr. W. Junk: Boston, 1982.
2. *Transuranic Elements in the Environment;* U.S. Department of Energy: Washington, DC, 1980; DOE/TIC-22800.
3. Eadie, B. J.; Chambers, R.; Gardner, W.; Bell, G. *J. Great Lakes Res.* **1984,** *10,* 307–321.
4. Karickhoff, S. W. *Chemosphere* **1981,** *10,* 833–846.
5. Bell, G. L.; Eadie, B. *J. Great Lakes Res.* **1983,** *9,* 559–567.
6. Strong, A.; Eadie, B. *Limnol. Oceanogr.* **1978,** *23,* 877–887.
7. Elzerman, A., in this book.
8. Kenaga, E. E.; Goering, C. *Proceedings of the 3rd Symposium on Aquatic Toxicology;* American Society for Testing and Materials: Philadelphia, PA, 1979; pp 78–115.
9. Means, J. C.; Hassett, J.; Wood, S.; Banwart, W. In *Polynuclear Aromatic Hydrocarbons;* Ann Arbor Science: Ann Arbor, MI, 1979; p 327.
10. Herbes, S. E. *Water Res.* **1977,** *2,* 493–496.
11. Karickhoff, S. W.; Brown, D.; Scott, T. *Water Res.* **1979,** *13,* 241–248.
12. Verschveren, K. *Handbook of Environmental Data on Organic Chemicals;* Van Nostrand Reinhold: New York, 1982.
13. Great Lakes Water Quality Board Report to the International Joint Commission; International Joint Commission: Windsor, Ontario, Canada, 1984.
14. Rice, C. P.; Eadie, B.; Erstfeld, K. *J. Great Lakes Res.* **1982,** *8,* 265–270.
15. Eadie, B. J.; Landrum, P.; Faust, W. In *PAH X;* Battelle: 1985.
16. Capel, P. D.; Eisenreich, S. J. *J. Great Lakes Res.* **1985,** *11,* 447–462.
17. O'Connor, D. J.; Connolly, J. *Water Res.* **1980,** *14,* 1517–1523.
18. Eadie, B. J.; Rice, C.; Frez, W. In *Physical Behavior of PCBs in the Great Lakes;* MacKay, D., Ed.; Ann Arbor Science: Ann Arbor, MI, 1982; p 213.
19. Ditoro, D. M. *Chemosphere* **1985,** *14,* 1503–1538.
20. Eadie, B. J.; Landrum, P. F.; Faust, W. *Chemosphere* **1982,** *11,* 847–858.
21. Landrum, P. F.; Reinhold, M.; Nihart, S.; Eadie, B. *Environ. Toxicol. Chem.* **1985,** *4,* 459–467.
22. Scavia, D. S. *Ecol. Modell.* **1979,** *8,* 49–78.
23. Vanderploeg, H. A. *Can. J. Fish. Aquat. Sci.* **1981,** *38,* 504–517.
24. Evans, M. Report to Great Lakes Environmental Research Laboratory/National Oceanic and Atmospheric Administration, Ann Arbor, MI, 1985.
25. Prahl, F. G.; Carpenter, R. *Geochim. Cosmochim. Acta* **1979,** *43,* 1959–1972.
26. Deuser, W. G.; Brewer, P.; Jickells, T.; Commeau, R. *Science (Washington, D.C.)* **1983,** *219,* 388–391.
27. Honjo, S. *J. Mar. Res.* **1978,** *36,* 469–492.
28. Tamura, T.; Jacobs, D. *Health Phys.* **1960,** *2,* 391.
29. Nelson, D. M.; Karttunen, J. ANL-83-100, Part III; Argonne National Laboratory: Argonne, IL, pp 41–44.

30. Nelson, D. M.; Karttunen, J.; Orlandini, K.; Larsen, R. ANL–80–115, Part III; Argonne National Laboratory: Argonne, IL, pp 19–25.
31. Nelson, D. M.; Karttunen, J.; Mehlhoff, P. ANL–81–85, Ecology; Argonne National Laboratory: Argonne, IL, pp 48–52.
32. Alberts, J. J.; Wahlgren, M.; Reeve, C.; Jehn, P. ANL–75–3, Part III; Argonne National Laboratory: Argonne, IL, pp 103–112.
33. Wahlgren, M. A.; Nelson, D. ANL–76–88, Part III; Argonne National Laboratory: Argonne, IL, pp 56–60.
34. Alberts, J. J.; Wahlgren, M. *Environ. Sci. Technol.* 1981, 5, 94–98.
35. Lerman, A.; Taniguchi, H. *J. Geophys. Res.* 1972, 77, 474–484.
36. Lerman, A. *J. Geophys. Res.* 1972, 77, 3256–3262.
37. Nelson, D. M.; Metta, D.; Karttunen, J. ANL–83–100, Part III; Argonne National Laboratory: Argonne, IL, pp 45–54.
38. Rossman, R. Special Report No. 108; Great Lakes Research Division; The University of Michigan: Ann Arbor, MI, 1984.
39. Hugel, E. A.; Robbins, J. A.; Eadie, B. Presented at the 28th Conference on Great Lakes Research, International Association for Great Lakes Research, Milwaukee, WI, 1985.
40. Hawley, N.; Robbins, J. A.; Eadie, B. *Geochim. Cosmochim. Acta* 1985, 50, 1127–1131.
41. *Transuranic Elements in the Environment;* U.S. Department of Energy: Washington, DC, 1980; DOC/TIC–22800.
42. Chambers, R. L.; Eadie, B. *Sedimentology* 1981, 28, 439–447.
43. "The Cycling of Toxic Organics in the Great Lakes: A 3-year Status Report"; Technical Memorandum ERL GLERL–45; National Oceanic and Atmospheric Administration: 1983.
44. Sandilands, R. G.; Mudroch, A. *J. Great Lakes Res.* 1983, 9, 190–200.
45. "Macrobenthos of Southern Lake Michigan"; Data Report ERL GLERL–28; National Oceanic and Atmospheric Administration: 1985.
46. Robbins, J. A.; Edgington, D. *Geochim. Cosmochim. Acta* 1975, 39, 285–304.
47. Robbins, J. A.; Krezoski, J.; Mozley, S. *Earth Planet. Sci. Lett.* 1977, 36, 325–333.
48. Fisher, J. B.; McCall, P.; Lick, W.; Robbins, J. A. *J. Geophys. Res.* 1980, 85, 3997–4006.
49. Robbins, J. A. In *Proceedings of the 2nd International Symposium on the Interactions between Sediments and Freshwater;* Sly, P. G., Ed.; Dr. W. Junk: Boston, 1982; p 611.
50. *Sediments of Southern Lake Huron: Elemental Composition and Accumulation Rates;* Ecological Research Series; U.S. Environmental Protection Agency. U.S. Government Printing Office: Washington, DC, 1980; EPA–600/3–80–080.
51. Keilty, T.; Robbins, J. A. *Abstract,* 46th Annual Meeting of the American Society of Limnology and Oceanography, 1983.
52. McCall, P. L.; Tevesz, M. *J. Great Lakes Res.* 1979, 5, 105–111.
53. Edgington, D. N.; Robbins, J. A. *Environ. Sci. Technol.* 1976, 10, 266–274.
54. Berner, R. A. *Early Diagenesis;* Princeton University: Princeton, NJ, 1980.
55. Robbins, J. A. *J. Geophys. Res.* 1986, 91, 8542–8558.
56. Karickhoff, S. W.; Morris, K. *Environ. Sci. Technol.* 1985, 19, 51–56.
57. Krezoski, J. R.; Robbins, J. A. *J. Geophys. Res.* 1985, 70, 11999–12006.
58. Sly, P. G.; Thomas, R. L.; Pelletier, B. R. In *Proceedings of the 2nd International Symposium on the Interactions between Sediments and Freshwater;* Sly, P. G., Ed.; Dr. W. Junk: Boston, 1982; pp 71–84.
59. Sly, P. G. In *Lakes: Chemistry, Geology, Physics;* Lerman, A., Ed.; Springer-Verlag: New York, 1978.

60. Frank, R.; Thomas, R. L.; Braun, H. E.; Gross, D. L.; Davies, T. T. *J. Great Lakes Res.* **1981**, *7(1)*, 42–50.
61. Frank, R.; Thomas, R. L.; Braun, H. E.; Rasper, J.; Dawson, R. *J. Great Lakes Res.* **1973**, *6*, 113–120.
62. Frank, R.; Thomas, R. L.; Holdrinet, M.; Kemp, A.; Braun, H. *J. Great Lakes Res.* **1968**, *5*, 18–27.
63. Cahill, R. A.; Shimp, N. F. In *Toxic Contaminants in the Great Lakes;* Nriagu, J., Ed.; Wiley: New York, 1984; pp 393–423.
64. Eisenreich, S. J.; Looney, B.; Thornton, J. *Environ. Sci. Technol.* **1981**, *15*, 30–38.
65. Edgington, D. N.; Robbins, J. A. In *Environmental Biogeochemistry;* Nriagu, J., Ed.; Ann Arbor Science: Ann Arbor, MI, p 705.
66. Edgington, D. N.; Benante, J.; Paddock, R.; Robbins, J. A. *Abstract,* Ocean Science Meeting of the American Geophysical Union and the American Society for Limnology and Oceanography, 1984.
67. Krishnaswami, S.; Lal, D.; Martin, J.; Meybeck, M. *Earth Planet. Sci. Lett.* **1971**, *11*, 407–414.
68. Robbins, J. A. In *Biogeochemistry of Lead in the Environment;* Nriagu, J., Ed.; Elsevier Scientific: Amsterdam, the Netherlands, 1978; p 285.
69. Robbins, J. A.; Edgington, D.; Kemp, A. *Quat. Res.* **1978**, *10*, 256–278.
70. Eisenreich, S. J.; Metzer, N.; Urban, N.; Robbins, J. A. *Environ. Sci. Technol.* **1986**, *20*, 171–174.
71. "Great Lakes Regional Fallout Source Functions"; Technical Memorandum ERL GLERL-56; National Oceanic and Atmospheric Administration: 1985.
72. "The Coupled-Lakes Model for Estimating the Long-Term Response of the Great Lakes to Time-Dependent Loadings of Particle-Associated Contaminants"; Technical Memorandum ERL GLERL-57; National Oceanic and Atmospheric Administration: 1985.
73. Tracy, B. L.; Prantl, F. *Water, Air, Soil Pollut.* **1983**, *19*, 15–27.
74. Thomann, R. V.; DiToro, D. *J. Great Lakes Res.* **1983**, *9*, 474–496.
75. Great Lakes Water Quality Board Report to the International Joint Commission; International Joint Commission: Windsor, Ontario, Canada, 1985.
76. Eadie, B. J. In *Toxic Contaminants in the Great Lakes;* Nriagu, J., Ed.; Wiley: New York, 1984; Vol. 14.

RECEIVED for review May 6, 1986. ACCEPTED October 2, 1986.

# Sediments as Archives
# of Environmental Pollution Trends

M. Judith Charles[1] and Ronald A. Hites[2]

School of Public and Environmental Affairs and Department of Chemistry,
Indiana University, Bloomington, IN 47405

*Retrospective measurements of dated sediment cores can be
used to determine the effect of regulations to control environmental inputs of hazardous chemicals. Sediment cores have been
used to reconstruct histories of environmental contamination by
mercury, lead, polycyclic aromatic hydrocarbons (PAHs), dioxins,
polychlorinated biphenyls (PCBs), DDT, octachlorostyrene, pH,
and carbon particles. These measurements have shown that most
anthropogenic chemicals first appeared in sediments at the turn
of the century at the time of the industrial revolution in most of
North America. Industrial growth after World War II also
resulted in inputs of pollutants such as PCBs and dioxins. Recent
decreases in the levels of several contaminants may have occurred because of environmental awareness and the onset of environmental regulations. Inputs of banned pollutants continue,
however, because of recycling of contaminants in the environment and long-range atmospheric transport.*

T O CONTROL THE INPUT OF HAZARDOUS CHEMICALS into the environment, many regulations have been implemented in the past 15 years.
As a result, industry and government have started monitoring programs
to prove or disprove compliance. Choosing which chemicals to monitor
depends on current technology, industrial processes, and health effects.
Advances in these areas are constantly occurring; thus, obscure contaminants discharged a decade ago are of grave concern today. Unfortunately, no data are available for such contaminants against which to

[1]Current address: California Public Health Foundation, P.O. Box 520, 2151 Berkeley
Way, Berkeley, CA 97404
[2]To whom correspondence should be addressed

0065-2393/87/0216-0365$07.25/0

judge the success of new regulations. Thus, a retrospective method is needed to answer questions about sources and inputs of contaminants after the fact. Historical sediment core measurements are such a method. This method relies on the collection of data from dated sediment cores and information about local and regional industries.

Obtaining accurate historical information from sediment cores is dependent on the rapid transfer of chemical contaminants from their sources to aquatic systems and then to the bottom sediments. Subsequent burial of these materials creates a stratigraphic record that reflects their time of deposition and environmental concentrations. Chemical or biological substances can be used as time markers, and they are usually associated with particulates in the water column. Ideally, these materials should not diffuse or undergo biological mixing. In cases where this does occur, the data must be cautiously interpreted and may require the use of mathematical models that consider these movements. Deposition histories of compounds that degrade or are altered in the environment can be reconstructed if their degradation products are stable, as in the case of the insecticide 1,1'-(2,2,2-trichloroethylidene)bis[4-chlorobenzene] (DDT), or by using analytical methods that convert altered species to their original form, as in the acid digestion of alkylated metals.

Dating of recent sediment core sections is usually done by measuring radionuclides (e.g., cesium-137 or lead-210) or by the presence of time markers (e.g., pollen) that reflect changes in the ecosystem or sources of input. After dating and chemical analyses of the core sections, the data are analyzed to establish chronologies of inputs or to determine fluxes of contaminants to the sediments. As part of this analysis, inferences can be made about sources. Once these inferences are identified and trends over time are known, environmental histories can be reconstructed.

In this chapter, the approaches and techniques used to conduct retrospective, sediment core measurements will be presented. Reconstructed histories will be discussed that have furthered our knowledge about environmental contaminants. Data from the Great Lakes will be emphasized. In this chapter, we will show that sediment core measurements are effective for establishing chronologies of environmental contaminants and for evaluating the effect of environmental regulations.

## Dating Techniques

Assigning dates to intervals of a sediment core is commonly done by interpreting vertical profiles of the natural radionuclide lead-210; the anthropogenic radionuclides cesium-137, plutonium-238, or plutonium-

239 and -240; pollen; or varves. Of the radiometric methods, dating by lead-210 is usually the most accurate method and is based on the specific activity of this nuclide in sediments as a function of depth. The other methods rely on identifying specific horizons. In all these cases, correctly interpreting the vertical distribution of the markers is the crux of accurate dating. Many studies evaluating the lead-210 dating technique provide examples of how this interpretation should be done and demonstrate the problems inherent to all the methods. For this reason, a discussion of the lead-210 method will be presented first with an emphasis on approaches used to obtain reliable dates.

**Lead-210.** The lead-210 technique has been used to date lake, estuary, and marine sediments (*1–5*). The half-life ($t_{1/2}$) of lead-210 (22.26 years) makes the technique attractive for retrospective studies of anthropogenic inputs over the last 150 years.

Lead-210, a member of the uranium-238 decay series, is produced from the decay of radium-226 ($t_{1/2} = 1622$ years) in soils to radon-222 ($t_{1/2} = 3.83$ days), a gas. Atmospheric lead-210 originates by disintegration of radon-222 after its diffusion to the air. In sediments, the primary source of lead-210 is atmospheric deposition. Other sources of lead-210 include indirect atmospheric fallout via the drainage basin and decay of radon-222 in the water column. These three sources are called the excess or "unsupported" lead-210. The "supported" lead-210 originates from the decay of radium-226 that reached sediments by soil erosion. The unsupported lead-210 is used for dating. The specific activity of the unsupported lead-210 is determined by the difference between the total and supported lead-210. The supported lead-210 is estimated from either the activity of lead-210 at sufficient core depths or by measuring the activity of radium-226 and then calculating the activity of lead-210, assuming secular equilibrium between these isotopes.

In dating, specific activity versus depth profiles are used to calculate sedimentation rates. Once these rates are obtained, dates are assigned to sections of the core. Ideally, the profile of the unsupported lead-210 decreases exponentially with depth; thus, a semilog plot of activity versus depth should be linear. In these cases, rates of sedimentation are calculated by using a simple model based on the assumptions that lead-210 rapidly deposits to the bottom sediment, that the flux of the unsupported lead to the sediment is constant, that the mass sediment accumulation rate is constant, and that no migration of lead-210 occurs after deposition (*4–7*). Difficulties in interpreting the profiles arise because of compaction, mixing by bottom currents, bioturbation, migration of lead-210 or polonium-210 (the alpha daughter of lead-210), or decreased activity of lead-210 at the surface due to dilution by organic matter or carbonates or due to slumps or erosion. Models developed by Robbins (8) consider these factors by expanding the simple model so that variables are included that consider changes in particle and lead-210 fluxes, sediment accumulation rates, and migration of lead-210 or polonium-210. These models will not be discussed in detail, but lead-210 sediment core profiles representing some of these situations will be discussed to aid in identifying anomalies.

Porosity changes in sediments usually do not cause obviously anomalous lead-210 profiles, but these changes must be considered. They result from the weight of overlying sediment such that porosity decreases in deeper layers. In assigning dates, compaction would be a source of error because a given time interval will span a greater depth in a top layer than in a deeper layer. By expressing depth in terms of cumulative dry weight of sediment, the effects of compaction on the lead-210 distribution are eliminated.

Mixing or variability in the sediment flux or accumulation rate is indicated by nonmonotonically decreasing lead-210 activity with depth. Identifying the cause of these anomalies is usually difficult unless other information is available. For example, deviations from exponential decay due to physical or biological mixing were observed in sediment cores taken from the Chesapeake Bay (2). Profiles for two of these cores are presented in Figure 1. Exponential decay was observed for one core (CHSP 7505-1416) at 3 cm/year, whereas an anomalous profile was observed in another core (CHSP 7505-1411) at 0.5 cm/year. The flat section of this profile may have been caused by an influx of sediment during Hurricane Agnes.

Slumping is indicated by a break in a linear profile. In Figure 2, lead-210 data showing a slump in two sediment cores from the Saguenay Fjord in Canada are depicted. The slumped region is indicated by the plateau and is characterized by a compacted layer without worm channels. Calculated sedimentation rates in these cores were based on two models, the constant flux model for dating the surficial layers and the constant initial concentration model to date the bottom layers. The constant flux model is used when the flux of lead-210 is constant, but the sediment accumulation rate varies with time. The constant initial concentration model is employed when the flux of lead-210 to the sediments and the sediment accumulation rate change proportionally so that a constant concentration of lead-210 is delivered to the sediments (9).

These examples show that interpreting lead-210 profiles is not always easy. Analyzing these profiles requires integrating information on meteorology, local discharges, and sedimentation processes. Often, sediment cores taken from a given lake or area will differ from each other because of at least one of these variables or because of sediment focusing, which favors sediment accumulation in deeper regions of the

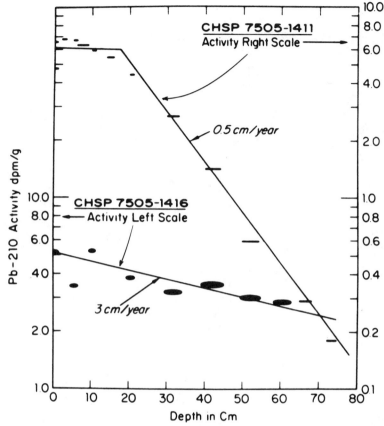

*Figure 1. Lead-210 activity in two Chesapeake Bay sediment cores as a function of depth. (Reproduced with permission from reference 2. Copyright 1978, Pergamon Press.)*

basin (*10*). Once anomalous profiles are recognized and their causes established, an appropriate model can be chosen that best represents the physical or biological phenomenon responsible for these deviations. Because these phenomena may affect the dating by any method, more than one dating technique should be used in assigning dates.

**Cesium and Plutonium Isotopes.**    Cesium-137 has been widely used to date sediment cores and to reconstruct histories of pollutants. This anthropogenic radionuclide was emitted into the atmosphere by nuclear weapons testing, and its deposition in sediments reflects the history of these tests. Cesium-137 first appeared in sediments in the early 1950s and decreased after 1963 because of the nuclear test ban treaty between the United States and the Soviet Union. Primary and

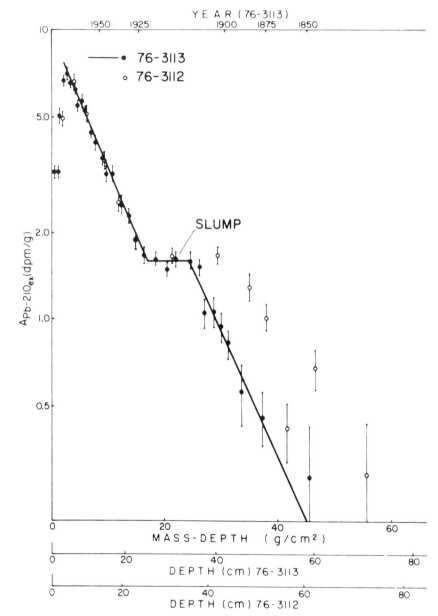

*Figure 2. Unsupported lead-210 activity in sediments from Saguenay Fjord, Quebec. The "slump" occurred in 1917 ± 6 years. (Reproduced with permission from reference 9. Copyright 1980, Pergamon Press.)*

secondary peaks of cesium-137, associated with intense fallout from the 1963 and 1958 test periods, are used as time markers. Usually, the assumption is made that sediment accumulation rates were constant from 1950 to 1958, from 1958 to 1963, and from 1963 to the present, and slopes of lines between these dates are used to calculate sedimentation rates. Clearly, the cesium-137 method is useful only over the last 35-40 years.

Ideal and anomalous cesium-137 profiles are presented in Figure 3. The ideal profile is presented as the solid line, and the anomalous profile is presented as the dashed line. In the ideal profile, primary and secondary peaks are associated with the years 1963 and 1958, respectively. A rapid decline in activity after 1963, the initial appearance of cesium-137 in 1950, and no tailing before 1950 indicate that mixing of cesium-137 is negligible. Less desirable profiles are obtained where physical or biological mixing or chemical diffusion occurs. In the anomalous profile presented in Figure 3, mixing is evident by the absence of distinct primary and secondary peaks and the lack of a rapid decline in cesium-137 activity after 1963. Cesium-137 and lead-210 dating methods sometimes do not agree because of the different geochemistries of the two elements; for example, lead binds to organic matter but cesium binds to clay minerals (*5, 9, 11*).

Plutonium is also an anthropogenic radionuclide that was emitted during weapons testing. Plutonium isotopes, except for plutonium-241, undergo alpha decay. The alpha energies emitted by plutonium-239 and 240 cannot be differentiated by alpha spectrometry, and hence, their activities are reported together. Plutonium-239 and -240 profiles parallel those of cesium-137 since major tests emitting plutonium in the United States began in 1951 at the Nevada test site, and peak emissions occurred in 1957-1958 and in 1961-1962. Other sources of plutonium provide additional markers. For example, the burn-up of the U.S. satellite SNAP-9A upon reentry emitted plutonium-238 into the southern hemisphere in 1964. This isotope was first detected in the northern hemisphere in 1966. Thus, an increase in the plutonium-238:plutonium-239 and -240 ratio in sediment cores serves as a marker for the year 1966. In addition, before the 1959 testing moratorium, the United States dominated weapons testing; after the moratorium, the Soviet Union dominated. This switch resulted in "signature" changes that reflect a change in the weapons tested. Ratios of cesium-137:plutonium-239 and -240 of less than 50 indicate the time interval of 1952-1959. Ratios of greater than 70 are associated with emissions during the early 1960s (*11*).

**Pollen.** The appearance of certain pollen types or changes in their distribution in sediment cores are indicators of environmental disturbances such as deforestation, agricultural development, or urbanization.

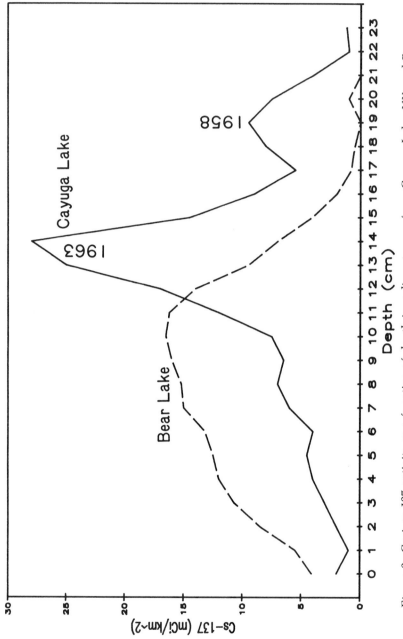

Figure 3. Cesium-137 activity as a function of depth in sediment cores from Cayuga Lake, NY, and Bear Lake, UT. (Data are from reference 11.)

Dating of sediments in the Great Lakes has frequently used horizons of *Ambrosia* (ragweed) or *Castanea* (chestnut) pollen as time markers for 1850 and 1930, respectively (*1, 12–16*). The abundance of *Ambrosia* in the midwestern United States rose because of deforestation and agricultural development of the Great Lakes basin in the mid-1800s. A time of 1850 ± 10 years has been suggested for the onset of *Ambrosia* in sediments because of time lapses between forest clearance and the buildup of ragweed pollen in the Great Lakes (*12, 15*). The decline of *Castanea* in profiles has also been used as a time marker for 1930–1935. During those years, chestnut blight killed most of the chestnut trees in northeastern forests (*9, 12*). Sedimentation rates may be calculated by assuming constant sediment accumulation above either the *Castanea* or *Ambrosia* horizons.

**Varves.**   Varves are annual deposits that are stratified in sediments to form distinct layers. Formation of varves occurs by seasonal inputs of mineral, organic, or chemical materials in temperate meromictic or deep dimictic lakes. Deposition of different materials (e.g., calcite, iron, or sulfur) in anoxic sediment induces varve formation (*17*). Dating by varves is accurate and straightforward because only counting of the layers is required, but one must verify that such laminations are indeed varves. Some type of microscopic examination of selected levels is often done to determine this. Unfortunately, most lakes are not varved.

## Interpreting Sources from Core Profiles

Data from sediment core studies can show concentration changes of a contaminant in the environment and can identify the source(s) of the chemical. This information is obtained by interpreting core profiles to distinguish between anthropogenic and natural inputs and between direct and diffuse sources.

**Anthropogenic versus Natural Sources.**   In general, anthropogenic and natural inputs are differentiated by comparing the flux of a contaminant in the younger and older strata. The flux in older strata (>150 years ago) is a measure of natural inputs, and the flux found in these layers can be subtracted from those in the top layers to calculate anthropogenic inputs. Other approaches, which depend on the type of contaminant and sources of interest, may be used to differentiate between preindustrial and postindustrial levels. For example, determining the source of a metal contaminant involves calculating enrichment factors or normalizing the data to a metal species that occurs naturally. For lead, isotopic ratios can also be used to indicate sources.

   Enrichment factors are ratios of the excess-to-background concentrations and are calculated by the following equation:

$$E = (C_m - C_o)/C_o$$

where $C_m$ is the concentration of a given metal at a depth $m$, and $C_o$ is the background or natural concentration of that metal, which is indicated by constant levels of the metal in the oldest segments of a sediment core. Enrichment factors greater than unity usually indicate anthropogenic sources (18–20). Normalizing concentrations or enrichment factors to a naturally occurring element such as Si, Al, Ca, Mg, Fe, or Na is another way to identify industrial sources.

Stable isotopes of lead that are found in the environment are lead-204, -206, -207, and -208. The isotopes that are formed by the decay of uranium and thorium are as follows: lead-206 from the decay of uranium-238, lead-207 from the decay of uranium-235, and lead-208 from the decay of thorium-235. These isotopes correspond to 23.6%, 22.6%, and 52.3% of all lead. Lead-204 accounts for a small percentage (1.5%) of all lead. Inferences about lead sources are commonly made by the isotopic ratios of lead-206:lead-204, lead-206:lead-207, and lead-206:lead-208 (21).

**Diffuse versus Direct Sources.**  Spatial patterns can sometimes be used to distinguish direct from diffuse sources. Direct or point sources originate from defined locations or events such as oil spills or sewage or industrial discharges. Spatial profiles in which a contaminant's concentration decreases with increasing distance from an area may suggest a point source. Nonpoint or diffuse sources stem from erosion, runoff, natural weathering of minerals, or long-range atmospheric transport and deposition of submicron-sized particles. Spatial patterns from nonpoint sources will frequently not show any distinct pattern or trend. The interpretation of spacial patterns, however, must be done cautiously. Some patterns have nothing to do with sources but rather reflect the mere accumulation of sediment at a given location.

### Environmental Histories

In this section, environmental histories reconstructed by the methods discussed above will be presented. These histories concern contamination by mercury, lead, polycyclic aromatic hydrocarbons (PAHs), dioxins, persistent chlorinated hydrocarbons [e.g., polychlorinated biphenyls (PCBs), DDT, and octachlorostyrene], acid, and carbon particles.

**Mercury.**  Histories of mercury contamination in Lake Ontario have been reconstructed by Thomas and by Breteler et al. (22, 23). The chronology of mercury inputs parallels the growth and practices of two electrochemical companies: Mathieson Alkali Corporation (now Olin Corporation) and Hooker Chemical (now Occidental Chemical Corporation). In 1893, Mathieson Alkali Corporation began using rocking-cell mercury electrodes. Mercury wastes from this process were directly discharged into the Niagara River. During 1945–1955, the rocking-cell electrodes were replaced by flat-bed mercury cells. The maximum levels of mercury in the sediments of Lake Ontario (Figure 4) correspond to this transition, presumably a disruptive process that caused an extra loss of mercury. In the early 1960s, Hooker Chemical also began operating a chloralkali plant and became the primary source of mercury in the Niagara River and Lake Ontario. The levels of mercury decreased during the late 1960s and early 1970s and seem to have reached background levels.

**Lead.** Sediment core histories that reconstruct the environmental contamination and sources of lead are noteworthy in evaluating the effect of environmental regulations. Regulations begun during the past decade have sought to control the industrial discharge and the automobile emission of lead. A survey of core histories shows inputs of lead from industrial discharges, mining operations, and from the combustion of leaded automobile fuel (*24–27*). In several studies, lead appears in the sediments above background levels around the middle

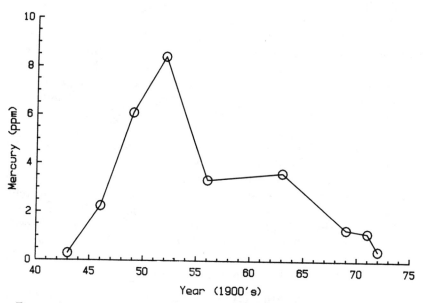

*Figure 4. Mercury concentration in a Lake Ontario sediment core as a function of year. (Data are from reference 23.)*

of the 19th century because of industrial or domestic activities and dramatically increases in the 1920s as a result of the increased use of leaded gasoline (28–31).

In areas remote from industrial inputs, atmospheric transport and deposition are responsible for lead contamination (18, 32–34). In Lakes Superior and Huron, as well as offshore marine environments, atmospheric inputs are a substantial proportion of the total input of lead (19). Studies by Chow et al. (35) and Edgington and Robbins (36) have contributed significantly to our understanding of source functions responsible for these inputs. In the study by Chow et al. (35), the rates of lead accumulation in sediments off the coast of southern California increased after the 1940s because of atmospheric transport of lead from automobile exhaust. Further evidence of this source was provided by the similarity between lead isotopic ratios in sediments and those in gasoline.

In contrast, the levels of lead in sediments of southern Lake Michigan were observed to increase in 1830 because of the combustion of coal for industrial and domestic purposes. Assuming that inputs of lead to these sediments were proportional to coal and gasoline emissions, Edgington and Robbins constructed a mathematical model that would predict lead inputs by using data from these emissions (36). This model also considered the decrease in contributions from coal after 1960 because of emission controls and processes affecting the deposition and distribution of lead in the sediments. For four core profiles, the predicted values obtained by using this model were in close agreement with the actual values. One of these profiles is presented in Figure 5. The solid line is the least-squares fit considering contributions from coal and gasoline emissions, and the dashed line represents the least-squares fit considering only contributions from gasoline. In this profile, the inclusion of coal emissions provides a better fit; thus, in recent years the contamination of these sediments by lead originates from both these sources.

Most studies show that the combustion of gasoline has been the primary source of lead in recent years. Thus, with the introduction of unleaded gasoline in the early 1970s, a decrease in lead would be expected in sediment cores starting at about that time. Unfortunately, most sediment core studies of lead were undertaken before the introduction of unleaded gasoline, and resolution in the sediment cores is not fine enough to show this recent effect. Cores taken over the next few years should show that emissions of lead are decreasing because of the use of unleaded gasoline. This issue is discussed further in Chapter 11.

**Polycyclic Aromatic Hydrocarbons.** The primary environmental source of PAHs is the incomplete combustion of carbonaceous materials. Early PAH papers debated whether these sources were natural

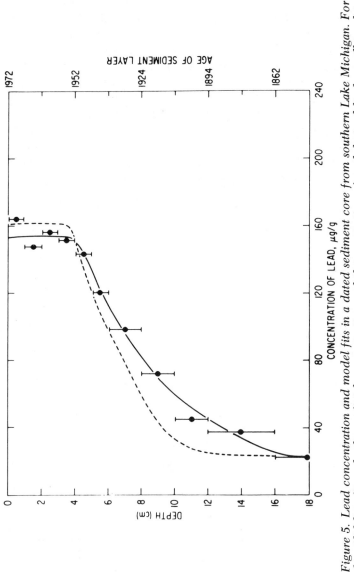

*Figure 5. Lead concentration and model fits in a dated sediment core from southern Lake Michigan. For the solid line fit, only the surficial concentration and the proportion of coal-derived lead are allowed to vary. The dashed line represents the least-squares fit assuming contributions from gasoline only. (Reproduced from reference 36. Copyright 1976, American Chemical Society.)*

events such as forest fires (*37*) or anthropogenic activities such as the combustion of coal and oil (*38*). Results of these studies showed that fossil fuels were the predominant source. Other historical sediment core studies have provided useful information about additional sources (e.g., direct oil spills, runoff from streets and roads, and effluent discharges) and the impact of fossil fuel combustion on PAH emissions. The relative contribution of these sources depends on whether the area of concern is in a remote or an urban region. In remote areas, atmospherically transported PAHs are the major, if not sole, source of PAHs in lakes. In urban areas, inputs by runoff or other effluents will be the primary source of PAHs. Furthermore, asphalt-associated PAHs in street dust are likely to affect these inputs to a greater extent than PAHs on particles originating from automobile tire wear or exhaust (*39*). In general, PAHs that have originated from fossil fuel combustion show an increase in sediment cores around the turn of the century (*40–43*) (Figure 6). This increase has been attributed to the increased use of coal as an energy source at that time (*40–43*).

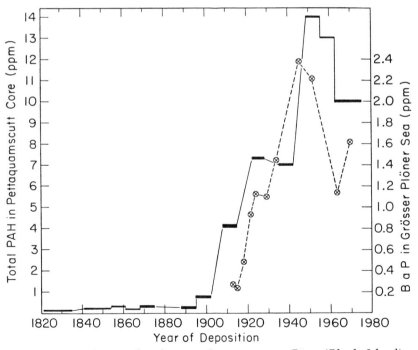

*Figure 6. Total PAH abundance in Pettaquamscutt River (Rhode Island) sediment core sections as a function of date of deposition (horizontal bars, left scale), and benzo[a]pyrene abundance in the Grösser Plöner Sea versus date of deposition (⊗, right scale). (Reproduced with permission from reference 40. Copyright 1980, Pergamon Press.)*

**Dioxins.** Concern about environmental sources of polychlorinated dibenzodioxins (PCDDs) and polychlorinated dibenzofurans (PCDFs) has been caused by the toxicity of 2,3,7,8-tetrachlorodibenzo-*p*-dioxin (TCDD). The PCDDs and PCDFs have been found as impurities in phenoxy herbicides [e.g., (2,4,5-trichlorophenoxy)acetic acid (2,4,5-T)], chlorophenols, and PCBs. They enter the environment as components of these chemicals or from the combustion of chlorinated organic materials. Czuczwa and Hites (*44*) analyzed the tetra-, penta-, hexa-, hepta-, and octachlorodioxins and furans in sediment cores from the Great Lakes. The resulting profiles all showed a predominance of octachlorodibenzodioxin (octa-CDD), with lesser amounts of hexachlorodibenzodioxin (hexa-CDD) and heptachlorodibenzofuran (hepta-CDF). Congener profiles from urban air particulate samples were also similar. The agreement among these profiles, the presence of dioxins and furans in remote lake sediments receiving only atmospheric inputs (Siskiwit Lake), and the occurrence of higher levels of dioxins and furans in urban versus remote sediments indicates atmospheric transport and deposition of these compounds.

Information on the source of sedimentary dioxins and furans was obtained from the history of congener profiles in the Great Lakes sediment cores (Figure 7). In all these profiles, an increase in dioxins and furans was observed after 1940. Comparisons among trends in coal consumption, production of chlorinated organic compounds, and levels of dioxins and furans in sediments have eliminated coal combustion as a major source of dioxins because the use of coal has remained relatively constant since about 1910. A strong relationship between dioxin and furan concentrations and the industrial production of chlorinated aromatic compounds suggests that the incineration of chlorinated compounds in municipal and chemical waste is the major source of these contaminants to the environment (*44, 45*).

**Polychlorinated Biphenyls.** PCBs were first commercially produced in 1930. They were widely used, particularly in the late 1960s, as plasticizers, dye solvents, hydraulic fluids, and dielectric fluids. In 1972, Monsanto, the sole U.S. producer, voluntarily banned the use of PCBs in open systems but continued to sell certain PCB mixtures for use in closed electrical systems until 1977. Currently, these compounds are introduced into the environment by transformer oil spills, incineration of wastes, municipal and industrial discharges, and recycling of carbonless copy paper (*46*). Historical measurements of PCBs have shown that both direct and indirect sources have caused PCB contamination.

Direct sources have contaminated sediments in southern California, Lake Erie, Lake Michigan, and the Hudson River. In oceanic sediments near the coast of southern California, the onset of PCB contamination

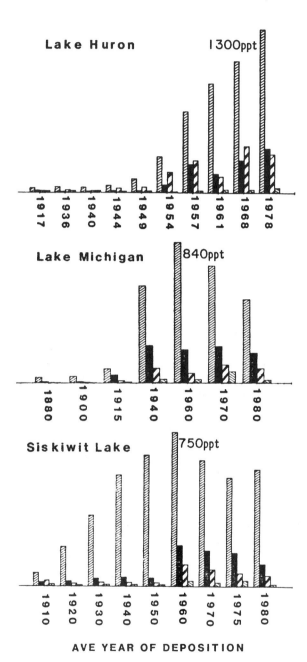

*Figure 7. Octachlorodibenzodioxin (▨), heptachlorodibenzodioxin (■), heptachlorodibenzofuran (▨), and octachlorodibenzofuran (▧) concentrations (pg/g) as a function of average year of deposition in several sediment cores.*

occurred around 1945. The time of their initial appearance reflects the increased use of PCBs during World War II. After this time, discharges of PCBs in sewage effluents increased (*47*). Similarly, increasing inputs of PCBs to Lake Erie after 1958 were thought to originate from Detroit (*48*). A study of PCBs in the Hudson River has shown that inputs from two General Electric plants contaminated the freshwater and tidal portions of the river. These sediments first became contaminated with PCBs in the mid-1950s and reached a maximum in the early 1970s. This maximum was generated by the removal of a dam below the source with subsequent dilution and burial of the contaminated sediments (*49*). Other contaminated waters such as the Fox River and Waukegon Harbor are sources of PCBs to Green Bay and Lake Michigan, respectively (*50*).

Several studies have investigated the atmospheric transport and deposition of PCBs. Such inputs contribute greatly to the continuing PCB contamination of fish, water, and sediments (*50–55*). The analysis of sediment cores taken from Lake Superior, Lake Michigan, and remote lakes in northern Wisconsin have substantiated this conclusion. Atmospheric sources were primarily responsible for PCB inputs to the sediments of Lake Superior and to remote lakes in Wisconsin; atmospheric and tributary sources contributed about equally to the levels of PCBs in Lake Michigan sediments (*50, 51, 55*). An average profile of PCBs in three sediment cores from Lake Superior is presented in Figure 8. A PCB maximum is evident in the early 1970s; after that time, decreases of PCBs in sediments parallel PCB sales. In remote lakes in Wisconsin receiving only atmospheric inputs, PCB inputs decreased from the mid-1970s to early 1980s by about 30% (*50*). Overall, the profiles of PCBs in Great Lakes sediment reflect the use and production of these chemicals and show that regulatory actions have been effective in controlling their release into the environment.

### 1,1′-(2,2,2-Trichloroethylidene)bis[4-chlorobenzene].

DDT is a chlorinated insecticide that was used extensively from 1942 to 1972. In the United States, maximum use and production of this compound occurred in the late 1950s until the early 1960s. Although its use was either banned or restricted in the United States, Canada, and Europe in the 1970s, DDT is still being used in Third World countries. In the environment, DDT is degraded primarily by microorganisms. Dechlorination and dehydrochlorination of DDT form compounds such as 1,1′-(dichloroethenylidene)bis[4-chlorobenzene] (DDE) in aerobic environments and 1,1′-(2,2-dichloroethylidene)bis[4-chlorobenzene] (DDD) in anoxic systems. Because of DDT's degradation, studies examining environmental contamination usually analyze for DDT, DDD, and DDE and report concentrations of DDT as the sum of these compounds, usually with the notation *t*-DDT for total DDT.

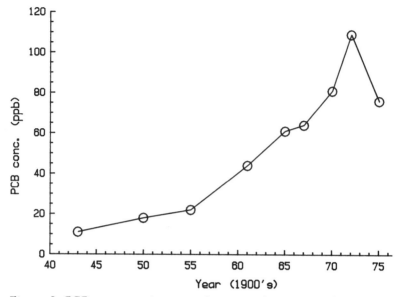

*Figure 8. PCB concentration as a function of year in Lake Superior sediment cores. Data were averaged from three cores. (Data are from reference 57.)*

In sediment cores taken off the coast of California and from Lake Erie, the initial appearance of *t*-DDT from the mid-1940s to the late 1950s reflects the time of DDT's initial production and use (Figure 9) (*47, 48, 56*). Concentrations of *t*-DDT in these sediments increased until the 1970s. In southern California, these inputs were primarily derived from discharges of DDT and its degradation products in sewage effluents (*47, 56*). As inputs of these compounds decreased in waste water discharges, the levels in the southern California sediments also decreased (*56*). Similar trends have been observed in peat bogs (*57, 58*). An example of this trend is shown in Figure 9. These cores are interesting because these peat bogs received only atmospheric inputs. As with other compounds, the appearance of *t*-DDT reflects usage patterns, and the levels of *t*-DDT decreased after the ban in 1972. Levels of DDT and ratios of DDT to DDE and DDD in the surface sediments are higher than expected in these bogs, considering that DDT has a half-life of 10 years. As a result, the long-range atmospheric transport of DDT from Mexico and Central America has been suggested to be responsible for continued inputs of DDT into the Great Lakes (*57, 58*).

**Octachlorostyrene.**   The chloralkali plants that contaminated Lake Ontario with mercury have also contaminated the lake with octachlorostyrene (OCS). The history of OCS provides an interesting story about

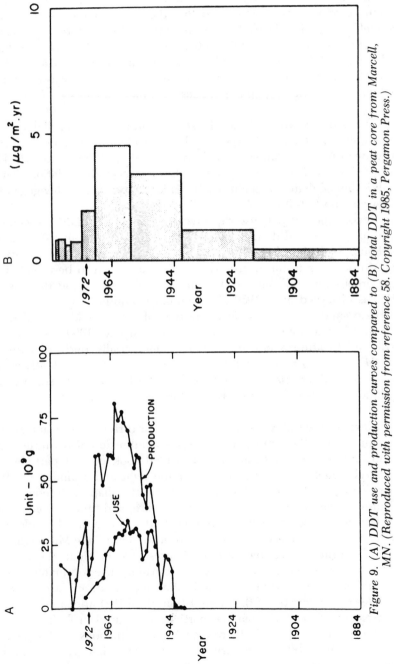

*Figure 9. (A) DDT use and production curves compared to (B) total DDT in a peat core from Marcell, MN. (Reproduced with permission from reference 58. Copyright 1985, Pergamon Press.)*

a contaminant that was never commercially produced. By comparing data for OCS in Lake Ontario cores with chlorine production data, Kaminsky and Hites (59) showed that the levels of OCS in the cores paralleled graphite-electrode chlorine production in the Great Lakes states (Figure 10). The hypothesis that the source of OCS was the "taffy tar" produced from these graphite anodes was verified by an analysis of a taffy tar sample. The decline of OCS in the 1970s was attributed to the replacement of graphite anodes with metal anodes.

**Diatoms as Indicators of pH.** Diatoms are unicellular algae commonly found in aquatic systems. Their species distribution is affected by water quality parameters such as nutrients, salinity, and pH. This group of algae has siliceous cell walls that resist degradation. These qualities make diatom remains useful as indicators of pH changes in the water column (60-62).

Reconstructing pH histories involves classifying diatom species in a sediment core and dating the cores by lead-210, cesium-137, or other methods. Such data have been used to reconstruct the pH history of Big Moose Lake in the Adirondacks, an area known to be affected by acid precipitation. From the diatom assemblages, a rapid decrease in pH was observed after 1950. The timing, magnitude, and rate of this decrease suggest atmospheric deposition of strong acids as the cause (63). In other diatom studies, lake acidification since 1900 has also been attributed to atmospheric deposition of industrially derived materials (61, 64). In one of these studies, this conclusion was substantiated by increases in oil soot particles in the sediment after 1950. The significance of these particles will be discussed in the next section.

**Carbon Particles.** Carbon particles of different sizes and shapes are emitted into the atmosphere from the combustion of fossil fuels and wood. Submicron-sized particles are called soot; coarser particles are called charcoal, coke, or char. Under the light and scanning electron microscopes, carbon particles have characteristic shapes, structures, and textures depending on the source, temperature, and time of combustion (65, 66). Such characteristics were used to determine the chronology of charcoal particles in a sediment core from southern Lake Michigan. The vertical profile of charcoal particles is presented in Figure 11. Prior to 1900, particles indicative of wood burning predominated; from 1900-1930, charcoal from coal combustion increased; and around 1930, particles characteristic of the combustion of oil first appeared. In 1960, the concentration of all charcoal particles decreased probably because of environmental controls to reduce fly-ash emissions. Although coal combustion was still the primary source of charcoal, a greater contribution from oil was observed after 1960.

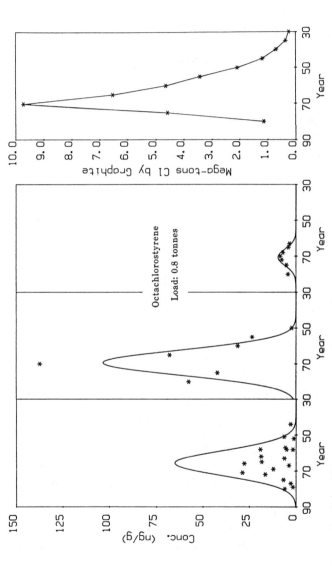

*Figure 10. Octachlorostyrene in six cores from Lake Ontario as a function of year. The three panels on the left represent each of the three sedimentation basins of the lake. The panel on the right is the U.S. production of chlorine gas by graphite electrodes as a function of year. The curves are Gaussian functions fitted to the core data; no input function is implied. (Reproduced from reference 59. Copyright 1984, American Chemical Society.)*

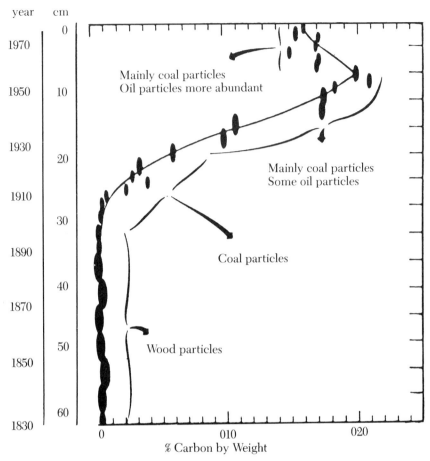

year　cm

*Figure 11. Charcoal particles (>38 μm) in a core from southeastern Lake Michigan. (Reproduced with permission from reference 66. Copyright 1981, Pergamon Press.)*

Profiles of particles less than 38 μm were similar to that shown in Figure 11, although their relative contribution decreased after 1900–1910. This decrease suggests that before the early 1900s, carbon particles originated from long-distance transport, but in more recent years, carbon particles were emitted from local sources. Historically, these changes reflect the development of the southern Lake Michigan area from a prairie to an agricultural region and then to an industrial center (66, 67).

In Sweden, soot particles in sediment cores also reflect coal and oil combustion (68, 69). Core profiles of soot particles correlate with diatom-inferred pH profiles; this correlation suggests fossil fuel combustion has been a source of strong acids in the atmosphere (61, 68). The concentration and type of charcoal particles in sediments can also be used to reconstruct the history of forest fires (*70, 71*).

## Acknowledgments

We acknowledge the U.S. Department of Energy (Grant 80EV-10449) and the U.S. Environmental Protection Agency (Grant R808865) for their support. We thank Deborah L. Swackhamer for comments on the manuscript.

## References

1. Bruland, K. W.; Koide, M.; Bowser, C.; Maher, L. J.; Goldberg, E. D. *Quat. Res.* **1975**, *5*, 89–98.
2. Goldberg, E. D.; Hodge, V.; Koide, M.; Griffin, J.; Gamble, E.; Bricker, O. P.; Matisoff, G.; Holden, G. R., Jr. *Geochim. Cosmochim. Acta* **1978**, *42*, 1413–1425.
3. Koide, M.; Bruland, K. W.; Goldberg, E. D. *Geochim. Cosmochim. Acta* **1973**, *42*, 1171–1187.
4. Krishnaswamy, S.; Lal, D.; Martin, J. M.; Meybeck, M. *Earth Planet. Sci. Lett.* **1971**, *11*, 407–414.
5. Robbins, J. A.; Edgington, D. N. *Geochim. Cosmochim. Acta* **1975**, *39*, 285–304.
6. Binford, M. W. *Proceedings of a Workshop on Paleolimnological Studies of the History and Effects of Acidic Precipitation*; U.S. Environmental Protection Agency: Corvallis, OR, 1984; pp 34–77; CR–811631–01–0.
7. Oldfield, F.; Appleby, P. G. In *Lake Sediments and Environmental History*; Haworth, E. Y.; Lund, J. W. G., Eds.; University of Minnesota: Minneapolis, MN, 1984; Chapter 3.
8. Robbins, J. A. In *The Biogeochemistry of Lead in the Environment: Part A*; Nriagu, J. O., Ed.; Biomedical: New York, 1978; Chapter 9.
9. Smith, J. N.; Walton, A. *Geochim. Cosmochim. Acta* **1980**, *44*, 225–240.
10. Lehman, J. T. *Quat. Res.* **1975**, *5*, 541–550.
11. Heit, M.; Miller, K. M.; Krey, P. W.; Begen, D.; Fealy, H. *Proceedings of a Workshop on Paleolimnological Studies of the History and Effects of Acidic Precipitation*; U.S. Environmental Protection Agency: Corvallis, OR, 1984; pp 34–77; CR–811631–01–0.
12. Kemp, A. L. W.; Anderson, T. W.; Thomas, R. L.; Mudruchova, A. *J. Sediment. Petrol.* **1974**, *44*, 207–218.
13. Nriagu, J. O.; Kemp, A. L. W.; Wong, H. K. T.; Harper, N. *Geochim. Cosmochim. Acta* **1979**, *43*, 247–258.
14. Thomas, R. L. *Can. J. Earth Sci.* **1972**, *9*, 636–651.
15. Robbins, J. A.; Edgington, D. N.; Kemp, A. L. W. *Quat. Res.* **1978**, *10*, 256–278.
16. Matthews, R. W.; D'Auria, J. M. *Can. J. Earth Sci.* **1982**, *17*, 2114–2125.
17. Renberg, I. *Proceedings of a Workshop on Paleolimnological Studies of the History and Effects of Acidic Precipitation*; U.S. Environmental Protection Agency: Corvallis, OR, 1984; pp 78–85; CR–811631–01–0.
18. Heit, M.; Tan, Y.; Klusek, C.; Burke, J. C. *Water, Air, Soil Pollut.* **1981**, *15*, 441–464.

19. Kemp, A. L. W.; Williams, J. D. H.; Thomas, R. L.; Gregory, M. L. *Water, Air, Soil Pollut.* **1978**, *10*, 381–402.
20. Muller, G.; Grimmer, G.; Bohnke, H. *Naturwissenschaften* **1977**, *64*, 427–431.
21. Matisoff, G. *Proceedings of a Workshop on Paleolimnological Studies of the History and Effects of Acidic Precipitation*; U.S. Environmental Protection Agency: Corvallis, OR, 1984; pp 201–233; CR–811631–01–0.
22. Thomas, R. L. *Can. J. Earth Sci.* **1972**, *9*, 636–651.
23. Breteler, R. J.; Bowen, V. T.; Schneider, D. L.; Henderson, R. *Environ. Sci. Technol.* **1984**, *18*, 404–409.
24. Christensen, E. R.; Chien, N. *Environ. Sci. Technol.* **1981**, *15*, 553–558.
25. Heit, M. H.; Klusek, C. S.; Volchok, H. L. *Environ. Int.* **1980**, *4*, 229–237.
26. Mudroch, A. *J. Great Lakes Res.* **1983**, *9*, 125–133.
27. Nriagu, J. O.; Wong, H. K. T.; Coker, R. D. *Environ. Sci. Technol.* **1982**, *15*, 551–560.
28. Christensen, E. R.; Scherfig, J.; Koide, M. *Environ. Sci. Technol.* **1978**, *12*, 1168–1173.
29. Crecelius, E. A.; Piper, D. Z. *Environ. Sci. Technol.* **1973**, *7*, 1053–1055.
30. Farmer, J. G. *Sci. Total Environ.* **1978**, *10*, 117–127.
31. Presley, B. J.; Trefry, J. H.; Shokes, R. F. *Water, Air, Soil Pollut.* **1980**, *13*, 481–494.
32. Bruland, K. W.; Bertine, K.; Koide, M.; Goldberg, E. D. *Environ. Sci. Technol.* **1984**, *8*, 425–431.
33. Davis, A. O.; Galloway, J. N. In *Atmospheric Pollutants in Natural Waters*; Eisenreich, S. J., Ed.; Ann Arbor Science: Ann Arbor, MI, 1981; Chapter 19.
34. Bertine, K. K.; Goldberg, E. D. *Science (Washington, D.C.)* **1971**, *16*, 233–235.
35. Chow, T. J.; Bruland, K. W.; Bertine, K.; Souter, A. *Science (Washington, D.C.)* **1973**, *10*, 551–552.
36. Edgington, D. N.; Robbins, J. A. *Environ. Sci. Technol.* **1976**, *10*, 266–274.
37. Youngblood, W. W.; Blumer, M. *Geochim. Cosmochim. Acta* **1975**, *39*, 1303–1314.
38. Laflamme, R. E.; Hites, R. A. *Geochim. Cosmochim. Acta* **1978**, *42*, 289–303.
39. Wakeham, S. G.; Schafner, C.; Giger, W. *Geochim. Cosmochim. Acta* **1980**, *44*, 403–413.
40. Hites, R. A.; Laflamme, R. E.; Windsor, J. G., Jr.; Farrington, J. W.; Deuser, W. G. Ibid., 873–878.
41. Prahl, F. G.; Carpenter, R. *Geochim. Cosmochim. Acta* **1979**, *43*, 1959–1972.
42. Gschwend, P. M.; Hites, R. A. *Geochim. Cosmochim. Acta* **1981**, *45*, 2359–2367.
43. Hites, R. A. *Proceedings of a Workshop on Paleolimnological Studies of the History and Effects of Acidic Precipitation*; U.S. Environmental Protection Agency: Corvallis, OR, 1984; pp 363–375; CR–811631–01–0.
44. Czuczwa, J. M.; Hites, R. A. *Environ. Sci. Technol.* **1984**, *18*, 444–450.
45. Czuczwa, J. M.; Hites, R. A. *Environ. Sci. Technol.* **1986**, *20*, 195–200.
46. *Polychlorinated Biphenyls*; National Academy of Sciences: Washington, DC, 1979.
47. Hom, W.; Risebrough, R. W.; Louter, A.; Young, D. R. *Science (Washington, D.C.)* **1974**, *184*, 1197–1199.
48. Frank, R.; Holdrinet, M.; Braun, H. E.; Thomas, R. L.; Kemp, A. L. W. *Sci. Total Environ.* **1977**, *8*, 205–227.
49. Bopp, R. F.; Simpson, H. J.; Olsen, C. R.; Tnier, R. M.; Kostyk, N. *Environ. Sci. Technol.* **1982**, *16*, 666–676.
50. Swackhamer, D. L. Ph.D. Thesis, University of Wisconsin, Madison, 1985.

51. Eisenreich, S. J.; Hollod, G. J.; Johnson, T. C.; Evans, J. In *Contaminants and Sediments*; Baker, R. A., Ed.; Ann Arbor Science: Ann Arbor, MI, 1980; Chapter 4.
52. Eisenreich, S. J.; Hollod, G. J.; Johnson, T. C. In *Atmospheric Pollutants in Natural Waters*; Eisenreich, S. J., Ed.; Ann Arbor Science: Ann Arbor, MI, 1981; Chapter 21.
53. Murphy, T. J.; Pokojowczyk, J. C.; Mullin, M. D. In *Physical Behavior of PCBs in the Great Lakes*; Mackay, D.; Patterson, S.; Eisenreich, S. J.; Simmons, M. S., Eds.; Ann Arbor Science: Ann Arbor, MI, 1983; Chapter 3.
54. Kihlstrom, J. E.; Berglund, E. *Ambio* 1978, 7, 175–178.
55. Eisenreich, S. J.; Looney, B. B.; Hollod, G. J. In *Physical Behavior of PCBs in the Great Lakes*; Mackay, D.; Patterson, S.; Eisenreich, S. J.; Simmons, M. S., Eds.; Ann Arbor Science: Ann Arbor, MI, 1983; Chapter 7.
56. Venkatesan, M. I.; Brenner, S.; Ruth, E.; Bonilla, J.; Kaplan, I. R. *Geochim. Cosmochim. Acta* 1980, 44, 789–802.
57. Eisenreich, S. J.; Rappaport, R. A.; Urban, N. R.; Capel, P. D.; Looney, B. B.; Baker, J. E. *Proceedings of a Workshop on Paleolimnological Studies of the History and Effects of Acidic Precipitation*; U.S. Environmental Protection Agency: Corvallis, OR, 1984; pp 387–408; CR–811631–01–0.
58. Rappaport, R. A.; Urban, N. R.; Capel, P. D.; Baker, J. E.; Looney, B. B.; Eisenreich, S. J.; Gorham, E. *Chemosphere* 1985, 14, 1167–1173.
59. Kaminsky, R.; Hites, R. A. *Environ. Sci. Technol.* 1984, 18, 275–279.
60. Charles, D. F. *Ecology* 1985, 66, 994–1011.
61. Battarbee, R. W. *Proceedings of a Workshop on Paleolimnological Studies of the History and Effects of Acidic Precipitation*; U.S. Environmental Protection Agency: Corvallis, OR, 1984; pp 275–307; CR–811631–01–0.
62. Renberg, I.; Hellberg, T. *Ambio* 1982, 11, 30–33.
63. Charles, D. F. *Verh. Int. Ver. Theor. Angew. Limnol.* 1984, 22, 559–566.
64. Flower, R. J.; Battarbee, R. W. *Nature (London)* 1983, 305, 130–133.
65. Griffin, J. J.; Goldberg, E. D. *Science (Washington, D.C.)* 1979, 206, 563–565.
66. Griffin, J. J.; Goldberg, E. D. *Geochim. Cosmochim. Acta* 1981, 45, 763–769.
67. Griffin, J. J.; Goldberg, E. D. *Environ. Sci. Technol.* 1983, 17, 244–245.
68. Renberg, I. *Proceedings of a Workshop on Paleolimnological Studies of the History and Effects of Acidic Precipitation*; U.S. Environmental Protection Agency: Corvallis, OR, 1984; pp 376–386; CR–811631–01–0.
69. Renberg, I.; Wir, M. *Verh. Int. Ver. Theor. Angew. Limnol.* 1984, 22, 712–718.
70. Tolonen, K. In *Handbook on Paleocology and Paleohydrology*; Berglund, B. B., Ed.; Wiley: New York, 1985.
71. Swain, A. M. *Quat. Res.* 1973, 3, 383–396.

RECEIVED for review May 6, 1986. ACCEPTED August 19, 1986.

# CASE STUDIES

# 13

# The Chemical Limnology of Nonpolar Organic Contaminants: Polychlorinated Biphenyls in Lake Superior

S. J. Eisenreich

Environmental Engineering Program, Department of Civil and Mineral Engineering, University of Minnesota, Minneapolis, MN 55455

*The detailed aquatic behavior of polychlorinated biphenyls (PCBs) in Lake Superior permits an evaluation of the chemical limnology and environmental fate of nonpolar organic contaminants having similar physicochemical properties in large lakes. The range of physicochemical properties [e.g., Henry's law constant ($K_H$) and the octanol-water partition coefficient ($K_{ow}$)] describing the fate of the 50-100 PCB congeners observed in the environment suggests that air-water and sediment-water transport processes dominate their aquatic behavior. PCBs likely enter the lake during intense episodes of precipitation (scavenging of particles) and are subsequently lost during longer periods of volatilization. Decreasing inputs and water column concentrations suggest that the atmosphere is now a sink for PCBs previously deposited. Resuspension of bottom sediment and pore water containing PCBs dominate internal cycling. Sorption and subsequent removal of particles to the bottom are important lake detoxification processes. The observed inverse relationship of log $K_p$, where $K_p$ is the partition coefficient, and log SS, where SS is suspended solids, may be explained by mixed-particle populations of differing organic carbon content and binding of PCBs to colloidal organic matter. The estimated residence time of PCBs in Lake Superior implies an efficient removal process. A dynamic mass balance model using a new PCB input function driven by atmospheric concentrations adequately predicts observed water column concentrations.*

T OXIC ORGANIC CHEMICALS DERIVED FROM ANTHROPOGENIC emissions are ubiquitous in the global environment (*1-3*). One general cate-

0065-2393/87/0216-0393$13.50/0
© 1987 American Chemical Society

gory of organic pollutants particularly troublesome to human and eco-system health is the nonpolar hydrophobic chemicals that partition readily into biotic compartments because of low activities in water and resist biological, chemical, and physical degradation. Study of the environmental fate of trace hydrophobic organic compounds in the interactive compartments of the atmosphere, water, sediment, and biota in lakes may be appropriately viewed as the chemical limnology of these compounds. Lakes are excellent environmental "test tubes" to study processes important in the environmental fate of organic pollutants, and the scientific knowledge gained may be applied to understanding their cycling in the more complex arenas of estuarine and oceanic systems. Large lakes such as the Laurentian Great Lakes and the Swiss lakes are especially suited to this challenge because steady-state approximations may still be possible in selected cases.

Polychlorinated biphenyls (PCBs) represent a class of nonpolar organic contaminants that are ubiquitous in the marine and freshwater environments of our planet (4, 5). The global distribution is demonstrated by their occurrence in fish products from as dissimilar environments as the Antarctic Ocean, North Atlantic Ocean, North Sea, Alpine Lakes (1, 6, 7), and Great Lakes (8, 9). Since 1929, approximately $5.7 \times 10^8$ kg of PCBs has been commercially produced, the majority in North America by Monsanto, and approximately $2 \times 10^8$ kg (35%) remains in mobile environmental reservoirs. The voluntary and legislative restrictions instituted since 1971 on the production, sale, and use of PCBs in closed systems and the banning of PCBs for future use suggest that the tropospheric and hydrospheric burdens should be decreasing. Indeed, decreasing concentrations of PCBs in certain fishes from the Great Lakes (9), sediment (10, 11), and atmosphere (12, 13) support this hypothesis. However, PCBs are largely refractory compounds resisting chemical and biological degradation and will remain active for decades (14, 15). This chapter summarizes the concentrations and burdens of PCBs in the various Lake Superior compartments, evaluates and quantifies several flux pathways, and reports on a dynamic steady-state model of PCB cycling in Lake Superior. This chapter is a result of nearly a decade of intensive study on the cycling of PCBs in Lake Superior.

Our approach to characterizing the cycling of nonpolar organic contaminants in aquatic systems is to use PCBs as tracers or surrogates. The physicochemical properties of PCBs, their relative resistance to biodegradation, and their ubiquitous distribution in the environment make them ideal as surrogates of hundreds of compounds emitted by anthropogenic sources. This approach is implemented by the construction of a model describing the inputs, outputs, compartmental burdens, and phase transformations in large lakes.

Anthropogenic hydrophobic organic chemicals such as PCBs enter the water column of large lakes from a variety of external sources (Figure 1). A mass balance constructed about the water column provides an understanding of the transport and distribution of PCBs in the lake and an estimate of the residence time that the pollutant remains in the ecosystem or in any individual compartment. Sources include riverine inputs, point discharges from municipal and industrial facilities, groundwater flow, and atmospheric deposition. Organic contaminants may be lost from the water column via sedimentation and incorporation into bottom sediments, volatilization, chemical and biological degradation, and riverine outflow. Chemical species in the sediments may be released into the water column by sediment and pore water resuspension and diffusional fluxes. Biological recycling of sediment-bound organic contaminants by benthic organisms serves to increase whole-lake chemical residence times. The major fluxes of organic species are restricted to particle settling from the water column, vapor-phase transfer across the air–water interface, and resuspension–diffusion across the sediment–water interface. The dominant loss processes of nonpolar organic contaminants in large lakes are sedimentation and volatilization, both of which are influenced by aqueous speciation. Thus, the fate and residence times of PCBs are largely determined by their affinity for abiotic and biotic particles and their net settling rate.

The conceptual framework of a mass balance model states that

$$\text{accumulation} = \text{inputs} - \text{outputs} \pm \text{internal processes} \qquad (1)$$

At steady state, no net accumulation in the lake occurs, and the rate of input equals the rate of output. For PCBs, the rate of PCB input inferred from production and sales has changed over time (*16–18*). Mathematically, the time-dependent change in PCB concentration in lakes may be expressed as

$$V \, d(C_t)/dt = \Sigma Q_i C_i - Q_o C_t - V_s C_d K_p / h - k_d A_{sl} C_d - k_v A_{sl} C_d \qquad (2)$$

where $C_t$ is the total PCB concentration in water, $Q_i$ is the inflow volume at influent concentration $C_i$, $V$ is the lake volume, $Q_o$ is the outflow volume, $V_s$ is the apparent net settling velocity, $A_{sl}$ is the lake surface area, $h$ is the lake depth, $k_d$ is the first-order rate constant for PCB degradation, $C_d$ is the dissolved PCB concentration, $k_v$ is the PCB volatilization rate, and $K_p$ is the PCB partition coefficient.

The $\Sigma Q_i C_i$ term signifies the summation of all inputs of PCBs. The outputs of PCBs are tributary outflow ($Q_o C_t$), sedimentation ($V_s C_d K_p / h$), and biological degradation and volatilization ($k_d C_d$). PCB burdens are expressed as mass accumulated in the atmosphere, water, sediment, and biota. The concentration of PCBs in Lake Superior over time may

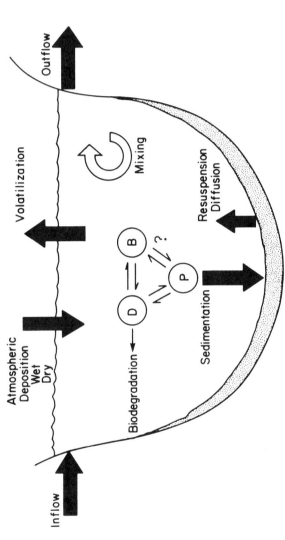

*Figure 1. A conceptual framework describing the major inputs, losses, and internal transfers for nonpolar organic contaminants in lakes. Key: D, dissolved phase; B, biotic phase; and P, particulate phase.*

be predicted from equation 2 using a time-dependent input function
(*18*) and specific chemical and physical relationships describing sorption
and volatilization. Predicted values may be compared to measured
compartmental concentrations and estimated fluxes in the field. Subsequent sections describe the physicochemical properties of PCBs and
their concentrations, and PCB burdens and behavior in air, water, and
sediment compartments. This description is followed by details of a
dynamic steady-state model for PCB cycling in Lake Superior.

## Limnological Setting

Lake Superior (Figure 2) has a large surface area ($8.21 \times 10^{10}$ m$^2$) and
volume ($1.21 \times 10^{13}$ m$^3$) containing 10% of the world's fresh surface
water (*19*). The lake is located in a subarctic temperate zone and experiences partial or full ice cover about 4–5 months per year. The lake
is dimictic and overturn occurs from May to June and again from
November to December. Although Lake Superior is deep [average
depth 145 m, maximum depth 405 m (*19*)], hydrodynamic activity at
overturn is able to resuspend and move bottom sediments (*20*). Bottom
currents are able to erode sediments at depths >200 m in some areas of
the lake, and no sediment accumulation occurs at depths <100 m, except
in sheltered areas (*21*). Summer surface currents move counterclockwise
around the perimeter of the lake and reach maximum speeds of greater
than 20 cm/s off the Keweenaw Peninsula (*22*). Hypolimnetic currents
have the same general movements as the surface water but have slower
velocities. A conservative substance released in the extreme western arm
(Duluth, MN) is predicted to reach the outflow of the lake in the
extreme eastern end in about 3 years; similar arrival times occur around
most of the perimeter of the lake, the center of the eastern basin, and
along the southern shore of Isle Royale. Recent excursions using a deep-
diving submersible (S. J. Eisenreich, unpublished data) show that currents are generally absent at depths greater than 25 m during the stratified period.

The lake is oligotrophic and has a suspended solids concentration
of 0.1–0.5 mg/L during the summer except in the extreme western
arm (0.1–4 mg/L) (*23*). The total organic carbon concentration averages
1.1 mg/L throughout the open lake (*24*), and particulate organic carbon
accounts for 10%–30% of the total. The higher suspended solids concentration in the western arm is due to the erosion of red clay soils along
Wisconsin's northern shore and contributes 60% of the total fine-sediment
inputs (*25*). Suspended solids in the eastern region consist mostly of
organic matter; this composition indicates little shoreline erosion or
riverine inputs. The central region has suspended solids concentrations
between these two extremes. The nepheloid layer, a zone of turbid

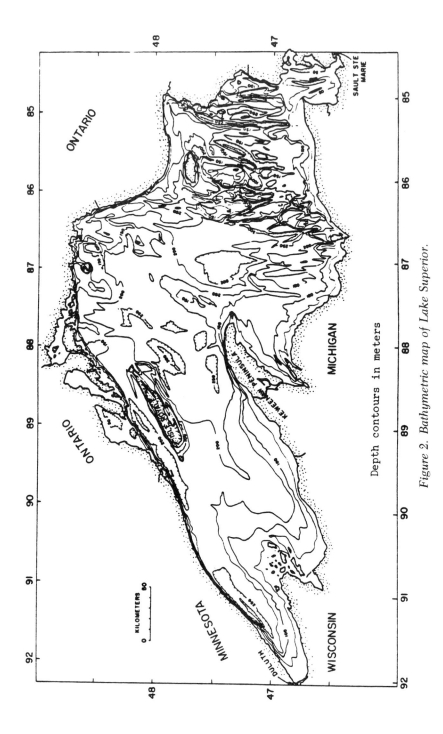

*Figure 2. Bathymetric map of Lake Superior.*

Depth contours in meters

water above the sediments containing higher suspended solids loads, is a prominent feature of Lake Superior during the stratified period (*26*).

The atmosphere exerts a significant influence on Lake Superior and contributes particles, nutrients, pollutants, and more than half of its water inputs. The remaining water comes from numerous tributaries throughout the drainage basin. The only outflow is the St. Mary's River (average 2100 m³/s) at the southeastern end, accounting for two-thirds of the water loss. The large lake volume and lack of other significant outflows result in an average water residence time of 112 years and a mean flushing rate of 177 years (*27*).

The lack of direct, localized impact of human activities and the surface area and volume of the lake imply that atmospheric deposition is a dominant input process for many chemical species (*12*). The basin has a small human population compared to the rest of the Great Lakes and little industrial activity except for Duluth, MN; Superior, WI; and Thunder Bay, Ontario. Because of this lack of activity, localized inputs of organic contaminants are minimal to the lake.

## Physicochemical Properties and Analytical Chemistry

PCBs are produced by the direct chlorination of biphenyl and result in 209 possible congeners, of which 50–100 are frequently observed in the environment. PCBs were marketed in the United States under the trade name Aroclor and contained a known average percentage of chlorine by weight (e.g., Aroclor 1242 contains an average 42% by weight chlorine). Aroclor mixtures employed for various commercial and industrial purposes were Aroclor 1221, 1242, 1016, 1248, 1254, 1260, and 1268. PCB congener distributions in the environment are frequently similar to linear combinations of Aroclor 1242, 1254, and 1260, although Aroclor 1248 is more important quantitatively in some aquatic systems [e.g., southern Lake Michigan (*19*)].

Analysis of PCBs in environmental samples requires a series of steps as outlined here:

Sample

Isolation

hexane:acetone     sediment; solvent extraction
DCM:MeOH     water and air; adsorbent resin

Cleanup–Fractionation
magnesia–silica gel     liquid–solid chromatography

Concentration of Extract
solvent volume reduction

*Continued on next page.*

Analysis
1. high-resolution glass capillary gas chromatography
   (GC) with electron-capture detection
2. confirmation by gas chromatography–mass spectrometry (GC–MS)

The analysis scheme used in our laboratory routinely involves the extraction of wet sediment and styrene–divinylbenzene copolymer (XAD–2) resins with 1:1 hexane:acetone followed by back extraction of the acetone with hexane. The combined hexane fractions are dried over anhydrous $Na_2SO_4$, concentrated by using a Kuderna–Danish (evaporator) apparatus and applied to a 1.25% deactivated magnesia–silica gel (13-g Florisil) column. The column is eluted with 60 mL of hexane and then 50 mL of 9:1 hexane:diethyl ether. PCBs elute in the hexane fraction. The eluate is concentrated as before to 0.5–1.0 mL and analyzed on a 25-m-high-resolution glass capillary column [SP–2100, SE–54, cross-linked phenyl methyl silicon (HP 19091 B)] by using a gas chromatograph (HP 5840 A) equipped with a [63]Ni electron-capture detector. Typical GC-run programs appear elsewhere (19, 20), and Figure 3 depicts the GC chromatographs of Aroclor 1242, 1254, and 1260 congeneric standards.

PCB congeners in Aroclor standards and environmental samples were initially identified by comparison to published chromatograms of Aroclor mixtures (28) and relative retention times of PCB congeners (29). Peak identifications were independently confirmed by using single congener standards and a combination of high-performance liquid chromatography (HPLC) and GC–MS (30, 31). Ballschmiter and Zell (28) have published capillary GC chromatograms in which all congeners were identified, and Mullin et al. (29) have reported relative retention times of all 209 PCB congeners. These data permitted the identification of all PCB congeners in Aroclor 1242, 1254, and 1260 mixtures of interest in our environmental studies.

Weight percentages of PCB congeners in Aroclor standard mixtures have been reported by Albro et al. (32, 33), but important discrepancies in identification of specific peaks and in mass balances warranted new determinations. Table I presents the weight percent and structures of 63 PCB congeners occurring at concentrations greater than 0.5 wt % as determined by GC–MS and reported by Capel et al. (31). Congener percentages estimated from the electron-capture relative response factors determined by Mullin et al. (29) showed close agreement to the GC–MS values. A least-squares multiple linear regression program (PCBQ) was employed to yield the total PCB concentration, the reconstructed contribution of Aroclor mixtures in the sample, and the individual PCB congener concentrations (31). The concentration of individual congeners in an Aroclor standard may be estimated from the total concentration of PCBs and the weight percent of that congener in the standard. In this way, a peak-to-peak comparison in the Aroclor standard and environmental sample yields the congener concentration. Details of the computerized least-squares multiple linear regression program based on a Lotus 1-2-3 spreadsheet format are given in Capel et al. (*31*).

The physicochemical properties of the most commonly observed PCB congeners in the environment are presented in Table I. Nonpolar organic compounds are characterized by their low solubility in water ($1 \times 10^{-4}$–$6 \times 10^{-6}$ mol/m$^3$) and low vapor pressures ($P_v = 1 \times 10^{-7}$–$1 \times 10^{-9}$ atm), the ratio of which ($P_v$/solubility) is equivalent to Henry's law constant ($K_H$). This term describes the equilibrium partitioning of a gas between the atmospheric (vapor) and aqueous (dissolved or unassociated) phases. The value of $K_H$ varies from $0.7 \times 10^{-4}$ to $9.7 \times 10^{-4}$

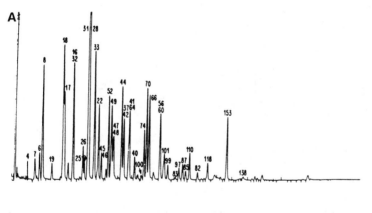

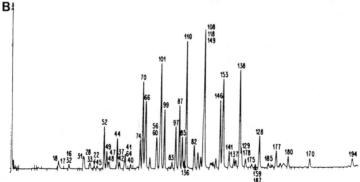

*Figure 3. Capillary column gas chromatograms of (A) Aroclor 1242 and (B) Aroclor 1254. (Reproduced with permission from reference 31. Copyright 1985, Pergamon Press.)* Continued on next page.

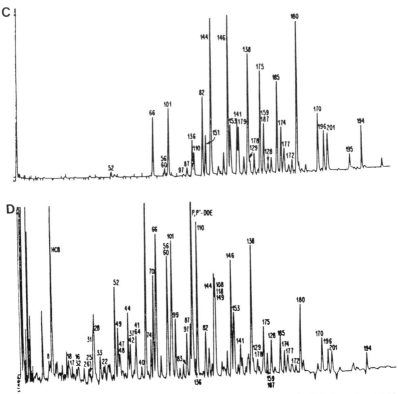

*Figure 3.* Continued. *Capillary gas chromatograms of (C) Aroclor 1260 and (D) an environmental sediment sample. (Reproduced with permission from reference 31. Copyright 1985, Pergamon Press.)*

atm · m³/mol, as predicted from the ratio of the liquid (or subcooled liquid) vapor pressure and aqueous solubility (34). The mean $K_H$ value for Aroclor mixtures at 25 °C is estimated to be $4 \times 10^{-4}$ atm · m³/mol. For PCB congeners, $K_H$ is independent of molecular weight and increases by 10 times with a 25 °C increase in temperature (34). The $P_v$ values and solubility of PCB congeners decrease in approximately the same ratio as the chlorination number ($N_{Cl}$) increases. This similarity leads to $K_H$ values that are relatively invariant. Physicochemical properties of PCB congeners have recently been reviewed by Shiu and Mackay (35).

Low aqueous solubilities and the hydrophobic nature of PCBs result in high partition coefficients to abiotic ($K_p$ and $K_{oc}$) and biotic particles ($K_{BCF}$). A commonly used measure of the ability of an organic compound to partition into aquatic particles having organic carbon content $\geq 0.1\%$ by weight is the octanol–water partition coefficient [$K_{ow}$ (36, 37)]. PCB congeners that have $N_{Cl} = 1$-8 have measured or calculated log $K_{ow}$

values from 4.89 to >7.71 (Table I) (*30*). The $K_{ow}$ values are high for organic compounds having low aqueous solubilities and have been correlated with several environmental parameters. These factors and the recalcitrant nature of PCBs to biological and chemical degradation result in observed hydrospheric lifetimes on the order of months to years. A summary of the ranges in physicochemical parameters for PCB congeners is presented in Table II.

## Concentrations and Processes

**Air Concentrations.** Air samples were collected over Lake Superior from 1978 to 1981 (*12, 38*) and analyzed for particulate and vapor-phase PCBs. The atmospheric sampler consisted of a high-volume air sampler (Universal) modified to include a backup adsorbent. The air sampler held a 7.8-cm bed of styrene–divinylbenzene copolymer macroreticular resin (XAD-2, 600 mL) in an aluminum standpipe atop a motor housing following a glass fiber filter. Recoveries of Aroclor 1221, 1242, and 1254 standards at flow rates of 0.4 m$^3$ min$^{-1}$ averaged 90%–95% in laboratory studies (*39*). The resins used provided good recovery, collection efficiency, and breakthrough behavior for airborne PCBs (*39–42*). The air sampler was affixed to the bow of the ship and collected sample only when the ship was underway (with one exception), and the wind was within a vector of 60° off the bow. Air sampling times of 5–10 h and collection volumes of 200–400 m$^3$ were typical. Flow rates of 0.4–0.5 m$^3$ min$^{-1}$ were used. Total PCB concentrations in air samples ($n = 38$) collected in 1978–1981 averaged 1.0 ng/m$^3$, with a range of 0.1–3.5 ng/m$^3$ and a standard deviation of 0.60 ng/m$^3$ (Table III). On average, 70% of the total PCBs occurred as Aroclor 1242. Because the air sampling system does not unambiguously differentiate between particle and vapor-phase PCB, reported concentrations are operationally defined as total values. However, PCBs were not detected on the glass fiber filters. For a sampling volume of approximately 300 m$^3$ and a total air particle concentration of about 7 μg m$^{-3}$ (*42*), the mean PCB concentration estimated from detection limits was less than 2.8 μg g$^{-1}$. This value compares to PCB concentrations measured in Lake Michigan aerosol of 4 (*43*) and 3 (*44*) μg g$^{-1}$. Based on these data and those summarized by Eisenreich et al. (*12*), more than 90% of the airborne PCBs exist as vapor over Lake Superior. These results are consistent with airborne PCBs collected over Lake Michigan (*43*), the North Pacific Ocean (*45*), the North Atlantic Ocean (*46*), and the calculated vapor–aerosol PCB distribution based on saturation vapor pressure ($P_v$) and available surface area (*47*).

Lowest air concentrations were observed following periods of mist or rain. These values suggest that rain scavenges PCBs from the atmosphere. Using an average concentration of PCBs in precipitation ($C_{rain}$)

**Table I. Physicochemical Properties of PCB Congeners and Mixtures**

| Congener (IUPAC No.) | Structure[a] | Congener wt % | | | $P_v$ (atm) | Solubility (mol/m³) | $K_H$ (×10⁴ atm·m³/mol) | log $K_{ow}$ |
| --- | --- | --- | --- | --- | --- | --- | --- | --- |
| | | Aroclor 1242 | Aroclor 1254 | Aroclor 1260 | | | | |
| 4 | 2,2'- | 0.64 | | | | | 2.2 | 4.89 |
| 7 | 2,4- | | | | | | | 5.30 |
| 6 | 2,3'- | | | | | | 2.1 | 5.02 |
| 8+ | 2,4'- | 10.73 | | | | | 2.2 | 5.10 |
| 19 | 2,2',6- | 11.91 | | | | | 2.1 | 5.48 |
| 18+ | 2,2',5- | 7.57 | | | | | 2.0 | 5.55 |
| 17+ | 2,2',4- | | | | | | | 5.76 |
| 24 | 2,3,6- | | | | | | | |
| 16,32+ | 2,2',3- | 6.52 | | | | | 3.0 | 5.31 |
| 26 | 2,3',5- | | | | | | | 5.76 |
| 25 | 2,3',4- | | | | | | 1.6 | |
| 31+ | 2,4',5- | 10.07 | | | | | | 5.69 |
| 28+ | 2,4,4'- | 10.26 | | | | | 2.0 | 5.69 |
| 33+ | 2',3,4- | 7.75 | | | | | 1.5 | 5.57 |
| 22+ | 2,3,4'- | 3.12 | | | | | 1.8 | 5.42 |
| 45 | 2,2',3,6- | | | | $0.39 \times 10^{-6}$ | $0.39 \times 10^{-3}$ | 1.0 | 6.09 |
| 46 | 2,2',3,6'- | | | | $0.27 \times 10^{-6}$ | $0.39 \times 10^{-3}$ | 6.9 | 6.22 |
| 52+ | 2,2',5,5'- | 4.19 | | | $0.19 \times 10^{-6}$ | $0.36 \times 10^{-3}$ | 5.3 | 6.29 |
| 49+ | 2,2',4,5'- | 2.89 | 3.17 | | $0.17 \times 10^{-6}$ | $0.36 \times 10^{-3}$ | 4.7 | 5.81 |
| 47,48+ | 2,2',4,4'- | 1.90 | | | $0.15 \times 10^{-6}$ | $0.35 \times 10^{-3}$ | 4.2 | 4.94 |
| 44+ | 2,2',3,5'- | 3.90 | | | $0.15 \times 10^{-6}$ | $0.45 \times 10^{-3}$ | 1.0 | 5.56 |
| 37,42+ | 3,4,4'-;2,2',3,4'- | 3.11 | | | $0.83 \times 10^{-7}$ | $0.12 \times 10^{-2}$ | 7.2 | |
| 41,64+ | 2,2',3,4-;2,3,4',6- | 3.57 | | | $0.12 \times 10^{-6}$ | $0.51 \times 10^{-3}$ | 2.4 | |
| 40 | 2,2',3,3'- | 0.46 | | | $0.11 \times 10^{-6}$ | $0.55 \times 10^{-3}$ | 2.0 | |
| 100 | 2,2',4,4',6- | | | | $0.81 \times 10^{-7}$ | $0.84 \times 10^{-4}$ | 9.6 | |
| 74+ | 2,4,4',5- | 1.76 | 1.55 | | $0.57 \times 10^{-7}$ | $0.34 \times 10^{-3}$ | 1.7 | 6.67 |
| 70+ | 2,3',4,5- | 3.93 | 7.55 | | $0.51 \times 10^{-7}$ | $0.34 \times 10^{-3}$ | 1.5 | 6.23 |
| 66+ | 2,3',4,4'- | 3.89 | 6.98 | 3.43 | $0.45 \times 10^{-7}$ | $0.34 \times 10^{-3}$ | 1.4 | 6.31 |
| 60,56+ | 2,3,4,4'- | 2.64 | 2.06 | | $0.42 \times 10^{-7}$ | $0.41 \times 10^{-3}$ | 1.0 | |
| 101+ | 2,2',4,5,5'- | | 11.32 | | $0.35 \times 10^{-7}$ | $0.11 \times 10^{-3}$ | 3.2 | 7.07 |
| 99+ | 2,2',4,4',5- | | 4.32 | 4.47 | $0.31 \times 10^{-7}$ | $0.11 \times 10^{-3}$ | 2.9 | 7.21 |

| No. | Chlorine positions[a] | | | | | | |
|---|---|---|---|---|---|---|---|
| 83 | 2,2',3,3',4- | | | $0.30 \times 10^{-7}$ | $0.14 \times 10^{-3}$ | 2.1 | 6.67 |
| 97+ | 2,2',3',4,5- | 2.40 | | $0.27 \times 10^{-7}$ | $0.13 \times 10^{-3}$ | 2.0 | 6.37 |
| 87+ | 2,2',3,4,5'- | 4.43 | | $0.26 \times 10^{-7}$ | $0.13 \times 10^{-3}$ | 2.0 | 6.61 |
| 85 | 2,2',3,4,4'- | 1.38 | 1.50 | $0.23 \times 10^{-7}$ | $0.13 \times 10^{-3}$ | 1.8 | 6.51 |
| 136 | 2,2',3,3',6,6'- | | | $0.37 \times 10^{-7}$ | $0.40 \times 10^{-4}$ | 9.2 | 5.62 |
| 110+ | 2,3,3',4',6- | 11.93 | | $0.23 \times 10^{-7}$ | $0.16 \times 10^{-3}$ | 1.7 | |
| 82+ | 2,2',3,3',4- | 0.61 | | $0.20 \times 10^{-7}$ | $0.16 \times 10^{-3}$ | 1.2 | |
| 151 | 2,2',3,5,5',6- | | 4.34 | $0.32 \times 10^{-7}$ | $0.37 \times 10^{-4}$ | 8.6 | |
| 144+ | 2,2',3,4,5',6- | | 1.56 | $0.34 \times 10^{-7}$ | $0.35 \times 10^{-4}$ | 9.7 | |
| 118,108 | 2,3',4,4',5- | 15.59 | 14.45 | $0.95 \times 10^{-7}$ | $0.10 \times 10^{-3}$ | 0.9 | 7.12 |
| 149+ | 2,2',3,4',5,5'- | | | $0.16 \times 10^{-7}$ | $0.35 \times 10^{-4}$ | 4.4 | 7.28 |
| 146+ | 2,2',3,4',5,5'- | 4.94 | 18.95 | $0.72 \times 10^{-8}$ | $0.39 \times 10^{-4}$ | 1.8 | 7.75 |
| 153+ | 2,2',3,3',5,5'- | 8.14 | | $0.65 \times 10^{-8}$ | $0.37 \times 10^{-4}$ | 1.8 | |
| 141+ | 2,2',3,4,5,5'- | | 2.17 | $0.27 \times 10^{-7}$ | $0.44 \times 10^{-4}$ | 6.1 | 8.13 |
| 179 | 2,2',3,3',5,6,6'- | | 2.57 | $0.14 \times 10^{-7}$ | $0.16 \times 10^{-4}$ | 8.8 | >7.71 |
| 137 | 2,2',3,4,4',5- | | | $0.24 \times 10^{-7}$ | $0.44 \times 10^{-4}$ | 5.4 | 7.44 |
| 138+ | 2,2',3,4,4',5'- | 9.49 | 11.68 | $0.48 \times 10^{-7}$ | $0.44 \times 10^{-4}$ | 1.1 | 7.32 |
| 129 | 2,2',3,3',4,5- | | | $0.21 \times 10^{-7}$ | $0.53 \times 10^{-4}$ | 3.9 | |
| 178 | 2,2',3,3',5,5',6- | | 0.20 | $0.65 \times 10^{-8}$ | $0.15 \times 10^{-4}$ | 4.4 | |
| 175+ | 2,2',3,3',4,5,6- | | 7.73 | $0.69 \times 10^{-8}$ | $0.14 \times 10^{-4}$ | 4.9 | |
| 187,159+ | 2,2',3,4',5,5',6- | | 2.03 | $0.59 \times 10^{-8}$ | $0.14 \times 10^{-4}$ | 4.2 | |
| 128 | 2,2',3,3',4,4'- | 1.26 | | $0.35 \times 10^{-8}$ | $0.52 \times 10^{-4}$ | 0.7 | 6.96 |
| 185+ | 2,2',3,4,5,5',6- | | 5.78 | $0.72 \times 10^{-8}$ | $0.16 \times 10^{-4}$ | 4.5 | |
| 174+ | 2,2',3,3',4,5,6'- | | 2.05 | $0.12 \times 10^{-7}$ | $0.16 \times 10^{-4}$ | 7.5 | |
| 177 | 2,2',3,3',4',5,6- | | 0.33 | $0.43 \times 10^{-8}$ | $0.17 \times 10^{-4}$ | 2.6 | |
| 172 | 2,2',3,3',4,5,5'- | | | $0.55 \times 10^{-8}$ | $0.17 \times 10^{-4}$ | 3.1 | |
| 180+ | 2,2',3,4,4',5,5'- | | 14.45 | $0.50 \times 10^{-8}$ | $0.17 \times 10^{-4}$ | 3.0 | |
| 170+ | 2,2',3,3',4,4',5- | | 3.80 | $0.37 \times 10^{-8}$ | $0.19 \times 10^{-4}$ | 1.9 | |
| 196+ | 2,2',3,3',4,4',5',6- | | 1.38 | $0.48 \times 10^{-8}$ | $0.68 \times 10^{-5}$ | 7.1 | |
| 201+ | 2,2',3,3',4',5,5',6- | | 1.50 | $0.45 \times 10^{-8}$ | $0.71 \times 10^{-5}$ | 6.4 | |
| 195 | 2,2',3,3',4,4',5,6- | | | $0.98 \times 10^{-9}$ | $0.78 \times 10^{-5}$ | 1.3 | |
| 194 | 2,2',3,3',4,4',5,5'- | | 0.83 | $0.38 \times 10^{-8}$ | $0.81 \times 10^{-5}$ | 4.7 | |
| Aroclor 1242 | | | | $0.51 \times 10^{-6}$ | $0.15 \times 10^{-2}$ | 3.4 | 4.5-5.8 |
| Aroclor 1248 | | | | $0.84 \times 10^{-7}$ | $0.19 \times 10^{-3}$ | 4.4 | 5.8-6.3 |
| Aroclor 1254 | | | | $0.26 \times 10^{-7}$ | $0.92 \times 10^{-4}$ | 2.8 | 6.1-6.8 |
| Aroclor 1260 | | | | $0.28 \times 10^{-8}$ | $0.81 \times 10^{-5}$ | 3.4 | 6.3-7.5 |

[a]The numbers in this column denote the chlorine atoms' location on each PCB molecule.

Table II. Physicochemical Properties of PCBs

| Property | Value |
|---|---|
| Solubility | $1 \times 10^{-4}$-$6 \times 10^{-6}$ mol/m$^3$ |
| Vapor pressure ($P_v$) | $1 \times 10^{-7}$-$1 \times 10^{-9}$ atm |
| Henry's law constants ($K_H$) | $0.7 \times 10^{-4}$-$1.0 \times 10^{-3}$ atm m$^3$/mol |
| Log octanol–water partition ($K_{ow}$) | 4.5->7.71 |
| Environmental concentrations | |
| water | 0.1–5.0 $\mu$g/m$^3$ |
| air | 0.05–5.0 ng/m$^3$ |
| Environmental partition coefficients | |
| ($K_p$, $K_{oc}$) | $10^4$-$10^7$ L/kg |
| Bioconcentration coefficients ($K_{BCF}$) | $5 \times 10^4$-$1 \times 10^6$ |
| Hydrospheric residence times | months to years |

Table III. PCB Concentrations in Air, Water, and Sediment in Lake Superior

| Time Period | Mean | Range | Standard Deviation | Number of Samples | Aroclor 1242 (%) |
|---|---|---|---|---|---|
| | | Air (ng/m$^3$) | | | |
| 1978 | 1.5 | 0.9-3.5 | 0.8 | 13 | 79 |
| 1979 | 0.9 | 0.4-1.4 | 0.4 | 8 | 54 |
| 1980[a] | 1.0 | 0.1-2.5 | 1.0 | 8 | 79 |
| 1981 | 1.0 | 0.1-3.5 | 0.14 | 9 | 61 |
| 1978-1981 | 1.0 | 0.1-35.5 | 0.6 | 38 | 70 |
| | | Water (ng/L) | | | |
| 1978 (July) | 1.3 | 0.4-7.4 | 1.3 | 28 | 49 |
| 1979 (June) | 3.8 | 0.3-8.4 | 1.9 | 35 | 39 |
| 1980 (August) | 0.9 | 0.3-2.1 | 0.4 | 63 | 56 |
| 1978-1980 | 1.8 | 0.3-8.4 | 0.2 | 126 | 50 |
| 1983[b] | 0.6 | 0.3-0.8 | 0.2 | 35 | 70 |
| | | Sediment (ng/g)[c] | | | |
| 1978, 1979, and 1982 | | | | | |
| (Whole-Lake) | 53 | 0.7-220 | 62 | 29 | 45 |
| (Nonimpacted) | 36 | 0.7-220 | 50 | 23 | 45 |

[a]Excludes value of 9.3 ng/L.
[b]Data are from reference 66.
[c]Sediment samples were between 0.0-0.5 cm.

over Lake Superior of 30 ng/L (*12, 48*) and an air concentration ($C_{air}$) of 1 ng/m$^3$, we estimated the mean scavenging coefficient ($W$) to be $3.0 \times 10^4$, which compares favorably to other field values (*12*).

$$W = \frac{C_{rain}}{C_{air}} = \frac{3.0 \times 10^{-6} \text{ g/m}^3}{1 \times 10^{-9} \text{ g/m}^3} = 3.0 \times 10^4 \tag{3}$$

Considering that $K_H$ for Aroclor mixtures and individual congeners is $3$-$4 \times 10^{-4}$ atm · m$^3$/mol (*see* Table I), then PCB concentrations in rain are likely due to efficient scavenging of particles. This inference is consistent with the model recently developed by Mackay et al. (*49*), which describes the transfer rates of organic chemicals between atmosphere and water. Thus, even though less than 10% of the atmospheric burden of PCBs is associated with particles, wet deposition is governed by this fraction. Over the 4 years of sampling, atmospheric PCB levels were not significantly correlated with location, distance from shore, wind direction or speed, or wave height. Between 1978 and 1981, mean PCB concentrations decreased from 1.5 to 0.3 ng/m$^3$. The concentrations were slightly higher in the western basin over the years of study, but the highest concentration (9.3 ng/m$^3$) was observed in the eastern basin in 1980.

**Air–Water Transfer Processes.**    AEROSOL-VAPOR PARTITIONING. Atmospheric organic chemicals exist in the vapor phase and are adsorbed to total suspended matter (TSM). The processes by which trace organic chemicals are removed from the atmosphere (wet particle, dry particle, and vapor deposition) and the quantity ultimately deposited on the water–land surface depend on the distribution between the vapor and particle phases. Theoretical considerations (*47*) and laboratory and field measurements (*12, 50, 51*) suggest that vapor–aerosol partitioning depends on contaminant vapor pressure, size and surface area of the aerosol, and the organic carbon content. The less volatile the compound, the more it tends to associate with the aerosol and the more important role played by particle scavenging. Junge (*47*) has estimated vapor–aerosol distributions of several chlorinated hydrocarbons based on saturation vapor pressure and available particle surface area for urban, rural, and clean air environments by using the following equation:

$$\phi = \frac{c \times \theta}{(P_v + c\ \theta)} \tag{4}$$

where $\phi$ is the ratio of adsorbed organic vapor on aerosol to the total amount of vapor in air, $\theta$ is the aerosol surface area, $P_v$ is the saturation vapor pressure, and $c$ is the constant that depends on heat of condensation and molecular weight (0.13 for many organic chemicals).

This simple model shows that $\phi$ in clean air environments (not in urban or industrial plumes) is small if $P_v < 10^{-6}$ torr. Therefore, in airsheds having low total suspended matter (1–20 $\mu$g/m$^3$), many PCB congeners, 1,1'-(2,2,2-trichloroethylidene)bis[4-chlorobenzene] (DDT), Hg, and low molecular weight polycyclic aromatic hydrocarbons (PAH) should exist primarily in the vapor phase. Organic compounds having $P_v > 10^{-6}$ torr should exist in the vapor phase, and those having $P_v \leq 10^{-8}$ torr should exist in the particulate phase. In reality, most high molecular weight organic compounds lie between these extremes, and their distribution and atmospheric half-lives depend largely on the particle concentration and composition in the atmosphere. PCBs, DDT, and other low molecular weight hydrocarbons and chlorinated hydrocarbons are found operationally to occur in the vapor phase in such diverse environments as the Great Lakes, remote marine systems, and urban air (12, 43, 45, 47, 50). Thus, both Junge's model calculations and field data support the hypothesis that many chlorinated hydrocarbons and PAHs, even those with low $P_v$ values, are transported in the vapor phase. Less than 10% of PCBs, DDT-group compounds, many PAHs, chlorinated benzenes, and even phthalate esters in the remote and rural atmosphere are on particles (see Chapter 2).

WET REMOVAL OF ORGANIC COMPOUNDS.    Wet removal of hydrophobic organic compounds from the atmosphere occurs by the scavenging of particles and by partitioning of vapor into rain and snow. The relative importance of these two processes depends on the distribution of the organic compound between vapor and aerosol, particle-size distribution, and Henry's law constant $(K_H)$. Wet removal rates of organic vapors are low if $K_H$ is small, and atmospheric residence times may be long. As a result, airborne organic species should be uniformly distributed through the troposphere away from sources.

An atmospheric vapor should attain equilibrium with a falling raindrop in 10 m (52, 53), and the washout ratio (W) defining the scavenging efficiency ($\alpha$) may be written as

$$W = \frac{RT}{K_H} = \alpha \tag{5}$$

where R is the gas constant (atm $\cdot$ m$^3$/mol $\cdot$ K), and T is the absolute temperature (K). Surface fluxes (F) may be written as

$$F_w = \alpha \times P \times C_a = W \times P \times C_a \tag{6}$$

where P is the rainfall amount (m/year), $C_a$ is the concentration of organic vapor in air (ng/m$^3$), and $\alpha$ is the solubility coefficient. Field-determined W values are generally larger than theoretical W values calculated from $K_H$. This difference suggests that particle scavenging by rain is an important flux term. A calculation of the expected organic concentrations in rain resulting from vapor scavenging is informative. The wet flux of an organic vapor scavenged by rain is

$$F_w = \alpha \times P \times C_a = P \times P_v/K_H \tag{7}$$

where P = 0.8 m/year rain, $C_a$ is in moles per cubic meter of organic vapor, $P_v$ is partial pressure (atm) of organic vapor, $K_H$ is in atmospheres cubic meters per mole, and $F_w$ is in moles per square meter per year. If $P_v$ is chosen as $1 \times 10^{-13}$ atm (1 ng/m$^3$ for a molecular weight of 250 g/mol), then $F_w$ may be calculated for different $K_H$ values. Washout ratios estimated from $K_H$ range from 0 to $10^4$. For PCBs having $K_H = 10^{-3}$–$10^{-4}$ atm m$^3$/mol, $F_w = 0.8$–$8 \times 10^{-10}$ mol/m$^2$ year. This value corresponds to rain concentrations of 0.025–0.25 ng/L by using a molecular weight of 250 g/mol. The rain PCB concentrations, which could not result from vapor scavenging, have been reported from 5 to 50 ng/L in the Great Lakes region (8, 12). More recent measurements suggest that PCB concentrations in rain collected in Minnesota are now lower, approximately 1–10 ng/L (S. J. Eisenreich, unpublished data). Mean PCB concentrations in wet-only precipitation collected in the Lake Superior basin in 1983 ranged from 1.4 to 16 ng/L on Isle Royale and from 0.6 to 48 ng/L on Caribou Island (54). If $K_H \geq 10^{-4}$ atm m$^3$/mol, vapor scavenging is unlikely to be important to the total flux.

Scavenging of particles with sorbed PCBs may be treated in a similar way by using equation 8.

$$F_w = W_p \times P \times C_p \tag{8}$$

where $p$ signifies particle-bound organic compounds and $W_p$ is the particle washout ratio. Depending on particle size, precipitation intensity, and type of meteorological event, $W_p = 10^4$–$10^6$ (55). The higher $W_p$ value implies that the aerosol has been readily incorporated into cloud water and is likely to be soluble. The lower $W_p$ value implies that the PCB is sorbed to particles in the 0.1–1.0-$\mu$m range and may be nonhygroscopic. Even though >90% of the atmospheric burden is in the vapor phase, field experiments support the hypothesis that particle scavenging dominates wet and total fluxes (49).

**DRY DEPOSITION.**   Transfer of organic vapors across the air–water interface is often predicted from a two-film diffusion model (12, 56–58). In this model, the rate of gas transfer between the well-mixed air and water reservoirs across the gas and liquid stagnant films at the interface is governed by molecular diffusion and is driven by the concentration gradient between the equilibrium concentrations at the interface and bulk reservoirs. For steady-state transfer, the flux ($F_v$) is given by equations 9 and 10.

$$F_v = K_{OL} \left[ C - (P_v/K_H) \right] \tag{9}$$

$$1/K_{OL} = 1/k_L + RT/k_g K_H \tag{10}$$

where $F$ is the flux (mol/m$^2 \cdot$ h); $K_{OL}$, $k_L$, and $k_g$ are the overall, liquid- and gas-phase mass-transfer coefficients (m/h); $C$ is the dissolved solute concentration in the liquid phase (mol/m$^3$); $P_v$ is the solute partial pressure (atm); $T$ is the absolute temperature $(K)$; and $R$ is the gas constant. The volatilization or absorption rate may be controlled by resistance to mass transfer in the liquid phase, gas phase, or a combination of the two. At typical values of $k_L$ and $k_g$ (20 and 2000 cm/h, respectively), resistance to mass transfer occurs >95% in the liquid phase for $K_H \geq 4.4 \times 10^{-3}$ atm m$^3$/mol and >95% in the gas phase for $K_H \leq 1.2 \times 10^{-5}$ atm m$^3$/mol. Considering the range of $K_H$ values for PCB congeners (Table I), 60%–90% of the resistance occurs in the liquid phase at 25 °C. This resistance extent implies that, in general, slightly soluble PCBs having $K_H > 10^{-4}$ atm m$^3$/mol tend to volatilize from water, but the direction and magnitude of the flux depends on the concentration gradient and mass-transfer coefficients. Examples of other compounds that are affected include many chlorinated hydrocarbons including pesticides, chlorinated benzenes, and tetrachloroethylene.

An alternate parameterization of vapor exchange involves the expression of concentration in fugacity $(f)$ terms. Thus, the driving force for vapor transfer is $f$, and a kinetic term is used for transport as follows $(49)$:

$$F_v = D_{AW} \, (f_W - f_A) \tag{11}$$

where

$$1/D_{AW} = 1/D_W + 1/D_A \tag{12}$$

$$D_A = k_A Z_A \qquad D_W = k_W Z_W \tag{13}$$

and where $D_A$ is the transport parameter for air, $D_W$ is the transport parameter for water, and $D_{AW}$ is the overall transport parameter. Fugacity is related to concentration by a fugacity capacity, $Z$ (mol/m$^3$ atm), and $Z_W = 1/K_H$ and $Z_A = 1/RT$. Also, $D_{AW}$, $D_A$, and $D_W$ have units of moles per square meter per day per atmosphere. Mackay et al. $(49)$ have shown that equations 9 and 10 are equivalent to equations 11–13. Again, large values of $K_H$ result in liquid-phase-controlled volatilization and can result in rapid vapor loss to the atmosphere given an appropriate concentration or fugacity gradient.

The critical parameters in determining vapor exchange at the air–water interface are $K_H$, $k_L$, $k_g$, and $K_{OL}$, or $K_H$, $D_A$, $D_W$, and $D_{AW}$. Henry's law constants may be calculated as the ratio of the subcooled liquid vapor pressure (atm) and the saturation aqueous solubility (mol/m$^3$) or as the ratio of atmospheric partial pressure of the vapor in equilibrium with the dissolved solute concentration, or may be measured directly by gas stripping.

Theoretical and experimental methods to estimate mass-transfer coefficients in the field for gas- and liquid-phase-controlled organic compounds have been developed. Mackay and Yuen (*56*) have developed equations to predict environmental mass-transfer coefficients based on Schmidt number $(S_c)$ and windspeeds at a reference height of 10 m $[U_{10} \ (m/s)]$.

$$k_g = (1.0 \times 10^{-3}) + (46.2 \times 10^{-3}) \ U^* \ S_{cG}^{-0.67} \tag{14}$$

$$k_L = (1.0 \times 10^{-6}) + (34.1 \times 10^{-4}) \ U^* \ S_{cL}^{-0.5} \text{ for } (U^* > 0.3) \tag{15}$$

$$k_L = (1.0 \times 10^{-6}) + (144 \times 10^{-4}) \ U^{*2.2} \ S_{cL}^{-0.5} \text{ for } (U^* < 0.3) \tag{16}$$

$$U^* = (6.1 + 0.63 \ U_{10})^{0.5} \ U_{10} \tag{17}$$

where $U^*$ is the air-side friction velocity (m/s); and $S_{cG}$ and $S_{cL}$ are the Schmidt numbers for gas- and liquid-phase-controlled organic compounds, respectively. These values correspond to still-air-transfer coefficients of $k_L = (1.0 \pm 0.5) \times 10^{-6}$ m/s (0.086 m/day) and $k_g = (1.0 \pm 0.5) \times 10^{-3}$ m/s (86.4 m/day). The effect of increasing wind speed is to increase the rate of mass transfer, especially for liquid-phase-controlled species. Values of $S_{cG}$ and $S_{cL}$ are about $2.0 \pm 0.2$ and $1000 \pm 200$, respectively, for organic solutes of interest. Results suggest that environmental mass-transfer coefficients will generally be lower than those measured in the laboratory. Annually integrated $k_L$ values for liquid-phase-controlled organic compounds determined from mass balance calculations on large lake systems are 0.2–0.25 m/day, or about 50%–100% smaller than predicted from laboratory experiments (Table IV). A physicochemical model of toxic substances in the Great Lakes (*62*) is in good agreement with open lake surface sediment data when a transfer coefficient of 0.1 m/day is used.

Table IV. Values of $k_L$ Determined from Mass Balance Calculations

| Location | Compound | $k_L$ (m/day) | Ref. |
|---|---|---|---|
| Lake Zurich | 1,4-dichlorobenzene | 0.24 | 59 |
| Saginaw Bay | PCBs | 0.2 | 60 |
| Lake Huron | (Aroclor 1242) | | |
| Lake Superior | PCBs | 0.24 | 61 |

Estimating volatilization of PCBs from lakes is complicated by our inability to accurately determine the concentration of the dissolved, unassociated species in equilibrium with the atmospheric vapor. This inability will be discussed later when partitioning between aqueous and solid phases is considered. An example of this phenomenon is presented in Figure 4, where Henry's law constants for several PCB congeners (Table I) are compared to apparent air–water partition coefficients $(H')$ determined from measurements of atmospheric PCBs in the vapor phase

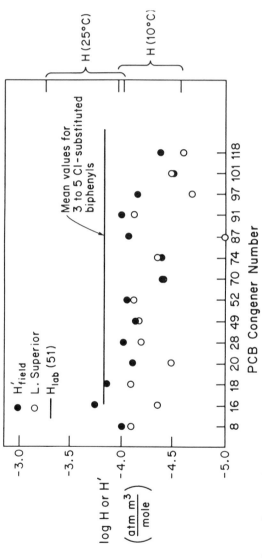

*Figure 4. Henry's law constants (H) for PCB congeners at 10 °C and 25 °C compared to air–water partition coefficients (H') measured in Lake Superior in 1980. (Data are from reference 34.)*

and dissolved PCBs in two surface water samples from Lake Superior. Henry's law constants are 50%–100% greater than air–water partition coefficients based on environmental measurements. Assuming equilibrium at the interface, we can calculate the fraction of nonfilterable PCBs that are dissolved in solution, and the fraction "associated" with other solution components such as dissolved organic matter or colloidal particles. For nonfilterable PCBs, 10%–15% are dissolved, and 55%–65% are associated. For filterable PCBs, 25%–30% are particulate.

Dry deposition of particle-bound organic chemicals depends on the type of surface, resistance to mass transfer in the deposition layer, particle size and concentration, and micro- and macrometeorology. Detailed discussions of the dry deposition process have been presented previously (*12, 63, 64*).

In the simplified case, the flux of particles to a receptor surface is given by

$$F_D = V_d \times C_p \tag{18}$$

where $V_d$ is the deposition velocity (cm/s), and $C_p$ is the contaminant concentration in the particle phase. $V_d$ depends strongly on near-surface turbulence, particle size distribution, and wind speed. Theoretical and experimental evaluations suggest that $V_d = 0.1$–1.0 cm/s for particles with mass median diameters of 0.1–1.0 $\mu$m (*63–65*). Much more research into the dry deposition process for particle input to lakes is needed before confidence can be placed in the results.

The relative importance of bubble stripping and bubble bursting to the net loss of PCBs from lakes has yet to be established. Ample evidence suggests that air bubbles are produced in natural waters, and bubble stripping models can estimate the loss due to this process. In the marine environment, bubble ejection at the air–water interface contributes significant quantities to the total burden of many chemical species in the lower atmosphere. At the present time, evaluating the net loss of organic chemicals from lakes is impossible by these processes.

AIR–WATER TRANSFER SCENARIO.  Figure 5 depicts our present understanding of the relative importance of air–water-transfer processes for hydrophobic organic chemicals and PCB cycling in large lakes distant from major sources. The dominant input pathway is the scavenging of particles from the atmosphere containing sorbed organic species; dry particle deposition is much less important. Particles thus reaching the lake equilibrate with the new aqueous environment and partition between the dissolved phase and other biotic and abiotic particle phases. PCBs in the dissolved phase equilibrate with the atmospheric gas phase. Several important observations can be made. The major input is through precipitation scavenging of particles, and the major loss is

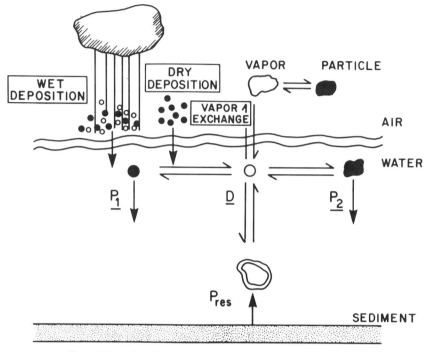

D = Dissolved - phase concentrations
P = Particle - phase concentrations

*Figure 5. Scenario for air–water transfer of PCBs in Lake Superior. Abbreviations are as follows: $P_1$ is atmospheric particles, $P_2$ is in-lake particles, and $P_{res}$ is resuspended particles.*

through volatilization and sedimentation. Conceivably, measurements of PCB flux to the atmosphere via volatilization may be more than counterbalanced by atmospheric inputs on particles (wet and dry inputs). Mackay et al. (*49*) describe this phenomenon as a dynamic, steady-state but nonequilibrium process whereby the input (particle scavenging) is connected to loss (volatilization) by the equilibrium partitioning of hydrophobic species. This model also suggests that airborne concentrations of hydrophobic organic chemicals over the lake are partially or wholly derived from in-lake processes. Compounds such as PCBs may cycle between water and air with intermittent periods of intense deposition (scavenging of PCB-laden particles) followed by slower but prolonged volatilization (*49*). The temperature dependence of $K_H$ for PCBs suggests that during periods of colder temperatures, gas-phase PCBs will partition (be absorbed) into the lake (*34*). For Lake Superior, this process may be important for 9 or 10 months per year.

## Water Column Concentrations and Processes

**Water Column Concentrations.** PCBs may be isolated from water by passage through styrene–divinylbenzene copolymer (XAD-2) macro-reticular resins that have been shown to be efficient at removing non-polar hydrophobic species (*61, 66*). In studies conducted in our labora-tory, PCBs in unfiltered water or PCBs passed through 0.6-$\mu$m glass fiber filters containing no bonding agent (292-mm diameter, stainless steel holder) were isolated by using a styrene–divinylbenzene copolymer resin (70–100 mL) held in a 2 × 15-cm glass column with a glass wool plug at each end. The water was drawn upward through the column at flow rates of 100–200 mL/min. From 90 to 300 L of water was passed through the glass fiber filters at a flow rate of 10–15 L/min. Filters and resins were extracted and analyzed as discussed previously. Although backup styrene–divinylbenzene copolymer columns showed no evi-dence of breakthrough, colloidal particles could pass through the glass fiber filter and the adsorbent resin column.

Bulk water or particle and filtered samples were obtained from Lake Superior during the consecutive summers of 1978–1980 (*23*) and in 1983 (*66*) and analyzed for total PCBs (*t*-PCBs), percentage of Aroclor mixtures, and 25–50 congeners. The range of PCB concentrations observed over the entire period was 0.3–8.4 ng/L (Table III), and annual PCB concentrations averaged 0.6–3.8 ng/L. With the exception of the summer of 1979, mean PCB concentrations have decreased from 1.3 to 0.6 ng/L over the years 1978–1983. This decrease represents a reduction of 54%. PCB congeners in the water column of Lake Superior may be best attributed to a 50:50 mixture of Aroclor 1242 and 1254 (1978–1980), in contrast to the dominance of Aroclor 1242 in the atmospheric PCB distributions. During the summer of 1983, the PCB congeners could best be explained by a 70:30 mixture of Aroclor 1242 and 1254 in the water column in response to the more efficient removal of the heavier PCB congeners by sedimentation in the western area and the decreasing input of PCBs from atmospheric sources. The areal and vertical trends were similar for the years 1978–1983; therefore, much of the discussion will be based on 1980 samples representing the most complete data set. Figure 6 depicts the overall data set including annual and areal distributions.

AREAL DISTRIBUTIONS. Concentrations of *t*-PCB in the surface (1 m) and bottom (5 m above the sediment) waters were relatively uniform throughout the open lake in 1980. The concentrations were 0.5–1.9 ng/L, (surface) and 0.3–2.1 ng/L (bottom). The lake was divided into four areas based on *t*-PCB and particle concentrations: the western arm, the central region between Isle Royale and the Keweenaw Peninsula, the eastern region, and the embayments (Figure 6). Waters of the western arm exhibited the lowest average PCB concentration (0.8 ± 0.2 ng/L) in

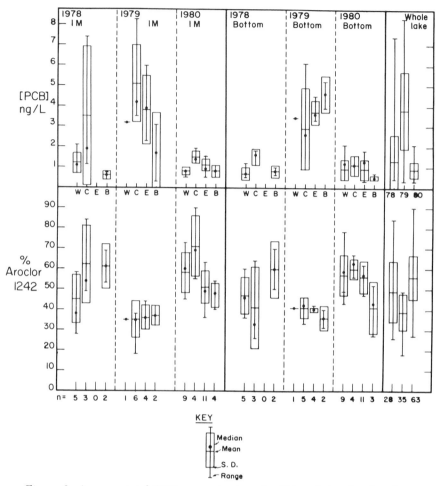

*Figure 6. Average total PCB concentrations (ng/L) and Aroclor distributions for Lake Superior water, 1978–1980. Abbreviations are as follows: n is the number of samples, W is west, C is central, E is east, and B is bays and harbors. (Reproduced with permission from reference 23. Copyright 1985, International Association of Great Lakes Research.)*

1980 surface water, the central region supported the highest concentration (1.5 ± 0.3 ng/L), and the eastern region showed an intermediate PCB level (1.1 ± 0.3 ng/L). This trend was also observed for the 1978 and 1979 samples (Figure 6) in those areas of the lake where similar samples were taken.

Because atmospheric deposition is a major source of PCBs to Lake Superior, and the PCB concentration in the air is relatively uniform over the entire lake (12), deposition to the surface waters should be approximately uniform. The surface waters might be expected to exhibit a uniform concentration, but instead, areas with consistently different average concentrations have been observed during all 3 years. These variations may be accounted for in terms of source, transport, and loss processes.

Observations of the lowest average t-PCB levels in the western arm were somewhat unexpected because the large volume of ship traffic, the relatively shallower depths for much of the area, and proximity to the Duluth–Superior urban area are likely to contribute PCBs to the water column. The low t-PCB concentrations can be attributed to a more rapid rate of sedimentation in this area of the lake (67). The western arm has the highest suspended solids concentration due mainly to erosion of red clay soils that also act as the major source of sediments to Lake Superior (68).

The surface waters of the central region showed the highest average t-PCB concentrations. This region also had the highest individual concentration of any open lake site (7.4 ng/L in 1978 and 8.4 ng/L in 1979). These concentrations suggest a local source of PCBs. Swain (8) measured elevated PCB water concentrations on and around Isle Royale and in fish from Siskiwit Lake (Isle Royale). Such levels are unexpected in this national park, which has stringent regulations aimed at maintaining a pristine environment. Because the prevailing winds are from the northwest more than 40% of the time, the industrial complex centered around Thunder Bay, Ontario, may be a significant source of the observed elevated PCB levels in the central region. Hites and co-workers (69, 70) have also found significant concentrations of atmospherically derived PAHs and chlorinated dioxins and furans on Isle Royale.

The eastern region had a t-PCB concentration that is intermediate between the two areas previously discussed. This value reflects the influence of lakewide atmospheric deposition. The only PCB removal mechanism unique to the eastern region is the outflow through the St. Mary's River, but this outflow is insignificant (<100 kg PCB/year, $Q = 2100$ m³/s, and t-PCB = 1 ng/L for 1980) compared to the total water burden of 11,000 ± 5400 kg PCBs estimated for 1980.

The embayments, harbors, and near-shore areas of Lake Superior represent environments far different from the open water and are susceptible to local sources and sediment resuspension. Some areas showed high suspended solids concentration (Duluth, 11.1 mg/L in 1979 and Black Bay, 5.0 mg/L in 1980), but most areas showed concentrations close to the open lake value of 0.1–0.5 mg/L. The measured t-PCB concentrations showed some bays with similar or lower concentrations (Black Bay and Thunder Bay in 1978 and 1980) as compared to the open water average. Anderson et al. (71) have observed elevated PCB levels along Harbor Beach, Lake Huron [154 ng/L (average) and 34–586 ng/L (range)] compared to the 1-ng/L concentration normally found in Lake Huron's open water. They attributed these high levels to resuspension of sediments by a major storm 2 days before sampling. Resuspension derived from storms, currents, and passing ships (72), and direct inputs from populated areas are significant mechanisms for the increased PCB levels in some near-shore areas of Lake Superior.

A sampling trip was conducted in early October, 1980, in and immediately outside Duluth–Superior Harbor, from which eight t-PCB samples were obtained from the surface waters. The average t-PCB concentration was 1.4 ± 1.2 ng/L (range 1.1–3.3 ng/L) and the Aroclor distribution was quantified as 35% ± 7% Aroclor 1242 (range 26%–45%). The elevated t-PCB concentration and the dominance of the higher chlorinated Aroclor mixture are both attributed to waste water discharges to the harbor and to resuspension of harbor sediments.

The Aroclor distributions within any 1 year were consistent throughout the surface waters (Figure 6), except for the central region in 1978 and 1980 where a few sites exhibited a higher Aroclor 1242 distribution. These sites also had elevated t-PCB concentrations compared to the rest of the area. For example, site 8 in 1978 had 7.4 ng/L t-PCB and 84% Aroclor 1242, and site 12 in 1980 had 1.9 ng/L t-PCB and 90% Aroclor 1242. This elevated concentration may reflect the direct influence of heavy rainfall at these sites prior to sampling because the atmospheric PCBs over Lake Superior contain 70% Aroclor 1242 (12).

VERTICAL DISTRIBUTIONS.    Concentrations of t-PCB were similar in the surface, intermediate, and bottom waters of Lake Superior within any year. The similarity in concentrations and Aroclor distributions (85% of the sites had a variability of <20% between the surface and the bottom) is attributed to physical mixing of the water column, upwelling in localized areas during times of stratification, transport by sedimenting particles, and most importantly, absence of local sources.

During the three summers of sampling, depth profiles were obtained at seven sites (Figure 7). Most of the fluctuations in t-PCB concentration occurred in the upper waters in close contact with the atmosphere. Many of the sites showed t-PCB concentrations below the thermocline that were nearly uniform (sites 5, 7, 11, and 24). Site 25 (1979) and site 14 (1980) had increases in concentration with depth. This increase may be attributed to resuspension of bottom sediments. Eisenreich and Looney (73) have used site 5 (1978) as evidence for the atmospheric input of PCBs to Lake Superior. This profile was taken after 3 days of extreme calm, reflecting the atmospheric flux of PCBs to the lake surface and their downward diffusion. Murphy (74) suggested that particulate PCBs depositing on the lake surface could yield the observed results. These concentration profiles provide evidence for periods when PCB deposition or volatilization is important.

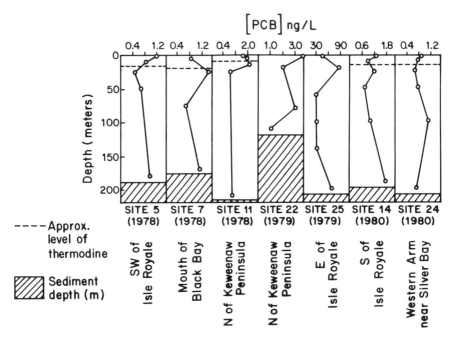

Figure 7. Total PCB profiles in the Lake Superior water column, 1978–1980. (Reproduced with permission from reference 23. Copyright 1985, International Association of Great Lakes Research.)

ANNUAL VARIATIONS. The three consecutive summers of sampling from 1978 to 1980 and again in 1983 in Lake Superior allowed for observations of the lake over time. The $t$-PCBs in 1979 exhibited elevated PCB levels compared to 1978, 1980, and 1983 throughout the lake. The surface waters of the central region provided the most consistent data set between 1978 and 1980. The $t$-PCB concentrations averaged 3.5 ± 3.4, 5.1 ± 1.9, and 1.5 ± 0.3 ng/L for 1978–1980. The Aroclor distributions for this area also reflect the overall trends for 3 years; [i.e., a smaller percentage of Aroclor 1242 in 1979 (35% ± 9%) compared to 1978 or 1980 (62% ± 19% and 71% ± 15%, respectively)]. Thus, water column PCBs in 1979 were dominated by the higher chlorinated congeners, and 1978 and 1980 data showed a predominance of the lower chlorinated congeners. The variations in $t$-PCB concentrations and Aroclor distributions are both unexpected results but appear to be real because analytical uncertainties were eliminated as a possibility. The changes in $t$-PCB levels and Aroclor distributions may be attributed to natural processes such as sediment resuspension.

From the water concentration, an estimate of the $t$-PCB water column burden can be made. Because the 1979 and 1980 data sets are the most complete, the best comparisons can be made between these two years. The volume-weighted averages based on the three open lake areas showed a total PCB burden of 40,000 ± 7300 kg for 1979 and 11,000 ± 5400 kg in 1980. The 14 months between the 1979 and 1980 cruises, therefore, represented a loss of 29,000 kg of PCBs from the water.

A qualitative as well as quantitative change took place in the PCB load during this time interval. A comparison between normalized concentration and normalized percentage Aroclor 1242 (23) showed a difference in the PCB composition between the 3 years. The samples from 1978 and 1980 exhibited a broader range of percentage Aroclor 1242, but a narrower PCB concentration range as compared to the 1979 samples. From an examination of the 1979 and 1980 data sets, 70% of the "additional" t-PCBs in the water column in 1979 (compared to 1978 and 1980) consisted of Aroclor 1254. The fraction of t-PCBs due to Aroclor 1242 was twice as high in 1979 as 1980, but those fractions due to Aroclor 1254 were 5 times as high. This fact strongly suggested a source that can contribute about 70% Aroclor 1254 to surficial sediments. Capel and Eisenreich (23) suggest that the length and severity of the previous winter explain these observations. The lake did not stratify in the summer of 1979, and this nonstratification permitted a prolonged period of sediment resuspension to occur.

Some evidence indicates that PCB concentrations exhibit seasonal variations in the water column as occur for nutrients, radionuclides, and metals (75–77). Richardson et al. (60) reported seasonal fluctuations in t-PCB concentrations for Saginaw Bay in Lake Huron. Rice et al. (78) reported t-PCB concentrations for the surface waters of Lake Michigan of 5.7 ± 1.1 ng/L for April 1978 and 2.9 ± 3.4 ng/L for August 1979. The elevated t-PCB levels during the spring in Lakes Huron and Michigan support our observations in this study. Data collected in May–June of 1984 and 1985 (isothermal period) by the Canadian surveillance program also showed higher PCB concentrations [1–4 ng/L (C. H. Chan, personal communication)]. The Lake Superior data for 1979 and 1980 represent the lake at near the ice-free seasonal extremes. The data set from 1979 showed elevated concentrations and an Aroclor distribution in which Aroclor 1242 represented <40% of the total. This value reflects the lake during the intensive spring mixing, whereas the 1980 samples showed late summer conditions with a distinct thermocline, lower t-PCB concentrations, and an average Aroclor 1242 distribution of >55%. This value implies a fast response of the lake during the ice-free months to the elevated spring concentrations and also the role of the benthic nepheloid layer (BNL) in distributing resuspended sediment throughout the water column during periods when the lake is not stratified.

**Benthic Nepheloid Layer.** The BNL is a well-characterized feature of oceanic systems (79, 80), but little information exists on its characteristics in freshwater environments, although the BNL may play an important role in recycling sedimentary contaminants. The BNL is a zone extending above the bottom sediments and containing measurable gradients of particles, heat, and chemical species. Particles in this region are small ($<$10 $\mu$m), have a high organic carbon content (*81, 83*), and generally have a similar organic composition to surficial sediments (*83*). Eadie and coworkers (*82, 84*) have studied the dynamics of BNL formation and destruction in Lake Michigan and its role in suspended sediment fluxes. Sandilands and Mudroch (*85*) and Baker et al. (*66*) indicate that the BNL is a common feature whose extent varies with depth in all of the Great Lakes. Evidence now suggests that seasonal variations in chemical species' concentrations (e.g., PCBs, cesium-137, and suspended solids) result from mixing of the BNL throughout the water column under nonstratified conditions. Eadie et al. (*81*) found that measured PCBs on particles collected in sediment traps were strongly influenced by the BNL in southern Lake Michigan. On the basis of sediment fluxes, particulate PCB concentrations, and sediment–water distribution coefficients, they estimated that reentrainment of surficial sediment may account for ≈14% of the particulate PCB inventory in the water column. Higher PCB concentrations observed in the water column of Lake Superior in 1979 and the similarity of PCB congener distributions to those observed in bottom sediments imply a major role for resuspension (*23*). Recently, Baker et al. (*66*) studied the dynamics and composition of the BNL in western Lake Superior with respect to PCB cycling. In this study, temperature and transmissivity depth profiles were measured at a grid of sampling sites during the stratified period of 1983. The BNL formed at the onset of stratification and was maintained through fall overturn. The BNL extended from the sediment to the thermocline (10–20 m thick) in water depths of 29–36 m. This layer was enriched in suspended solids and PCBs, and deficient in dissolved organic carbon relative to overlying waters. During this period, mean PCB concentrations were 0.56 ng/L in the surface waters and 0.87 ng/L in the BNL (different at $p < .001$). Average surface and BNL suspended solids concentrations were 1.1 and 4.8 mg/L, respectively.

To assess temporal variations in the mass loadings of PCBs to western Lake Superior, an integrated water column burden was calculated on the basis of in situ temperature and transmissivity profiles and PCB concentrations at depth. Figure 8 shows the water column burdens of suspended solids, PCBs, and 1,1'-(2,2-dichloroethylidene)bis[4-chlorobenzene] (*p,p'*-DDE) at one site over the study period from June to October 1983. The increase in PCB burden in mid-June coincides with elevated suspended solids concentrations in the BNL and is due mostly to increases of the heavier chlorinated congeners (e.g., Aroclor 1254). The increased PCB burdens were likely due to seiche-induced resuspen-

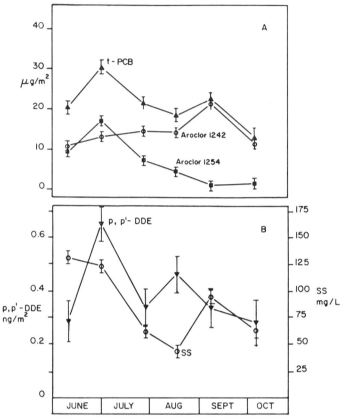

*Figure 8. Water column burdens of PCBs, p,p'-DDE, and suspended solids (SS) at site 10 (A, B) from June to October 1983. Error bars represent ±1 standard deviation based on propagated analytical uncertainties. Suspended solids burden is expressed in g/m². (Reproduced from reference 66. Copyright 1985, American Chemical Society.) Continued on next page.*

sion of surficial sediments in the lake, and this conclusion is supported by an increased p,p'-DDE burden, which is a tracer of sediment resuspension. In early September, another peak in PCB burden was observed amounting to 3.8 μg/m² above the summer level of ≈18.5 μg/m². The net increase in PCB burden consisted of an increase in the dissolved Aroclor 1242 concentration of 7.2 μg/m² and a decrease in the Aroclor 1254 concentration of 3.4 μg/m². The conclusion was that colder open lake water of higher PCB concentration and enriched in Aroclor 1242 contributed to the increase.

An interesting feature of PCB cycling in this study was the selective loss of higher chlorinated PCB congeners (dissolved and particulate Aroclor 1254). By assuming a first-order removal process, the mean loss

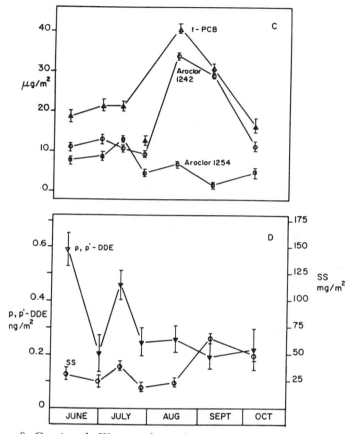

*Figure 8.* Continued. *Water column burdens of PCBs, p,p'-DDE, and suspended solids (SS) at site 12 (C, D) from June to October 1983. Error bars represent ±1 standard deviation based on propagated analytical uncertainties. Suspended solids burden is expressed in g/m². (Reproduced from reference 66. Copyright 1985, American Chemical Society.)*

rate of Aroclor 1254 at two sites, 0.23 and 0.10 $\mu g/m^2 \cdot$ day, corresponds to half-lives of $\approx$17 and 28 days, respectively. A Stokes settling velocity of 14 m/day is required (5-$\mu m$ particles) to yield the resulting profiles. Thus, higher chlorinated congeners are rapidly and efficiently removed from the water column under stratified conditions, probably by processes of differential settling, coagulation–flocculation, and incorporation into fecal pellets. The increase in dissolved Aroclor 1242 to maintain constant total PCB concentrations may result from diffusion of PCBs from pore water or mineralization of organic matter in the BNL followed by release of PCBs to solution. In any case, sorption or uptake of the more insoluble PCBs on biotic or abiotic particles must account for the rapid response. In 1985 and 1986, an instrumented four-person submersible was used to carefully examine the type and composition of particles in the BNL and the flocculant "fluff" layer on the sediment surface. No quantitative results have been available as yet.

**Sorption to Aquatic Particles.** The dominant fluxes of PCBs are largely restricted to particle transport and deposition, resuspension and diffusion, and vapor-phase transfer across the air–water interface. The dominant loss processes for PCBs are particle transport and sedimentation in quiescent areas of the lake and volatilization. The fate and residence times of PCBs are largely determined by their affinity for abiotic and biotic particles and the net settling rate of those particles.

Sorption of slightly soluble organic compounds to suspended particles and their subsequent settling is a dominant removal mechanism in lakes, estuaries, and oceans (*66, 81, 86, 87*). The magnitude of such fluxes should simply be the product of the contaminant concentration on the particles and their settling rates. This calculation is complicated by difficulties in estimating net fluxes of heterogeneous particle populations and by selective enrichment of certain organic compounds on particles. The more insoluble organic species may be removed more rapidly and efficiently than the more soluble ones by their incorporation into large particles such as fecal pellets, flocculated clays, or organic aggregates (*66*). Biologically active elements are preferentially scavenged from the water column by rapidly settling particles (*88, 89*). These elements may be subsequently released when the particles are degraded. Alternatively, abiotic formation of large (10–100-$\mu$m) flocculated clays may account for rapid organic chemical fluxes (*90*).

PCBs entering the lake will be distributed between the dissolved phase and the biotic, or detrital, phase. In most instances, the dissolved and particulate phases are assumed to be in local equilibrium and respond quickly to changes in pool levels. However, this assumption may not always be valid (*91*). The availability of the chemical and its removal from the water column by sedimentation and volatilization is dependent on the extent of partitioning and settling rate of particles. Biota may assimilate nonpolar organic compounds from the dissolved phase by direct partitioning or from the particulate phase by ingestion. The pathways are not independent in that the dissolved phase supports concentrations in both. The fraction of the chemicals in the particulate phase, coupled with net sediment fluxes, control water concentrations.

CONTROLLING FACTORS. Sorption of nonpolar organic chemicals from water by sediment consists primarily of partitioning into the sediment organic phase; adsorption onto or into inorganic mineral phases is relatively unimportant. The relevant factors governing partitioning to suspended and settled particles are attributed to the sediment (particle size, organic carbon content, concentration, and surface area) and organic contaminant (polarity, ionic behavior, solubility, octanol–water partitioning, and kinetics of adsorption and desorption). In general, sorption of organic compounds to particles increases with decreasing particle size and increasing organic carbon content of the particles and with decreasing ionic character and solubility of the organic compound (86). The organic pollutant concentration in the aqueous phase is related to the sorbed-phase concentration by an equilibrium isotherm. At low aqueous concentrations typical of the ambient environment ($<1$ $\mu$g/L), sorption isotherms are linear and the partition coefficient ($K_p$) can be expressed as

$$K_p = \frac{C_p/\text{SS}}{C\,(\text{aq})}\ (\text{L/kg}) \tag{19}$$

where $C_p$ is the concentration on particles ($\mu$g/kg), $C$ (aq) is the dissolved concentration ($\mu$g/L); and SS is suspended solids (kg/L).

Several researchers (36, 92–94) have noted that the partition coefficient is strongly related to the organic carbon content of the particle. Thus, Karickhoff et al. (36) expressed the partition coefficient of a hydrophobic organic compound in terms of the organic carbon content of the sorbent.

$$K_{oc} = K_p/f_{oc} \tag{20}$$

where $K_{oc}$ is the partition coefficient of the compound between water and a hypothetical natural sorbent containing 100% organic carbon, and represents the natural organic matter of the particle, and $f_{oc}$ is the fractional organic carbon content by weight. The partition coefficient, $K_{oc}$, is related substantially to the organic carbon content of the particle for nonpolar or hydrophobic organic compounds, and these substances are sorbed independently. The latter phenomenon and related sorption behavior strongly support a sorption mechanism whereby the nonpolar organic compound partitions into the organic matter on the particle.

Partitioning of nonpolar organic compounds between sediment organic matter and water are treated through linear free energy relationships as between an organic solvent and water. Thus, the partitioning of various chlorinated and nonchlorinated hydrocarbons between natural particles and water is strongly correlated with octanol–water partition coefficients ($K_{ow}$) and aqueous solubility ($S$). Because $K_{ow} \propto S^{-1}$, the aqueous solubility of the nonpolar organic compounds appears to be the dominant factor in determining $K_{oc}$. The role of different types of natural organic matter as it affects partitioning needs to be assessed. The $K_{oc}$ relationships should not be used where $f_{oc} < 0.001$ (37), and caution is recommended in directly applying various correlations to the environmental system of interest.

From both thermodynamic principles and experimentation, $K_p$ ($K_{oc}$) should be independent of solids concentration at solute concentrations much below solubility. However, nonlinear partitioning is observed in both laboratory and field systems (i.e., $K_p$ decreases with increasing suspended solids concentration). The reasons proposed for the observed nonlinear partitioning include experimental artifacts, nonlinear sorption isotherms at low aqueous concentration, slow desorption kinetics, and solute complex formation with aqueous organic matter. Voice et al. (92) proposed that organic microparticles break off from larger solids in laboratory sorption experiments and result in an overestimate of $C$ (aq), and thus a reduced $K_p$ or $K_{oc}$. This mechanism does not, however, explain many field observations (e.g., PCB partitioning in Lake Superior). Gschwend and Wu (93) suggest that slow sorption–desorption kinetics or filtration artifacts explain the phenomenon. Elzerman (see Chapter 10) further demonstrates the slow kinetics of desorption at high suspended solids concentrations. Regardless of the explanation, the implications on partitioning in natural aquatic systems have yet to be determined.

The rate of approach to equilibrium for partitioning of hydrophobic PCBs has been investigated (91, 93, 94). A two-component model is proposed whereby an aqueous pollutant ($P$) rapidly achieves sorptive equilibrium with phase $S_1$ and achieves equilibrium much more slowly with phase $S_2$.

$$P \underset{\longleftarrow}{\overset{X_1 K_p}{\longrightarrow}} S_1 \overset{k_d}{\longrightarrow} S_2 \qquad (21)$$

where $X_1$ is a fraction of the total sorptive capacity, and $k_d$ is a rate constant. In this case, $S_1$ and $S_2$ are distinguished only by the kinetics of transport and probably do not represent different sorbents. The time to achieve equilibrium for $S_1$ is at most minutes and represents $X_1$ (20%–60%) of the total sorptive capacity. Pollutant sorbed in $S_2$ accesses the remaining sorptive capacity over much longer time frames (hours or weeks). Note that $K_p$ (the equilibrium constant) describes partitioning in $S_1$ and $S_2$. Karickhoff (86) stated that the time required to achieve equilibrium varies significantly with suspended solids concentration. The higher the suspended solids level, the longer time needed to reach equilibrium. At low suspended solids concentration (a few milligrams per liter), even in high organic content particles, hydrophobic organic chemicals achieve equilibrium rapidly, and $K_{oc}$ may be predicted from $K_{ow}$. Equilibrium may not be achieved for months at very high solids concentrations, such as in pore water regimes of sediments. The nonlinearity of $K_p$ with increasing suspended solids may be explained by considering that kinetics of sorption and desorption are different at different levels of particle aggregation, which is influenced by suspended solids. Conceptually, $X_1$ likely corresponds to surface-bound organic species, and the remaining fraction corresponds to the slow diffusion of solute into the organic matrix. This relationship is consistent with the observation that desorptive release is dependent on pollutant molecular size and thus molecular diffusion.

In the natural aquatic environment, nonpolar solutes and sorbing particles are in contact for considerable periods, and "local" equilibrium is assumed. This equilibrium is favored, especially for sorption–desorption reactions, in systems where suspended solids concentrations are only a few milligrams per liter.

An estimate of the fraction of the total aqueous burden of hydrophobic organic chemicals in the dissolved phase ($F_d$) may be estimated as follows:

$$F_d = \frac{1}{[SS]K_p + 1} \tag{22}$$

where [SS] is the suspended solids concentration (kg/L), and $K_p$ is the equilibrium partition coefficient (L/kg). Applying the relationship developed by Schwarzenbach and Westall (37) for estimating $K_p$,

$$\log K_p = 0.72 \log K_{ow} + \log f_{oc} (s) + 0.49 \tag{23}$$

where $F_d$ may be calculated for a variety of $K_{ow}$ values ($10^3$–$10^6$), $f_{oc}$ (s) (0.01–0.4) values, and SS concentrations (1–1000 mg/L).

Figure 9 demonstrates that most of the organic component exists in the dissolved phase for suspended solids concentrations less than 10 mg/L, even where $K_{ow}$ values ($10^6$) and organic carbon content ($f_{oc} = 0.4$) are large. These conditions may be met in a productive lake. In the absence of nonequilibrium effects, the majority of the organic burden is dissolved in aqueous solution. Colloidal or dissolved natural organic matter binds hydrophobic organic compounds, effectively competing with particulate organic matter for the dissolved species and increasing $F_d$.

Gschwend and Wu (93), Means and co-workers (95-97), and Caron et al. (98) reported that colloidal organic matter binds compounds such as chlorinated pesticides and PCBs in freshwater and estuarine environments. Baker et al. (99) provided field data to support the hypothesis that a sizeable fraction of the dissolved PCBs in Lake Superior is bound to colloids.

In an estuarine environment, natural organic matter ("colloids") from drainage of peaty soils and in situ production influences the speciation and thus transport and bioavailability of incoming PCBs. In Lake

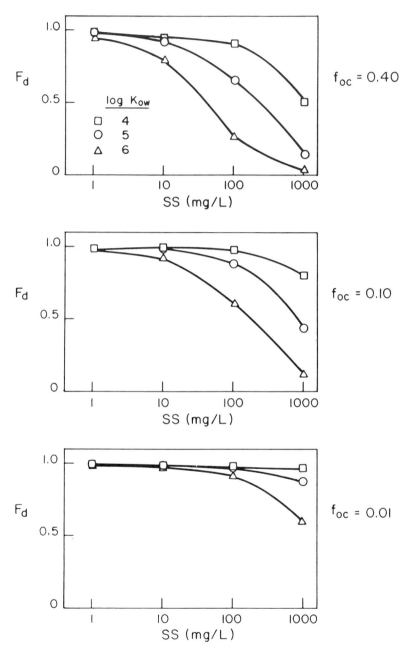

*Figure 9. Fraction of solute concentration in the dissolved phase* ($F_d$) *as a function of* $K_{ow}$, $f_{oc}$, *and SS.*

Superior, where total organic carbon values of 1–2 mg/L are common, this influence is less clear.

OBSERVATIONS. Capel and Eisenreich (23) measured PCBs (1980) in filtered and nonfiltered samples ($N = 24$) collected in the surface waters of Lake Superior to gain a better understanding of their distribution between particulate and nonparticulate (dissolved) fractions. The range in particulate PCB concentrations (30–2770 ng/g) far exceeded the range in $t$-PCB concentrations (0.5–1.9 ng/L). The particulate PCB concentrations were higher than had previously been reported for surficial sediments (18). The particulate PCB fraction was arbitrarily defined by the efficiency of the glass fiber filter; colloidal particles could pass through the filter and contribute to the dissolved-phase concentration. The distribution coefficient, $K_D$ (L/kg), can be calculated as

$$K_D = \frac{[PCB]_{filter}/[SS]}{[PCB]_{filtrate}/\text{water volume}} \qquad (24)$$

The distribution coefficient, $K_D$, is not a thermodynamic constant but is useful in quantifying distribution of PCBs and potential loss processes. The $K_D$ values varied from $10^4$ to $10^7$ L/kg, and 27% ± 12% of the water column PCBs were in the particulate phase. This value range for $K_D$ is in general agreement with the predicted distribution using the equation presented by Schwarzenbach and Westall (37) applied to Lake Superior conditions. The $K_D$ values exhibited an inverse log relationship to suspended solids (Figure 10).

$$\log K_D = -1.2 \log [SS] + 5.4 \qquad r^2 = 0.75, n = 24 \qquad (25)$$

This behavior may be due to analytical artifacts derived from colloidal organic interactions (92, 93, 96, 99), particle–particle interactions (100–102), slow desorption kinetics (91, 93), or exchangeable and nonexchangeable sorption sites (101). This behavior has been observed in other laboratory (100) and field (78, 103) studies. For example, the relationships observed in Lake Michigan from references 78 and 103, respectively, are

$$\log K_D = -1.19 \log [SS] + 6.1 \qquad (26)$$

$$\log K_D = -0.94 \log [SS] + 5.89 \qquad (27)$$

Variations in particulate organic carbon and different populations of particles having different sorption characteristics (i.e., $f_{oc}$) explain much of this variation. When particulate PCB concentrations (ng/g) are plotted against suspended solids concentration (mg/L) (Figure 11), high

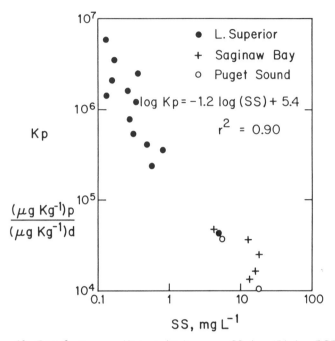

*Figure 10. Distribution coefficient ($K_p$) versus SS (mg/L) for PCBs in the Lake Superior water column, 1980. (Reproduced with permission from reference 23. Copyright 1985, International Association of Great Lakes Research.)*

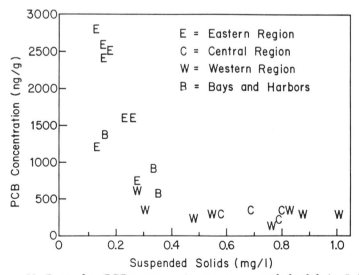

*Figure 11. Particulate PCB concentration versus suspended solids for Lake Superior water, 1980. (Reproduced with permission from reference 23. Copyright 1985, International Association of Great Lakes Research.)*

particulate PCB levels (PCBs) were found in the eastern basin as compared with the western basin, and the central basin was in transition. Solids in the eastern basin are largely organic particles of biotic origin and have little mineral matter (high $f_{oc}$ values), and the suspended solids in the western basin are a mixture of biotic particles and red clay particles from bluff and soil erosional inputs. Thus, much of the variation in $K_D$ is explained by high PCB concentrations in regions of low suspended solids concentration (<0.3 mg/L) and high $f_{oc}$ values. Filtrate concentrations remain relatively constant at about 75% of t-PCB levels.

Baker et al. (99) measured the distribution of PCBs between dissolved and particulate fractions in western Lake Superior throughout the ice-free period in 1983. Values of log $K_D$ for t-PCBs and 28 individual PCB congeners varied inversely with log [SS], and only 34% of this variation could be attributed to variations and controls exerted on partitioning by $K_{ow}$, $f_{oc}$, and SS. Laboratory-derived correlations between $K_D$, $f_{oc}$, and $K_{ow}$ (Table V) overestimate variation in $K_D$ observed in Lake Superior. Baker et al. modified a three-phase model proposed earlier by Gschwend and Wu (93) that incorporated binding of PCBs to nonfilterable microparticles and macromolecular organic matter and concluded that colloidal-bound PCBs may be a dominant component in waters of low suspended solids concentrations. The relationship of $K_D$ and SS for 1980 and 1983 is

$$\log K_D = -1.09 \ (\pm 22\%) \ \log [SS] + 5.21 \tag{28}$$

The 1980 and 1983 data, respectively, yield relationships given by equations 29 and 30.

**Table V. Correlative Relationships between Solid–Water Distribution Coefficients and Octanol–Water Partition Coefficients**

| Correlative Relationship | Matrix | Ref. |
|---|---|---|
| $\log K_{oc} = 1.00 \log K_{ow} - 0.21$ | sediments | 143 |
| $\log K_{oc} = 0.904 \log K_{ow} - 0.779$ | soils | 144 |
| $\log K_p = 0.72 \log K_{ow} + \log f_{oc} + 0.49$ | soils | 37 |
| $\log K_{oc} = 0.72 \log K_{ow} + 0.49$ | soils | 37 |
| $\log K_p = 0.748 \log K_{ow}$ $- 0.648 \log SS + 0.131 \log f_{oc}$ $+ 0.346$ | sediments | 92 |
| $\log K_{om} = 1.03 \log K_{ow} - 0.49$ | aquatic humic substances | 145 |

SOURCE: Data are from reference 131.
NOTE: $K_{oc} = K_p/f_{oc}$.

$$\log K_D = -1.2 \log [\text{SS}] + 5.4 \tag{29}$$

$$\log K_D = -0.85 \ (\pm 21\%) \log [\text{SS}] + 5.11$$

$$\text{range} - [\text{SS}] = 0.35 - 7.15 \text{ mg/L} \tag{30}$$

## Sedimentary Concentrations and Processes

**Factors Controlling Sedimentary Accumulation.** Lacustrine sediments act as ultimate sinks for hydrophobic organic compounds in large lakes, but the processes responsible for net sedimentary accumulation of hydrophobic organic compounds are only now being unravelled (Figure 12). As discussed previously, biotic and abiotic particles have a strong affinity for hydrophobic organic compounds that have $\log K_{ow} > 4$. Particles having higher organic carbon content ($f_{oc} > 0.001$) preferentially scavenge these compounds from the water column relative to inorganic particles (e.g., clays, sand, and calcite). Suspended particles are delivered to the sediment by gravitational settling, by being packaged into fecal pellets by zooplankton and subjected to the downward "fecal pellet express", and by coagulation followed by settling of larger aggregates. O'Melia (*104*) provides a case for efficient particle removal from the water column in large lakes by the latter process. Although deposition of fecal pellets from the marine pelagic zone is important, the importance of this process in freshwaters is unclear. For example, Heuschele (*105*) found

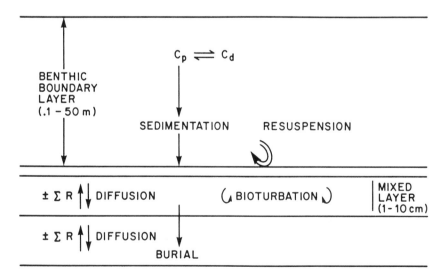

R = ADSORPTION/DESORPTION, DEGRADATION, ION EXCHANGE, DECAY

*Figure 12. Conceptual framework describing sediment–water interactions for nonpolar organic contaminants in large lakes.*

that fecal pellets in surficial Lake Superior sediments were derived from benthic organisms, and not those organisms from the euphotic zone. In this case, particles deposited onto the benthic zone may be repackaged by benthic organisms.

Particles reaching the sediment are focused into quiescent, deeper areas where resuspension and bottom currents are not sufficient to cause further transport. This process makes lakewide deposition estimates from one or a few cores tenuous. Because lake residence time of fine particles is often less than 1 year, even in large lakes (*see* Chapter 11), sediment profiles may still be interpreted as historical inputs to the ecosystem. Apparent settling velocities of 100–300 m/year are typical for large lakes. Sedimentary particles and solutes are subjected to diffusive and advective processes that result in the generation of a steady-state mixed zone of variable depth containing constant pollutant concentrations, and a deeper layer corresponding to permanent particle burial. Until recently, diffusional and advective transport were considered to be the primary processes responsible for water–sediment interactions, but now an increased awareness of solute and solid transport resulting from resuspension (*66, 82, 84, 85*) and bioturbation (*106*) has evolved. Robbins and co-workers (*107–112*) have been especially successful in using radionuclide tracers (e.g., $^{210}$Pb, $^{137}$Cs, and $^{7}$Be) in studying the effects of burrowing organisms on the rate of bioturbation and the chemical profiles that result. Mixing of the surficial sediments has the net effect of increasing the time over which sedimentary pollutants may be recycled into the ecosystem.

Sources of sediment deposition are either autochthonous (derived from in-lake processes, usually primary production) or allochthonous (erosion or riverine inputs), and the organic particles derived from photosynthesis in the water column have strong affinity for hydrophobic organic chemical species. Lakes receiving large amounts of organic and mineral particles have their water columns scavenged of dissolved organic pollutants, which are deposited in the bottom sediments. The particles deposited in deep waters have small grain sizes and higher organic carbon content than the sands and clays of nondepositional areas. One of the best examples of the influence of lake morphology and sediment focusing on the accumulation of organic pollutants is the accumulation of *t*-PCBs in Great Lakes sediment [Figure 13 (*113*)]. Distribution of sediment types was determined from echo sounding, and the bottom sediment was sampled at approximately 1200 locations for analysis of grain size and a variety of inorganic and organic species. Nondepositional zones in shallow areas of lakes were characterized by the occurrence of bedrock, rare sand deposits, lag sands, and gravels that veneer the exposed surfaces of glacial tills and glacial lacustrine clays. Depositional zones in deeper, offshore areas of the lake were

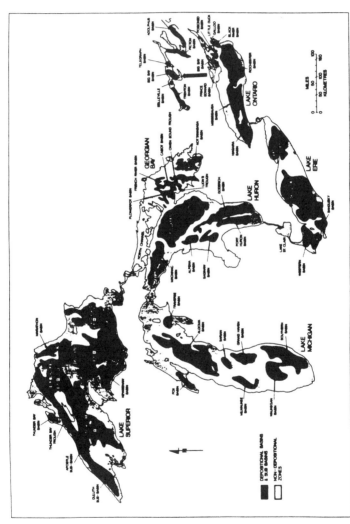

*Figure 13A. Distribution of the depositional basins in the Great Lakes. (Reproduced with permission from reference 113. Copyright 1983, Ann Arbor Science Publishers.)*

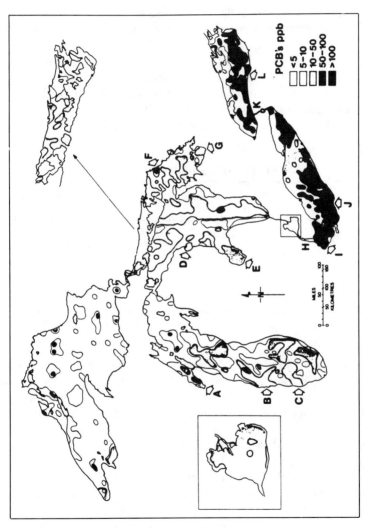

*Figure 13B. PCBs (ng/g) in the surface (0-3-cm) section of sediment in the Great Lakes. (Reproduced with permission from reference 113. Copyright 1983, Ann Arbor Science Publishers.)*

characterized by silty clays and clays rich in organic matter. The lower part of Figure 13 clearly shows that PCBs, typical of hydrophobic organic species, accumulate in areas receiving heavy local loads and in depositional basins. Eisenreich and Johnson (11) calculated a conservative burden of PCBs in the Great Lakes ecosystem and the fraction in the sediment compartment (0.0–3.0 cm) on the basis of the preceding data. Table VI shows that of the 442–504 metric tons in the Great Lakes, 83%–94% occur in "active" surficial sediments, and the largest burdens occur in the urban (industrial) lower lake areas and those areas with the highest mean sedimentation of fine particles. PCB burdens may be overestimated by a factor of 2 for Lakes Superior and Huron but are likely underestimated in the other lakes by a factor of 2 or 3. These totals do not include PCBs deposited in contaminated harbors (e.g., 500 metric tons in Waukegan Harbor in Lake Michigan) or embayments (e.g., Green Bay in Lake Michigan).

**Table VI. PCB Burden in the Great Lakes**

| Lake | Total Lake Burden ($\times$ $10^3$ kg) | % Accumulation in Sediment |
|---|---|---|
| Superior | 32–44 | 53 |
| Michigan | 47–67 | 74 |
| Huron | 62–76 | 84 |
| Erie | 183–193 | 97 |
| Ontario | 119–125 | 94 |
| Total | 442–505 | |

SOURCE: Reproduced with permission from reference 11. Copyright 1983, Butterworth Publishers.

**Sedimentary Fluxes and Diagenesis.**   Sedimentation and burial is usually the most important removal pathway for hydrophobic organic compounds in large lakes because of the strong sorptive processes and long water residence times. The burial process, and therefore the chemical profile in the sediment, is influenced by the presence of benthic organisms that mix sediments differentially with depth. Macrobenthos such as *Pontoporeia* mix surficial sediment in a manner analogous to particle diffusion, and tubificid worms ("conveyor-belt" species) mix sediment in a manner best modeled by advection (114). The conveyor-belt species feed with their heads extending downward into sediments and tails protruding above the sediment surface. Ingested material passing through the gut is redeposited on surface sediments. Advective mixing decreases in intensity with increasing sediment depth. Robbins (114) concludes that diffusive and advective processes derived from bioturbation combine with sorptive equilibrium and degradation to yield sedimentary profiles observed in the Great Lakes.

Sedimentation rates in the Great Lakes vary from 0.01 to 0.2 cm/year in depositional basins of Lake Superior to as much as 2 cm/year in Lake Erie. Bioturbation rates ($D_B$) vary from $10^{-7}$ to $10^{-11}$ cm$^2$/s depending on type and integrated number of organisms. Solute diffusion coefficients in water ($D_w$) are in the range of $10^{-6}$ cm$^2$/s, and apparent diffusion coefficients ($D_{app}$) for hydrophobic species are in the range $10^{-9}$–$10^{-13}$ cm$^2$/s. These values suggest that biological mixing has the greater opportunity to affect the resulting sedimentary profiles. Mixed depths on the order of 1–10 cm have been observed in large and small lakes, and the depth of mixing is dependent on bulk and organic carbon sedimentation rates as well as the total number of organisms.

The combined effect of the above physical and biological processes on the sedimentary profile can best be demonstrated by using the hypothetical situation in which an organic pollutant is discharged to a lake followed by deposition and mixing. Assuming diffusion is negligible and sedimentation is constant, the resulting profile depends on $D_B$. If the rapid steady-state mixing model is used (115), the pulse input yields an undistorted profile if $D_B = 0$ or is very small, but the sediment is mixed vertically in both directions and the historical pattern is destroyed if $D_B \geq 3$ cm$^2$/year. Christensen (116) has mathematically modeled the mixing process by assuming that the intensity of bioturbation decreases exponentially with depth and with the shape of the chemical profile in the mixed zone corresponding to half of a Gaussian curve. Robbins (114) interprets mixing in Great Lakes sediments similarly by interpreting that depth-dependent mixing results from the activity of diffusive-type organisms, which burrow down 1–2 cm, and advective-type organisms such as conveyor-belt species, which can burrow 10 cm. Downward perturbations of the chemical profile may be 10%–20% of the upward mixing.

**Sedimentary Concentrations and Fluxes.**   Sediments represent the ultimate sink for PCBs in the Great Lakes. As a result, numerous studies have been conducted on the distribution of PCBs in surficial sediments as an indication of environmental degradation. The most ambitious study to date was conducted from 1968 to 1976 by the Canadian government in which more than 1200 surficial sediment samples from the top 3 cm were collected and analyzed for PCBs and organochlorine pesticides. The intent of this study was to establish baseline levels and identify sources to enable future trends to be assessed. The surficial sediment data have been summarized by Thomas and Frank (113) and are presented for comparison in Table VII. The lower lakes, especially Erie and Ontario, have the highest mean and range of concentrations, averaging 50–60 ng/g. As expected, Lake Superior has the lowest PCB concentration, averaging 3.3 ± 5.7 ng/g followed by Lakes Huron, Michigan, Ontario, and Erie.

Table VII. PCB Concentrations in Great Lakes Sediments

| Location | Whole-Lake | Nondepositional Zones | Depositional Zones | Range |
|---|---|---|---|---|
| Superior | 3.3 | 3.9 | 4.8 | <2.5–57 |
| | (5.7) | (2.1) | (5.5) | |
| Michigan | 9.7 | 6.8 | 17.3 | <2–190 |
| | (15.7) | (8.1) | (23.9) | |
| Huron | 12.8 | 10.7 | 15.4 | <3–90 |
| | (10.3) | (7.3) | (12.8) | |
| Georgian Bay–North Channel | 11.2 | 11.2 | 11.1 | <3–43 |
| | (10.7) | (13.2) | (8.1) | |
| Erie | 94.6 | 64.0 | 115 | 4–800 |
| | (114) | (105) | (115) | |
| Lake St. Claire | 9.9 | | | <2–28 |
| | (6.3) | | | |
| Ontario | 57.5 | 28.1 | 85.3 | 5–280 |
| | (56.2) | (34.7) | (57.0) | |
| Bay of Quinte | 48.0 | 47.0 | | <2–260 |
| | (43) | (45) | | |

SOURCE: Data are from reference 113.
NOTE: All values are in units of nanograms per gram. Parentheses denote standard deviation.

The highest concentrations in Lake Superior were observed in the vicinity of Thunder Bay, Marathon Bay, and in the western arm near Duluth–Superior. These are all areas near or downwind of urban and industrial areas and may be expected to exhibit elevated concentrations. Actual surficial concentrations are somewhat higher because of dilution of surface PCBs by uncontaminated sediment in 0–3-cm sections. On the basis of these data and preliminary sediment cores collected in 1977, Eisenreich et al. (117) concluded that these accumulations could be accounted for by atmospheric deposition.

Dated sediment cores from aquatic ecosystems have the potential for providing detailed chronologies of contaminant input. The depositional history of PCBs in sediments has been documented for relatively few dated cores, including those from the Great Lakes (10, 11, 117–119), the Hudson River and estuary (120), and the Santa Barbara Basin off the southern California coast (121). Here, the chronology of PCB input to Lake Superior as recorded in bottom sediment cores dated with lead-210 is reported. These cores exhibit little mixing in the surface sediments due to low benthic activity (10, 67). Lake Superior receives in excess of 75% of its PCB burden from the atmosphere (12), and the sedimentary record should provide a detailed chronology of atmospheric inputs and the response time of a large lake to changes in loading.

The character of the bottom sediments in Lake Superior has been strongly influenced by Pleistocene glaciation, changes in lake conditions since the last ice age, and present-day sedimentary processes. Glacial till is exposed locally and underlies younger sediment elsewhere throughout most parts of the lake. Varved, glacial-lacustrine sediment overlies till in most deep areas of the lake, and this sediment is overlain by more uniform postglacial clays and silts where conditions have allowed the accumulation of sediments during the past 9000 years. The sedimentation rates of postglacial sediments throughout much of the open lake are 0.2–0.4 mm/year, and higher rates are observed in embayments such as Thunder Bay (north central) and Whitefish Bay (extreme east). Sedimentation rates for Lake Superior are reported in Evans et al. (67). Large surface waves generated by storms prevent the accumulation of sediment on topographical highs shallower than 110 m, except in protected areas of the western arm and embayments (21). Bottom currents located off the Keweenaw Peninsula measured in late 1985 by R. Flood approach 40 cm/s for periods of several days at water depths of about 150 m (Lamont Doherty Geological Observatory, personal communication). Other factors complicating sedimentation are bottom currents in some deep valleys (>200 m) that prevent deposition or erode sediment, and some slumping occurs in the northeast–southwest trough located off Minnesota. A pronounced nepheloid layer also is a prominent feature when Lake Superior is stratified (66). Particulate organic carbon, as a tracer of fine sediment and a carrier of nonpolar organic contaminants, is concentrated in the depositional zones but rarely exceeds 4% by weight of sediment (122).

A variety of sediment cores have been taken with an oceanographic box corer in the years 1977–1979 and in 1982. These sediment box cores provide good coverage of the depositional and nondepositional areas of the lake bottom and provide the data necessary to construct chronological inputs and sediment burdens. Figure 14 notes the locations of box cores taken in 1977–1979 and 1982, and Table VIII presents detailed information on sedimentation rates, PCB concentrations in the top 0.0–0.5 cm, and PCB accumulation rates. Each of these cores was dated using lead-210 geochronology (67).

Figure 14 and Table VIII show the locations of sediment coring sites in Lake Superior to be relatively concentrated in the central and western basins of the lake. The sedimentation rate of the 16 cores collected in 1978 and 1979 ranged from 0.01 to 0.17 cm/year, or from 7 to 83 mg/cm$^2$ year. This range represents the low end of measured sedimentation rates in the Great Lakes. Additional data on cores collected in other years may be found in reference 18.

Surficial sediment concentrations (0.5 cm) of *t*-PCBs in 1978 ranged from 3 to 147 ng/g of dry sediment; this value corresponds to estimated recent accumulation rates of 0.02–5.4 ng/cm$^2$ year (Table VIII). Surficial

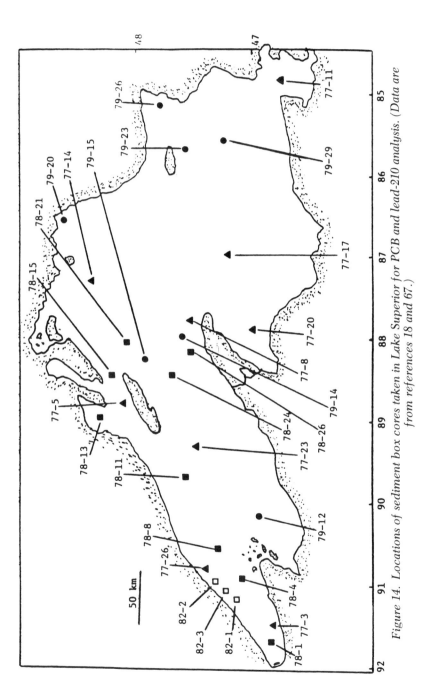

*Figure 14. Locations of sediment box cores taken in Lake Superior for PCB and lead-210 analysis. (Data are from references 18 and 67.)*

Table VIII. Sediment and PCB Accumulation Rates in Lake Superior

| Core[a] | Water Depth (m) | Sedimentation Rate | | PCB Conc. (ng g⁻¹) | PCB Accum. (ng cm⁻² year⁻¹) | PCB Burden (ng cm⁻²) |
|---|---|---|---|---|---|---|
| | | w (mg cm⁻² year⁻¹) | R (cm year⁻¹) | | | |
| 78-1 | 48 | 45 | 0.11 | 120 | 5.4 | 80 |
| 78-4 | 140 | 16 | 0.04 | 147 | 2.4 | 47 |
| 78-8 | 86 | 22 | 0.03 | 7 | 0.15 | 6.0 |
| 78-11 | 184 | 10 | 0.02 | 50 | 0.50 | 22 |
| 78-13 | 77 | 53 | 0.17 | 35 | 1.9 | 45 |
| 78-15 | 186 | 24 | 0.07 | 88 | 2.1 | 45 |
| 78-21 | 201 | 6.9 | 0.03 | 28 | 0.19 | 9.0 |
| 78-24 | 175 | 7.4 | 0.01 | 3 | 0.02 | 6.0 |
| 78-26 | 180 | 22 | 0.05 | 34 | 0.75 | 31 |
| 79-12 | 115 | 16 | 0.04 | 8.7 | 0.14 | 5.4 |
| 79-14 | 245 | 15 | 0.04 | 11 | 0.17 | 7.0 |
| 79-15 | 265 | 27 | 0.03 | 16 | 0.43 | 16 |
| 79-20 | 136 | 83 | 0.17 | 19 | 1.6 | 37 |
| 79-23 | 203 | 14 | 0.05 | 8.5 | 0.12 | 8.5 |
| 79-26 | 200 | 33 | 0.12 | 18 | 0.59 | 10 |
| 79-29 | 143 | 14 | 0.01 | 2.6 | 0.04 | 0.9 |

SOURCE: Data are from references 11 and 12.
[a] All cores are identified from the data in Figure 14.

concentrations in 1979 ranged from 2.6 to 19 ng/g, and accumulation rates ranged from 0.04 to 1.6 ng/cm$^2$ year. Highest concentrations and fluxes of $t$-PCBs occur in the western arm of the lake nearest the urban areas of Duluth–Superior and the major source of sediment to the lake: erosion of red clay along the Wisconsin shoreline. Levels are also generally elevated in the central region of the lake in the vicinity of Isle Royale. Swain (8) also found elevated PCB concentrations in water, precipitation, and fish near Isle Royale. This area may be impacted by atmospheric transport from the upwind urban and industrial center of Thunder Bay, Ontario. The congener distribution approximated an Aroclor 1242:1254 ratio of 6:4, which is somewhat higher than observed in later sediment analysis employing high-resolution GC (18). In 1979 cores, Aroclor 1242:1254 ratios of 4:6 were observed. Surficial $t$-PCB concentrations are higher than those generally observed by Frank et al. (88), who reported values integrated over 3 cm and generally lower than those observed for the other lakes (113).

Figure 15 shows the $t$-PCB concentration–depth profiles for the nine sediment box cores taken in 1978 in Lake Superior. Sectioning the cores into 0.25-cm segments in the top few centimeters corresponds to a time resolution of about 6 years and a linear sedimentation rate of 0.04 cm/year (approximate lakewide average) to 2.5 years for a 0.1-cm/year rate. Lead-210 analyses of these sediment cores (67) show mixing depths of 0–4 cm, and most cores had values of 0–2 cm. A sediment accumulation rate of 0.04 cm/year and a mixing depth of 1 cm yields a time resolution of 25 years. This time resolution is taken to mean that an event occurring more recently than 25 years ago should not be observed in the sediment profile. However, application of biological mixing models (114–116) using lead-210 as the tracer and the depth-integrated numbers of benthic organisms [*Ponteporeia hoyi* and oligochaete worms (105)] suggest a low rate of mixing. Unpublished data from the hi-sed (high sediment) study [Robbins et al., Great Lakes Environmental Research Laboratory, National Oceanic and Atmospheric Administration (GLERL–NOAA)] generated in our laboratory demonstrate that mixing by benthic organisms may not destroy the historical profile for a transient input tracer. The detailed shape of the PCB profile in the top few centimeters argues against significant mixing.

The $t$-PCB profiles exhibit a decrease in concentration with depth ($<1$ ng/g for values occurring at depths of 2–5 cm). In seven of the nine cores, the top 0.25-cm increment has concentrations 20%–100% less than the next 0.25-cm increment. This behavior suggests either a mechanical loss of the uppermost flocculant material enriched in $t$-PCBs or provides evidence for a recent decrease in sedimentary PCB flux in response to decreased loadings. An alternate explanation is that PCBs are being lost from surface sediments in response to decreasing water column concen-

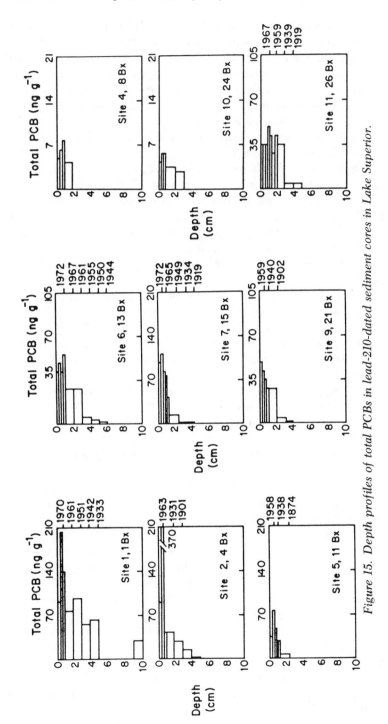

*Figure 15. Depth profiles of total PCBs in lead-210-dated sediment cores in Lake Superior.*

trations and whole-lake degassing. Several of the cores exhibit a depth of penetration for PCBs to sediment older than the first production date. On average, this depth corresponds to 1900. If corrected for downward mixing by one mixed depth, the average approaches 1930. In the case of core 78-1, sediment mixing by storm-induced wave action may account for the PCBs penetrating to 10 cm. Recent data from our laboratory suggest that lower chlorinated congeners diffusing in pore waters also contribute to downward movement (unpublished results).

Figure 16 shows the historical profiles of $t$-PCBs in three sediment cores—78-1 (site 1), 78-13 (site 6), and 78-15 (site 7)—differing in mass accumulation rates by more than a factor of 2. In nearly all cores from 1978, the onset of elevated concentrations occurs near 1950, peaks in 1972–1973, and decreases in the last decade. The date at which $t$-PCB levels increase most rapidly (1950) is about the same in all the Great Lakes (S. J. Eisenreich and J. A. Robbins, unpublished data) and may be related to the average lifetime of electrical transformers containing PCB fluids.

The sedimentary profiles of $t$-PCBs in Lake Superior are in close agreement with the decrease in PCB residues measured in Lakes Michigan and Superior coregonids and chubs (*124, 125*) and the U.S. sales of PCBs (*4*). The decrease in PCB concentrations in major fish species demonstrates reduced loading to Lake Superior, probably as a result of improved disposal practices. The difference in time between the peak in PCB sales and the peak in sedimentary concentrations corresponds fortuitously to the average residence time of PCBs and DDT determined from mass balance calculations [2–4 years (*11, 127*)]. We conclude that the $t$-PCB profiles in Lake Superior sediments record the historical input pattern. Furthermore, the sedimentary burden responds rapidly to decreases in loading on the order of the chemical residence time of the organic compound. Of course, the PCB input to the upper Great Lakes has decreased over the last decade. For PCBs, sales and production in the United States appear to be adequate predictors of the shape of the PCB input function.

The total mass of PCBs stored in the sedimentary compartment of Lake Superior was estimated by dividing the lake into zones impacted to varying extents, as evidenced by their accumulation of PCBs and DDT (*18*). Zone I is minimally affected by anthropogenic sources and consists of primarily open lake and deep depositional areas. Zone II is influenced by activities in marginal bays and includes the Thunder Bay and Marathon Bay areas. Zone III is the extreme western end of Lake Superior impacted by the Duluth–Superior urban complex. The mean accumulation in each zone was multiplied by the zone's area and the values were then summed. The "atmospheric" contribution to PCB accumulation in zones II and III was estimated from zone I; excess accumulation was

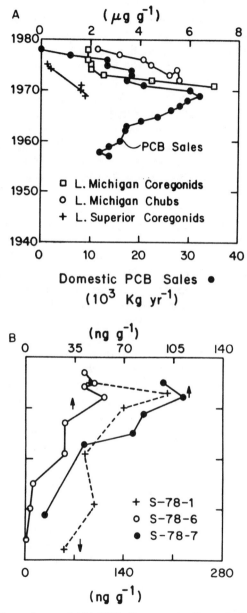

Figure 16. Chronology of (A) PCB decrease in Great Lakes fish species
and (B) sedimentary accumulation in Lake Superior.

attributed to local sources. On the basis of these data, the total quantity of PCBs in Lake Superior sediments was estimated to be approximately 10,000 kg, and 78% was atmospherically derived. Similarly, 2400 kg of total DDT (*t*-DDT) was in Lake Superior sediments, and nearly all was derived from atmospheric deposition.

Another method of estimating the atmospheric contribution of PCBs to Lake Superior is suggested by the work of Rapaport (*17*). Figure 17 presents total sediment burdens of PCBs in Lake Superior in comparison to PCB burdens in small northern Wisconsin lakes (*103*) and regional peatlands (*17*) receiving all of their hydrologic and chemical input from the atmosphere. Excluding the higher PCB burdens in Lake Superior sediment cores from "impacted" zones II and III, the atmospheric signal is about 80% of the total burden. Eisenreich et al. (*12*) concluded that PCB input to Lake Superior is dominated by atmospheric deposition.

**Pore Water Concentrations.**  PCBs are scavenged from the water column by settling biotic and abiotic particles (i.e., detritus) and are delivered to the sediment. Estimated values of $K_p$ and $K_{oc}$ for PCB species in the water column at low suspended solids concentrations are approximately $5.0 \times 10^4 – 10^6$ L/kg (*23*). However, similar distribution coefficients estimated from empirical correlations at high suspended solids concentrations (*100, 101*) and measured values (*81, 127*) are on the order of $10^2 – 10^3$ L/kg. Chapra and Reckhow (*128*) pointed out that diffusive recycling of PCBs may be important in reducing net settling velocities of particle-bound contaminants. In addition, diffusion of PCBs in pore waters may assist in redistributing sedimentary accumulations. PCB input profiles may be altered by this diagenetic process as demonstrated by Looney (*18*) in Great Lakes cores and Brownawell and Farrington (*129*) in coastal marine sediment near New Bedford Harbor (Acushnet

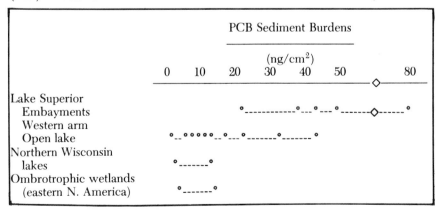

*Figure 17. PCB burdens in Lake Superior, northern Wisconsin lakes, and peat (ombrotrophic wetlands) sediments. (Data are from references 17 and 103.)*

River estuary). Baker et al. (*66*) have argued that the diffusive flux of the lower chlorinated PCB congeners (i.e., Aroclor 1242) serves to support a dissolved PCB concentration in the BNL of western Lake Superior.

These data require that PCB concentrations in pore water exceed overlying water column concentrations. Duinker and Hildebrand (*130*) have measured PCBs and several chlorinated hydrocarbon pesticides in pore waters and sediments of the Rhone–Meuse estuary and adjacent coastal sea. Eadie et al. (*81*) measured PCB concentrations in Lake Michigan surficial sediments and pore waters. Concentrations of $t$-PCB in pore waters ranged from 159 to 342 ng/L and in sediments from 64 to 160 ng/g. Distribution coefficients, defined as $[PCB_{sed}]/[PCB_{water}]$, averaged 400–750 L/kg and were not related to the organic carbon content of the sediments. Brownawell and Farrington (*129*) reported concentrations of PCBs in New Bedford Harbor sediments and pore water of 20–40 × 10³ ng/g and 1–22 × 10³ ng/L, respectively. Concentrations of dissolved organic carbon (DOC) ranged from 14 to 87 milligrams of carbon per liter in these reducing sediments and were thought to control PCB diagenesis. Distribution coefficients typically were 1–5 × 10³ L/kg. Elevated PCB concentrations in pore waters, low $K_D$ values, and correlation of these properties with DOC suggested a strong PCB interaction with colloidal organic carbon in the pore waters.

To investigate the role of diffusion and mixing in the diagenesis of PCBs in lake sediments, several box cores collected in Lakes Erie, Huron, and Superior were segmented into 2-cm-depth increments immediately after being brought on board ship. The sediment was centrifuged at 900 $g$ for 30 min, and the supernate filtered through 0.6-$\mu$m glass fiber filters. The filtrate contained dissolved PCBs and any microparticulate or colloidal PCBs surviving the process. The centrifuged solids and the filtrate were analyzed for PCBs. Table IX shows the PCB concentrations in sedimentary pore waters to be 10–100 times the levels in overlying waters. As observed in other laboratory and field studies conducted at high sediment concentrations, $K_D$ values were approximately 10³ L/kg. In general, log $K_D$ increases with increasing log $K_{ow}$ values for individual PCB congeners. This relationship deteriorates when log $K_{ow}$ > 6 (when pore water DOC levels are high).

Core profiles of pore water and sedimentary PCBs were developed for a sediment core collected in western Lake Superior in 1982 (Figure 18). The PCB profiles (*131*) exhibited for pore water and sediment were similar in the top 5 cm and ranged from 12–35 ng/L, as compared to <1 ng/L in the overlying water, and the sedimentary concentrations ranged from 9 to 19 ng/g. Pore water PCBs in the top 10 cm were dominated by the lower chlorinated congeners (e.g., Aroclor 1242), and sedimentary PCBs were dominated by the higher chlorinated species. This

Table IX. PCBs in Pore Waters of Great Lakes Sediments

| Site and Depth | Pore Water Concentration (ng/L) | Sediment Concentration (ng/g) | log $K_D$ (L/kg) |
|---|---|---|---|
| Lake Erie, 1981 | | | |
| 0–2 cm | 72 | 76 | 3.03 |
| 2–4 cm | 35 | 40 | 3.06 |
| Lake water | 1–2 | | |
| Lake Huron, 1981 | | | |
| 0–2 cm | 39 | 46 | 3.07 |
| 2–4 cm | 26 | 39 | 3.18 |
| 4–6 cm | 307 | 21 | 1.85 |
| Lake water | 0.5–2.0 | | |
| Lake Superior (0–2 cm) | | | |
| 81–9 | 26 | 5.0 | 2.28 |
| 81–19 | 18 | 7.9 | 2.64 |
| 81–31 | 12 | 7.8 | 2.81 |
| 82–2 | 9–35 | 9–19 | 2.7–3.0 |
| (core profile) | | | |
| Lake water | 0.5–1.0 | 300–3000 | 5.0–6.0 |

behavior is in agreement with the decreasing solubility and increasing $K_{ow}$ of PCB compounds with increasing degree of chlorination.

The diffusive flux of PCBs out of the sediments may be estimated (*132*).

$$F = - \phi D_m \frac{dC}{dz} \qquad (31)$$

where $F$ is the diffusive flux (ng/cm² year), $\phi$ is the sediment porosity, $D_m$ is the molecular diffusion coefficient (cm²/s), $C$ is the concentration of PCB in pore water (ng/cm³), and $z$ is the depth (cm). The pore water profile for PCBs in Figure 18 yields a concentration gradient of 4–5 ng/L cm from 5 cm below the sediment–water interface. A flux of approximately 4 ng PCB/m² · day can be calculated by using a molecular diffusion coefficient of $10^{-6}$ cm²/s (*131*). Baker et al. (*66*) found that this flux was insufficient to support the elevated Aroclor 1242 concentrations observed in the BNL in this area in 1983. Eadie et al. (*81*) suggested that advection of pore water PCBs through the resuspension process might well account for the remainder.

The increased concentrations of PCBs in pore water and reduced $K_D$ values in sediments may be the result of PCB binding to dissolved and colloidal organic matter as a competing third phase. Nonpolar organic compounds have been shown to exhibit higher binding constants to colloidal organic matter than to sedimentary organic matter (*95–98*),

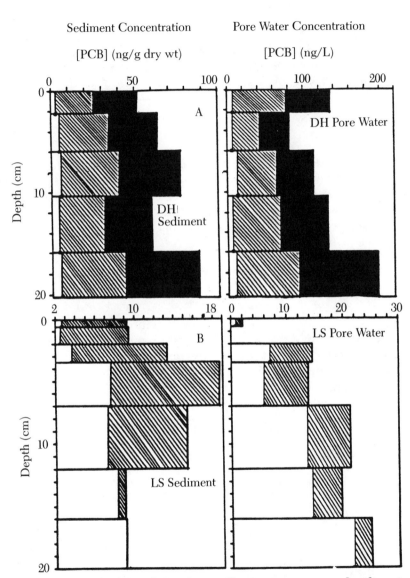

*Figure 18. Total PCB and Aroclor profiles in pore water and sediment from western Lake Superior, 1982. The darkened area represents Aroclor 1260, the shaded area represents Aroclor 1254, and the nonshaded area represents Aroclor 1242. (Reproduced with permission from reference 131. Copyright 1986, P. D. Capel.)*

but more recent data do not support this conclusion ($131$). The general affinity of natural organic matter for organic compounds having log $K_{ow} > 3$ has been known for some time ($36, 37, 134$–$137$). Brownawell and Farrington ($129$) developed a modified version of the three-phase model proposed by Voice et al. ($92$) and Gschwend and Wu ($93$) in which the measured or apparent distribution coefficient $K'_D$ is calculated from

$$K'_D = \frac{f_{oc}(s)K_{oc}(s)D}{D + f_{oc}(c)K_{oc}(c)D} \tag{32}$$

where $f_{oc}(s)$ and $f_{oc}(c)$ are the fractional organic carbon contents of the sediment and pore water colloids, respectively; $K_{oc}(s)$ and $K_{oc}(c)$ are the organic-carbon-normalized distribution coefficients for the sediment and pore water colloids, respectively; and $D$ is the dissolved (uncomplexed) species concentration. If $D$ is small compared to $f_{oc}(c)K_{oc}(c)D$, then the equation reduces to

$$K'_D = \frac{f_{oc}(s)K_{oc}(s)}{f_{oc}(c)K_{oc}(c)} \tag{33}$$

If $K_{oc}(s) = K_{oc}(c)$, then $K'_D$ is simply the ratio of the organic carbon concentrations in the sediment and pore water. Neglecting non-equilibrium effects influenced by slow desorption kinetics ($79, 82$), the relative distribution of PCBs between sediment and pore water is the ratio of the organic carbon distribution between the two phases. For the New Bedford sediments, this model adequately explained the PCB sediment–pore water profiles.

## Modeling of PCB Cycling in Lake Superior

**Dynamic Mass Balance Model.** A dynamic steady-state model was constructed ($18$) on the basis of the approaches of Bierman and Swain for DDT cycling in Lakes Michigan and Superior ($126$), Rogers and Swain for PCB loading trends in Lake Michigan ($138$), a general toxic substances model used by Chapra and Reckhow ($128$), and a physico-chemical model of toxic substances in the Great Lakes by Thomann and DiToro ($62$). The overall structure of the dynamic mass balance as shown in Figure 19 differentiates the well-mixed lake conditions of fall, winter, and spring in Lake Superior from the stratified conditions of summer. The model assumes that PCB concentrations in the lake are homogeneous on an areal scale but may vary from year to year. The data and model output were assumed to represent seasonal values. Factors such as the amount of ice cover, suspended solids, and DOC may be

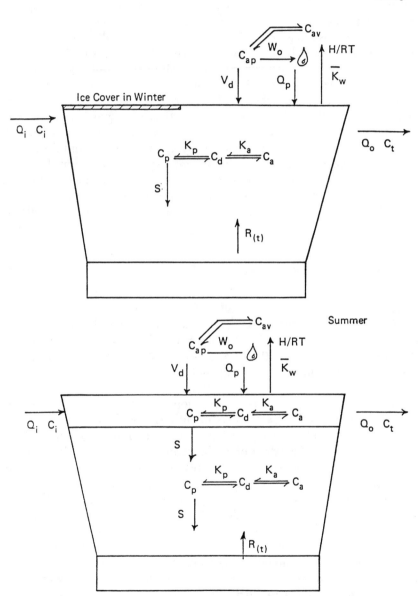

*Figure 19. Structure of dynamic mass balance model describing PCB cycling in Lake Superior.*

varied on a seasonal basis, and resuspension rates may be varied monthly. These physical factors are assumed to be constant from year to year. The model constructed in PASCAL describes mathematically the pertinent mass balance relationships as well as chemical and physical relationships in the lake. These mathematical relationships are solved over time by using a numerical integration. Overall input processes included in the model are inflow and atmospheric deposition and resuspension; loss processes included are volatilization, outflow, and sedimentation. Biological degradation was not included because turnover rates are a decade or more. However, internal loss processes not included explicitly are implicitly contained in rate coefficients describing other losses. Model inputs include the various partition coefficients measured or calculated, mass-transfer rates at air–water and sediment–water interfaces, and a time-dependent atmospheric input function. Model output includes dissolved, particulate, and total concentrations of PCBs in the water column and sediment concentrations over time.

The time-dependent equation describing the dynamic mass balance model is

$$\frac{dC_t}{dt}(V) = Q_iC_i - Q_oC_t + F_t - k_dC_dV - SA_sC_pD + RA_sC_{sed} \qquad (34)$$

where $Q_i$ and $Q_o$ are the inflow and outflow, respectively ($m^3$/month); $C_t$, $C_d$, $C_i$, and $C_p$ are the total, dissolved, inflow, and particulate PCB concentrations in water ($\mu g/m^3$); $F_t$ is the total atmospheric flux ($\mu g$/month); $V$ is the lake volume ($m^3$); $k_d$ is the degradation constant (month$^{-1}$); $S$ and $R$ are the sedimentation and resuspension rates, respectively (m/month); $A_s$ is the surface area of the deposition zone ($m^2$); $C_{sed}$ is the concentration in sediment ($\mu g/m^3$); and $D$ is the fraction of water column particulate organic carbon (POC) reaching sediment (dimensionless).

The historical trends and overall magnitude of atmospheric concentrations determine the extent of PCB accumulation in the Lake Superior ecosystem. These concentrations represent long-range or regional PCB transport, and the shape of the historical curve may be assumed to be representative of (and represented by) the material in the "mobile environment reservoir" (MER) (4, 16, 139). This reservoir includes all compartments of the biosphere from which contaminants may be released or move. In general, the atmosphere and hydrosphere, as well as portions of the lithosphere, freshwater sediments, and sewage sludge are considered to be part of the MER. Ocean sediments, deep ocean water, and other environmental compartments represent areas that accumulate material lost from the MER. The mass of material in the MER is determined by source strength and residence time. The sources are a function of manufacturing rate, handling losses, and uses and disposal methods, and the sources represent the total mass of material that enters the environment. The residence time is determined by physicochemical processes that move the material to areas where it is sequestered from further interaction with the MER.

Conway et al. (*16*) have developed a complex model that estimates the mass of PCBs released to the environment. The resulting prediction is inconsistent with sediment profiles (*18*), peat profiles (*17*), and fish concentrations (*138*) in rural environments. Thus, a detailed model may not be justified for predicting MER concentrations based on uncertainties in the necessary input parameters.

For this work, a simple model based on the description of the MER as a well-mixed reactor was developed to estimate the input function of PCBs over time to Lake Superior. In this model, the shape of the U.S. production and use curve, the residence time calculated for the MER, and measured air concentrations (1978–1981) were the determining parameters. The model assumes that a fraction of U.S. manufactured PCBs is released to the environment. This material makes its way through the compartments of the MER into nonmobile environmental compartments at a rate consistent with the residence time. The air concentrations over Lake Superior as a function of time (representing national use and release trends) were estimated by fitting the calculated input curve shape to the observed air concentrations (1978–1981).

The residence time of PCBs in the MER ($\gamma$) was calculated from the environmental burdens of PCBs in 1979 (*4*) by assuming a pulse input to the environment at the mass-weighted-mean input year (1965). The fraction of the environmental burden residing in the MER in 1979 can be used to calculate $\gamma$ by using equation 35.

$$\gamma = - \left[ \ln \left( \frac{C}{C_o} \right) \right] (t) \tag{35}$$

where $\gamma$ is the residence time (years), $C/C_o$ is the fraction of environmental burden in MER, and $t$ is the time during which removal occurred (14 years).

The residence time of PCBs in the MER ranges from 6 years [assuming that $C/C_o = 0.9$, which means that the ocean water; ocean sediment; and 50% of the lithosphere, freshwater sediments, and sewage sludge are removed from the MER (*9*)] to 5 years (assuming that $C/C_o = 0.94$, which means that the ocean water; ocean sediment; and 100% of the lithosphere, freshwater sediments, and sewage sludge are removed from the MER). The estimated $\gamma$ (5.5 years) may then be used to calculate a normalized MER burden based on equation 36.

$$B_{(t)} = \frac{\sum_{i=0}^{t} P_i \, e^{-\left(\frac{t-i}{\gamma}\right)}}{\sum_{j=0}^{t} P_j \, e^{-\left(\frac{t-j}{\gamma}\right)}}$$
(36)

where $B_{(t)}$ is the normalized MER burden as a function of time and ranges from 0 to 1; $t$ is the time; and $P_i$ and $P_j$ are the production of PCBs at times $i$ and $j$, respectively. Finally, the atmospheric concentration over Lake Superior can be estimated by using equation 37.

$$C_{(t)} = (B_{(t)}) \, (C_{(n)}/B_{(n)})$$
(37)

where $C_{(t)}$ is the concentration at time $t$, $B_{(t)}$ is the normalized burden at time $t$, $C_{(n)}$ is the measured concentrations over lake at time $n$, and $B_{(n)}$ is the normalized burden at time $n$.

The resulting numerically smoothed PCB production curve and the estimated atmospheric input function are shown in Figure 20. The shape of the input function based on production, sales, and the MER is similar to that derived by Rapaport (17) from atmospherically driven peat profiles and loading trends estimated for Lake Michigan (138). The input function predicts an onset of PCB loading in the late 1940s, a maximum in 1968–1970, and a rapid decrease from 1970 to the present.

Figure 20 also shows the close agreement of predicted and measured atmospheric PCB concentrations in the period 1978–1981. The input function and the corresponding model calculations closely predicted both the shape and the magnitude of PCB variations in the water column over the period 1978–1986 with the exception of 1979. As stated previously, Lake Superior did not stratify in 1979, and this phenomenon permitted resuspension of bottom sediments (and PCBs) to occur throughout the year. The close correspondence of independently predicted and measured PCB concentrations in the water column is consistent with a particle settling velocity of 125 m/year and a sediment resuspension rate of 0.0012 m/year. The predicted sediment burden of $1.5 \times 10^4$ kg is in agreement with that estimated from measurement ($1 \times 10^4$ kg). Trends in mass balance calculations as driven by atmospheric deposition show a net flux of PCBs to Lake Superior during times of increasing input (100 kg in 1940 and 500 kg in 1969) and a net loss in times of decreasing input (300 kg in 1980). Water concentrations are predicted to decrease from a lakewide mean of 3.6 ng/L in 1969 to 0.9 ng/L in 1980 and 0.2 ng/L in 1990 (Figure 21). The PCB concentrations measured in a limited number of samples in 1985 support the decreasing trend.

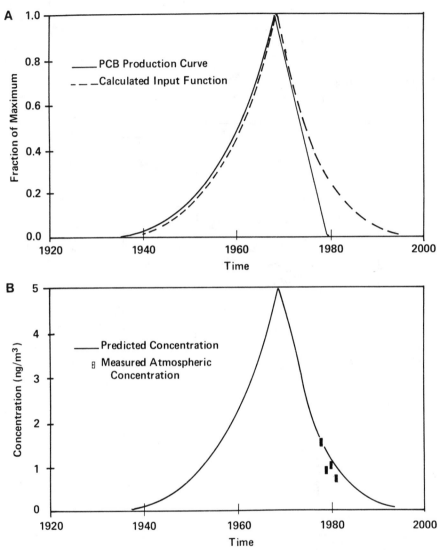

*Figure 20. (A) PCB production and input functions. (B) Comparison of measured and predicted atmospheric PCB concentrations derived from the dynamic mass balance model.*

**Mass Balance Model.** The simple mass balance model has been used by many researchers as both a conceptual and practical framework for viewing the cycling of PCBs through a "box" version of a lake (*12*). Murphy and Rzeszutko (*44*) and Weininger and Armstrong (*139*) previously used field measurements in conjunction with mass transport the-

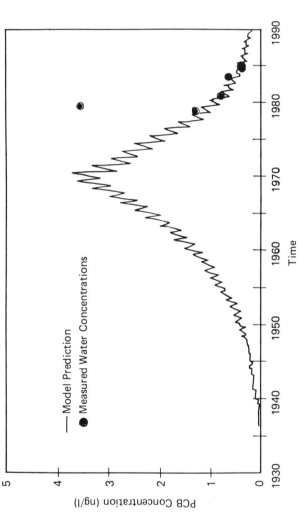

*Figure 21. Comparison of model output to measured PCB concentrations in the waters of Lake Superior.*

ory to estimate inputs and outputs. These early attempts were later criticized for overestimating the magnitude of atmospheric inputs. For example, atmospheric loading rates of 5000–9000 kg/year of PCBs to Lake Superior have been reported (12). Table X lists our recent best estimates of PCB inputs to the lake. Strachan (54) calculated PCB wet-only inputs of about 300 kg/year for 1983 based on rain measurements conducted on in-lake islands.

Table X. Precipitation Inputs of PCBs to Lake Superior

| Mode | Input (kg/year) | Reference |
|------|-----------------|-----------|
| Rain | 300 | Strachan (54) |
| Rain | 100–300 | C. H. Chan (unpublished data) |
| Rain and snow | 150–400 | Eisenreich (unpublished data) |
| Peat (net input) | 100–200 | Rapaport (17) |

Rain measurements conducted by us in 1984 in Minnesota (unpublished data) yielded input estimates of 100–300 kg/year of PCBs to Lake Superior. These data combined with snow core measurements made in north central Minnesota from 1981 to 1985 (17) yielded loading estimates of 150–400 kg/year. Rapaport (17) estimates that recent net PCB inputs derived from ombrotrophic peat profiles in the vicinity of Lake Superior are 100–200 kg/year. Excluding intense localized deposition in industrial areas, Lake Superior today receives 150–400 kg/year of PCBs from atmospheric wet inputs.

Table XI places the atmospheric inputs into perspective by comparison to other input sources and sinks and to a PCB budget for Lake Michigan (141). On the basis of these data, the Lake Superior PCB budget is in approximate balance, whereas Looney and Eisenreich (13) estimated a net loss to the lake of 500 kg/year. Tributary inputs to Lake Michigan appear to dominate inputs and represent a continuing problem for the ecosystem.

**Residence Time.** The residence time $(t_R)$ of a chemical species in a lake at steady state can be calculated as

$$t_R = \frac{\text{mass of lake (kg)}}{\text{input rate (kg/year)}} = \frac{\text{mass of lake (kg)}}{\text{output rate (kg/year)}} \tag{38}$$

A lake not at steady state with respect to a transient input (e.g., PCBs) would have a $t_R$ between the input and output rate values. Lake Superior has a hydrologic residence time on the order of 170–180 years. Conservative chemical species such as $Na^+$ or $Cl^-$ would be expected to have $t_R$ on the order of the $t_R$ for $H_2O$ ($t_{R, H_2O}$). Nonconservative species such as non-

Table XI. PCB Mass Balance in Lakes Superior and Michigan

| Source | Lake Superior (kg/year) | Lake Michigan (kg/year) |
|---|---|---|
| Inputs | | |
| Atmosphere | | |
| Wet deposition—measured | 150–400 | |
| Wet deposition—washout (vapor) | | 5 |
| Wet deposition—rainout (part.) | | 270–510 |
| Dry deposition—particles | 35–70 | 25–50 |
| Net peat profiles | 100–200 | |
| Tributary inflow | 1000 | 270–2600 |
| Municipal and industrial discharges | 60 | |
| Total | 1350–1700 | 570–2600 |
| Outputs | | |
| Water-to-air transport | 500 | 150–410 |
| Outflow | 100 | 60 |
| Sedimentation | 300–800 | 440 |
| Total | 900–1400 | 650–910 |

SOURCE: Data are from references 13, 17, and 140.

polar organic contaminants will have much lower $t_R$ due to in-lake removal processes such as volatilization and sedimentation. On the basis of a water column concentration of 0.6 ng/L (1983) and a PCB sedimentation rate (based on lead-210-dated cores) of 1.0 ng/cm$^2$ year, the $t_R$ for PCB ($t_{R,\,PCB}$) is calculated as

$$t_{R,\,PCB} = \frac{7200 \text{ kg}}{820 \text{ kg/year}} = 9 \text{ years} \tag{39}$$

Looney and Eisenreich (13) estimate PCB residence time based on a dynamic mass balance model of about 3 years. Bierman and Swain (126) have estimated the residence time of DDT in Lake Superior (2.7 years) based on its disappearance in fish. Apparent settling velocities of particle-associated species such as plutonium (369 m/year) and DDT (150 m/year) as compared to a mean lake depth of 144 m and standing crops suggest a $t_R < 1$ year. Baker et al. (66) found that the more highly chlorinated PCB congeners typical of Aroclor 1254 were removed from the water column, and the $t_R$ value was on the order of weeks. These data (see also Chapter 11) lead to the conclusion that sorption of nonpolar organic contaminants and other strongly sorbing species to suspended solids and their subsequent settling is a dominant mechanism in the detoxification of the lakes. Rates of particle removal inferred by short $t_R$ values cannot be explained by Stokes settling behavior. Selective removal of nonpolar organic compounds from the water column must involve incorporation into larger particles such as fecal pellets, flocculated clays, organic aggregates, or precipitated carbonate solids.

Santschi (*141*) proposed that a simple relationship exists between $t_R$ and particle flux through the water column for those chemical species with $K_p > 10^4$ L/kg. In its simplest form,

$$t_R^{-1} = S \times K_D/h \qquad (40)$$

where $S$ is the particle flux or sedimentation rate (g/cm$^2$ year), $K_D$ is the distribution coefficient (cm$^3$/g), and $h$ is the mean depth of the lake (cm).

In log-transformed form, $\log t_R = -\log K_D - \log (S/h)$. A plot of $\log t_R$ versus $\log (S/h)$ should yield a line with a slope equal to $-1$ and an intercept equal to $-\log K_D$. The model assumes $K_D$ describes the equilibrium condition and that SS $\leq 20$ mg/L. The model was verified using Zn from the freshwater environment and Th isotopes from the marine system. Figure 22 shows a plot of $\log t_{R, \text{PCB}}$ versus $\log (S/h)$ for a variety of freshwater and marine systems (*143*). The PCB $t_R$ values were estimated from both sediment trap and sediment accumulation data applied to various aquatic systems. The data are best described by the following equation:

$$\log t_{R, \text{PCB}} = -4.54 - 0.980 \log (S/h) \qquad r^2 = 0.74 \qquad (41)$$

Regression analysis on the model output due to variations in data input obtained from the literature showed that the slope was not significantly different from $-1$ at the 50% confidence interval. The 95% confidence intervals for the slope and $\log K_D$ were $-1.3$–$-0.7$ and 2.24-6.8, respectively. This simple model demonstrates that the residence time in aquatic systems is a direct function of the affinity of PCBs for aquatic particles and the rate of particle removal from the system. The $\log K_D$ calculated from the regression line (4.54) is similar to those values observed in numerous field studies (*23, 66, 103*).

## Summary

The detailed aquatic behavior of PCBs in Lake Superior obtained from scientific studies over the last decade permits an evaluation of the chemical limnology and environmental fate of nonpolar organic contaminants having similar physicochemical properties in large lakes. The range of physicochemical constants describing the environmental partitioning of 50-100 PCB congeners observed in the environment suggests that air-water and sediment-water interactions dominate their aquatic behavior. These ranges were as follows: from $1 \times 10^{-4}$ to $6 \times 10^{-6}$ mol/m$^3$ for solu-

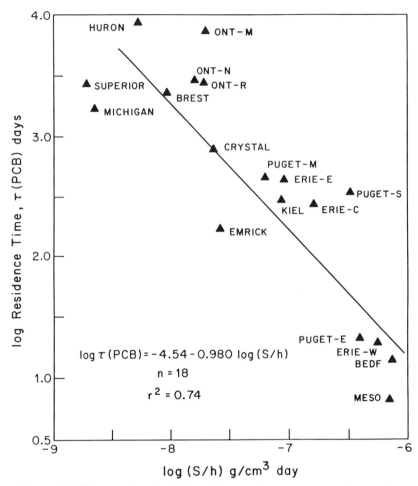

*Figure 22. Relationship of PCB residence time to mean sedimentation rates (in aquatic systems).*

bility, from $1 \times 10^{-7}$ to $1 \times 10^{-9}$ atm for $P_v$ (vapor pressure), from $0.7 \times 10^{-4}$ to $1 \times 10^{-3}$ atm m$^3$/mol for $K_H$ (Henry's law constant), and from 4.5 to >7.71 for log $K_{ow}$ (octanol–water partition coefficient).

PCBs likely enter the lake during intense episodes of precipitation (particle scavenging) and are subsequently lost during longer periods of volatilization. Although the atmosphere has served as a long-term source of PCBs to the lake, decreasing inputs and water column concentrations suggest that the atmosphere is now a sink for PCBs previously deposited. The ecosystem burden of PCBs is about 20,000 kg, which is about evenly distributed between the water column and bottom sediments (Table XII). Although only about 25% of the water column PCBs occur in

**Table XII. PCBs in the Lake Superior Ecosystem**

| Ecosystem Component | Time Period | PCBs | | $t_R$ |
| | | Concentration | Burden (kg) | |
|---|---|---|---|---|
| Atmosphere | 1978–1981 | 0.3–1.5 ng/m$^{3a}$ | 27[a] | 1–5 days |
| Water | 1978–1983 | 0.6–3.8 ng/L[b] | 10,000[b] | 2–6 years |
| Sediment | 1977–1982 | 0.7–220 ng/g[c] | 10,000[c] | 10–100 years |

[a] Of the PCBs, >90% was vapor and >50%–80% was Aroclor 1242.
[b] Of the PCBs, >70% was dissolved and 40%–60% was Aroclor 1242.
[c] Of the PCBs, >50% was Aroclor 1254.

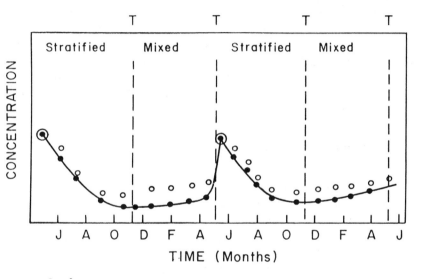

T: Turnover

*Figure 23. Seasonal contaminant cycle in large lakes. The letters along the time scale denote months.*

the particulate phase, sorption and subsequent removal of particles to the bottom is an important lake detoxification process. PCB fluxes to the sediments range from 0.02 to 5.4 ng/cm$^2$ year and have a typical open lake range of 1.5–2.0 ng/cm$^2$ year. The integrated burdens of PCBs in Lake Superior sediments are similar to those of small isolated lakes and peatlands in the region receiving only atmospheric inputs.

The estimated residence time of PCBs in Lake Superior based on mass balance modeling is about 2–6 years, in contrast to hydraulic residence times of 170–180 years. This short residence time implies efficient removal processes.

The internal cycling of PCBs is dominated by sediment resuspension and seasonal inputs. The seasonal cycle for PCBs (Figure 23) shows that PCBs entering the lake through spring flows and sediment resuspension during nonstratified periods are removed to bottom sediments and lost to the atmosphere over the summer. PCB concentrations are highest in the late spring and reduce over the stratified period.

A dynamic mass balance model using a new PCB input function driven by atmospheric concentrations and air–water and sediment–water interactions adequately predicts observed water column concentrations from 1978 to 1985. PCB concentrations should reach about 0.2 ng/L by 1990.

## Abbreviations and Symbols

| | |
|---|---|
| $\alpha$ | scavenging efficiency |
| $\gamma$ | residence time |
| $\phi$ | sediment porosity; ratio of adsorbed organic vapor on aerosol to the total amount of vapor in air |
| $\theta$ | aerosol surface area |
| $A_s$ | surface area of the deposition zone |
| $B_{(n)}$ | normalized burden at time $n$ |
| $B_{(t)}$ | normalized burden as a function of time |
| $C$ | constant that depends on heat of condensation and molecular weight |
| $C_a$ | associated PCB concentration; concentration of organic vapor in air |
| $C_d$ | dissolved PCB concentration |
| $C_p$ | particulate PCB concentration |
| $C_{sed}$ | sediment PCB concentration |
| $C_t$ | total PCB concentration |
| $D$ | dissolved species concentration; fraction of water column organic carbon reaching sediment |
| $D_B$ | bioturbation rate coefficient |
| $D_M$ | molecular diffusion coefficient |

| | |
|---|---|
| $f$ | fugacity |
| $f_{oc}(s)$ | fractional organic carbon in sediment |
| $f_{oc}(c)$ | fractional organic carbon in pore water colloids |
| $F$ | surface flux |
| $F_d$ | total aqueous burden of hydrophobic organic carbon in the aqueous phase |
| $F_D$ | flux of particles to a receptor surface |
| $h$ | lake depth |
| $H$ | Henry's law constant |
| $H'$ | apparent air–water partition coefficient |
| $k_d$ | first-order rate constant for PCB degradation |
| $k_g$ | gas-phase mass-transfer coefficient |
| $k_L$ | liquid-phase mass-transfer coefficient |
| $k_v$ | PCB volatilization rate |
| $K_{BCF}$ | bioconcentration coefficient |
| $K_D$ | distribution coefficient |
| $K'_D$ | ratio of the organic carbon concentrations in sediment and pore water |
| $K_H$ | Henry's law constant |
| $K'_H$ | air–water partition coefficient |
| $K_{oc}$ | partition coefficient between water and a natural sorbent |
| $K_{ow}$ | octanol–water partition coefficient |
| $K_p$ | equilibrium partition coefficient |
| $N_{Cl}$ | chlorination number |
| $p$ | particle-bound organic compounds |
| $P$ | aqueous pollutant; rainfall amount |
| $P_i, P_j$ | production of PCB at times $i$ and $j$, respectively |
| $P_v$ | vapor pressure |
| $Q_i$ | inflow volume |
| $Q_o$ | outflow volume |
| $S$ | solubility; sediment rate |
| $S_c$ | Schmidt number |
| SS | suspended solids |
| $t$ | time |
| $t_m$ | time of maximum production |
| $t_R$ | residence time of lake |
| $T$ | absolute temperature |
| $U^*$ | air-side, frictional velocity |
| $V$ | lake volume |
| $V_d$ | deposition velocity |
| $V_s$ | apparent net settling velocity |
| $W$ | mean scavenging coefficient |
| $W_p$ | particle washout ratio |
| $X_1$ | fraction of total sorptive capacity |

$z$       depth
$Z$       fugacity capacity

## Acknowledgments

I am indebted to former students (G. Hollod, B. Looney, and R. Rapaport) and present students (P. Capel and J. Baker) for their contributions to this article. In addition, stimulating conversations with T. C. Johnson, J. Robbins, R. Hites, A. Andren, D. Armstrong, D. Swackhamer, B. Eadie, D. Mackay, and A. Elzerman over the last several years have greatly helped to shape my ideas about the cycling of hydrophobic organic chemicals. The research described here has been supported in part by grants from the U.S. Environmental Protection Agency, the National Science Foundation, the National Oceanic and Atmospheric Administration (NOAA), the Minnesota Sea Grant College as part of NOAA and the Department of Commerce, and the University of Minnesota.

## References

1. Ballschmiter, K.; Zell, M. *Int. J. Environ. Anal. Chem.* **1980,** *8,* 15–35.
2. Woodwell, G. M.; Craig, P. P.; Johnson, H. A. *Science (Washington, D.C.)* **1971,** *174,* 1101–1107.
3. Rapaport, R. A.; Urban, N. R.; Capel, P. D.; Baker, J. E.; Looney, B. B.; Eisenreich, S. J.; Gorham, E. *Chemosphere* **1985,** *14,* 1167–1173.
4. *Polychlorinated Biphenyls;* National Academy of Sciences; National Academy Press: Washington, DC, 1979; p 182.
5. Nisbet, I. C. T.; Sarofim, A. F. *Environ. Health Perspect.* **1972,** *1,* 21–38.
6. Zell, M.; Ballschmiter, K. *Fresenius' Z. Anal. Chem.* **1980,** *304,* 337–349.
7. Ballschmiter, K.; Buchert, H.; Bihler, S.; Zell, M. *Fresenius' Z. Anal. Chem.* **1981,** *306,* 323–339.
8. Swain, W. J. *J. Great Lakes Res.* **1978,** *4,* 398–407.
9. Water Quality Board Report to the International Joint Commission, Great Lakes Regional Office; International Joint Commission: Windsor, Ontario, Canada, 1985.
10. Eisenreich, S. J.; Hollod, G. J.; Johnson, T. C.; Evans, J. E. In *Contaminants and Sediments;* Baker, R. A., Ed.; Ann Arbor Science: Ann Arbor, MI, 1980; Vol. 1, pp 67–94.
11. Eisenreich, S. J.; Johnson, T. C. In *PCBs: Human and Environmental Hazards;* D'Itri, F. M.; Kamrin, M. A., Eds.; Butterworth: Boston, 1983; Chapter 4, pp 49–75.
12. Eisenreich, S. J.; Looney, B. B.; Thornton, J. D. *Environ. Sci. Technol.* **1981,** *15,* 30–38.
13. Looney, B. B.; Eisenreich, S. J., submitted for publication in *Environ. Sci. Technol.*
14. Ballschmiter, K.; Zell, M.; Neu, H. J. *Chemosphere* **1978,** *1,* 173.
15. *PCBs: Human and Environmental Hazards;* D'Itri, F. M.; Kamrin, M. A., Eds.; Butterworth: Boston, 1983; p 443.
16. Conway, R. A.; Whitmore, F. C.; Hanson, N. J. In *Environmental Risk*

*Analysis for Chemicals;* Conway, R. A., Ed.; Van Nostrand Reinhold: New York, 1982; pp 61–84.

17. Rapaport, R. A. Ph.D. Thesis, University of Minnesota, Minneapolis, 1985.
18. Looney, B. B. Ph.D. Thesis, University of Minnesota, Minneapolis, 1984, p 315.
19. Munawar, M. *J. Great Lakes Res.* 1978, *4*, 554.
20. Thomas, R. L.; Dell, C. I. *J. Great Lakes Res.* 1978, *4*, 264–275.
21. Johnson, T. C. *Quat. Res.* 1980, *13*, 380–391.
22. Lam, D. C. *J. Great Lakes Res.* 1978, *4*, 343–349.
23. Capel, P. D.; Eisenreich, S. J. *J. Great Lakes Res.* 1985, *11*, 447–461.
24. Maier, W. J.; Swain, W. R. *Water Res.* 1978, *12*, 403–412.
25. Kemp, A. L. W.; Dell, C. I.; Harper, N. S. *J. Great Lakes Res.* 1978, *4*, 276–287.
26. Baker, J. E.; Eisenreich, S. J.; Johnson, T. C.; Halfman, B. M. *Environ. Sci. Technol.* 1985, *19*, 854–861.
27. Bennett, E. B. *J. Great Lakes Res.* 1978, *4*, 331–342.
28. Ballschmiter, K.; Zell, M. *Fresenius' Z. Anal. Chem.* 1980, *302*, 20–31.
29. Mullin, M. D.; Pochini, C. M.; McCrindle, S.; Romkes, M.; Safe, S. H.; Safe, L. M. *Environ. Sci. Technol.* 1984, *18*, 468–476.
30. Rapaport, R. A.; Eisenreich, S. J. *Environ. Sci. Technol.* 1984, *18*, 163–170.
31. Capel, P. D.; Rapaport, R. A.; Eisenreich, S. J.; Looney, B. B. *Chemosphere* 1985, *14*, 439–450.
32. Albro, P. W.; Parker, C. E. *J. Chromatogr.* 1979, *169*, 161–166.
33. Albro, P. W.; Corbett, J. T. *J. Chromatogr.* 1981, *295*, 103–111.
34. Burkhard, L. P.; Armstrong, D. E.; Andren, A. W. *Environ. Sci. Technol.* 1985, *19*, 590–596.
35. Shiu, W. Y.; Mackay, D. *J. Phys. Chem. Ref. Data* 1986, *15*, 911–929.
36. Karickhoff, S. W.; Brown, D. S.; Scott, T. A. *Water Res.* 1979, *13*, 241–248.
37. Schwarzenbach, R. P.; Westall, J. *Environ. Sci. Technol.* 1981, *15*, 1360–1367.
38. Eisenreich, S. J.; Looney, B. B.; Hollod, G. J. In *Physical Behavior of PCBs in the Great Lakes;* Mackay, D.; Paterson, S.; Eisenreich, S. J.; Simmons, M. S., Eds.; Ann Arbor Science: Ann Arbor, MI, 1983; Chapter 7, pp 115–125.
39. Hollod, G. J.; Eisenreich, S. J. *Anal. Chim. Acta* 1981, *124*, 31.
40. Doskey, P. V.; Andren, A. W. *Anal. Chim. Acta* 1979, *110*, 129.
41. Billings, W. N. T.; Bidleman, T. F. *Environ. Sci. Technol.* 1980, *14*, 679.
42. Hollod, G. J. Ph.D. Thesis, University of Minnesota, Minneapolis, 1979, p 247.
43. Doskey, P. V.; Andren, A. W. *J. Great Lakes Res.* 1981, *7*, 15.
44. Murphy, T. J.; Rzeszutko, C. P. *J. Great Lakes Res.* 1977, *3*, 305.
45. Atlas, E. L.; Giam, G. S. *Science (Washington, D.C.)* 1981, *211*, 163.
46. Harvey, G. R.; Steinhauer, W. G. *Atmos. Environ.* 1974, *8*, 77.
47. Junge, C. E. In *Fate of Pollutants in the Air and Water Environment;* Suffet, I. H., Ed.; Wiley-Interscience: New York, 1977; pp 7–25.
48. Strachan, W. M. J.; Huneault, H. J. *J. Great Lakes Res.* 1979, *5*, 61–68.
49. Mackay, D.; Paterson, S.; Schroeder, W. H. *Environ. Sci. Technol.* 1986, *20*, 810–816.
50. Bidleman, T. F.; Christensen, E. J.; Harder, H. W. In *Atmospheric Pollutants in Natural Waters;* Eisenreich, S. J., Ed.; Ann Arbor Science: Ann Arbor, MI, 1981; Chapter 24, pp 481–508.
51. Bidleman, T. F.; Foreman, W. T., in this book.
52. Scott, B. C. In *Atmospheric Pollutants in Natural Waters;* Eisenreich, S. J., Ed.; Ann Arbor Science: Ann Arbor, MI, 1981; Chapter 1, pp 3–21.

53. Ligocki, M. P.; Levenberger, C.; Pankow, J. F. *Atmos. Environ.* **1985**, *19*, 1609–1617.
54. Strachan, W. M. *J. Environ. Toxicol. Chem.* **1985**, *4*, 677–683.
55. Ligocki, M. P.; Leuenberger, C.; Pankow, J. F. *Atmos. Environ.* **1985**, *19*, 1619–1626.
56. Mackay, D.; Yuen, A. T. K. *Environ. Sci. Technol.* **1983**, *17*, 211.
57. Liss, P. S.; Slater, P. G. *Nature* **1974**, *247*, 181–184.
58. Liss, P. S. In *Air–Sea Exchange of Gases and Particles;* Liss, P. S.; Slinn, W. G. N., Eds.; NATO Advanced Studies Institute Series; Reidel: Boston, 1983; pp 241–298.
59. Schwarzenbach, R. P., Molnar-Kubica, E.; Giger, W.; Wakeham, S. *Environ. Sci. Technol.* **1979**, *13*, 1367.
60. Richardson, W. L.; Smith, V. E.; Wethington, R. In *Physical Behavior of PCBs in the Great Lakes;* Mackay, D.; Paterson, S.; Eisenreich, S. J.; Simmons, M. S., Eds.; Ann Arbor Science: Ann Arbor, MI, 1983; Chapter 18, pp 329–366.
61. Capel, P. D. M.S. Thesis, University of Minnesota, Minneapolis, 1983.
62. Thomann, R. V.; DiToro, D. M. *J. Great Lakes Res.* **1983**, *8*, 695–699.
63. Slinn, W. G. N.; Hasse, L.; Hicks, B. B.; Hogan, A. W.; Lal, D.; Liss, P. S.; Munnich, K. O.; Sehmal, G. A.; Vittori, O. *Atmos. Environ.* **1978**, *12*, 2055–2087.
64. Slinn, S. A.; Slinn, W. G. N. In *Atmospheric Pollutants in Natural Waters;* Eisenreich, S. J., Ed.; Ann Arbor Science: Ann Arbor, MI, 1981; Chapter 2, pp 23–54.
65. Slinn, W. G. N. In *Air–Sea Exchange of Gases and Particles;* Liss, P. S.; Slinn, W. G. N., Eds.; NATO Advanced Studies Institute Series; Reidel: Boston, 1983; pp 299–406.
66. Baker, J. E.; Eisenreich, S. J.; Johnson, T. C.; Halfman, B. M. *Environ. Sci. Technol.* **1985**, *19*, 854–861.
67. Evans, J. E.; Johnson, T. C.; Alexander, E. C.; Lively, R. S.; Eisenreich, S. J. *J. Great Lakes Res.* **1981**, *7*, 299–310.
68. Kemp, A. L. W.; Dell, C. I.; Harper, N. S. *J. Great Lakes Res.* **1978**, *4*, 276–287.
69. Gschwend, P. M.; Hites, R. A. *Geochim. Cosmochim. Acta* **1981**, *45*, 2359–2367.
70. Czuczwa, J. M.; McVeety, B. D.; Hites, R. A. *Science (Washington, D.C.)* **1984**, *226*, 568–569.
71. Anderson, M. L.; Rice, C. P.; Carl, C. C. *J. Great Lakes Res.* **1982**, *8*, 196–200.
72. Stortz, K. R.; Sydor, M. *J. Great Lakes Res.* **1980**, *6*, 223–231.
73. Eisenreich, S. J.; Looney, B. B. In *Physical Behavior of PCBs in the Great Lakes;* Mackay, D.; Paterson, S.; Eisenreich, S. J.; Simmons, M. S., Eds.; Ann Arbor Science: Ann Arbor, MI, 1983; pp 141–156.
74. Murphy, T. J. In *Toxic Contaminants in the Great Lakes;* Nriagu, J. O.; Simmons, M. S., Eds.; Wiley: New York, 1984; 53–79.
75. Dolan, D. M.: Bierman, V. J., Jr. *J. Great Lakes Res.* **1982**, *8*, 676–694.
76. Tisue, T.; Fingleton, D. In *Toxic Contaminants in the Great Lakes;* Nriagu, J. O.; Simmons, M. S., Eds.; Wiley: New York, 1984; Chapter 5, pp 105–125.
77. Wahlgren, M. A.; Robbins, J. A.; Edgington, D. N. In *Transuranic Elements in the Environment;* Hanson, W. C., Ed.; U.S. Department of Energy: Washington, DC, 1980; pp 659–683.
78. Rice, C. P.; Eadie, B. J.; Erstfield, K. M. *J. Great Lakes Res.* **1982**, *8*, 265–270.

79. *The Benthic Boundary Layer;* McCave, I. N., Ed.; Plenum: New York, 1976; p 323.
80. *Suspended Solids in Water;* Gibbs, R. J., Ed.; Plenum: New York, 1974.
81. Eadie, B. J.; Rice, C. P.; Frez, W. A. In *Physical Behavior of PCBs in the Great Lakes;* Mackay, D.; Paterson, S.; Eisenreich, S. J.; Simmons, M. S., Eds.; Ann Arbor Science: Ann Arbor, MI, 1983; pp 213–228.
82. Eadie, B. J.; Chambers, R. L.; Gardner, W. S. *J. Great Lakes Res.* 1984, *10*, 307–321.
83. Meyers, P. A.; Leenher, M. J.; Eadie, B. J. *Geochim. Cosmochim. Acta* 1984, *48*, 443–452.
84. Chambers, R. L.; Eadie, B. J. *Sedimentology* 1981, *28*, 439–447.
85. Sandilands, R. G.; Mudroch, A. *J. Great Lakes Res.* 1983, *9*, 190–200.
86. Karickhoff, S. W. *J. Hydraul. Div. Am. Soc. Civ. Eng.* 1984, *110*, 707–735.
87. Imboden, D. M.; Schwarzenbach, R. P. In *Chemical Processes in Lakes;* Stumm, W., Ed.; Wiley–Interscience: New York, 1985; Chapter 1, pp 1–30.
88. Gardner, W. D.; Southard, J. B.; Hollister, C. D. *Mar. Geol.* 1985, *65*, 199–242.
89. Noriki, S.; Ishimori, N.; Harada, K.; Tsunogai, S. *Mar. Chem.* 1985, *17*, 75–89.
90. Gibbs, R. J. *J. Geophys. Res.* 1985, *90*, 3249–3251.
91. Coates, J. T.; Elzerman, A. W. *J. Contam. Hydrol.* 1986, *1*.
92. Voice, T. C.; Rice, C. P.; Weber, W. J. *Environ. Sci. Technol.* 1983, *17*, 513–518.
93. Gschwend, P. M.; Wu, S.-C. *Environ. Sci. Technol.* 1985, *19*, 90–96.
94. Karickhoff, S. W. In *Contaminants and Sediments;* Baker, R. A., Ed.; Ann Arbor Science: Ann Arbor, MI, 1980; Vol. 2, pp 193–205.
95. Means, J. C.; Wijayaratne, R. D. *Bull. Mar. Sci.* 1984, *35*, 449–461.
96. Means, J. C.; Wijayaratne, R. D.; Boynton, W. R. *Can. J. Fish. Aquat. Sci.* 1983, *40*, 337–345.
97. Means, J. C.; Wijayaratne, R. D. *Science (Washington, D.C.)* 1982, *215*, 968–970.
98. Caron, G.; Suffet, I. H.; Belton, T. *Chemosphere* 1985, *14*, 993–1000.
99. Baker, J. E.; Capel, P. D.; Eisenreich, S. J. *Environ. Sci. Technol.* 1986, *20*.
100. O'Connor, D. J.; Connolly, J. P. *Water Res.* 1980, *14*, 1571.
101. DiToro, D. M.; Mahoney, J. D.; Kirchgraber, P. R.; O'Byrne, A. L.; Pasquale, L. R.; Piccirilli, D. C. *Environ. Sci. Technol.* 1986, *20*, 55–61.
102. DiToro, D. M.; Horzempa, L. M.; Casey, M. M.; Richardson, W. *J. Great Lakes Res.* 1982, *8*, 336–349.
103. Swackhamer, D. Ph.D. Thesis, University of Wisconsin, Madison, 1985.
104. O'Melia, C. R. In *Chemical Processes in Lakes;* Stumm, W., Ed.; Wiley–Interscience: New York, 1985; Chapter 10, pp 207–224.
105. Heuschele, A. S. *J. Great Lakes Res.* 1982, *8*, 603–613.
106. *Animal–Sediment Relations: The Biogenic Alteration of Sediments;* McCall, P. L.; Tevesz, M. J. S., Eds.; Plenum: New York, 1982; p 336.
107. Fisher, J. B.; Lick, W. J.; McCall, P. J.; Robbins, J. A. *J. Geophys. Res.* 1980, *85*, 3997–4006.
108. Robbins, J. A. *Hydrobiologia* 1982, *92*, 611–622.
109. Robbins, J. A.; Husby-Copeland, K.; White, D. S. *J. Great Lakes Res.* 1984, *10*, 335–347.
110. Robbins, J. A.; Krezoski, J. R.; Moseley, S. C. *Earth Planet. Sci. Lett.* 1977, *36*, 325–333.
111. Robbins, J. A.; McCall, P. L.; Fisher, J. B.; Krezoski, J. R. *Earth Planet. Sci. Lett.* 1979, *42*, 277–287.
112. Krezoski, J. R.; Robbins, J. A. *J. Geophys. Res.* 1985.

113. Thomas, R. L.; Frank, R. In *Physical Behavior of PCBs in the Great Lakes;* Mackay, D.; Paterson, S.; Eisenreich, S. J.; Simmons, M. S., Eds.; Ann Arbor Science: Ann Arbor, MI, 1983; Chapter 14, pp 245–267.
114. Robbins, J. A. *J. Geophys. Res.* **1986**.
115. Schink, D. R.; Guinasso, N. L. *Mar. Geol.* **1977**, *23*, 133–154.
116. Christensen, E. R. *J. Geophys. Res.* **1982**, *87*, 566–572.
117. Eisenreich, S. J.; Hollod, G. J.; Johnson, T. C. *Environ. Sci. Technol.* **1979**, *13*, 569.
118. Durham, R. W.; Pliver, B. G. *J. Great Lakes Res.* **1983**, *9*, 160–168.
119. Armstrong, D. E.; Swackhamer, D. L. In *Physical Behavior of PCBs in the Great Lakes;* Mackay, D.; Paterson, S.; Eisenreich, S. J.; Simmons, M. S., Eds.; Ann Arbor Science: Ann Arbor, MI, 1983; pp 229–244.
120. Bopp, R. F.; Simpson, H. J.; Olsen, C. R.; Trier, R. M.; Kostyk, N. *Environ. Sci. Technol.* **1982**, *16*, 666–676.
121. Hom, W.; Risebrough, R. W.; Soutar, A.; Young, D. *Science (Washington, D.C.)* **1974**, *184*, 1197–1199.
122. Johnson, T. C.; Evans, J. E.; Eisenreich, S. J. *Limnol. Oceanogr.* **1982**, *27*, 81–91.
123. Frank, R.; Thomas, R. L.; Braun, H. E.; Rasper, J. *J. Great Lakes Res.* **1980**, *6*, 113–120.
124. Hartig, J. H. In *Highlights of Water Quality and Pollution Control in Michigan;* Publication No. 4833-9804; Department of Natural Resources: East Lansing, MI, 1981.
125. Swain, W. R. In *PCBs: Human and Environmental Consequences;* D'Itri, F. M.; Kamrin, M. A., Eds.; Ann Arbor Science: Ann Arbor, MI, 1983; pp 11–48.
126. Bierman, V. J., Jr.; Swain, W. R. *Environ. Sci. Technol.* **1982**, *16*, 572–579.
127. Oliver, B. G. *Chemosphere* **1985**, *14*, 1087–1106.
128. Chapra, S. C.; Reckhow, K. H. *Engineering Approaches to Lake Management;* Butterworth: Boston, 1983; Vol. 2, pp 378–401.
129. Brownawell, B. J.; Farrington, J. W. *Geochim. Cosmochim. Acta* **1986**, *50*, 157–169.
130. Duinker, J. C.; Hildebrand, M. T. J.; Neth, J. *Sea Res.* **1979**, *13*, 256–281.
131. Capel, P. D. Ph.D. Thesis, University of Minnesota, Minneapolis, 1986.
132. Lerman, A. *Geochemical Processes: Water and Sediment Environments;* Wiley–Interscience: New York, 1979; p 340.
133. Poirrier, M. A.; Bordelon, B. R.; Laseter, J. L. *Environ. Sci. Technol.* **1972**, *6*, 1033–1035.
134. Hassett, J. P.; Anderson, M. A. *Environ. Sci. Technol.* **1979**, *13*, 1526–1529.
135. Hassett, J. P.; Anderson, M. A. *Water Res.* **1982**, *16*, 681–686.
136. Landrum, P. F.; Nihart, S. R.; Eadie, B. J.; Gardner, W. S. *Environ. Sci. Technol.* **1984**, *18*, 187–192.
137. Rodgers, P. W.; Swain, W. R. *J. Great Lakes Res.* **1983**, *9*, 548–558.
138. Whitmore, F. C. *A First Order Mass Balance Model for the Sources, Distribution and Fate of PCBs in the Environment*, U.S. Environmental Protection Agency: Washington, DC, 1977; EPA-560/6-77-006.
139. Weininger, D.; Armstrong, D. E. In *Restoration of Lakes and Inland Waters;* U.S. Environmental Protection Agency: Washington, DC, 1981; pp 364–372; EPA-444/5-81-010.
140. Swackhamer, D. L.; Armstrong, D. E. *Environ. Sci. Technol.* **1986**, *20*, 879–883.
141. Santschi, P. H. *Limnol. Oceanogr.* **1984**, *29*, 1100–1108.
142. Eisenreich, S. J. *Evaluation of a Simple Residence Time Model for PCBs in*

*Lakes;* Environmental Engineering Program Report; University of Minnesota: Minneapolis, 1985.
143. Karickhoff, S. W.; Brown, D. S.; Scott, T. A. *Water Res.* **1979,** *13,* 241–248.
144. Chiou, C. T.; Porter, P. E.; Schmedding, D. W. *Environ. Sci. Technol.* **1983,** *17,* 227–231.
145. McCarthy, J. F.; Jimenez, B. D. *Environ. Sci. Technol.* **1985,** *19,* 1072–1076.

RECEIVED for review May 6, 1986. ACCEPTED October 22, 1986.

# 14

# Fate of Some Chlorobenzenes from the Niagara River in Lake Ontario

B. G. Oliver

Environmental Contaminants Division, National Water Research Institute, Canada Centre for Inland Waters, Burlington, Ontario L7R 4A6, Canada

*Pollution of the Niagara River with chemical wastes has led to severe contamination of Lake Ontario. This chapter discusses the behavior of five chlorobenzenes in the Niagara River, in the river plume, and in Lake Ontario. The importance of sedimentation and volatilization on chemical pathways in the lake has been assessed. Bottom sediments are shown to contain the bulk of the chlorobenzenes that remain in the lake. The physical processes that affect the concentration distribution of chlorobenzenes in bottom sediments are explained, together with processes that affect the desorption rates of chlorobenzenes from sediments and biouptake of chlorobenzenes by benthic invertebrates. The contamination of fish and other biota is reported, and contaminant trends in sediment cores and biota are discussed.*

T HE CONTAMINATION OF LAKE ONTARIO BY THE NIAGARA RIVER has been well documented. The papers presented at the 1982 Niagara River Symposium at the Great Lakes Conference have been published as a special volume (1) and accurately present the current status of the Niagara River pollution problem. In addition to direct discharges from chemical manufacturers along the river, more than 200 chemical waste dumps can be found in the river's vicinity (2). Many of these waste dumps leak into the river. Elder et al. (3) demonstrated the presence of high concentrations of chlorobenzenes, chlorotoluenes, polychlorinated biphenyls (PCBs), and many other chlorinated and fluorinated chemicals in sediments from creeks near dumpsites along the river. Oliver and Nicol (4) showed elevated concentrations of chlorobenzenes in water samples below dumpsites and chemical manufacturing discharges along the river. Yurawecz (5) identified some unusual chlorotoluenes

0065-2393/87/0216-0471$06.00/0

and chlorinated trifluorotoluenes in fish from the river. These chemicals are produced in the river's vicinity.

In addition to direct contamination of the river, considerable evidence indicates that the sediments of Lake Ontario have been severely polluted. Elevated concentrations of organochlorine insecticides and PCBs (6), mirex (7), chlorobenzenes (4), and octachlorostyrene (8) have been observed in Lake Ontario sediments. Unusual fluorinated compounds, which seem to result from the chemical reaction of wastes in the dumpsites, have also been shown to be dispersed widely throughout the lake (9).

In this chapter, the pathways of several chlorobenzenes in Lake Ontario will be discussed for which the Niagara River appears to be the major source. These compounds were chosen for discussion because they are present in significant concentrations in all compartments of both the river and the lake. They also are chemicals that exhibit a wide range of physical and chemical properties. Some of the properties of chlorobenzenes are presented in Table I and some useful characteristics of Lake Ontario are listed on page 473. A diagram of the general study area is shown in Figure 1. Analytical methods, dual column capillary gas chromatography with electron-capture detectors, and extraction and cleanup procedures were described previously (10).

**Table I. Chlorobenzene Properties**

| Chemical | Abbreviation | $\log K_{ow}$ (Ref.) | Water Solubility at 25 °C[a] (mg/L) |
|---|---|---|---|
| 1,4-Dichlorobenzene | 1,4-DCB | 3.4 (11) | 90 |
| 1,2,4-Trichlorobenzene | 1,2,4-TCB | 4.0 (12) | 30 |
| 1,2,3,4,-Tetrachlorobenzene | 1,2,3,4-TeCB | 4.5 (13) | 4.3 |
| Pentachlorobenzene | QCB | 4.9 (11) | 0.56 |
| Hexachlorobenzene | HCB | 5.5 (12) | 0.005 |

[a]Data are from reference 14.

## The Niagara River

The Niagara River is one of North America's larger rivers (flow is 6400 $m^3/s$), and the presence of Niagara Falls and the Whirlpool Rapids below the falls makes much of the river unnavigable. Thus, sampling along many of the river reaches is difficult if not impossible. A comparison of the concentration of five chlorobenzenes at site 1 (Figure 1), the start of the river (Fort Erie); at site 2, below a chemical dumpsite and a chemical manufacturing discharge just above Niagara Falls, New York (a helicopter-collected sample); and site 3, a sample at the mouth

```
┌──────────────────────────────────────────┐
│         Lake Ontario Characteristics       │
│                                            │
│                   Water                    │
│          Area: 18,500 km²                  │
│          Length: 311 km                    │
│          Mean depth: 86 m                  │
│          Maximum depth: 244 m              │
│          Volume: 1640 km³                  │
│          Niagara R. inflow: 6400 m³/s      │
│          St. Lawrence R. outflow: 6700 m³/s│
│          Mean residence time: 7.8 years    │
│                                            │
│      Sedimentation Basins and Their Areas  │
│                                            │
│          Niagara: 1600 km²                 │
│          Mississauga: 2700 km²             │
│          Rochester: 3800 km²               │
│          Kingston: 560 km²                 │
└──────────────────────────────────────────┘
```

of the river [Niagara-on-the-Lake (NOTL)], is shown in Table II. As can be seen from the table, significant sources of chlorobenzenes are entering the river. Contributions from both chemical dumpsite leachates and direct discharges appear to be the major sources (*15*).

The turbulence in the river produces a reasonably homogeneous water mass at the river's mouth. Transect sampling at NOTL has shown virtually the same concentrations of organic chemicals on the Canadian side, in the middle, and on the American side of the river. On the basis of this data, a water quality monitoring station has been established at NOTL (*16*). The intake line for this station is located 30 m from the Canadian shore, about 6 m above the river bottom, and 13 m below the river surface. Originally, samples of suspended solids from the river were collected by centrifugation and analyzed for organic compounds in an attempt to estimate loadings to Lake Ontario (*16*). This estimation was done on the basis of the assumption that most hydrophobic organic chemicals such as PCBs would be found mainly in the particulate phase. However, subsequent studies have shown that this procedure vastly underestimated chemical loadings (in the low turbidity river, suspended solids concentrations were 3–10 mg/L) because the bulk of the organic compounds was found to be present in the dissolved phase (*17*).

Subsequently, weekly 16-L whole-water samples were collected from the river at the NOTL station for a 2-year period (1981–1983) to estimate concentrations, to assess concentration variability, and to approximate loading for several chlorinated organic compounds (*18*).

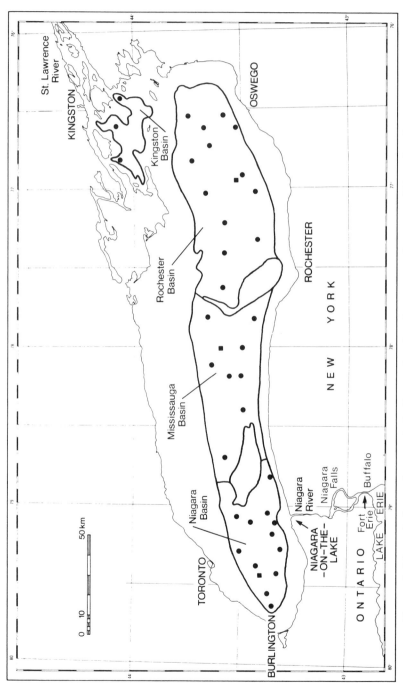

*Figure 1. Map of the study area including surficial sediment sampling sites (●) and sediment core sampling sites (■) in Lake Ontario.*

Table II. Chlorobenzene Concentrations in the Niagara River in 1981

| Site | 1,4-DCB | 1,2,4-TCB | 1,2,3,4-TeCB | QCB | HCB |
|------|---------|-----------|--------------|-----|-----|
| 1 | 1.7 | 0.5 | 0.06 | 0.05 | 0.05 |
| 2 | 94 | 110 | 130 | 22 | 1 |
| 3 | 29 | 12 | 3.8 | 1.2 | 0.6 |

NOTE: All values are in units of nanograms per liter.

Figure 2 shows a plot of the concentration of 1,2- and 1,4-dichlorobenzenes (1,2- and 1,4-DCB, respectively) over this 2-year period. A fairly steady background concentration was observed in the river, likely due to steady leaching from waste disposal sites along the river. Superimposed on this background are large concentration spikes that are likely due to direct industrial discharges to the river. To illustrate the magnitude of the concentration spikes, on May 17, 1982, the concentration of 1,2-DCB was measured to be 240 ng/L compared to 29 ng/L recorded the previous week. The 1,2-DCB remained at an elevated concentration for 4 weeks. The estimated total quantity of chemical in this slug was about 2000 kg, or about 50% of the total 1982 river loading for this chemical. Many other chlorinated chemicals showed similar concentration profiles to these chlorobenzenes, although concentration spikes for the other chemicals were not as large (*18*).

A crude estimate of loadings for some chlorobenzenes to Lake Ontario from the river can be made by using median river concentrations from this 2-year study. The loadings are 5800 kg/year for 1,4-DCB, 2400 kg/year for 1,2,4-trichlorobenzene (1,2,4-TCB), 760 kg/year for 1,2,3,4-tetrachlorobenzene (1,2,3,4-TeCB), 240 kg/g for pentachlorobenzene (QCB), and 120 kg/year for hexachlorobenzene (HCB). Earlier studies provided chlorobenzene concentrations in some sewage treatment plant (STP) effluents flowing into Lake Ontario (*4*). By using this data and a total STP discharge of $10^9$ m$^3$/year (*19*), the loadings of these chlorobenzenes from STPs are roughly 660 kg/year for 1,4-DCB, 11 kg/year for 1,2,4-TCB, 2 kg/year for 1,2,3,4-TeCB, 1 kg/year for QCB, and 2 kg/year for HCB. Thus, STPs appear to be only minor contributors of chlorobenzenes (with the possible exception of 1,4-DCB) to Lake Ontario. Also, sediment samples near the mouths of other major rivers entering Lake Ontario were found to contain very low concentrations of chlorobenzenes. Thus, the dominant chlorobenzene loading to the lake appears to originate in the Niagara River.

## Niagara River Plume and Near-River Lake Ontario

The distinct plume of the Niagara River has been shown to extend 10–15 km into Lake Ontario. The direction of the plume can vary from

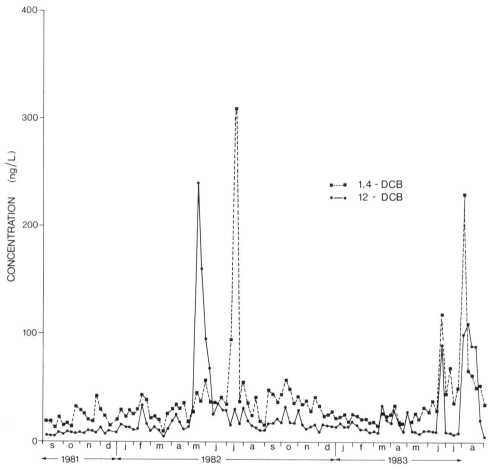

*Figure 2. Concentrations of 1,2-DCB and 1,4-DCB in the Niagara River at NOTL (Reproduced with permission from reference 18. Copyright 1984, Elsevier.)*

northerly towards Toronto (Figure 1) to easterly depending mainly on wind direction and speed (20). Because of the prevailing wind direction, the most frequent direction of the plume is bent over to the east along the south of the lake where it is caught up in the strong eastward coastal current (21). The plume can be tracked by using physical measurements such as temperature, turbidity, and conductivity. Because of the higher surface-to-volume ratio of Lake Erie compared with Lake Ontario, for much of the year the water temperature in the plume will be higher than that in Lake Ontario. Another method that has been used to track the plume is with radiotransmitting drogues. These drifters, which are propelled by underwater sails (2.4 × 3 m), move in the direction of the underwater currents and are minimally affected by surface winds (21).

In addition to tracking by physical techniques, some measurements have also been made on contaminant concentrations in the plume. These measurements were made at a 1-m depth on centrifuged, large-volume (200-L) samples over a period of 24 h. Sampling occurred in the same water mass as indicated by the drogues. The concentrations of chlorobenzenes and other chlorinated contaminants were virtually constant (within experimental error) over the course of the sampling. This constant level indicates that processes that influence the concentration of these chemicals occur over a much longer time frame. Contaminant concentrations in surrounding lake water were much lower than in the plume, so a region with a concentration gradient between the plume and the lake water must exist.

In addition to following the chlorobenzene concentration in the water column in the plume, the partitioning between suspended sediments and the water was also studied. The sediment–water partition coefficient ($K_{oc}$) normalized to organic carbon is defined as

$$K_{oc} = C_{ss}/(C_{H2O}f_{oc}) \qquad (1)$$

where $C_{ss}$ and $C_{H2O}$ are the chemical concentrations in the suspended sediment and water phases, respectively, and $f_{oc}$ is the organic carbon fraction of the suspended sediments. In the plume on September 22, 1984, the suspended sediment concentration averaged 3 mg/L, and the mean organic carbon content of the sediments was 6.6% ($f_{oc} = 0.066$). In Table III, the range and mean log $K_{oc}$ values for the five samples in the plume are compared to values calculated with Karickhoff's empirical equation (22), which relates $K_{oc}$ to $K_{ow}$ (the octanol–water partition coefficient).

$$K_{oc} = 0.411 \, K_{ow} \qquad (2)$$

The $K_{oc}$ values were fairly consistent within the plume and varied by a maximum of a factor of 3. With the exception of 1,4-DCB, for

**Table III. Sediment–Water Partition Coefficients in the Niagara River Plume and from Karickhoff's Empirical Equation**

| log K$_{oc}$ Value | 1,4-DCB | 1,2,4-TCB | 1,2,3,4-TeCB | QCB | HCB |
|---|---|---|---|---|---|
| log $K_{oc}$ (range) | 5.3–5.6 | 4.8–5.3 | 4.9–5.4 | 5.5–5.9 | 6.0–6.5 |
| log $K_{oc}$ (mean) | 5.5 | 5.0 | 5.1 | 5.7 | 6.3 |
| Calculated $K_{oc}$ | 3.0 | 3.6 | 4.1 | 4.6 | 5.1 |

which we suspect analytical difficulties in the suspended solids fraction, the $K_{oc}$ value increases systematically with $K_{ow}$ as predicted theoretically. But the absolute values of the field $K_{oc}$ values are more than 1 order of magnitude greater than those predicted by Karickhoff's empirical equation. Recent studies show that the partition coefficient for organic compounds changes considerably with changing suspended sediment concentration (23). Weber et al. (24) showed that partition coefficients increased by 1 order of magnitude for each 2.5 order-of-magnitude decrease in solids concentration. The suspended sediment concentrations used by Karickhoff to generate equation 2 were about 1000 mg/L (25). This phenomenon of changing partition coefficient with changing suspended solids concentration is the likely explanation for the apparent discrepancy between the field samples (suspended solids concentration was 3 mg/L) and the empirical equation. This result shows that considerable care must be taken before applying laboratory-derived measurements to field situations.

The amount of chemicals entering the lake that were sequestered to settling particles and eventually became bottom sediments was estimated by placing three sediment traps around the edge of the plume (26). The sediment trap assembly consisted of five plexiglass tubes (7 cm in diameter and 106 cm in length) fitted at the bottom with removable cups (27). Samples in the traps were collected every month during the field season from May to November in 1981 and in 1982. The mean contaminant values in the down-fluxing material were determined. At the same time, the same contaminants and suspended solids levels were quantified in weekly samples of Niagara River water. The loading of contaminants to the bottom sediments was calculated from the mean concentration in the sediment trap material by assuming that all the suspended sediments in the Niagara River were down-fluxing to the bottom of Lake Ontario. The data for several chlorobenzenes is shown in Table IV. The data in the table clearly show that only a very small percentage of the chlorobenzene input from the river deposits to the bottom sediments. As expected, the compounds with higher $K_{ow}$ values and lower water solubilities exhibit a greater tendency to become associated with particulate matter in the water column (22).

**Table IV. Amount of Chlorobenzene Sedimentation in Lake Ontario**

| Source | 1,4-DCB | 1,2,4-TCB | 1,2,3,4-TeCB | QCB | HCB |
|---|---|---|---|---|---|
| Niagara R. loading (g/day) | 23,000 | 8800 | 2900 | 830 | 510 |
| Down-fluxing material (g/day) | 300 | 93 | 61 | 33 | 75 |
| % of input down-fluxing | 1 | 1 | 2 | 4 | 15 |

## Lakewide Processes

A block diagram presenting the major processes of contaminant losses from Lake Ontario is shown in Figure 3. The Niagara River input to the lake and the proportion of this loading that settles to the lake bottom have already been discussed. Some measurements have been made at the eastern end of the lake to quantify the export of chemicals from the lake via the St. Lawrence River. Table V gives a brief summary of the yearly input and the losses from the lake that we have been able to determine. These data show that our mass balance is very poor. More than 80% of all the chlorobenzenes are unaccounted for when only loadings to the sediment and losses via the St. Lawrence River are considered.

Volatilization losses from the lake are the likely cause for the mass balance discrepancies in Table V. Biodegradation and photodegradation are other processes that must be considered. The lower chlorobenzenes (DCB and TCB) have been shown to be slowly degraded by high concentrations of bacteria in the laboratory (28). Whether or not such processes occur in the lake environment that has much lower bacterial populations is not known. The higher chlorobenzenes are definitely recalcitrant to biodegradation. Photodechlorination at solar wavelengths of TeCBs, QCBs, and HCBs was reported in acetonitrile–water mixtures

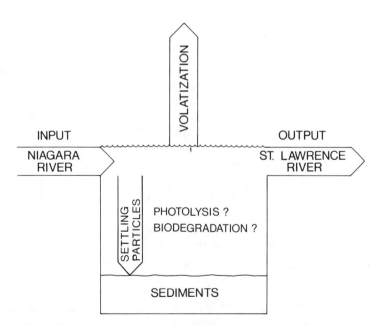

*Figure 3. Block diagram of major loss processes for Lake Ontario.*

Table V.  Yearly Chlorobenzene Inputs and Losses from Lake Ontario

| Source | 1,4-DCB | 1,2,4-TCB | 1,2,3,4-TeCB | QCB | HCB |
|--------|---------|-----------|--------------|-----|-----|
| Niagara R. loading | 5800 | 2400 | 760 | 240 | 120 |
| Sediment loading | 76 | 25 | 16 | 10 | 18 |
| St. Lawrence R. loading | 180 | 74 | 15 | 6 | 6 |
| Unaccounted material | 5500 | 2300 | 730 | 220 | 96 |

NOTE: All values are in units of kilograms per year.

(29). Also, sensitized photodechlorination and photoisomerization were demonstrated with acetone (29). These studies were conducted at much higher concentrations than are encountered in the environment. Because the light absorbance of these chemicals in the solar region is very weak, indirect photosensitized photoreactions may be more important than direct photolysis in the environment. But the occurrence and significance of photodegradation of chlorobenzenes in the environment has not been demonstrated.

These considerations lead to the conclusion that volatilization is probably the major mechanism of loss of chlorobenzenes from Lake Ontario. This result is in good agreement with earlier studies on Lake Zurich, Switzerland (30), where volatilization was shown to be the major process for loss of 1,4-DCB from the lake. Volatilization has also been shown to be important for 1,4-DCB and 1,2,4-TCB in sea water mesocosms (31). This process may provide a partial explanation for the widespread distribution of chlorinated contaminants such as PCBs and HCB in rainwater (32).

During the last few years, samples from most compartments of the Lake Ontario ecosystem have been analyzed. The map in Figure 1 shows the surficial sediment sampling network in the sedimentation basins of the lake and the location of the three cores (one from each major basin) that have been collected and analyzed. From these data, a crude estimate can be made of the masses of the various chlorobenzenes in the lake's bottom sediment (33). Similarly, by analysis of the lake water from many diverse lake locations in the fall after turnover and in the spring before stratification, an estimate of the mass of chlorobenzenes in the water column of the lake was made. The approximate mass of contaminants in suspended solids in the lake was estimated by analysis of material collected with a continuous flow centrifuge and with sediment traps throughout the lake. A very crude estimate of the mass of chlorobenzenes in biota was made by using biomass estimates (34) combined with analysis of algae, zooplankton, benthic organisms, and fish from the lake. The masses of chemicals in the lake compartments are shown in Table VI. The bulk of chlorobenzenes remaining in Lake Ontario are present in bottom sediments. Surprisingly, very little

Table VI. Masses of Chlorobenzenes in Lake Ontario Compartments

| Compartment | 1,2,4-TCB | 1,2,3,4-TeCB | QCB | HCB |
|---|---|---|---|---|
| Bottom sediments | 11,000 | 3300 | 4100 | 8500 |
| Lake water | 700 | 210 | 90 | 90 |
| Suspended sediments | 10 | 4 | 4 | 9 |
| Biota | 2 | 2 | 2 | 8 |

NOTE: All values are in units of kilograms.

mass of chlorobenzenes is found in other lake compartments. Fish and other biota contain significant concentrations of chlorobenzenes and other chlorinated chemicals, and this problem is serious. But, because of the low biomass in the lake ($\approx 10$ g dry weight/$m^2$), the total mass of chlorobenzenes in this compartment is small. Because of the importance of the bottom sediments as both a sink and a possible future source for contaminants, the sediments in the lake will be discussed in detail in the next section.

## Sediment Processes

Three main sedimentation basins occur in Lake Ontario: the Niagara, the Mississauga, and the Rochester, which are separated by sils. One minor basin occurs, the Kingston (35) (Figure 1). These basins are the only areas in the lake where net accumulation of sediment occurs over the year. Outside the basins, a net erosion of sediment occurs. The sampling grid in Figure 1 shows that samples were collected only in the sedimentation basins because past measurements showed that the concentration of chemicals in erosional-zone bottom material is near zero. The mean total chlorobenzene concentrations (di- through hexa-) are 610 $\mu$g/kg dry weight in the Niagara Basin, 560 $\mu$g/kg in the Mississauga Basin, 480 ppb in the Rochester Basin, and 120 $\mu$g/kg in the Kingston Basin. The data for individual samples as well as the basin means show that no strong plume of contaminated sediment extends from the Niagara River. The major basins have roughly the same chlorobenzene concentrations, and somewhat lower values occur in the Kingston Basin. This observation contrasts with studies on other rivers and lakes (e.g., reference 36, where elevated sediment concentrations are observed near point sources).

To explain these observations, several sediment traps were placed at various locations and depths in the lake. A series of sediment traps at roughly 20-m-depth intervals were used at each site; the lowest trap was placed 2 m from the bottom. Trap catches for a typical year for an offshore site near the center of the Niagara Basin are shown in Table VII. One observation that can be made from this table is that the

Table VII. Sedimentation Rate from May 1981 to May 1982 in the Middle of
the Niagara Basin

| Trap Depth (m) | May | June | July | Aug. | Sept. | Oct. | Nov. | (Dec. to May) |
|---|---|---|---|---|---|---|---|---|
| 20 | 2.80 | 2.29 | 3.12 | 1.65 | 1.11 | 0.94 | 0.35 | 2.60 |
| 40 | 2.78 | 2.09 | 2.76 | 1.53 | 1.23 | 0.91 | 0.74 | 2.99 |
| 60 | 2.84 | 2.05 | 3.06 | 1.56 | 1.22 | 1.32 | 0.67 | 3.17 |
| 80 | 2.74 | 2.03 | 3.25 | 1.65 | 1.41 | 2.34 | 1.41 | 4.29 |
| 90 | 2.91 | 1.65 | 3.54 | 2.07 | 1.80 | 3.67 | 1.83 | 4.30 |
| 98 | 3.18 | 2.07 | 4.98 | 2.81 | 2.75 | 5.73 | 3.53 | 5.71 |

NOTE: All sedimentation rates are in units of grams per square meter per day.

bottom traps (particularly the trap 2 m from the bottom) collect more material than the higher traps throughout much of the year. The largest differences between the bottom and upper traps occur when the lake is stratified in the summer and early fall. The higher catches in the lower trap indicate that significant sediment resuspension occurs even in the deepest basins of Lake Ontario. This result agrees with optical measurements in the lake that have shown the presence of a nepheloid layer (a layer of high turbidity) near the lake bottom (37). The thickness of the layer, which was present over the entire lake, was variable but could extend up to 45 m from the bottom (37).

The analysis of contaminant profiles in the sediment trap material, Niagara River suspended sediments, and Lake Ontario surficial bottom sediments provided interesting evidence as to the source of the additional material in the bottom traps. Mirex in bottom sediments of the Niagara Basin averaged 48 $\mu$g/kg, whereas recent suspended sediments from the Niagara River at NOTL contained about 5 $\mu$g/kg mirex. Also, the ratio of 1,2,4,5-TeCB to 1,2,3,4-TeCB in bottom sediments was 1.64, whereas this ratio was 0.54 in recent suspended sediments from the river. These contaminant concentration changes appear to be due to reduced loading of mirex and 1,2,4,5-TeCB in recent years (38). The mirex concentration and the ratio of 1,2,4,5-TeCB to 1,2,3,4-TeCB in bottom trap material were much higher than that in the 20- and 40-m-depth traps. This observation indicates that the major source of the additional catch in the lower traps was probably resuspended bottom sediments.

Another observation that can be made from Table VII is that the trap catches are much higher in early spring and in winter when the lake is isothermal. During this time of the year, violent storms occur on the lake, and wind-driven currents penetrate deeply (because of isothermal conditions) and cause significant sediment resuspension.

The data in Table VII and information generated from traps near the Niagara River indicate that sediments and detritus from the river adsorb contaminants from the river water and settle temporarily in the vicinity of the river. Currents in Lake Ontario, especially during spring and winter, resuspend the material and redistribute it to the various sedimentation basins in the lake. A strong counterclockwise circulation pattern occurs in the lake, and current speeds average 5 cm/s in the winter and 2 cm/s in the summer (39). The Kingston Basin is out of the main circulation pattern of the lake; this fact probably explains why the sediment chlorobenzene concentrations are lower in this basin. The sediment chlorobenzene concentrations in the other three basins are similar because of the dynamic nature of sedimentary processes within the lake's major circulation regime.

The mean total (di- through hexa-) chlorobenzene concentration in Great Lakes surficial sediments is 560 $\mu$g/kg for Lake Ontario, 25 $\mu$g/kg for Lake Erie, 33 $\mu$g/kg for Lake Huron, and 5.5 $\mu$g/kg for Lake Superior. Thus, Lake Ontario sediments are more than 1 order of magnitude more contaminated than the other three Great Lakes located partially in Canada. We have no data for Lake Michigan.

What is the significance of this sediment contamination? Sediment-associated contaminants can influence the concentrations in both the water column and in lake biota if they are desorbed or are available to benthic organisms. Some of the chlorobenzenes in Lake Ontario bottom sediment would be desorbed if the sediments were resuspended (40). Laboratory studies showed that chlorobenzene desorption half-lives in clean lake water averaged about 60 days at 4 °C, 40 days at 20 °C, and 10 days at 40 °C (40). Temperature was a much more important variable than chemical structure in governing desorption rates. Field observations in Lake Ontario have indicated that a sediment layer about 1 mm thick in the sedimentation basins ($\approx$8700 km$^2$) is in a constant state of flux. This sediment is almost continuously being resuspended and then resettled to the bottom. The near-bottom temperature in the lake is 4 °C, and the active layer contains about 2% solids. By using these approximations, the half-life data above, and the mean chlorobenzene concentrations in lake surficial sediments, a crude estimate of loading rates from suspended sediments to the water column can be made. These loadings are compared to loadings from the Niagara River (Table VIII). The table shows that, for chemicals with large active sources such as the lower chlorinated chlorobenzenes, the contribution from sediment desorption is low. But, for compounds with low current loadings such as HCB, desorption from sediments could play a significant role in controlling lake water concentrations.

## Biological Component

Chemicals in bottom sediments may also be taken up directly from sediments and detritus by benthic organisms. Laboratory studies (41)

Table VIII. Loadings to Lake Ontario Water Column from Suspended Sediments and the Niagara River

| Loadings Source | 1,4-DCB | 1,2,4-TCB | 1,2,3,4-TeCB | QCB | HCB |
|---|---|---|---|---|---|
| Resuspended sediments | 11 | 19 | 7 | 7 | 19 |
| Niagara River | 5800 | 2400 | 760 | 240 | 120 |

NOTE: All values are in units of kilograms per year.

have demonstrated that oligochaete worms, which are found in significant numbers in Lake Ontario sediments, bioaccumulate many chlorinated compounds from these sediments. Field studies near the Niagara River have shown a strong correlation between HCB concentration in oligochaete worms and in the sediments in which they lived (42). These benthic organisms are at the lower end of the food chain and can influence chemical concentrations in sport and commercial fish through this food chain link.

Benthic organisms also enhance the rate of release of pollutants from sediments by the process of bioturbation. Karickhoff and Morris (43) showed that sediment reworking by oligochaetes enhanced the flux of QCB and HCB from a bed of sediment in a laboratory microcosm by a factor of 4-6. This process was mitigated to a certain degree by fecal pelletization of the sediment by the worms, which actually caused a reduction in the desorption rates of the chemicals from the sediment particles (43). Worms feed at depths of 8-10 cm and defecate this material at the surface (44). By this mechanism, the worms can recycle more contaminated deeper sediments (in locations where control measures have been implemented) to the sediment surface. Thus, the presence of benthic organisms may increase the time required, after implementation of controls, to observe dramatic reductions in contaminant levels in the ecosystem. This induction period will be greater in locations that have low sedimentation such as Lake Ontario [sedimentation rates are 1-5 mm/year (45)].

The chlorobenzene concentrations in biota, such as algae and zooplankton, were studied briefly in plankton net (125 μm) catches from the Niagara River (42). From purely physical considerations for inert material, the smallest particles would be predicted to have the highest concentration because of their higher surface–mass ratio. But in general, chlorobenzene concentrations in plankton net catches were higher in the larger size fractions. Qualitative examination of the material showed that smaller size fractions (<175 μm) contained mainly algae, whereas the larger size fractions comprised copepod zooplankton and their skeletons. Thus, chlorobenzene association with this biotic material is consistent with partitioning based on lipid content in each fraction (46).

The chemical residues in Lake Ontario fish are related to chemicals in the water column (47) and to chemical contamination in benthic and pelagic food chains (48). Arguments abound in the literature as to the relative importance of water-borne and food chain uptake for contaminants. Recent studies clearly demonstrate that, in Lake Ontario and probably in other aquatic systems, the relative importance of these exposure routes depends mainly on chemical structure (47). For chlorobenzenes, the usual strong correlation between bioconcentration factor (BCF) and the octanol–water partition coefficient ($K_{ow}$) was demonstrated. For example, the BCF for chlorobenzenes gradually increases from about 400 for 1,4-DCB to 13,000 for QCB (47). Using these laboratory-derived BCFs to predict the chlorobenzene residue levels in field populations of rainbow trout is possible. These predicted values are compared to measured mean values for 10 field trout in Table IX. For the lower chlorinated benzenes, good agreement was found between the predicted and measured residue levels. The half-lives of di- through pentachlorobenzene in fish are of the order of several days (49), so equilibrium between fish and water concentrations can be attained rapidly for these chemicals. In contrast, the half-life of HCB in fish is greater than 7 months (50), so equilibrium with the water cannot be established, and residue HCB concentrations in the fish will continue to increase over time. The much higher measured value for HCB than the predicted value (Table IX) shows that HCB water concentrations play only a minor role in controlling HCB fish concentrations, and that food chain accumulation is the more important exposure vector for this chemical. This work was extended to other classes of chemicals (51) and showed that, as a general rule, residue levels of compounds with short half-lives are controlled mainly by chemical concentrations in the water, whereas residue levels for chemicals with long half-lives in fish depend mainly on fish food contamination and food consumption rates. Because of the extremely low concentration of chlorobenzenes in the Lake Ontario water column (<1 ng/L), the previous discussion shows that HCB is probably the only chlorobenzene that could pose a problem in the lake through fish bioconcentration or bioaccumulation.

Fish in Lake Ontario contain higher concentrations of many

**Table IX. Predicted and Measured Chlorobenzene Concentrations in Lake Ontario Rainbow Trout**

| Value Basis | 1,2,4-TCB | 1,2,3,4-TeCB | QCB | HCB |
|-------------|-----------|--------------|-----|-----|
| Predicted | 0.8 | 0.5 | 2.6 | 0.7 |
| Measured | 0.6 ± 0.3 | 1.0 ± 0.4 | 3.4 ± 1.3 | 33 ± 15 |

NOTE: All values are in units of micrograms per kilogram.

chlorinated compounds (including HCB) than fish from the other Great Lakes (52), and fish consumption advisories have been issued (53). By using some of the concentration data presented in this chapter, the potential human exposure routes can be compared for chemicals such as HCB. The concentration of HCB in Lake Ontario is about 0.06 ng/L. For dose calculations, water consumption is usually taken as 2 L/day, or about 700 L/year. This value yields an HCB dose for consumers of Lake Ontario drinking water of 0.04 $\mu$g/year (assuming no removal at the treatment plant). If one consumed a meal (about 125 g) of rainbow trout from the lake (HCB concentration 33 $\mu$g/kg) in 1 year, one would be exposed to a dose of 4 $\mu$g/year. In other words, drinking the lake water for 100 years is necessary to obtain the same HCB dose obtained from consuming one fish meal from the lake. Similar dose calculations for other bioaccumulated chemicals such as PCBs yielded similar results and showed the minimal impact of drinking water as compared to fish or food consumption.

## Historical Chemical Inputs from the Niagara River

Monitoring of organic contaminants in the Niagara River has occurred only since the late 1970s (16). Thus, the only way to obtain information about historical inputs is by the analysis of sediment cores from Lake Ontario. A diagram of the total sediment chlorobenzene concentration versus sediment depth is shown in Figure 4 for a core collected off the mouth of the Niagara River in Lake Ontario in 1981. The top portion of the graph shows the corresponding $^{210}$Pb dating of the core and CB production figures for those years. The peak discharge of chlorobenzenes to Lake Ontario occurred in the 1960s. This observation correlates well with U.S. production figures for chlorobenzenes that also were greatest in this period. A significant biological impact of the high discharges of chlorobenzenes and other chlorinated chemicals during the 1960s was observed in fish-eating bird populations in the region. Population declines and reproductive failures were documented in the Common Tern (54) and the Herring Gull (55) in Lake Ontario colonies.

In recent years, the sediment core data shows significant reduction in levels of contamination. Bird populations and reproductive success in the region are returning to normal (56). However, recent studies show that PCB concentrations in fish from Lake Ontario, after many years of decline, are beginning to increase again (57). Considerable concern has been raised about apparent increased leakage from chemical dumpsites close to the river (58). As long as these chemical dumpsites—which contain many metric tons of toxic chemicals (15)—are present, people in the Lake Ontario region will be justifiably concerned. The only long-term viable solution to the problem appears to be removal and destruc-

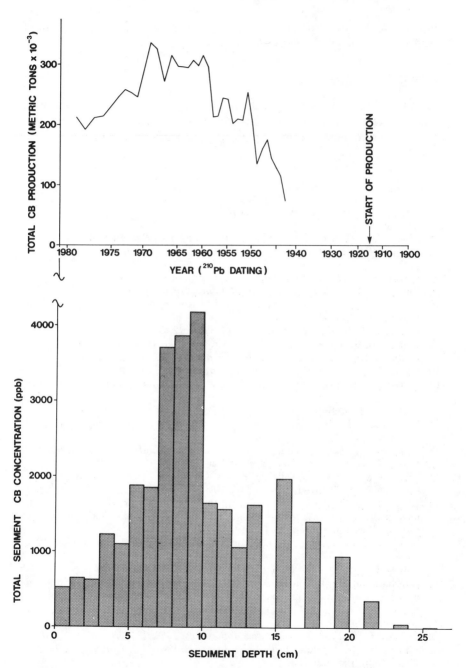

Figure 4. Total concentration of chlorobenzene (CB) versus depth in sediment core and corresponding $^{210}Pb$ dating of core sections, and U.S. production figures for CBs.

tion of the chemicals in these waste sites, and the termination of direct persistent chemical discharges to the Niagara River.

## Acknowledgments

I thank A. Niimi, M. Charlton, R. Durham, M. Fox, and J. Carey, who collaborated with me on several aspects of the research described in this chapter. I thank Karen Nicol and Lee Durham for their technical assistance.

## References

1. Allan, R. J.; Mudroch, A.; Munawar, M., Eds. *J. Great Lakes Res.* **1983**, *9*, 109–340.
2. Allan, R. J.; Mudroch, A.; Sudar, A. *J. Great Lakes Res.* **1983**, *9*, 111.
3. Elder, V. A.; Proctor, B. L.; Hites, R. A. *Environ. Sci. Technol.* **1981**, *15*, 1237.
4. Oliver, B. G.; Nicol, K. D. *Environ. Sci. Technol.* **1982**, *16*, 532.
5. Yurawecz, M. P. *J. Assoc. Off. Anal. Chem.* **1979**, *62*, 36.
6. Frank, R.; Thomas, R. L.; Holdrinet, M.; Kemp, A. L. W.; Braun, H. E. *J. Great Lakes Res.* **1979**, *5*, 18.
7. Kaiser, K. L. E. *Environ. Sci. Technol.* **1978**, *12*, 520.
8. Kaminsky, R.; Hites, R. A. *Environ. Sci. Technol.* **1984**, *18*, 275.
9. Jaffe, R.; Hites, R. A. *Environ. Sci. Technol.* **1985**, *19*, 736.
10. Oliver, B. G.; Nicol, K. D. *Chromatographia* **1982**, *16*, 336.
11. Banerjee, S.; Yalkowsky, S. H.; Valvani, S. C. *Environ. Sci. Technol.* **1980**, *14*, 1227.
12. Chiou, C. T.; Schmedding, D. W. *Environ. Sci. Technol.* **1982**, *16*, 4.
13. Könemann, H.; Zelle, R.; Busser, F.; Hammers, W. E. *J. Chromatogr.* **1979**, *178*, 559.
14. Horvath, A. L. *Halogenated Hydrocarbons Solubility-Miscibility with Water*; Marcel Dekker: New York, 1982.
15. "Report of the Niagara River Toxics Committee"; Inland Waters Directorate, Canada Centre for Inland Waters: Burlington, Ontario, Canada, 1984.
16. Kuntz, K. W. "Toxic Contaminants in the Niagara River, 1975–1982"; Technical Bulletin No. 134; Inland Waters Directorate, Canada Centre for Inland Waters: Burlington, Ontario, Canada, 1984.
17. McCrea, R. C.; Fischer, J. D.; Kuntz, K. W. *Water Pollut. Res. J. Can.* **1985**, *20*, 67.
18. Oliver, B. G.; Nicol, K. D. *Sci. Total Environ.* **1984**, *39*, 57.
19. International Joint Commission. "Inventory of Major Municipal and Industrial Point Source Dischargers in the Great Lakes Basin"; Great Lakes Water Quality Board, Great Lakes Regional Office: Windsor, Ontario, Canada, 1979.
20. Murthy, C. R. *Proc.—Conf. Great Lakes Res.* **1969**, *12*, 635.
21. Murthy, C. R.; Simons, T. J.; Lam, D. C. L. *Proc. Int. Counc. Explor. Sea Conf.* **1986**, *186*, 150–164.
22. Karickhoff, S. W. *Chemosphere* **1981**, *10*, 833.
23. O'Connor, D. J.; Connolly, J. P. *Water Res.* **1980**, *14*, 1517.
24. Weber, W. J., Jr.; Voice, T. C.; Pirbazari, M.; Hunt, G. E.; Ulanoff, D. M. *Water Res.* **1983**, *17*, 1443.

25. Karickhoff, S. W.; Brown, D. S.; Scott, T. A. *Water Res.* **1979**, *13*, 241.
26. Oliver, B. G.; Charlton, M. N. *Environ. Sci. Technol.* **1984**, *18*, 903.
27. Charlton, M. N. *J. Great Lakes Res.* **1983**, *9*, 201.
28. Davis, E. M.; Murray, H. E.; Liehr, J. G.; Powers, E. L. *Water Res.* **1981**, *15*, 1125.
29. Choudhry, G. G.; Hutzinger, O. *Environ. Sci. Technol.* **1984**, *18*, 235.
30. Schwarzenbach, R. P.; Molnar-Kubica, E.; Giger, W.; Wakeham, S. G. *Environ. Sci. Technol.* **1979**, *13*, 1367.
31. Wakeham, S. G.; Davis, A. C.; Karas, J. L. *Environ. Sci. Technol.* **1983**, *17*, 611.
32. Strachan, W. M. J. *Environ. Toxicol. Chem.* **1985**, *4*, 677.
33. Oliver, B. G. *Water Pollut. Res. J. Can.* **1984**, *19*, 47.
34. Borgmann, U. *Sci. Total Environ.* **1985**, *44*, 111.
35. Thomas, R. L.; Kemp, A. L. W.; Lewis, C. F. M. *J. Sediment. Petrol.* **1972**, *42*, 66.
36. Bopp, R. F.; Simpson, H. J.; Olsen, C. R.; Kostyk, N. *Environ. Sci. Technol.* **1981**, *15*, 210.
37. Sandilands, R. G.; Mudroch, A. *J. Great Lakes Res.* **1983**, *9*, 190.
38. Durham, R. W.; Oliver, B. G. *J. Great Lakes Res.* **1983**, *9*, 160.
39. Pickett, R. L.; Bermick, S. *Limnol. Oceanogr.* **1977**, *22*, 1071.
40. Oliver, B. G. *Chemosphere* **1985**, *14*, 1087.
41. Oliver, B. G. *Can. J. Fish. Aquat. Sci.* **1984**, *41*, 878.
42. Fox, M. E.; Carey, J. H.; Oliver, B. G. *J. Great Lakes Res.* **1983**, *9*, 287.
43. Karickhoff, S. W.; Morris, K. R. *Environ. Sci. Technol.* **1985**, *19*, 51.
44. Fisher, J. B.; Lick, W. J.; McCall, P. L.; Robbins, J. A. *J. Geophys. Res.* **1980**, *85*, 3997.
45. Kemp, A. L. W.; Anderson, T. W.; Thomas, R. L.; Mudrochova, A. *J. Sediment. Petrol.* **1974**, *44*, 207.
46. Clayton, J. R., Jr.; Pavlov, S. P.; Breitner, N. F. *Environ. Sci. Technol.* **1977**, *11*, 676.
47. Oliver, B. G.; Niimi, A. J. *Environ. Sci. Technol.* **1983**, *17*, 287.
48. Borgmann, U.; Whittle, D. M. *Can. J. Fish. Aquat. Sci.* **1983**, *40*, 328.
49. Könemann, H.; Van Leeuwen, K. *Chemosphere* **1980**, *9*, 3.
50. Niimi, A. J.; Cho, C. Y. *Can. J. Fish. Aquat. Sci.* **1981**, *38*, 1350.
51. Oliver, B. G.; Niimi, A. J. *Environ. Sci. Technol.* **1985**, *19*, 842.
52. Whittle, D. M.; Fitzsimons, J. D. *J. Great Lakes Res.* **1983**, *9*, 295.
53. "Guide to Eating Ontario Sport Fish. Southern Ontario Great Lakes 1984–85"; Ontario Ministry of the Environment: Toronto, Canada, 1984.
54. Gilbertson, M.; Reynolds, L. M. *Bull. Environ. Contam. Toxicol.* **1972**, *7*, 371.
55. Gilbertson, M. *Chemosphere* **1983**, *12*, 357.
56. Mineau, P.; Fox, G. A.; Norstrom, R. J.; Weseloh, D. V.; Hallett, D. J.; Ellenton, J. A. *Adv. Environ. Sci. Technol.* **1984**, *14*, 425.
57. International Joint Commission. Great Lakes Water Quality Board Report; Great Lakes Regional Office: Windsor, Ontario, Canada, 1985; p 104.
58. Anderson, E. G. Presented at the 27th Conference on Great Lakes Research, St. Catharines, Ontario, Canada, 1984.

RECEIVED for review May 6, 1986. ACCEPTED September 12, 1986.

# 15

# Cycles of Nutrient Elements, Hydrophobic Organic Compounds, and Metals in Crystal Lake

## Role of Particle-Mediated Processes in Regulation

David E. Armstrong, James P. Hurley, Deborah L. Swackhamer,[1] and Martin M. Shafer

Water Chemistry Program, University of Wisconsin–Madison, Madison, WI 53706

*Measurements of the chemical composition and fluxes of particulate matter were used to assess the particle-mediated cycling of selected nutrient elements, hydrophobic organic compounds, and metals in Crystal Lake, located in north central Wisconsin. The absence of surface water input simplified the analysis of in-lake cycles. Sediment incorporation and accumulation fluxes were calculated on the basis of an assumption of negligible sediment focusing. Removal of $^{210}Pb$ was rapid with negligible recycling, and $^{210}Po$ was partly recycled in the water column. The nutrient elements (C, N, P, and Si) contained in deposited particles were partly recycled (~50%) into the water column, but most of the P recycled was subsequently redeposited through interaction with Fe(III) formed near the sediment–water interface. Hydrophobic organic compounds such as polychlorinated biphenyls (PCBs) were also removed rapidly to the sediment–water interface by particle deposition but apparently returned partly to the water column during particle incorporation into surface sediments. Although PCBs were partly returned to the sediments by redeposition, recycling from bottom sediments increased the residence time of PCBs in the water column. Differences among chemical constituents in particle-mediated fluxes were regulated by differences in biogeochemical processes.*

[1]Current address: School of Public Health, Environmental and Occupational Health, University of Minnesota, Minneapolis, MN 55455

CHEMICAL LIMNOLOGY IS FUNDAMENTALLY CONCERNED with the processes regulating the chemical composition of lakes. Advances in the field involve improved qualitative and quantitative understanding of the controlling processes. Regulation occurs through both external and internal processes. External processes include transport of materials by air and water, input of light and heat, and water circulation by wind energy. Internal processes involve changes in chemical forms by biological and chemical reactions and transport of materials by advection and diffusion.

Models based on quantitative input–output relationships, which treat most internal processes empirically, have been very useful tools in lake management (1). However, element-specific internal processes often play a major role in controlling chemical composition (2–4), and prediction of lake composition and response to changes in external factors requires an understanding of internal processes and cycles.

Particle-mediated processes play a major role in the internal regulation of the chemical composition of lakes. Essential elements are incorporated and released by particles through photosynthesis and respiration. Similarly, the uptake and release of other chemical substances by particles occurs through adsorption–desorption and precipitation–dissolution reactions. Combined with particle transport by settling, these particle-mediated reactions have a major influence on chemical composition.

Although laboratory experiments have provided major contributions toward understanding the mechanisms of particle-mediated reactions in lakes, simulation of the complex interactions among physical, chemical, and biological processes is difficult. Thus, whole-lake measurements and experiments are useful in resolving the role of particle-mediated processes in controlling lake composition. Our analysis focuses on Crystal Lake, located in the Northern Highlands area of north central Wisconsin (Vilas County). Crystal Lake is an oligotrophic lake that has no surface water inlets or outlets. Direct precipitation accounts for about 90% of the water input (Table I). The remainder occurs through groundwater inflow. Consequently, external inputs are low, and Crystal Lake is well suited for investigation of the regulation of chemical composition by internal processes.

Three groups of chemical constituents are examined in this chapter: the nutrient elements (C, N, P, and Si), hydrophobic organic compounds [represented by the polychlorinated biphenyls (PCBs)], and metals ($^{210}$Pb and $^{210}$Po). The role of particle-mediated processes in regulating biogeochemical cycles is emphasized.

## Nutrient Elements

The biogeochemical cycles of nutrient elements in lakes are complex. Some processes are common to all nutrient elements, including incorporation of inorganic forms from lake water into cells (particles) through primary production. Other processes are unique to specific elements or chemical forms, such as adsorption or precipitation reactions and the biological transformations among the inorganic forms of nitrogen (nitrification and denitrification). Biogenic particles link the cycles of nutrient elements together, and similarities in behavior result. Primary production incorporates the elements into particles in characteristic proportions, and particle settling leads to similar vertical transport rates. However, element-specific reactions can uncouple the nutrient elements as particles flow through the lake system. This uncoupling leads to enrichment or depletion of the relative concentration of the nutrient element in the lake water. Thus, particle-mediated cycling influences both the absolute and relative supply of nutrients to the photic zone of lakes.

**Biological Regulation.** For the nutrient elements, active uptake from lake water by phytoplankton has a major influence. Although bacteria and higher plants also assimilate inorganic forms of nutrients, phytoplankton play a dominant role in the pelagic zones of lakes. The influence of phytoplankton on the nutrient elements is related to their requirements for growth and organic matter production. As ultimately determined by the chemical composition of cell components, phytoplankton remove nutrients from lake water in specific proportions. The resulting average stoichiometry of the phytoplankton is the basis of the Redfield model (equation 1) of the influence of nutrient assimilation and mineralization on nutrient elements dissolved in lake water (5).

$$106CO_2 + 16NH_3 + 15H_4SiO_4 + H_2PO_4^- + H^+ + 106H_2O \underset{R}{\overset{P}{\rightleftharpoons}}$$

$$[(CH_2O)_{106}(NH_3)_{16}(H_4SiO_4)_{15}H_3PO_4] + 106O_2 \tag{1}$$
$$\text{phytoplankton}$$

The nutrient elements are incorporated into biogenic particles during photosynthesis $(P)$ and regenerated during cell decomposition, represented as respiration $(R)$. The stoichiometry in equation 1 $(C_{106}N_{16}Si_{15}P_1)$ is representative of diatoms. For nonsiliceous algae, the C:N:P stoichiometry is similar $(C_{106}N_{16}P)$. The Redfield model is an approximation because the nutrient stoichiometry of the phytoplankton varies with species and nutrient concentrations in lake water (6). Nevertheless, stoichiometry is extremely useful in modeling the biological cycling of the nutrient elements.

**Chemical Regulation.** Chemical processes also control nutrient element concentrations in particulate phases. These chemical reactions may result in deviations from the stoichiometry predicted by the Redfield model. In general, the important processes include (1) the precipitation of inorganic carbon as calcium carbonate, (2) adsorption of the ammonium ion by cation-exchange reactions, (3) the solubility and dissolution kinetics of amorphous silica contained in diatom frustules and the formation of silicate minerals, and (4) adsorption and precipitation reactions involving phosphate. The alkalinity, calcium concentration, and pH of Crystal Lake preclude calcium carbonate formation (Table I). However, chemical reactions may play a role in retaining ammonium ion, silica, and phosphate in particles, particularly near the sediment–water interface.

### Table I. Limnological Characteristics of Crystal Lake

| Characteristic | Value |
|---|---|
| Surface area | 36 ha |
| Volume | $3.8 \times 10^6$ m$^3$ |
| Terrestrial drainage area | negligible |
| Direct precipitation | $346 \times 10^3$ m$^3$ year$^{-1}$ |
| Groundwater inflow | $30 \times 10^3$ m$^3$ year$^{-1}$ |
| Groundwater outflow | $127 \times 10^3$ m$^3$ year$^{-1}$ |
| Phosphorus | |
|     Total dissolved P | 2.7 $\mu$g L$^{-1}$ |
|     Total particulate P | 2.5 $\mu$g L$^{-1}$ |
| Nitrogen | |
|     NO$_3$ (N) | 18 $\mu$g L$^{-1}$ |
|     NH$_4$ (N) | 28 $\mu$g L$^{-1}$ |
|     Particulate N | 34 $\mu$g L$^{-1}$ |
| Silicon | |
|     Dissolved reactive Si | 16 $\mu$g L$^{-1}$ |
|     Particulate biogenic Si | 23 $\mu$g L$^{-1}$ |
| Calcium | 1.1 mg L$^{-1}$ |
| Chlorophyll | 1.5 $\mu$g L$^{-1}$ |
| Conductivity | 12 $\mu$S cm$^{-1}$ |
| Alkalinity | 15 $\mu$equiv L$^{-1}$ |
| pH | 5.85 |
| Mass sedimentation rate | 8 mg cm$^{-2}$ year$^{-1}$ |
| Sediment mixed-layer depth | 4 cm |

SOURCE: Data are from references 19, 20, 24, and 27.
NOTE: Concentrations represent volume-weighted mean annual values.

Sediments of lakes in the Northern Highlands area, including Crystal Lake, are generally rich in iron (7). Thus, phosphate tends to be retained in sediments through interactions with iron. The removal of phosphate from lake waters by iron is linked to the iron reduction–oxidation cycle, which mobilizes Fe(II) in anoxic zones and forms Fe(III) in oxic zones. This "ferrous wheel" (8) scavenges phosphate from lake waters above anoxic–oxic boundaries. If Fe(III) is formed in the presence of phosphate, a basic iron phosphate [Fe$_2$(OH)$_3$PO$_4$] with a Fe:P stoichiometry of 2:1 is apparently formed (9). This reaction can be represented as a combined Fe(II) oxidation–basic iron phosphate precipitation reaction:

$$2Fe^{2+} + \tfrac{1}{2}O_2 + H_2PO_4^- + 2H_2O \longrightarrow Fe_2(OH)_3PO_4(s) + 3H^+ \quad (2)$$

Alternatively, if Fe(III) is formed and hydrolyzed before interaction with phosphate, adsorption of phosphate by $Fe(OH)_3(s)$ results in Fe:P ratios $\geq 5:1$ (*10*). However, either reaction is efficient in removing phosphate from solution if the Fe:P ratio is sufficiently high. Below the oxic–anoxic boundary (e.g., after incorporation into surficial sediments), Fe(III) reduction in association with organic matter oxidation by sediment bacteria tends to solubilize the iron-bound phosphate. For the basic iron phosphate, the reaction can be depicted as

$$Fe_2(OH)_3PO_4(s) + \tfrac{1}{2}CH_2O + 3H^+ \longrightarrow$$

$$2Fe^{2+} + H_2PO_4^- + \tfrac{1}{2}CO_2 + \tfrac{5}{2}H_2O \quad (3)$$

Similarly, reduction of $Fe(OH)_3(s)$ results in solubilization of adsorbed phosphate. The removal of phosphate from lake waters by iron is also related to reactions that control $Fe^{2+}$ concentrations and transport from the anoxic zone into the lake water, especially reactions that result in $FeCO_3(s)$ and $FeS(s)$ formation (*11*). In lakes of the Northern Highlands, the Fe:S and Fe:C (inorganic) ratios are high and lead to a high mobility of $Fe^{2+}$ in anoxic waters (*12*). Although relatively selective for phosphate, $Fe(OH)_3(s)$ also adsorbs smaller proportions of other nutrient elements including silica, organic N, and organic C (*13*).

Amorphous silica is typically undersaturated in lake waters, and this undersaturation results in a tendency for dissolution of biogenic (diatom) silica. Because of the relative rates of silica dissolution and diatom sinking, dissolution in lakes occurs mostly after diatoms sink to the sediment–water interface (*14, 15*). Consequently, some of the amorphous silica is buried and retained in the lake sediments (*14, 16*). In addition, silica subsequently released by dissolution in sediments may be incorporated into aluminosilicates (*17*).

Although sediment particles usually possess a net negative charge and the capacity to adsorb $NH_4^+$ and other cations, the exchange capacity is typically small relative to the organic N content (*18*). This capacity indicates that most of the $NH_4^+$ formed by mineralization of biogenic particles will not be retained by adsorption to the particulate phase.

In summary, chemical regulation of nutrient elements in particulate matter in Crystal Lake may be important for inorganic phosphorus and silica. Chemical reactions, especially involving $Fe(OH)_3$, may also retain organic forms of carbon and nitrogen.

## Particle-Mediated Nutrient Cycling in Crystal Lake

In this section, data from Crystal Lake (19) is used to examine links in nutrient element behavior through particle-mediated processes. Specifically, the processes investigated include the removal of nutrient elements by particle production and settling, the regeneration of nutrient elements through mineralization of particulate organic matter, and the uncoupling of nutrient elements through element-specific interactions occurring in association with regeneration.

**Particulate Matter Compartments.** The biogeochemical cycling of nutrient elements in association with particulate matter is evaluated by dividing the lake into four vertically segregated particulate-matter compartments; suspended particulate matter, recently deposited material collected by sediment traps, the surficial sediment layer mixed by bioturbation, and the underlying buried sediment layer. This approach views the particulate matter as originating as suspended particulate matter in the water column. Settling of the suspended particulate matter forms a thin layer of recently deposited material at the sediment–water interface. This material is subsequently incorporated into the surficial sediment by advection and mixing (bioturbation in Crystal Lake) and eventually leaves the surface layer by burial as additional sediment is deposited at the surface.

Nutrient regeneration (removal from the particulate phase) as the particulate matter moves through these four compartments can be evaluated by measuring the composition of the particulate matter in each compartment and the rate of gain or loss from each compartment. For the nutrient elements (19), suspended particulate matter was obtained by filtration (0.4 μm), and recently deposited particulate matter was collected by sediment traps. The rate of deposition of an element was calculated as the product of the particulate matter mass flux (g cm$^{-2}$ year$^{-1}$) and the element concentration in the deposited particulate matter (mmol g$^{-1}$). The sedimentation rate and mixed-layer depth were calculated from $^{210}$Pb profiles (20).

**Temporal Changes in Composition and Fluxes.** The composition and flux of suspended particulate matter in Crystal Lake reflect the annual primary production cycle (Figures 1 and 2). Details of the chemical and biological composition of the particulate matter are given elsewhere (19). Phytoplankton are an important component of the suspended particulate matter. Diatoms (e.g., *Asterionella*) are prevalent in early spring and late fall, but nondiatom algae, some containing silica (e.g., *Dinobryon*), are also relatively abundant throughout the ice-free season. Nonalgal components of the suspended particulate matter and

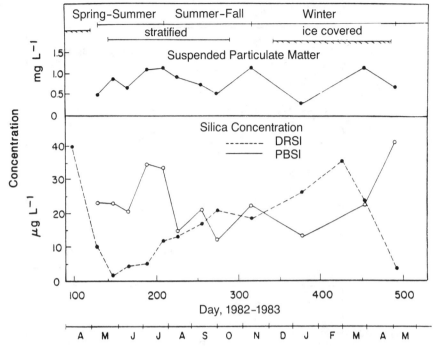

*Figure 1. Particulate matter and silica cycling in Crystal Lake. Suspended particulate matter concentration versus day of the year (top), and concentrations of dissolved reactive silicon (DRSl) and particulate biogenic silicon (PBSl) versus day of the year (bottom). The scale represents the months of the year.*

trap particulate matter include detritus, zooplankton (mainly copepods) and their fecal matter, and pollen.

The particulate matter concentration and flux rise from low values in the early spring to a maximum in the midsummer and decline to low levels during the winter ice-covered period (Figures 1 and 2). This fluctuation reflects the importance of in-lake primary production as a source of particulate matter in Crystal Lake. Erosional transport to the lake of soil from the surrounding landscape is unimportant. Atmospheric input and groundwater are the main external sources of chemical components. However, atmospheric deposition of particulate matter (*21*) corresponds to less than 2% of the in-lake depositional flux of autochthonous particulate matter.

The deposition rates of the nutrient elements (C, N, Si, and P) generally reflect the pattern for total particulate matter abundance, and maximum values occur in the midsummer (Figure 2). However, fluxes of P are higher in fall than in spring. In a related manner, Fe was not

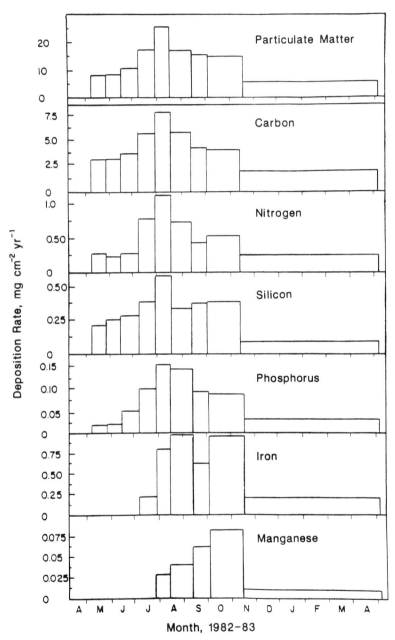

*Figure 2. Deposition rates of particulate matter, nutrient elements (C, N, Si, and P), iron, and manganese in Crystal Lake.*

detected in sediment traps in the spring, but high deposition rates of Fe were observed in midsummer and fall. Mn was also deposited during this period, but in small amounts in comparison with Fe. These results suggest a primary link of N, Si, and P deposition to biogenic organic matter deposition and a secondary link between P and Fe deposition.

Even though the particulate material is derived mostly from autochthonous primary production, chemical composition varies considerably with time (*19*). This variability reflects, in part, differences in phytoplankton and zooplankton assemblages, alteration by secondary biological processes (phytoplankton autolysis and grazing by zooplankton, and bacterial decomposition), varying proportions of biogenic detritus, and chemical processes (biogenic silica dissolution and precipitation of Fe and Mn hydrous oxides). Variations in concentration of C, N, Si, and P are illustrated in Figure 3. The data are grouped according to the three seasonal periods depicted in Figure 1. In general, variations within a season (as shown by the 95% confidence limits) are of the same magnitude as variations between seasons. Seasonal concentrations of nutrient elements in the particulate matter were usually within 80%–120% of the mean annual concentration. However, some seasonal patterns are apparent. Higher Si concentrations in the suspended particulate matter during spring–summer are associated with diatom production. High P concentrations in the deposited particulate matter during summer–fall and winter reflect the increased rate of P deposition during these seasons.

Changes in composition and deposition rates of the nutrient elements over the annual cycle (Figures 2 and 3) provide some insight into the importance of particle production and removal events in controlling these nutrient elements. The major event appears to be primary production, which leads to large relative differences in deposition rates with season, smaller average changes in composition with season, and relatively large within-season variations in composition. Superimposed on this pattern are the production of specific types of phytoplankton. The production of diatoms and other Si-containing algae is reflected in both the seasonal and within-season variations in amorphous Si concentration. Changes in phytoplankton have a smaller influence on C and N. Variations in P within season are relatively high and are consistent within the well-established range in P concentration of algal cells as influenced by species and external supply (*6*). Deposition of P in association with Fe appears as a secondary pattern in addition to the influence of primary production. Biological alteration of organic matter during deposition no doubt causes variation in both concentrations and deposition rates but does not emerge as "temporal" patterns in the nutrient concentration data sets.

In the following sections, the nutrient composition of the particulate

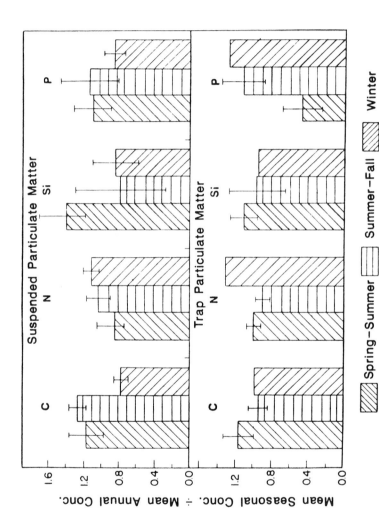

*Figure 3. Ratios of mean seasonal to mean annual nutrient element concentrations in particulate matter.*

matter compartments and intercompartmental fluxes are used to quantify nutrient cycling among compartments in Crystal Lake. To enable comparison of the bottom sediments with other compartments, annual weighted average composition and fluxes are used. This approach is necessary because short-term (seasonal) changes in composition of the surface sediments cannot be resolved. The validity of this approach is supported by the similarity among seasons in major element composition of the suspended and trap particulate matter. Although fluxes are highly seasonal, composition varies as much within season as between seasons (Figure 3). Thus, using an annual time scale should not bias the analysis.

**Changes in Stoichiometry during Deposition and Burial.** Changes in stoichiometry as particulate matter moves through the lake from suspended particulate matter to sediments are shown in Figure 4. The average stoichiometry reflects a general similarity to the composition of phytoplankton ($C_{106}N_{16}Si_{15}P$ for diatoms). The suspended particulate matter is somewhat depleted in P and is consistent with the expected lower P concentration in P-limited algal cells (6). Also, Si is lower than

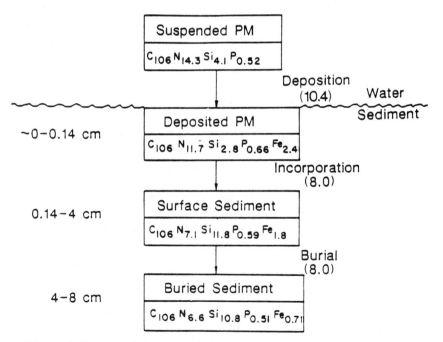

*Figure 4. Nutrient element stoichiometry of particulate matter in Crystal Lake. Compartments (boxes) show composition relative to carbon on a molar basis. Values in parentheses denote the mass fluxes of particulate matter (mg cm$^{-2}$ year$^{-1}$).*

expected for diatoms and reflects a mixed population of diatoms and nonsiliceous algae.

Comparisons of the particulate matter compartments show that the major changes in stoichiometry are depletion of C and N during incorporation of recently deposited particulate matter into surficial sediments, apparent enrichment of Si during incorporation, and enrichment of P and Fe in the recently deposited particulate matter over the other compartments. Although the stoichiometry suggests a partial depletion of N and Si during deposition, the concentrations (mmol/g) show this results in part from an increase in the carbon content of the trap particulate matter (Figure 5). The concentrations of N and Si in suspended particulate matter and deposited particulate matter are similar. The lower C and N levels in the active sediments indicate that organic matter mineralization occurs mainly after deposition and that N is mineralized to a greater extent than C. Presumably, P is mineralized similarly to N during incorporation. However, the inorganic phosphate formed by mineralization is partly retained as Fe-bound P. Although P contained in plankton is organic, about 45% of the P in Crystal Lake sediment is inorganic (7). Some of the inorganic P is released from the anoxic sediments but converted to particulate P near the sediment–water interface by association with Fe(III) oxides formed from soluble Fe(II) also released from the sediments. This ferrous wheel accounts for the enrichment of P in the recently deposited particulate matter.

In contrast to the other elements, Si shows a marked enrichment in the surface sediments in comparison to suspended and deposited particulate matter (Figure 4). The source of this Si is unclear (*see* discussion later in the chapter). The increase in Si concentration in the sediments could arise in part from a corresponding loss of organic matter. However, comparing the deposited particulate matter and surface sediment compartments shows that a loss of 9 mmol/g of organic C from the trap particulate matter would increase the Si content to only about 1 mmol/g, as compared to the observed concentration of 2 mmol/g (Figure 4). Also, this "concentrating" effect of organic matter loss would be partly offset by the increased Al content of the sediments (21).

Changes in nutrient element composition during burial were minor. This fact no doubt reflects the relatively high mixed-zone thickness (4 cm or 0.35 g cm$^{-2}$), low sedimentation rate (0.008 g cm$^{-2}$ year$^{-1}$), and corresponding long residence time of particles (~40 years) in the mixed zone of Crystal Lake sediments. Most alterations in composition involving mineralization of biogenic particles likely occur during this time period. Although the stoichiometry (Figure 4) shows some decreases in N, Si, and P relative to C during burial, the concentration data (Figure 5) show these changes are due mainly to a higher C content in the buried sediment zone. One notable exception to this trend is shown by Fe. The

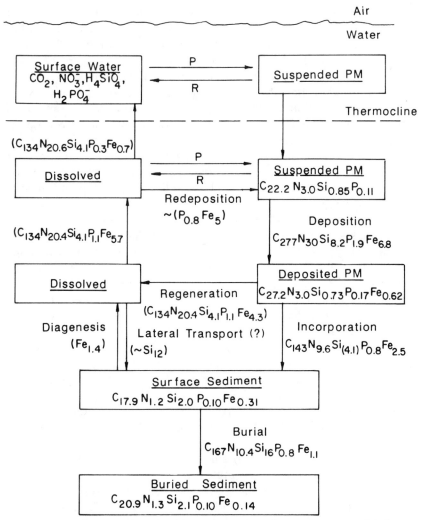

*Figure 5. Particle-mediated cycling of nutrient elements in Crystal Lake. Compartments show composition (mmol $g^{-1}$) and arrows denote fluxes ($\mu mol\ cm^{-2}\ year^{-1}$). Values in parentheses were estimated by difference. Fluxes associated with photosynthesis (P) and respiration (R) were not estimated.*

Fe concentration in the surface sediment (0.31 mmol/g) was twice that in the buried sediment. This relationship suggests a tendency for post-burial migration of Fe from the reduced zone toward the surface. The Fe content of the recently deposited particulate matter in late summer indicates that Fe migrates into the water column, precipitates, and redeposits in the surface sediment compartment.

**Nutrient Element Fluxes during Deposition and Burial.** The data on element concentrations and mass fluxes can be combined to obtain estimates of nutrient fluxes associated with particulate matter deposition, incorporation, and burial (Figure 5). Depositional fluxes are based on sediment trap-measured sedimentation rates and element concentrations in the deposited particulate matter collected over approximately 3-week periods. Incorporation and burial fluxes are based on the mean annual sedimentation rate (0.008 g cm$^{-2}$ year$^{-1}$) and the nutrient concentrations in the two sediment layers. Element concentration data are based on direct measurements and are subject mainly to analytical, sampling, and averaging errors or biases. The flux data, in addition, may include errors associated with trapping efficiency and sediment focusing and must be regarded as less certain.

One potential pitfall in comparing water column and sediment fluxes in lakes is sediment focusing (22). Particles initially deposited in shallow (usually near-shore) waters may gradually migrate down-slope, and higher accumulation rates in the deeper zones result. If focusing is appreciable in Crystal Lake, the mass sedimentation rate measured by using midlake sediment cores will not represent the mean lakewide value, and element fluxes calculated from these cores may overestimate lakewide fluxes.

Data collected on $^{210}$Pb and stable Pb in Crystal Lake (20, 23) provide a basis for evaluating the importance of sediment focusing. Measurements were made of the atmospheric input, the flux from the water column, and the flux into midlake sediments. Water column fluxes were measured by using sediment traps and also were calculated from $^{210}$Pb by box modeling based on the atmospheric flux and changes in water column concentrations.

Comparison of these measurements indicates sediment focusing has a minor influence on Pb accumulation in midlake sediments (20). Atmospheric, water column, and bottom sediment fluxes were similar. These fluxes were about 1.79–1.86 dpm cm$^{-2}$ year$^{-1}$ for $^{210}$Pb and 1–2 $\mu$g cm$^{-2}$ year$^{-1}$ for stable Pb. The flux comparisons also indicate the sediment traps were ~100% efficient in collecting sedimentation of Pb.

In marked contrast to the results for Pb, data on Al indicate sediment incorporation fluxes greatly exceed deposition fluxes as a result of higher Al concentration in the sediments than in the trap particulate matter (20). Thus, the importance of focusing is uncertain. In the analysis presented later in the chapter, focusing is assumed to be unimportant on the basis of the $^{210}$Pb data.

The measured fluxes (deposition, incorporation, and burial) indicate a loss of C, N, and P from particulate matter occurs during incorporation into the sediments (Figure 5). Incorporation and burial fluxes are similar because both composition and mass fluxes are similar for the two sediment compartments. The loss of C, N, and P during incorporation presumably reflects a partial mineralization of biogenic particulate matter during the residence time of particles in the surficial sediment.

Comparison of the deposition (8.2 $\mu$mol cm$^{-2}$ year$^{-1}$) and incorporation (16 $\mu$mol cm$^{-2}$ year$^{-1}$) fluxes for Si indicates another source of silica to the sediments. Three factors may account for the silica discrepancy: analytical problems, particle focusing, and precipitation of dissolved silica transported to the depositional zone sediments.

Our results are based on measurements of amorphous or biogenic particulate silica (*19, 24*). If the alkaline reagent used partially dissolves other silicate minerals (e.g., aluminosilicates) in the sediments, sediment incorporation and burial fluxes of biogenic silica would be overestimated. However, the contribution from these sources should be small (*25*).

As discussed above, focusing is not evident in $^{210}$Pb data. The extent of silica focusing necessary to account for the silica flux discrepancy can be estimated as follows: On the basis of mass balance calculations, at least 50% of the particulate biogenic silica deposited in Crystal Lake is regenerated annually (*24*). This amount corresponds to equal incorporation and regeneration fluxes of about 4.1 $\mu$mol cm$^{-2}$ year$^{-1}$ (Figure 5). Comparison to the measured incorporation flux (16 $\mu$mol cm$^{-2}$ year$^{-1}$) indicates focusing by a factor of 4 would be required to account for the observed silica accumulation. This factor corresponds approximately to the focusing factor calculated as the ratio of the midepilimnion area to the minimum depositional area (17-m plane) for Crystal Lake.

The stoichiometry of the particulate matter (Figure 4) shows that enrichment of nutrient elements in surface sediments over deposited particulate matter occurs only for silica. Thus, if lateral transport (focusing) accounts for the enrichment of silica, then either the process is selective for silica, or regeneration during incorporation substantially exceeds lateral transport for C, N, and P. This comparison combined with the $^{210}$Pb data indicates that focusing is probably not the source of the silica excess.

An alternative source of the excess silica in surface sediments is transport and precipitation of dissolved silica. Sediment pore waters are near saturation with respect to amorphous aluminosilicates (*26*). Although groundwater is enriched in dissolved silica and is the major source of silica to the lake (*24*), groundwater apparently does not enter the lake through the depositional zone sediments (*27*). To account for the excess particulate silica by precipitation of dissolved silica, lateral advection of dissolved silica to the depositional zone sediments must be invoked. It seems unlikely that this would account for the relatively large flux of excess silica. Thus, we were unable to resolve the excess silica flux. Analysis of silica in sediment cores distributed over the entire lake bottom would be helpful in resolving the silica source.

The differences between deposition and incorporation fluxes (Fig-

ure 5) provide estimates of regeneration fluxes occurring during incorporation. These fluxes correspond to about 70% and 50% of the N and Si
deposited (Table II). Although P is also regenerated during incorporation, comparison of the P concentrations in suspended particulate matter
and deposited particulate matter shows that the P regenerated during
incorporation was partly gained during deposition. Fe also increased in
the deposited particulate matter during the summer–fall period (Figure
2). We concluded that Fe and P are released from the surface sediment
and are partly redeposited. Gross regeneration of P, calculated as the
difference between deposition and incorporation, amounts to 1.1 mol
$cm^{-2}$ $year^{-1}$, or 60% of the P deposited to the surface sediment. Assuming
the gain in P relative to N in the deposited particulate matter as compared to the suspended particulate matter is due to redeposition of P,
the redepositional flux is about 0.8 $\mu mol$ $cm^{-2}$ $year^{-1}$, and net regeneration is 0.3 $\mu mol$ $cm^{-2}$ $year^{-1}$, or 15% of the P deposited.

Table II. Cycling Characteristics of Crystal Lake

| Element | Regeneration[a] (%) | Residence Time[b] with Respect to | |
| --- | --- | --- | --- |
| | | Deposition (years) | Regeneration (years) |
| Nitrogen | 70 | 0.41 | 0.60 |
| Silicon | 50 | 0.35 | 0.71 |
| Phosphorus | | | |
| Gross | 60 | 0.19 | 0.32 |
| Net | 15 | | 1.2 |

NOTE: Focusing is assumed to be unimportant.
[a] Fraction of deposited nutrient returned to the water column.
[b] Residence times are calculated as mean annual areal concentrations ($\mu mol$ $cm^{-2}$) divided
by the fluxes shown in Figure 4 ($\mu mol$ $cm^{-2}$ $year^{-1}$). Areal concentrations are based on
volume-weighted mean annual concentrations (Table I) and the area of the 12-m plane.

Particle-mediated processes play a major role in nutrient element
cycling in Crystal Lake. Estimated residence times for total N, Si, and P
in the water column show that the amounts removed annually by deposition exceed the quantity in the lake water by factors ranging from
about 2.5 to 5 (Table II). (A residence time of 0.5 years for deposition
means that annual deposition is twice the mean annual quantity in the
lake water column.) Regeneration at the sediment–water interface is a
major source of nutrients to the water column. If the lake is at steady
state on an annual time scale, then total removal is equal to removal by
deposition, input consists of external input and regeneration, and

$$deposition = external\ input + regeneration \qquad (4)$$

Thus, regeneration is >50% of deposition (Table II), and gross regeneration therefore accounts for more than one-half of the N, Si, and P loadings to the water column. The residence times presented in Table II are approximations because steady-state conditions are assumed (concentrations actually vary considerably over the annual cycle) and some fluxes are ignored (e.g., nitrogen fixation and denitrification). Nevertheless, the residence times demonstrate that regeneration is one of the major factors regulating flux of nutrients to the water column of Crystal Lake.

The chemical uncoupling of P from the cycle of biological regulation by organic matter production, settling, and decomposition has a major influence on the cycling of P. Although gross regeneration accounts for about 60% of the P deposited, most of the P regenerated is redeposited through interaction with Fe, which is also released from the sediment. Thus, net regeneration replaces P in the water column only about 0.8 times per year (Table II) and corresponds to a regeneration residence time of about twice that for N and Si. This fact and the N:P ratios in the suspended particulate matter (27) and lake water (35) indicate Crystal Lake is P-limited. The N:P ratio in the gross regeneration flux is about 19. Thus, the selective retention of phosphate by Fe(III), which removes about 75% of the P regenerated, may account for the high N:P ratio in the lake water and conditions indicative of P limitation.

## Polychlorinated Biphenyls

Polychlorinated biphenyls (PCBs) are ubiquitous contaminants in the environment. This mixture of compounds contains 209 theoretically possible congeners. Because of the wide range of physicochemical properties encompassed by this family of compounds, PCBs serve as important indicators of the environmental behavior of other hydrophobic organic compounds.

The transport of PCBs in lakes is largely controlled by their distribution between water and suspended particulate matter. Association of PCBs with particulate matter regulates the flux to the sediments and influences rates of biodegradation, volatilization, photolysis, and uptake by biota. Thus, the ultimate fate of PCBs in aquatic systems is linked closely to particle-mediated processes.

**Chemical and Biological Regulation.** The distribution of a hydrophobic compound between the dissolved and particulate phases in aqueous systems can be treated as an equilibrium process. In dilute systems, the distribution can be described by a dimensionless particle–water partition coefficient ($K_p$):

$$K_p = C_s/C_w \tag{5}$$

where $C_s$ and $C_w$ are the concentrations in the solid and aqueous phases, respectively. Partitioning is controlled by both particle and sorbate properties. The partition coefficient varies inversely with sorbate solubility (33) and directly with $K_{ow}$ (29, 30). Thus, for PCBs, steric configuration and degree of chlorination affect $K_p$. The value of $K_p$ may vary inversely with particle size (28). Also, $K_p$ may depend directly on the fractional organic carbon content of the particulate matter (29). Consequently, a related partition coefficient, $K_{oc}$, the particle–water partition coefficient normalized to the weight fraction of organic carbon (OC), can be defined as

$$K_{oc} = K_p/OC \tag{6}$$

An empirical relationship between $K_{oc}$ and the octanol–water partition coefficient ($K_{ow}$) was found for a wide range of hydrophobic organic compounds in sediments (29).

$$K_{oc} = 0.63 \, K_{ow} \tag{7}$$

However, for PCBs and suspended particulate matter in lake water, a different relationship was observed (30).

$$\log K_{oc} = 0.34 \log K_{ow} + 4.2 \tag{8}$$

Thus, $K_p$ can be estimated from the $K_{ow}$ of the organic compound and the organic carbon content of the sediment or suspended particulate matter.

Although equilibrium theory indicates that $K_p$ should be independent of suspended particulate matter concentration, field and laboratory data have shown an inverse relationship of $\log K_p$ to the log of suspended particulate matter. Several explanations have been advanced to explain this observation (see Chapter 13). Our own data on Lake Michigan (30) are consistent with the hypothesis that a fraction of the PCBs in lake water associates with colloidal material (31). The colloidal PCB fraction may be either inadvertently included in the dissolved fraction or not collected at all.

For a constant $K_p$, the fraction of PCBs associated with the particulate phase should depend on the suspended particulate matter concentration in the water column. This relationship can be shown by first defining the concentration terms as follows: $C_w$ is grams of dissolved PCBs per gram of $H_2O$, $C_p$ is grams of particulate PCBs per gram of $H_2O$, $C_T = C_w + C_p$, and $C_{SPM}$ is grams of suspended particulate matter per gram of $H_2O$.

The partition coefficient can be expressed as

$$K_p = \frac{C_p/C_{\text{SPM}}}{C_w} \tag{9}$$

and the fraction of compound in the particulate phase, $C_p/C_T$, is related to $K_p$ and $C_{\text{SPM}}$ by the equation

$$C_p/C_T = \frac{C_p}{C_p + C_w} = \frac{1}{1 + (C_{\text{SPM}}K_p)^{-1}} \tag{10}$$

Thus, $C_p/C_T$ increases with $C_{\text{SPM}}$ and $K_p$. Even when $K_p$ is large, $C_p/C_T$ will be small if $C_{\text{SPM}}$ is low (32).

Particle processes affecting PCBs in water also involve the biological community. Because of their relatively high lipid content, algae are thought to accumulate PCBs from the surrounding water (34). Zooplankton and fish may remove dissolved PCBs by direct uptake or ingest PCBs associated with food particles. The biological community also cycles PCBs through excretion and release processes. Fecal pellets (35) and detrital biogenic particles are important in removal of PCBs from the water column.

**Field Observations.** Water samples were collected from Crystal Lake four times between June and October 1983 at 1 m below the surface and at midhypolimnion (36). Particulate matter was isolated by continuous flow centrifugation, and dissolved PCBs were extracted by using styrene–divinylbenzene copolymer (XAD-2) cartridges. Sediment traps were deployed from March 1983 to May 1984. Sediment cores were obtained from a depositional zone in the center of the lake in summer 1982. All samples except sediments were Soxhlet extracted and cleaned by silica gel and alumina column chromatography. Sediments were steam distilled and cleaned with alumina. All extracts were analyzed by capillary column gas chromatography with electron-capture detection. Total PCB concentration was determined by multiple linear regression analysis (37). Full details of sampling and analytical methodology can be found elsewhere (36).

The mean concentrations and standard deviations of PCBs in water, suspended and trap particulate matter, and sediments are shown in Table III. In the water column, 95% of the measured PCBs was in the dissolved phase and independent of sampling time. Temporal variations in the dissolved concentrations were large and ranged from 66 to 720 pg/L in surface water and from 1150 to 2630 pg/L in the hypolimnion. After fall overturn, the average dissolved concentration was 1590 pg/L. Surface

Table III. Polychlorinated Biphenyl Concentrations in Crystal Lake

| Lake Compartment | Concentration[a] | Standard Deviation |
|---|---|---|
| Lake water (dissolved) | | |
| 2 m below surface | 480 pg/L | ±320 pg/L |
| Midhypolimnion | 1890 pg/L | ±740 pg/L |
| Lake water PM | | |
| 2 m below surface | 100 ng/g | ±95 ng/g |
| Midhypolimnion | 73 ng/g | ±20 ng/g |
| Sediment trap PM | | |
| Midhypolimnion | 170 ng/g | ±38 ng/g |
| 2 m above bottom | 330 ng/g | ±140 ng/g |
| Lake sediments at midlake | 15 ng/g | ±0.55 ng/g |

[a]Particulate matter and sediment concentrations are expressed on a dry-weight basis.

and hypolimnetic concentrations in particulate matter ranged from 17 to 240 and 52 to 94 ng/g, respectively. The average concentration after turnover was 45 ng/g. The suspended particulate matter was size fractioned into >120 (primarily zooplankton), 50–120 (primarily algae), and <50 (algae and detritus) groups. Average PCB concentrations in surface-suspended particulate matter were 330, 520, and 58 ng/g, and 340, 160, and 35 ng/g in hypolimnion-suspended particulate matter for the >120, 50–120, and <50-$\mu$m size fractions. During turnover, the average concentrations were 350, 140, and 30 ng/g.

Sediment trap particulate matter was retrieved three times during trap deployment. Mean PCB concentrations in trap particulate matter at the top and bottom of the hypolimnion were 170 and 330 ng/g, respectively. Greater variability was seen in the PCB concentrations in bottom trap particulate matter.

Surface sediments in Crystal Lake are mixed, probably by bioturbation, and this mixing results in essentially constant PCB concentration with depth. PCBs were detected down to 9 cm at concentrations ranging from 15 to 16 ng/g.

Partition coefficients ($K_p$) for suspended particulate matter were calculated for each sample by dividing the PCB concentration in suspended particulate matter (ng/kg) by the dissolved concentration (ng/L). The average log $K_p$ value was 4.9 ± 0.46. No temporal variations in $K_p$ were observed. However, log $K_p$ was higher for surface (5.3) samples than for hypolimnion samples (4.6). The relationship of $K_p$ to particle and sorbate properties was examined in order to explain the observed variations. Factors considered included organic carbon content, particle size, particle type, and differences in congener composition of total PCB.

A dependence of $K_p$ on the organic carbon content of the suspended particulate matter was not observed. This observation is in contrast to observations for soil and sediment (*29*), possibly because Crystal Lake suspended particulate matter is dominated by plankton, and humic substances are major components of the organic carbon content in soil and sediment.

Partition coefficients were significantly different between the larger size fractions (>120 and 50–120 $\mu$m) and the smaller size material (<50 $\mu$m). Average log $K_p$ values for the three size fractions, largest to smallest, were 5.6, 5.4, and 4.7. Thus, differences in surface area do not account for the change in $K_p$ with particle size. For all samples within a given size fraction, surface $K_p$ values were greater than hypolimnion values.

Differences in particle type could account for the observed differences in $K_p$ among size fractions and with depth. Zooplankton, contained in the largest size fraction, accumulate PCBs by both adsorption and consumptive uptake. The higher PCB concentrations in this fraction are reflected in higher $K_p$ values. The middle size fraction also contained some zooplankton and exhibited high PCB concentrations and $K_p$ values. The algae–detrital fraction had lower PCB concentrations and low $K_p$ values. A relationship of log $K_p$ versus log suspended particulate matter was not observed for this data set.

Sorbate properties may also control $K_p$. If the PCB composition of samples was different, then the different physicochemical properties of the different congeners would be reflected in the resulting $K_p$. To examine this possibility, $K_p$ values were calculated for 15 congeners having a wide range of physicochemical properties. Congener $K_p$ values were found to be directly related to molecular weight, degree of chlorination, and $K_{ow}$. Thus, the relative congener composition of a sample influences the composite $K_p$. This factor may explain the different $K_p$ values found between surface and hypolimnion waters. For all size fractions, the hypolimnion contained higher proportions of the lower chlorinated congeners, and this situation possibly resulted in the lower $K_p$ values.

In summary, the field data indicate that the distribution between the dissolved and particulate phases, indicated by $K_p$, is controlled in part by particle (particle type) and sorbate properties (physicochemical properties of the different congeners). In the water column where the suspended particulate matter contains living biota, particle type becomes a more important factor than the organic carbon content of suspended particulate matter.

## Cycling of Polychlorinated Biphenyls in Crystal Lake.

The particle-mediated cycling of PCBs in Crystal Lake is depicted in Figure 6. This interpretation is based in part on measurements or analysis of certain fluxes, including rates of deposition by particle settling measured by using

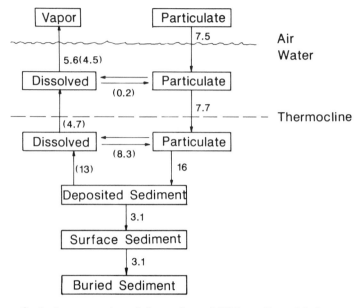

*Figure 6. An interpretation of the cycling of PCBs in Crystal Lake. Arrows denote fluxes ($\mu g\ m^{-2}\ year^{-1}$); values in parentheses were calculated by difference. This figure is intended to illustrate only the relative magnitude of the fluxes. Because the range in PCB concentrations in the sediment trap particulate matter was relatively large (Table III), the uncertainty in the corresponding depositional fluxes is similarly high.*

sediment traps (36), accumulation rates in sediments derived from analysis of sediment cores (38), atmospheric deposition (39), net atmospheric deposition, and net vapor exchange (38). The other fluxes were calculated by assuming steady-state conditions on an annual time scale and balancing fluxes to and from each PCB compartment. Although Crystal Lake was assumed to be at steady state, an imbalance is shown between the net atmospheric deposition (1.9 $\mu g\ m^{-2}\ year^{-1}$) and sediment accumulation rates (3.1 $\mu g\ m^{-2}\ year^{-1}$). This imbalance arises because net atmospheric accumulation was calculated from the mean sediment PCB accumulation rate for several remote lakes in Wisconsin (38), and the rate for Crystal Lake was higher than the mean.

The relative concentrations of PCBs in the dissolved and particulate compartments are likely controlled in part by equilibrium partitioning. Interchange between these compartments must occur because PCBs enter the lake in association with particles in precipitation and are found mostly in the dissolved phase (>95%) in the lake. Although the gross flux is unknown, the net flux between the dissolved and particulate compartments can be calculated by balancing the fluxes. If atmospheric particulate deposition is 7.5 $\mu g\ m^{-2}\ year^{-1}$, and particulate deposition to the hypolimnion is 7.7 $\mu$g m$^{-2}$ year$^{-1}$, the particulate compartment gains 0.2 $\mu$g m$^{-2}$ year$^{-1}$ from the dissolved compartment in the epilimnion.

PCBs are removed rapidly from the epilimnion by particle settling. The particles include both phytoplankton and fecal matter produced by zooplankton grazing on the phytoplankton. However, the rate of burial in bottom sediments is apparently about one-half the rate of deposition from the epilimnion.

The measured fluxes of particle-associated PCBs from the epilimnion to the hypolimnion (7.7 $\mu$g m$^{-2}$ year$^{-1}$), to the sediment–water interface (16 $\mu$g m$^{-2}$ year$^{-1}$), and into the sediments (3.1 $\mu$g m$^{-1}$ year$^{-1}$) play a major role in our interpretation of the cycle. To account for these observations, we postulated substantial recycling of PCB from the sediment surface to the hypolimnion (13 $\mu$g m$^{-2}$ year$^{-1}$). Transport probably occurs by diffusion as $K_p$ values decrease in surface sediment (*40*) possibly because of high dissolved organic matter concentrations and the high solid-to-liquid ratios in surficial sediments. Alternatively, the imbalance between deposition and sediment accumulation rates could be attributed to PCB degradation in sediments (*41*). However, this scenario would not account for the high rates of deposition that substantially exceed the rate of input to the lake. Note that vapor loss was calculated under the assumption that PCB degradation in sediments was negligible (*38*). However, even if the vapor loss was eliminated, recycling of PCBs from sediments must be invoked to explain the observed deposition rates.

Recycling from sediments could occur through resuspension as well as diffusion (*40, 42*). However, the mass fluxes of particulate matter collected in the upper and lower traps were similar. This fact indicates that resuspension was unimportant. Furthermore, $^{210}$Pb and Al data do not reflect resuspension (*21*). Thus, we postulated that dissolved PCBs are transported from the sediment surface by diffusion and mixing, are partly readsorbed by settling particulate matter, and are transported back to the sediment–water interface by particle deposition. This explanation is consistent with the higher dissolved PCB concentration in bottom waters (2000 pg L$^{-1}$) than in surface waters (500 pg L$^{-1}$) and the enrichment of the lower chlorinated congeners in the hypolimnion (Table III).

Importantly, our analysis may underestimate particle-mediated transport of PCBs from the water column. We used poisoned traps and discarded particles >120 $\mu$m in diameter to eliminate zooplankton that were possibly over-trapped. The measured particulate matter fluxes of about 5 mg cm$^{-2}$ year$^{-1}$ (*36*) were about one-half the total fluxes measured in other years (*19, 21*).

**Residence Times.**  Residence times are useful in relating PCB fluxes or transport rates to the total amount of PCBs in the lake. Values are calculated as the areal concentration of total PCBs in the lake ($\mu$g m$^{-2}$) divided by the PCB flux ($\mu$g m$^{-2}$ year$^{-1}$) for the respective process. Thus, a residence time corresponds to the time required for transport by a particular process of a quantity of PCBs equal to the quantity in the lake.

The residence time for total PCBs in Crystal Lake is estimated to be about 2 years (Table IV). Considering the accuracy of the flux estimates, vapor loss and burial in bottom sediments are of similar importance in controlling the overall residence time. Our analysis indicates that recycling at the sediment–water interface substantially increases the residence time with respect to PCB incorporation into bottom sediments. The burial residence time is about 6 years, and the residence time for transport to the sediment–water interface is about 1–2 years. Apparently, permanent incorporation of PCBs into bottom sediments in Crystal Lake is a relatively slow process. Because recycling at the sediment–water interface may play a major role in retarding PCB removal from lakes, the factors and mechanisms controlling recycling deserve thorough investigation. Similarly, in view of the apparent importance of vapor loss, attention also should be focused on vapor-phase transport at the air–water interface.

Table IV. Rates of Polychlorinated Biphenyl Transport in Crystal Lake

| Process | Estimated Flux ($\mu g \ m^{-2} \ year^{-1}$) | Calculated Residence Time[a] (years) |
|---|---|---|
| Deposition | | |
| Midhypolimnion | 7.7 | 2.5 |
| Bottom | 16 | 1.2 |
| Incorporation into surface sediment | 3.1 | 6.4 |
| Total removal[b] | 8.7 | 2.3 |

[a] Areal concentration divided by flux. Areal concentration (20 $\mu$g m$^{-2}$) was calculated as total PCB content of the lake ($\mu$g) divided the area of the 12-m plane.
[b] Total removal was calculated as sediment incorporation plus vapor loss.

## Lead-210 and Polonium-210

Lead-210 is a useful tracer of the geochemical cycling of metals in lakes because the input rate and in-lake concentrations can be accurately quantified. The links between atmospheric input (major source), water column concentrations, and sediment accumulation were investigated recently by Talbot and Andren (21, 23) to explore the particle-mediated cycling of $^{210}$Pb and $^{210}$Po (a radioactive decay daughter) in Crystal Lake. The results are summarized briefly for comparison to other elements.

Measurements of $^{210}$Pb and $^{210}$Po in precipitation and aerosol were used to estimate the atmospheric flux (21). Vertical and temporal changes in the concentrations of dissolved and particulate $^{210}$Pb and $^{210}$Po in the water column were measured over a 1-year period (23). The vertical distribution of $^{210}$Pb in sediment cores was also determined. Mass balancing was used to estimate the rates of removal from the water column.

The atmospheric flux of $^{210}$Pb to Crystal Lake was estimated to be 0.70 pCi cm$^{-2}$ year$^{-1}$ (21). The wet flux ranged from 0.24 to 1.2 pCi cm$^{-2}$ year$^{-1}$ over the sampling intervals, and the mean dry flux was 0.07 pCi cm$^{-2}$ year$^{-1}$. The atmospheric flux of $^{210}$Po was about 0.06 pCi cm$^{-1}$ year$^{-1}$. Radioactive decay of $^{210}$Pb in the lake was the main $^{210}$Po source.

Concentrations of $^{210}$Pb and $^{210}$Po in the water column varied considerably among the sampling periods (23). The mean total $^{210}$Pb concentration was 5.1 pCi/100 L (Table V). Dissolved $^{210}$Pb averaged 1.6 pCi/100 L and varied within a factor of 2. Particulate $^{210}$Pb concentrations varied more widely, especially in surface water, and tended to parallel changes in suspended particulate matter concentration. The mean $^{210}$Po concentration was 4.3 pCi/100 L. Although proportions of particulate $^{210}$Po and $^{210}$Pb were similar overall, wide differences were observed at specific times. This observation suggested some specificity in the particle-mediated behavior of the two elements.

**Table V. Mean Concentrations, Distributions, and Fluxes of $^{210}$Po and $^{210}$Pb in Crystal Lake**

| Property | $^{210}Pb$ | $^{210}Po$ |
|---|---|---|
| Concentration (pCi/100 L) | 5.1 | 4.3 |
| Particulate (dissolved) concentration | | |
| SML[a] | 2.4 | 3.2 |
| DL[b] | 2.6 | 3.3 |
| Apparent log $K_p$ | | |
| SML | 6.6 | 6.8 |
| DL | 6.6 | 6.7 |
| Removal rate (pCi cm$^{-2}$ year$^{-1}$) | 0.82 | 0.13 |
| Residence time (years) | 0.095 | 0.26 |

SOURCE: Data are from references 20, 21, and 23.
NOTE: Temporal variations are large (*see* text for discussion).
[a]SML is surface mixed layer.
[b]DL is deep layer.

Mass balancing between the atmospheric input and changes in the water column content of $^{210}$Pb and $^{210}$Po was used to calculate rates of removal from the water column (23). Removal from the upper waters and bottom waters was estimated separately. The mean removal rate for $^{210}$Pb ranged from 0.31 to 2.1 pCi cm$^2$ year$^{-1}$ over the annual period, but the average rate (0.82 pCi cm$^{-2}$ year$^{-1}$) was essentially equal to the atmospheric input and sediment accumulation rates (0.70 and 0.79 pCi cm$^{-2}$ year$^{-1}$, respectively). These values correspond to a mean residence time of 0.095 years (Table V) and suggest a dependence of $^{210}$Pb removal on particle type and abundance, but little recycling. In contrast, the mean removal rate of $^{210}$Po was 0.13 pCi cm$^{-2}$ year$^{-1}$, and this value corresponds to a water column residence time of about 0.26 years.

Differences in rates of $^{210}$Pb and $^{210}$Po removal did not appear to be controlled by differences in distribution between dissolved and particulate phases. Although the ratio of particulate-to-dissolved concentrations and the apparent log $K_p$ values were higher for $^{210}$Po than $^{210}$Pb (Table V), the mean removal rate was 6.3 times lower and the residence time 2.7 times longer for $^{210}$Po. The slower removal rate was thus attributed to greater recycling of $^{210}$Po within the lake.

The recycling of $^{210}$Po may occur either within the water column or at the sediment–water interface. In fact, short-term increases in the $^{210}$Po content of the water column indicated that some release from bottom sediment occurs. However, the rate of particle-mediated removal from the water column was also apparently slower for $^{210}$Po as shown by both mass balance (23) and sediment trap measurements (20).

The comparison of $^{210}$Po and $^{210}$Pb is in contrast to the expected direct dependence of the −log metal residence time on log $K_p$ and log suspended particulate matter flux (43). Under the influence of the same suspended particulate matter flux, the element with the higher $K_p$ value ($^{210}$Po) is removed more slowly. Two explanations may account for this behavior: (1) $^{210}$Po and $^{210}$Pb may be associated in part with different particle types or phases, and the particles containing $^{210}$Po may settle more slowly. Scavenging of $^{210}$Pb occurs mostly through adsorption and precipitation reactions (44–46). The important particles in Crystal Lake may include plankton, detritus, and iron hydroxide-coated particles. In contrast, $^{210}$Po is biologically concentrated to a greater extent than $^{210}$Pb by phytoplankton and zooplankton (47). (2) $^{210}$Po may be biologically recycled within the water column (e.g., Scheme I). The behavior of $^{210}$Po is thus more similar to nutrients (e.g., P) or "nutrient-type" metals such as Cd (46, 48). Recycling would tend to retain $^{210}$Po in the water column even though the proportion associated with particles is high. Thus, the comparison of $^{210}$Pb and $^{210}$Po illustrates the dependence of particle-

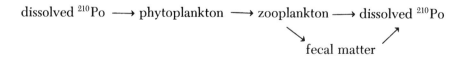

*Scheme I. Biologic recycling of $^{210}$Po.*

mediated behavior on the specific chemical properties of the element and the resulting differences in their chemical and biological cycles.

## Abbreviations and Symbols

$C_s$      concentration in solid phase
$C_T$      total concentration in solid and aqueous phases
$C_{SPM}$      concentration of suspended particulate matter
$C_w$      concentration in aqueous phase
$K_{oc}$      particle–water partition coefficient normalized to the weight fraction of organic carbon
$K_{ow}$      octanol–water partition coefficient
$K_p$      particle–water partition coefficient

## Acknowledgments

We thank H. Grogan and J. Schneider for typing the manuscript. This work was supported in part by the National Science Foundation, Long-Term Ecological Research Program (LTER), Contract DEB8012313, and by the University of Wisconsin Sea Grant Program under grants from the Office of Sea Grants, National Oceanic and Atmospheric Administration; the U.S. Department of Commerce; and the state of Wisconsin.

## References

1. Reckhow, K. H.; Chapra, S. C. *Engineering Approaches for Lake Management;* Butterworth: Boston, 1983; Vol. 1, p 340.
2. Sholkovitz, E. R.; Copland, D. *Geochim. Cosmochim. Acta* **1982**, *46*, 393–410.
3. Cook, R. B.; Kelly, C. A.; Schindler, D. W.; Turner, M. A. *Limnol. Oceanogr.* **1986**, *31*, 134–148.
4. Stauffer, R. E.; Armstrong, D. E. *Geochim. Cosmochim. Acta* **1986**, *50*, 215–229.
5. Redfield, A. C.; Ketchum, B. H.; Richards, F. A. In *The Sea;* Hill, M. N., Ed.; Interscience: New York, 1963; pp 26–77.
6. Reynolds, C. S. *The Ecology of Freshwater Phytoplankton;* Cambridge University: Cambridge, MA, 1984; p 384.
7. Williams, J. D. H.; Syers, J. K.; Armstrong, D. E.; Harris, R. F. *Soil Sci. Soc. Am. Proc.* **1971**, *35*, 556–561.
8. Mayer, L. M.; Biotta, F. P.; Norton, S. A. *Water Res.* **1982**, *16*, 1189–1196.
9. Tessenow, V. U. *Arch. Hydrobiol. Suppl.* **1974**, *47*, 1–79.
10. Lyklema, L. In *Interactions Between Sediments and Fresh Water;* Gollerman, H. L., Ed.; Dr. W. Junk: The Hague, 1977; pp 313–317.
11. Cook, R. B. *Can. J. Fish Aquat. Sci.* **1984**, *41*, 286–293.
12. Armstrong, D. E. In *Lake Restoration;* U.S. Environmental Protection Agency: 1979; pp 169–175.
13. Tipping, E.; Woof, C.; Cooke, D. *Geochim. Cosmochim. Acta* **1981**, *45*, 1411–1419.

518 SOURCES AND FATES OF AQUATIC POLLUTANTS

14. Wollast, R. In *The Sea;* Goldberg, E. D., Ed.; Wiley: New York, 1974; Vol. 5, Chapter 11, pp 359–392.
15. Schelske, C. L.; Eadie, B. J.; Krause, G. L. *Limnol. Oceanogr.* **1984**, *29*, 99–110.
16. Schelske, C. L.; Conley, D. L.; Warwick, W. F. *Can. J. Fish. Aquat. Sci.* **1985**, *42*, 1401–1409.
17. Paces, T. *Geochim. Cosmochim. Acta* **1978**, *42*, 1487–1493.
18. Keeney, D. R. *J. Environ. Qual.* **1973**, *2*, 15–29.
19. Hurley, J. M.S. Thesis, University of Wisconsin, Madison, 1984.
20. Talbot, R. W. Ph.D. Dissertation, University of Wisconsin, Madison, 1981.
21. Talbot, R. W.; Andren, A. W. *J. Geophys. Res.* **1983**, *88*, 6752–6760.
22. Hilton, J. A. *Limnol. Oceanogr.* **1984**, *29*, 99–110.
23. Talbot, R. W.; Andren, A. W. *Geochim. Cosmochim. Acta* **1984**, *48*, 2053–2063.
24. Hurley, J. P.; Armstrong, D. E.; Kenoyer, G. J.; Bowser, C. J. *Science (Washington, D.C.)* **1985**, *227*, 1576–1578.
25. Krause, G. L.; Schelske, C. L.; Davis, C. O. *Freshwater Biol.* **1983**, *13*, 13–81.
26. Machesky, M. L. M.S. Thesis, University of Wisconsin, Madison, 1982.
27. Kenoyer, G. J. Ph.D. Dissertation, University of Wisconsin, Madison, 1986.
28. Hiraizumi, Y.; Takahashi, M.; Nishimura, H. *Environ. Sci. Technol.* **1979**, *13*, 580–584.
29. Karickhoff, S. W.; Brown, D. S.; Scott, T. A. *Water Res.* **1979**, *13*, 241–248.
30. Swackhamer, D. L.; Armstrong, D. E. *J. Great Lakes Res.* **1987**, *13*.
31. Gschwend, P. M.; Wu, S. *Environ. Sci. Technol.* **1985**, *19*, 90–96.
32. Voice, T. C.; Weber, W. J. *Water Res.* **1983**, *17*, 1433–1441.
33. Kenaga, E. E.; Goring, C. A. I. Special Technical Publication No. 707; American Society for Testing and Materials: Philadelphia, PA, 1980; pp 78–115.
34. Wang, K.; Rott, B.; Korte, F. *Chemosphere* **1982**, *11*, 525–530.
35. Elder, D. L.; Fowler, S. W. *Science (Washington, D.C.)* **1979**, *197*, 459–461.
36. Swackhamer, D. L. Ph.D. Dissertation, University of Wisconsin, Madison, 1985.
37. Burkhard, L. P.; Weininger, D. *Anal. Chem.* **1987**, *59*.
38. Swackhamer, D. L.; Armstrong, D. E. *Environ. Sci. Technol.* **1986**, *20*, 879–883.
39. Andren, A. W. In *Physical Behavior of PCBs in The Great Lakes;* Mackay, D.; Paterson, S.; Eisenreich, S. J.; Simmons, M. S., Eds.; Ann Arbor Science: Ann Arbor, MI, 1983; pp 127–140.
40. Eadie, B. J.; Rice, C. P.; Frez, W. A. Ibid, pp 213–218.
41. Brownawell, B. J.; Farrington, J. W. *Geochim. Cosmochim. Acta* **1986**, *50*, 157–169.
42. Capel, P. D.; Eisenreich, S. J. *J. Great Lakes Res.* **1985**, *11*, 447–461.
43. Santschi, P. H. *Limnol. Oceanogr.* **1984**, *29*, 1100–1108.
44. Bacon, M. P.; Brewer, P. G.; Spencer, D. W.; Murray, J. W.; Goddard, J. *Deep-Sea Res.* **1980**, *27A*, 119–135.
45. Spencer, D. W.; Bacon, M. P.; Brewer, P. G. *J. Mar. Res.* **1981**, *39*, 119–138.
46. Bruland, K. W. In *Chemical Oceanography;* Riley, J. P.; Chester, R., Eds.; Academic: New York, 1983; Vol. 8, Chapter 45, pp 157–220.
47. Shannon, L. V.; Cherry, R. D.; Orren, M. J. *Geochim. Cosmochim. Acta* **1970**, *34*, 701–711.
48. Jones, C. J.; Murray, J. W. *Limnol. Oceanogr.* **1984**, *29*, 711–720.

RECEIVED for review June 10, 1986. ACCEPTED September 4, 1986.

# Element Cycling in Wetlands: Interactions with Physical Mass Transport

Harold F. Hemond, Thomas P. Army, William K. Nuttle, and Diane G. Chen

Department of Civil Engineering, Massachusetts Institute of Technology, Cambridge, MA 02139

*In peatlands and other waterlogged ecosystems, a severely restricted physical transport regime may be regarded as a rate-controlling factor for many biogeochemical processes. Transport in the liquid phase by advection, diffusion, and dispersion, and vapor-phase transport by ebullition and desaturation are the abiotic transport processes in wetland sediment. Biotic transport may also occur. Quantitative formulations for several of these transport processes exist, and specific, quantitative, useful examples of transport-oriented biogeochemical cycles are presented for two wetland ecosystems, namely, Belle Isle marsh and Thoreau's Bog.*

## Mass Transport Processes

**M**ASS TRANSPORT IS A LIMITING ECOLOGICAL FACTOR in wetland ecosystems. Wetlands (represented in this chapter by two Massachusetts peatlands) contain large reservoirs of reduced carbon in close proximity to an oxidizing atmosphere. The wetlands generate a steep gradient of redox potential. The above-ground plant parts carry out photosynthesis and release $O_2$ to the atmosphere. Photosynthesis also reduces and deposits carbon by leaf-litter fall, by below-ground litter deposition, and perhaps by direct exudation of reduced carbon from the below-ground parts (Figure 1). The long-term persistence or accretion of the reducing organic sediments is, in large part, testimony to the fact that physical transport of oxidants into the sediment is highly restricted by a waterlogged sediment.

0065–2393/87/0216–0519$06.00/0

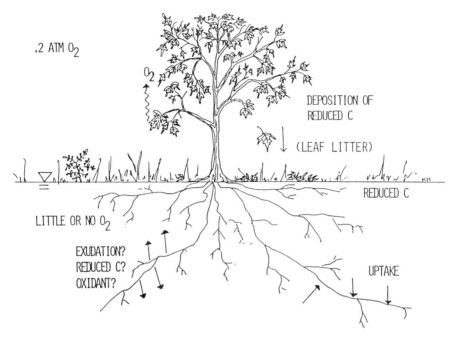

*Figure 1. Deposition of reduced carbon in a wetland ecosystem. The pro-
duction of a steep gradient of redox status results from the carbon-
reducing activity of wetland vegetation in close proximity to the
atmosphere.*

**Liquid-Phase Transport.**   The nature of the transport mechanisms
that operate in the waterlogged environment of wetland sediments will
be considered first. Because the sediments are typically close to water
saturation and contain at most a few percent air (except perhaps in
fibrous peats during drought events), we will consider transport from
the perspective of solutes contained in the aqueous phase (pore water).
Given the complex, highly variable, and poorly understood physical and
hydrologic characteristics of peaty sediments in general, assigning absolute magnitudes to the several transport processes is difficult. Wetland
peats range in texture from highly porous, undecomposed *Sphagnum*
remains typical of the upper horizons of ombrotrophic bogs to nearly
impermeable, structureless, saprobic peat and to clayey deposits of
rather high mineral content. Therefore, assigning parameters that are
generally applicable to all peatlands is not possible.

The first mass transport mechanism considered, molecular diffusion, operates in all cases. This mechanism is modeled as a Fickian process [i.e., the mass flux $(\overline{J})$ is proportional to the concentration gradient
$(\overline{\nabla}C)$ of a solute].

$$\overline{J} = D_m^{\,\circ}\,\overline{\nabla}C \tag{1}$$

The constant of proportionality $(D_m^{\,\circ})$ is the diffusion coefficient for the solute in water and is decreased by the effect of the tortuous paths the solute must follow to move through a porous sediment. Significantly, $D_m^{\,\circ}$ is small, on the order of $10^{-5}$ cm$^2$ s$^{-1}$.

Transport of solutes also occurs via the net movement, or advection, of pore waters through a sediment. Although generally slow by surface flow standards, pore water flow may greatly increase the flux of solute at a point in the sediment. The flux of solute due to advection is proportional to both velocity and concentration and is given by

$$\overline{J} = C\overline{V} \tag{2}$$

where $\overline{V}$ is the pore water velocity. Pore water velocities are generally low; average lateral flux velocity in Belle Isle marsh is highest at creek banks and is of the order of $2 \times 10^{-4}$ cm s$^{-1}$ (*1*). Elsewhere in the marsh, fluxes driven by evapotranspiration ($3.5 \times 10^{-6}$ cm s$^{-1}$) dominate. However, in special circumstances, much higher lateral velocities may occur in areas of steep hydraulic gradients and very conductive peats. Hofstetter (*2*) has observed water to move locally as rapidly as 1–2 cm s$^{-1}$ in the upper 15 cm of peat in certain Minnesota peatlands. In general, lateral flow in *Sphagnum* bog peats is largely restricted to an upper, high-conductivity acrotelm layer (*3*). Tritium studies (*4*) have suggested that deeper bog peats generally have stagnant pore water. However, coupling with regional groundwater, which implies vertical flow through a catotelm layer, has been shown to occur in large peatland complexes (*5*). Such zonation cannot be expected to exist in all wetlands. Indeed, in one Massachusetts salt marsh, a pattern of low hydraulic conductivity at the top and bottom of the peat profile and distinctly higher values at middepth have been documented (*6*).

A third mechanism of mass transport may also occur where pore water movement exists. This mechanism is mechanical dispersion and results from the fact that a pore water velocity averaged over a bulk, macroscopic volume is actually the net effect of local microscopic-scale flows that vary greatly in both magnitude and direction. Because different parcels of water may thus have different travel times associated with a given net displacement, the result is a mixing process that can cause mass transport. The modeling of mechanical dispersion is a very complex problem (*7*) and is an area of active current research. Although mechanical dispersion is scale-dependent and is demonstrably non-Fickian under some conditions, it is often treated as a Fickian process for simplicity. A mass flux due to dispersion in an isotropic sediment is described in Fickian terms as

$$\overline{J} = - \overline{\overline{D}} \, \overline{\nabla} C \qquad (3a)$$

where $\overline{\overline{D}}$ is a dispersion coefficient tensor and reflects the fact that even in an isotropic sediment, the dispersivity is greater in the direction of average pore water flow than in a direction perpendicular to the flow. In one dimension, a Fickian dispersive flux formulation becomes analogous to that for a diffusive flux:

$$J = - D \, \frac{dC}{dz} \qquad (3b)$$

Significantly, however, the dispersion coefficient may be much larger than the molecular diffusion coefficient if pore water velocities have nonzero values. The magnitude of $D$ is of the order of the magnitude of $Va$, where $V$ is the average pore water velocity and $a$ is a representative pore size.

Pore sizes in peatland sediments may approach zero in highly decomposed amorphous granular peats ($8, 9$) but may be of the order of millimeters in loose, raw *Sphagnum* material. Hayward and Clymo ($10$) indicate that the pores of hyaline cells of one species of *Sphagnum* may have typical radii of the order of $4$–$8 \times 10^{-6}$ m, although gaps between leaves may range from $10^{-4}$ to $10^{-3}$ m. The larger characteristic void sizes in such material tend to enhance water velocity as well as macrodispersivity. In the absence of highly porous material such as undecomposed *Sphagnum*, high mechanical dispersion coefficients could potentially result from the existence of preferred flow channels (macropores). For example, Chen ($11$) has shown that vertical flow in Belle Isle sediments is in large part associated with root channels and results in a large variability in local pore water velocities and, by inference, considerable mechanical dispersion.

**Vapor-Phase Transport.** Although wetland sediments are by definition mostly water-saturated, gas phases may exist. Although volumetrically small, gas volumes in the sediment may be associated with fluxes of volatile species that far exceed the corresponding fluxes in the pore water. Two major modes of vapor-phase transport can be important. One mode is the well-known phenomenon of bubble ebullition ($12, 13$). Ebullition can occur whenever the sum of the concentrations of solutes times their Henry's law constants exceeds the hydrostatic pore pressure:

$$\Sigma \, C_i H_i > P \qquad (4)$$

where $H_i$ is Henry's law constant for species $i$, and $P$ is the pore water pressure. (Surface tension effects, which may be important for very small bubbles, are neglected, and pore water pressure is assumed to equal gas pressure.) Under steady-state conditions where a net gas source (e.g., biota) exists, such bubbles must eventually escape to the atmosphere unless some sink process in the sediment periodically acts to reduce the sum of partial pressures.

A second gas-phase process is associated with desaturation. Initially, the removal of water from an organic peatland sediment leads to a compression of the bulk sediment structure. However, when pore pressures drop below a certain level, called the air-entry threshold, air may also begin to enter the sediment to partially replace the water that is removed. Although the water storage ability of peats is high (14), little is known in general about the relative role of desaturation as opposed to other water-release processes such as compression in peat sediments. Although critical peatland phenomena associated with aeration are attributed to desaturation (15, 16), the desaturation process is quite poorly quantified for wetland sediments. Standard soil-moisture characteristic measurement apparatus must be modified for use with peat soils, where large volumes of water may be released at rather low suctions and where bulk soil volume change can be large and must be accounted for. The principle of measurement of desaturation is illustrated in Figure 2; actual data from Belle Isle marsh sediment are shown in Figure 3.

The amount of water storage associated with peat compression must be determined in order that the amount of desaturation may be calculated by difference. Other techniques, such as those involving lysimeters (1, 17) or water content measurement by nuclear methods (10), are also possible if peat compression is monitored during the measurement procedures. Mass transport is associated with advection of gas into the sediment as desaturation proceeds and with molecular diffusion along the gas-filled channels. The latter phenomenon is important because the molecular diffusion coefficient for a vapor in air is about 4 orders of magnitude larger than for a solute in water. Thus, a few percent desaturation may increase mass transport of volatile compounds by 2 or 3 orders of magnitude. The process of desaturation will not be discussed further in this chapter because a sufficient body of quantitative field data is not yet available. This limitation should in no way imply that desaturation is considered less important than the other processes represented in the examples in this chapter.

**Biotic Transport.** Biotic processes must also be considered in any expression for mass conservation within wetland pore waters. The biotic processes provide the chemical energy that ultimately drives much of the chemical cycling in the system. In a mass conservation equation, these processes typically appear as sources or sinks. Biotic processes may be divided into microbial processes, which are in general highly versatile

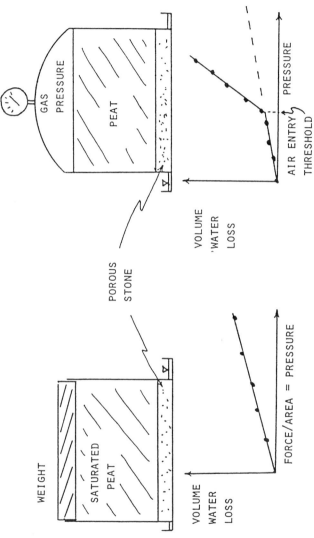

*Figure 2. Schematic representation of the measurement of air-entry (desaturation) thresholds for sediments by comparing water yield of a sample in a standard compressibility test (left) against water yield when the corresponding sample loading is applied by gas pressure (right). Observed compressibility is nearly constant over the range of environmentally relevant pressures.*

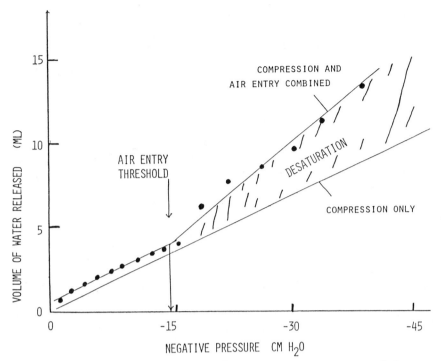

*Figure 3. Results of compression and air-entry tests on a peat sample from Belle Isle marsh. Solid circles correspond to actual water release as a function of pore pressure in the porous plate apparatus [as in Figure 2 (right)]. The upper line is fitted to the aforementioned data points by eye. Water release under bulk peat compression [as in Figure 2 (left)] is shown by the lower line, which closely fits the observed data and exhibits no inflection point. Desaturation is represented by the differences between the lines at any given value of pore pressure.*

and ubiquitous in wetland sediments, and higher plant processes, which usually provide most of the reduced carbon. Higher plants also provide additional mass transport mechanisms via active uptake–translocation–exudation and via passive transport facilitated by the physical structure of the plants. Gas transport via aerenchymous root and rhizome tissues, and gas or water movement in former root channels are two prominent possibilities where subsurface plant organs are present. Gas transport by plants has been shown to be important in certain cases (*18*). Present knowledge, however, does not generally permit the quantification of plant transport processes in wetlands.

## Quantitative Formulations for Transport Processes

From the perspective of the pore waters, both transport and transformation via gas phases or biological entities may be considered as a source or sink, as summarized here.

| Process | Application | Source or Sink |
|---|---|---|
| Movement in gas-filled channels | $P < P_{\text{air entry}}$ | Either |
| Ebullition | $\Sigma C_i H_i > P_{\text{hydro}}$ | Sink |
| Plant transport | Plant-specific | Either |
| Microbial processes | Ubiquitous | Either |

When the sum of these gas-phase and biotic processes is represented as a net source or sink term $(r)$, an expression can be written for the conservation of mass of any substance $i$ at a point in a sediment. This expression is the well-known advection–dispersion–reaction (a–d–r) equation:

$$\frac{dC_i}{dt} + \overline{V} \cdot \overline{\nabla} C_i = \nabla \cdot \overline{\overline{D}} \overline{\nabla} C_i + D_m^{\circ} \, \nabla^2 C_i + r_i \tag{5a}$$

In the general case, none of the quantities needs to be constant. Because wetland sediments often exhibit their strongest gradients in the vertical direction, consideration of the simpler one-dimensional expression in the vertical $(z)$ direction is often of interest.

$$\frac{dC_i}{dt} + V_z \frac{dC_i}{dz} = \frac{d}{dz} (D + D_m^{\circ}) \frac{dC_i}{dz} + r_i \tag{5b}$$

The above expressions are well-recognized in environmental fluid mechanics and groundwater hydrology. The issue, from the perspective of biogeochemistry, is to be able to apply these quantitative expressions of mass conservation in a useful way to the study of chemical cycling in wetland sediments. The following examples represent efforts in this direction. The first example, involving a conservative nonvolatile solute, is seemingly very simple, although the biogeochemical problem (salinity regulation) is of profound importance to salt marsh systems and has not yet been satisfactorily resolved. The second example includes vapor-phase transport, is pertinent to understanding internal functioning of a wetland, and also offers a new methodology for assessing the roles of such ecosystems as sources of trace gases (e.g., $CH_4$, $H_2S$, and $H_2$) to the atmosphere.

## Salt Balance of a Coastal Salt Marsh

Salinity controls the general nature of salt marsh vegetation and restricts the flora to a few species of halophytes, notably *Spartina* and *Distichlis spicata* in northeastern marshes. Salinity also strongly affects net primary productivity within a given stand of marsh grass. The nature of the salt balance is not understood, however. Salt enters the sediment in infiltrating sea water. Although the plants are known to have a salt transport capability, salt export also might occur by drainage to the tidal creeks that run through the marsh as well as by mixing into the regional groundwater system.

Results of a hydrologic analysis of interior sites at the Belle Isle marsh in Boston simplify this picture. Belle Isle marsh is a tidal salt marsh of Boston Harbor, encompasses about 100 ha, and experiences tidal heights of the order of 3 m. This marsh was described in more detail by Nuttle (1). The interior of this marsh is shown schematically in Figure 4. On the basis of measured peat properties, average horizontal pore water velocities are minimal at sites sufficiently removed from tidal creeks (1). This statement was confirmed in Belle Isle by extensive field observations. Vertical movement through the bottom of the sediments is minimal because of the low permeability of a clayey layer about 2 m below the marsh surface. Consequently, far from the tidal creeks, annual average infiltration into the surface, which occurs during tidal flooding of the marsh as well as during precipitation events, must equal annual average evaporation plus transpiration if no long-term change in storage is assumed.

Because horizontal velocities are small, horizontal advective transport of salt is small. Because horizontal dispersive plus diffusive salt fluxes are also small (horizontal salt gradients are minimal in the interior marsh), the annual average salt flux into the sediment associated with the infiltration of water must be balanced by vertical export of salt. The two

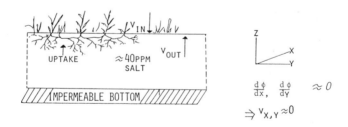

1) HORIZONTAL, BOTTOM WATER FLUXES $\approx 0$

2) $\bar{V}_{IN}$ = UPTAKE + $V_{OUT}$

3) HORIZONTAL SALT GRADIENTS $\approx 0$

4) VERTICAL SALT <u>EXPORT</u> OCCURS

*Figure 4. A simplified representation of the interior of the Belle Isle marsh in Boston, MA. Absence of appreciable horizontal head gradients ($\partial\phi/\partial x$ and $\partial\phi/\partial y$) and horizontal velocities ($V_x$ and $V_y$) is one result of a detailed hydrologic field study.*

possible mechanisms, plant export and export via a possible salt-crusting mechanism, are represented in Figure 5. Although studies to determine which is the dominant process are still ongoing, the definitive experiments are made feasible by the foregoing analysis and the elimination of confounding alternative pathways for salt transport.

## Dinitrogen and Carbon Gas Fluxes at Thoreau's Bog

A second example was taken from work at Thoreau's Bog in Concord, MA. This system is an ombrotrophic, floating-mat *Sphagnum* bog of about 0.4 ha and is described in more detail elsewhere (*19, 20*). As part of biogeochemical studies at this site, a high priority task was to quantify the export of gases, which include methane and carbon dioxide as well as hydrogen, carbon monoxide, and reduced sulfur gases, to the atmosphere.

Figure 6 is a schematic representation of the floating mat in Thoreau's Bog. The mean vertical and horizontal velocities are taken to be zero, and short-term fluctuations in vertical velocity associated with

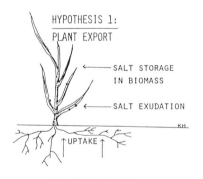

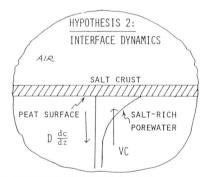

*Figure 5. Two salt export processes, plant transport (hypothesis 1) and crusting (hypothesis 2), collectively account for substantial net vertical export of salt from the interior of Belle Isle salt marsh. The plant mechanism entails substantial salt uptake by marsh halophytes together with above-ground salt exudation (hypothesis 1). Salt export by crusting (hypothesis 2) is driven by an upward flow of pore water at a velocity (V) corresponding to surface evaporation. The resulting crust is periodically removed by tidal flood water. An opposing diffusive downward flux of salt (D dc/dz) also occurs as a result of the high salt concentrations at the peat-atmosphere interface. Abbreviations are as follows: D is a dispersion coefficient, and C is salt concentration in pore water.*

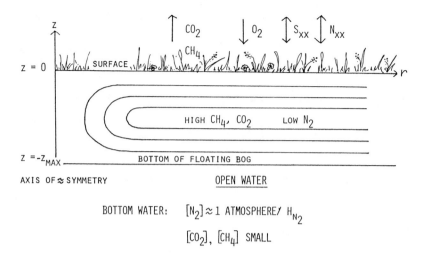

Figure 6. *Floating mat of Thoreau's Bog in Concord, MA. A radial section that has relatively mixed water near the partially open center is shown. Because the bog floats, desaturation is of minor importance at this site.*

precipitation and evapotranspiration are accounted for by a vertical dispersion coefficient $(D)$ of initially unknown magnitude. The value of $D$ is assumed to be constant with depth in the floating bog. In this steady-state analysis, the one-dimensional (vertical) a–d–r equation was used to quantify the transport parameters for $N_2$, $CO_2$, and $CH_4$. The presence of ebullition was evident from informal field evidence and from the depletion of $N_2$ in the sediment. The latter could only occur to the extent observed if bubble stripping occurred. In situ rates of $CH_4$ and $CO_2$ production $(P_{(CH4)}$ and $P_{(CO2)}$, respectively) are inferred from sediment geochronology (17, 18) and from the assumption that the carbon oxidation state is nearly constant with depth. The latter assumption was supported by elemental analysis of the peat. $P_{(CH4)}$ and $P_{(CO2)}$ were assumed to be equal and independent of depth below the aerobic zone. The a–d–r equations for $N_2$, $CO_2$, and $CH_4$ are

$$D \frac{d^2[N_2]}{dz^2} - B[N_2] = 0 \qquad (6a)$$

$$D \frac{d^2[CH_4]}{dz^2} - B[CH_4] + P_{(CH4)} = 0 \qquad (6b)$$

$$D \frac{d^2[CO_2]}{dz^2} + P_{(CO2)} = 0 \qquad (6c)$$

For simplicity, Henry's law constants for $N_2$ and $CH_4$ are assumed equal and are absorbed into the bubble sink term $(B)$. The bubble export of each gas is thus proportional to $B$ and to the aqueous concentration of the gas. The ebullition term for $CO_2$ is neglected because of the low Henry's law constant for $CO_2$, and aqueous concentrations of all gases are assumed in equilibrium with the atmosphere at the top and bottom ($z = 0$ and $z_{max} = -2$ m, respectively) of the sediment profile. Additionally, the sum of partial pressures of $CH_4$ and $N_2$ is set equal to 1 atm. This condition is approximated by actual field observations (unpublished data). When $P_{(CH_4)}$ and $P_{(CO_2)}$ are equated and replaced by $P$, the above system of equations (6a–6c) corresponds to the following solution (7a–7c):

$$[CO_2] = \frac{-P}{2D} (z \, z_{max} + z^2) \tag{7a}$$

$$[CH_4] = [1 - \frac{(e^{z^*} -1) \, e^{\frac{z^*}{z_{max}} z} + (1-e^{-z^*}) \, e^{\frac{-z^*}{z_{max}} z}}{e^{z^*} - e^{-z^*}} ] \, H^* \tag{7b}$$

$$[N_2] = \frac{(e^{z^*} -1) \, e^{\frac{z^*}{z_{max}} z} + (1-e^{-z^*}) \, e^{\frac{-z^*}{z_{max}} z}}{e^{z^*} - e^{-z^*}} H^* \tag{7c}$$

where $H^*$ is the approximate solubility of either $N_2$ or $CH_4$, which are considered equal under 1-atm partial pressure. $H^* = 1$ mol/m$^3$ and is considered invariant with temperature, and

$$z^* = z_{max} \, (P/DH^*)^{\frac{1}{2}} \tag{8}$$

When equations 7a–7c are plotted, extrema are observed at the middepth of the sediment. In particular, the maxima for $CO_2$ and $CH_4$ may be expressed in terms of the biotic production rate plus physical properties of the gases and ecosystem.

$$[CO_2]_{max} = \frac{3P \, Z_{max}^2}{8D} \tag{9a}$$

$$[CH_4]_{max} = [1 - \frac{2}{e^{-z^*/2} + e^{z^*/2}} ] \, H^* \tag{9b}$$

$$[N_2]_{min} = \frac{2H^*}{e^{-z^*/2} + e^{z^*/2}} \tag{9c}$$

Examination of equations 9a–9c leads to several interesting hypotheses regarding the controls on dissolved gas profiles in this peatland ecosystem. Of particular interest is the appearance of the quantity $P/D$ in each equation. The appearance of this quantity suggests that for any given sediment thickness, the observed gas profiles are controlled equally by biotic production rates and by the magnitude of vertical dispersion in the peat. A corollary is that the gas profiles are not sufficient to establish the rate of biotic gas production; an independent estimate of either $P$ or $D$ must be obtained. Geochronological techniques (to establish $P$) or use of an additional tracer substance (to obtain $D$) appear to be required.

Also of interest is the dependence of gas profiles on sediment thickness. Although $CO_2$ maxima in this model increase with the square of sediment thickness, $CH_4$ maxima increase in a strongly nonlinear manner to an asymptotic value equal to $H^\circ$ (atm L mol$^{-1}$), the reciprocal of Henry's law constant.

The predictions of this model are compared to the single "snapshots" of dissolved gas profiles of Thoreau's Bog in Figures 7–9. This comparison was carried out by adjusting the $P/D$ value until the measured and modeled $CO_2$ maxima were approximately equal (Figure 7). The resulting value of $P/D$ (3.3 mmol L$^{-1}$ m$^{-2}$) was used in equations 7b and 7c to yield the predictions for $CH_4$ and $N_2$, respectively, shown in Figures 8 and 9.

The approximate agreement between data and prediction reinforces our belief that equations 6a–6c express a large component of the mass balance for $CO_2$, $CH_4$, and $N_2$ in the bog ecosystem. At the same time, the differences between modeled and observed data underscore the need to compare this coupled transport model against a larger data set having explicit and sufficiently narrow confidence limits on the data. Differences may also arise from simplifications contained in equations 6a–6c and their solutions. These differences include: (1) boundary conditions (in particular, dissolved $N_2$ actually approaches 0.8 atm at the top and bottom of the sediment, but neither $CO_2$ nor $CH_4$ actually equals zero at these boundaries), (2) the neglect of possible vertical advective transport, (3) the aforementioned equating of Henry's law constants for $CH_4$ and $N_2$ (the actual solubilities are about 30% higher than $H^\circ$ for $CH_4$ and about 30% lower for $N_2$), and (4) the assumption that $P$ and $D$ are constant with depth. In addition, seasonal variations may be expected to alter $P$, $D$, and $H$ through changes in hydrology, temperature, and ice formation, and the effect of a narrow oxidized zone of sediment at the upper surface of the mire has not been accounted for. Further work is required to quantify these effects and modify the application of the coupled mass transport equations if indicated.

Estimating the fluxes of other dissolved gases from the sediment to the atmosphere is possible on the basis of the preceding (or a similar) model of physical transport, insofar as the flux of other gases is controlled by the same processes of diffusion–dispersion and ebullition

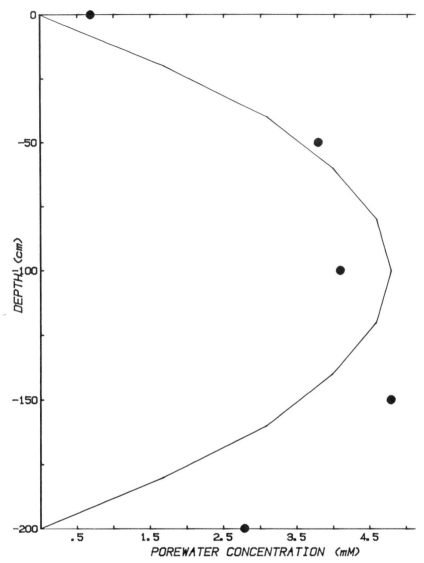

*Figure 7. Modeled CO$_2$ profile in Thoreau's Bog as fitted to observations by adjusting the vertical dispersivity.*

that control CH$_4$ and CO$_2$ fluxes (vegetation-mediated fluxes are also possible but are not considered here). Several gases of major interest (H$_2$, CO, and H$_2$S) have Henry's law constants similar to those of CH$_4$ and N$_2$; hence, according to this hypothesis, their efflux will be controlled largely by ebullition and will be scaled to the CH$_4$ efflux by their concentrations relative to that of CH$_4$. By contrast, the efflux of gases

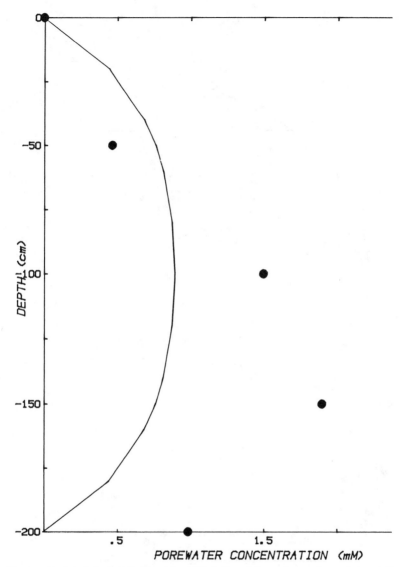

*Figure 8. Modeled $CH_4$ profile (solid lines) based on the value of D determined in Figure 7 compared with an actual "snapshot" observation of $CH_4$ concentration during the growing season (circles).*

such as $N_2O$ and $NH_3$, which have low Henry's law constant values, may be mostly controlled by diffusion and dispersion. Their efflux rates may be scaled by their concentrations relative to that of $CO_2$ if they have a net source within the sediment that is similarly distributed to that of $CO_2$. Order-of-magnitude estimates of the bubble component of these

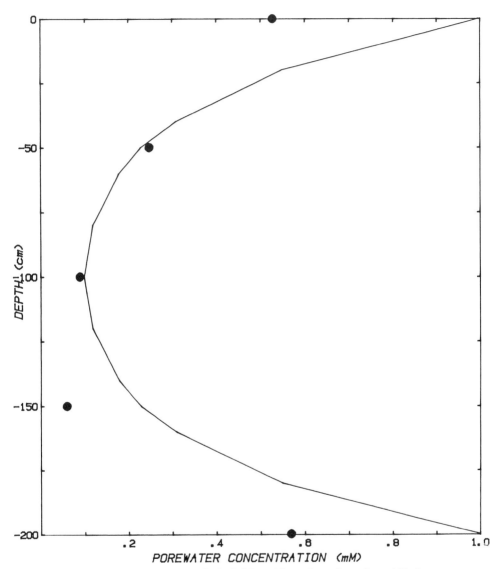

*Figure 9. Modeled $N_2$ profile (solid lines) based on the value of D determined in Figure 7 compared with an actual "snapshot" observation of $N_2$ concentration during the growing season (circles).*

fluxes are presented in Table I. The efflux of $CO_2$ and $CH_4$ is assumed to equal half the carbon mineralization rate as calculated from [210]Pb dating. Bubble passage through the upper fibrous peat layers is assumed to be sufficiently rapid to preclude appreciable oxidation (especially of $H_2S$) in oxic pore waters near the atmospheric interface.

Table I. Order-of-Magnitude Estimates of Ebullition of Selected Trace Gases at Thoreau's Bog

| Compound | Pore Water Concentration[a] | $H_{gas}/H_{CH_4}$ | Flux $(mol/m^2\ year)$ |
|---|---|---|---|
| $CH_4$ | $1 \times 10^{-3}$ M | 1.0 | 3 |
| $H_2$ | $60 \times 10^{-9}$ M | 2.3 | $4 \times 10^{-4}$ |
| CO | $20 \times 10^{-9}$ M | 1.9 | $1 \times 10^{-4}$ |
| $H_2S$ | $1 \times 10^{-6}$ M | 1.9 | $6 \times 10^{-3}$ |
| $N_2O$ | 5 nM | 0.04 | $6 \times 10^{-7}$ |
| $NH_3$ | 1 nM | $2.9 \times 10^{-5}$ | $1 \times 10^{-10}$ |

[a]Depth-averaged value on 1 day at one multilevel well during the growing season.

The advantages of being able to base atmospheric flux estimates on a defensible physical transport model are considerable. The theoretical difficulties of flux boxes, problems due to surface-related artifacts, and the great difficulties of measuring often-reactive gases at trace levels in the air are avoided. These difficulties have been replaced by the more tractable problem of analysis of millimolar to micromolar levels of gases in pore water. Analysis of micromolar levels can be accomplished by gas chromatographic techniques or potentially by an in situ probe.

The approach presented may be of general utility in many mire ecosystems, including those having nonfloating sediments and perhaps a relatively impermeable (zero flux) lower boundary such as the clay layer underlying Belle Isle marsh. The same mass conservation equations should apply and yield somewhat different solutions by virtue of a different lower boundary condition. Also, inclusion of an advective term may be necessary in some systems. By arguments of symmetry, the upper half of the floating-mat system corresponds to a system having a zero flux lower boundary. (Although they physically rise through the sediment, bubbles are assumed to do so rapidly once they start to move, and so are mathematically treated as point sinks at their location of formation.) This correspondence suggests that stationary sediment systems of comparable $P/D$ ratios may exhibit $CO_2$ maxima 4 times higher and exhibit $CH_4$ maxima higher by a factor that varies with the $P/D$ ratio in comparison with floating-mat bog systems.

## Conclusions

Considerable work, theoretical as well as experimental, remains to be done to understand physical transport within wetland ecosystems. In particular, the difficult problem of desaturation requires advances, especially in measurement methodology. Seasonal or episodic desaturation may be extremely important, especially to biomass accretion and element cycling in both fresh and saltwater wetlands. It is hoped that the foregoing examples will encourage further work by illustrating the considerable potential such analysis may offer to the understanding of wetland biogeochemistry.

## Abbreviations and Symbols

| | |
|---|---|
| $\overline{\overline{\nabla}}$ | concentration of gradient of solute |
| $a$ | pore size |
| $B$ | bubble sink term |
| $\underline{D}$ | vertical dispersion coefficient |
| $\overline{\overline{D}}$ | dispersion coefficient tensor |
| $D_m^{\,\circ}$ | diffusion coefficient for a solute in water |
| $C$ | concentration |
| $H^{\,\circ}$ | approximate solubility |
| $H_i$ | Henry's law constant for species $i$ |
| $P$ | pore water pressure |
| $P_{(CH_4)}$ | in situ rate of $CH_4$ production |
| $P_{(CO_2)}$ | in situ rate of $CO_2$ production |
| $r$ | sink term |
| $\overline{V}$ | pore water velocity |
| $z$ | vertical direction |

## Acknowledgments

Support for this research was provided by the Massachusetts Water Resources Research Center; National Oceanic and Atmospheric Administration, Office of Sea Grant; the Massachusetts Division of Water Pollution Control; and National Science Foundation Grant BSR 8306433.

## References

1. Nuttle, W. K. M.S. Thesis, Massachusetts Institute of Technology, 1982.
2. Hoffstetter, R. H. Ph.D. Thesis, University of Minnesota, 1969.
3. Verry, E. S. *Proceedings of the Seventh International Peat Congress*, Dublin, Ireland, 1986.
4. Gorham, E.; Hoffstetter, R. H. *Ecology* **1971**, *52*, 898–902.
5. Siegel, D. I. Water Resources Investigations No. 81-24; U.S. Geological Survey: 1981.
6. Knott, J. F.; Nuttle, W. K.; Hemond, H. F. *Hydrol. Proc.*, in press.
7. Bear, J. *Hydraulics of Groundwater;* McGraw–Hill: Jerusalem, Israel, 1979; p 567.
8. Landva, A. O.; Pheeney, P. E. *Can. Geotech. J.* **1980**, *17*, 416–435.
9. Levesque, M. P.; Dinel, H. *Soil Sci.* **1982**, *133*, 324–332.
10. Hayward, P. M.; Clymo, R. S. *Proc. R. Soc. London* **1982**, *B215*, 299–325.
11. Chen, D. G. M.S. Thesis, Massachusetts Institute of Technology, 1986.
12. Klump, J. V.; Martens, C. S. *Geochim. Cosmochim. Acta* **1981**, *45*, 101–121.

13. Martens, C. S.; Kipphut, G. W.; Val Klump, J. *Science (Washington, D.C.)* **1980**, *208*, 285–288.
14. Boelter, D. H.; Verry, E. S. Technical Report No. NC-31; U.S. Forest Service: 1976.
15. Boggie, R. *Plant Soil* **1977**, *48*, 447–454.
16. Schwintzer, C. T. *Am. Midl. Nat.* **1978**, *100*, 441–451.
17. Dacey, J. W. H.; Howes, B. L. *Science (Washington, D.C.)* **1984**, *224*, 487–489.
18. Dacey, J. W. H.; Klug, M. J. *Science (Washington, D.C.)* **1979**, *203*, 1253–1255.
19. Hemond, H. F. *Ecol. Monogr.* **1980**, *50(4)*, 507–526.
20. Hemond, H. F. *Ecology* **1983**, *64(1)*, 99–109.

RECEIVED for review May 6, 1986. ACCEPTED September 4, 1986.

# INDEXES

# AUTHOR INDEX

# AFFILIATION INDEX

# SUBJECT INDEX

*Copy editing and indexing by Keith B. Belton*
*Production by Cara Aldridge Young*
*Jacket design by Carla L. Clemens*
*Managing Editor: Janet S. Dodd*

*Typesetting of text by McFarland Company, Dillsburg, PA*
*Typesetting of front matter and index by Hot Type Ltd., Washington, DC*
*Printing and binding by Maple Press Company, York, PA*

# Recent ACS Books

*Personal Computers for Scientists: A Byte at a Time*
By Glenn I. Ouchi
288 pp; clothbound; ISBN 0-8412-1001-2

*Writing the Laboratory Notebook*
By Howard M. Kanare
145 pp; clothbound; ISBN 0-8412-0906-5

*The ACS Style Guide: A Manual for Authors and Editors*
Edited by Janet S. Dodd
264 pp; clothbound; ISBN 0-8412-0917-0

*Chemical Demonstrations: A Sourcebook for Teachers*
By Lee R. Summerlin and James L. Ealy, Jr.
192 pp; spiral bound; ISBN 0-8412-0923-5

*Phosphorus Chemistry in Everyday Living, Second Edition*
By Arthur D. F. Toy and Edward N. Walsh
342 pp; clothbound; ISBN 0-8412-1002-0

*Pharmacokinetics: Processes and Mathematics*
By Peter G. Welling
ACS Monograph 185; 290 pp; ISBN 0-8412-0967-7

*New Directions in Electrophoretic Methods*
Edited by James W. Jorgenson and Marshall Phillips
ACS Symposium Series 335; 275 pp; ISBN 0-8412-1021-7

*Biotechnology in Agricultural Chemistry*
Edited by Homer M. LeBaron, Ralph O. Mumma, Richard C. Honeycutt,
and John H. Duesing
ACS Symposium Series 334; 367 pp; ISBN 0-8412-1019-5

*High-Energy Processes in Organometallic Chemistry*
Edited by Kenneth S. Suslick
ACS Symposium Series 333; 336 pp; ISBN 0-8412-1018-7

*Nucleophilicity*
Edited by J. Milton Harris and Samuel P. McManus
Advances in Chemistry Series 215; 494 pp; ISBN 0-8412-0952-9

*Organic Pollutants in Water*
Edited by I. H. Suffet and Murugan Malaiyandi
Advances in Chemistry Series 214; 796 pp; ISBN 0-8412-0951-0

---

For further information and a free catalog of ACS books, contact:
American Chemical Society
Distribution Office, Department 225
1155 16th Street, NW, Washington, DC 20036
Telephone 800-424-6747